Fundamentals of Inorganic and Organic Chemistry

Fundamentals of Inorganic and Organic Chemistry

By

Ilia Manolov

Cambridge
Scholars
Publishing

Fundamentals of Inorganic and Organic Chemistry

By Ilia Manolov

This book first published 2023

Cambridge Scholars Publishing

Lady Stephenson Library, Newcastle upon Tyne, NE6 2PA, UK

British Library Cataloguing in Publication Data
A catalogue record for this book is available from the British Library

ISBN (10): 1-5275-4061-8
ISBN (13): 978-1-5275-4061-3

ANNOTATION

The textbook is intended mainly for participants in chemistry and environmental protection competitions, national and international chemistry Olympiads, chemistry candidates and students in the initial courses of chemistry, medicine, dentistry and pharmacy in all European and other countries of the world.

The development of the material is in line with the chemistry and environmental science curricula in these areas of human knowledge. Since the programs of universities with a chemistry major are practically no different from those of medical universities, the book can be used by a wider range of high school students interested in chemistry, as well as by students in their initial years of study.

Appropriately developed material can be the basis of good preparation not only for the chemistry and environmental science matriculation examinations, for national competitions, for national and international chemistry Olympiads, but also for gaining the knowledge to qualify in chemistry for competitive examinations in medical and other universities in the respective countries.

Sample problems and solutions are proposed for a significant number of the topics and, I hope, will be a useful and interesting tool for developing skills of analysis, comparison, generalisation, searching for relationships and dependencies, and will facilitate the preparation of young people for successful competitive participation.

Serious attention is paid to the redox processes taking place not only in the respective subject but in all cases with inorganic and organic objects. In this way, the reader will be able to determine the degrees of oxidation of the individual constituent atoms of molecules, will learn to determine correctly the oxidant and reductant, and the changes in the degrees of oxidation at electronic transitions.

The book also includes qualitative reactions for identifying the most important ions and elements, as well as characteristic reactions for determining the functional groups and the membership of a molecule in a particular class of organic compounds.

I hope the proposed material will appeal to readers and users, and will be useful and promising reading for their future careers.

Ilia Manolov

Periodic Table of the Elements

Atomic mass values reflect the IUPAC accepted values as of 09/2013.

Masses expressed in [a;b] format show the lower and upper limit of atomic mass depending on the physical and chemical history of the element.

Masses expressed in < > format are the mass numbers of the longest-lived isotope for elements with no stable nucleus.

Key: Atomic Number / Symbol / Name / Atomic Mass

1 IA 1A	2 IIA 2A	3 IIIB 3B	4 IVB 4B	5 VB 5B	6 VIB 6B	7 VIIB 7B	8 VIII 8	9 VIII 8	10 VIII 8	11 IB 1B	12 IIB 2B	13 IIIA 3A	14 IVA 4A	15 VA 5A	16 VIA 6A	17 VIIA 7A	18 VIIIA 8A
1 H Hydrogen [1.00784;1.00811]																	2 He Helium 4.002602(2)
3 Li Lithium [6.938;6.997]	4 Be Beryllium 9.0121831(5)											5 B Boron [10.806;10.821]	6 C Carbon [12.0096;12.0116]	7 N Nitrogen [14.00643;14.00728]	8 O Oxygen [15.99903;15.99977]	9 F Fluorine 18.998403163(6)	10 Ne Neon 20.1797(6)
11 Na Sodium 22.98976928(2)	12 Mg Magnesium [24.304;24.307]											13 Al Aluminum 26.9815386(8)	14 Si Silicon [28.084;28.086]	15 P Phosphorus 30.973761998(5)	16 S Sulfur [32.059;32.076]	17 Cl Chlorine [35.446;35.457]	18 Ar Argon 39.948(1)
19 K Potassium 39.0983(1)	20 Ca Calcium 40.078(4)	21 Sc Scandium 44.955908(5)	22 Ti Titanium 47.867(1)	23 V Vanadium 50.9415(1)	24 Cr Chromium 51.9961(6)	25 Mn Manganese 54.938045(5)	26 Fe Iron 55.845(2)	27 Co Cobalt 58.933194(4)	28 Ni Nickel 58.6934(4)	29 Cu Copper 63.546(3)	30 Zn Zinc 65.38(2)	31 Ga Gallium 69.723(1)	32 Ge Germanium 72.630(8)	33 As Arsenic 74.921595(6)	34 Se Selenium 78.971(8)	35 Br Bromine [79.901;79.907]	36 Kr Krypton 83.798(2)
37 Rb Rubidium 85.4678(3)	38 Sr Strontium 87.62(1)	39 Y Yttrium 88.90584(2)	40 Zr Zirconium 91.224(2)	41 Nb Niobium 92.90637(2)	42 Mo Molybdenum 95.96(1)	43 Tc Technetium <98>	44 Ru Ruthenium 101.07(2)	45 Rh Rhodium 102.90550(2)	46 Pd Palladium 106.42(1)	47 Ag Silver 107.8682(2)	48 Cd Cadmium 112.414(4)	49 In Indium 114.818(1)	50 Sn Tin 118.710(7)	51 Sb Antimony 121.760(1)	52 Te Tellurium 127.60(3)	53 I Iodine 126.90447(3)	54 Xe Xenon 131.293(6)
55 Cs Cesium 132.90545196(6)	56 Ba Barium 137.327(7)	57-71	72 Hf Hafnium 178.49(2)	73 Ta Tantalum 180.94788(2)	74 W Tungsten 183.84(1)	75 Re Rhenium 186.207(1)	76 Os Osmium 190.23(3)	77 Ir Iridium 192.217(3)	78 Pt Platinum 195.084(9)	79 Au Gold 196.966569(5)	80 Hg Mercury 200.592(3)	81 Tl Thallium [204.382;204.385]	82 Pb Lead 207.2(1)	83 Bi Bismuth 208.98040(1)	84 Po Polonium <209>	85 At Astatine <210>	86 Rn Radon <222>
87 Fr Francium <223>	88 Ra Radium <226>	89-103	104 Rf Rutherfordium <267>	105 Db Dubnium <268>	106 Sg Seaborgium <271>	107 Bh Bohrium <272>	108 Hs Hassium <270>	109 Mt Meitnerium <276>	110 Ds Darmstadtium <281>	111 Rg Roentgenium <280>	112 Cn Copernicium <285>	113 Uut Ununtrium unknown	114 Fl Flerovium <289>	115 Uup Ununpentium unknown	116 Lv Livermorium <293>	117 Uus Ununseptium unknown	118 Uuo Ununoctium unknown

Lanthanide Series	57 La Lanthanum 138.90547(7)	58 Ce Cerium 140.116(1)	59 Pr Praseodymium 140.90766(2)	60 Nd Neodymium 144.242(3)	61 Pm Promethium <145>	62 Sm Samarium 150.36(2)	63 Eu Europium 151.964(1)	64 Gd Gadolinium 157.25(3)	65 Tb Terbium 158.92535(2)	66 Dy Dysprosium 162.500(1)	67 Ho Holmium 164.93033(2)	68 Er Erbium 167.259(3)	69 Tm Thulium 168.93422(2)	70 Yb Ytterbium 173.054(5)	71 Lu Lutetium 174.9668(1)
Actinide Series	89 Ac Actinium <227>	90 Th Thorium 232.0377(4)	91 Pa Protactinium 231.03588(2)	92 U Uranium 238.02891(3)	93 Np Neptunium <237>	94 Pu Plutonium <244>	95 Am Americium <243>	96 Cm Curium <247>	97 Bk Berkelium <247>	98 Cf Californium <251>	99 Es Einsteinium <252>	100 Fm Fermium <257>	101 Md Mendelevium <258>	102 No Nobelium <259>	103 Lr Lawrencium <262>

TABLE OF CONTENTS

1. STRUCTURE OF THE ATOM

ATOMIC NUCLEUS. CHEMICAL ELEMENT. ISOTOPES. ISOBARS. STATE OF ELECTRON IN THE ELECTRON SHELL OF AN ATOM. STRUCTURE OF THE ELECTRON SHELL. GROUND AND EXCITED STATES OF ATOMS. EXAMPLE PROBLEMS AND SOLUTIONS.

The idea of the existence of atoms as the basic building blocks of matter was already enshrined in the ideas of Leucippus and Democritus, but it was not until the late seventeenth and early nineteenth centuries, following the works of Dalton, that it became the basis of modern chemistry. Until the late 19th century, the atom was thought to be indivisible and the smallest building block of matter. With the rapid development of natural science in the 19th and early 20th centuries, sufficient experimental data accumulated to show that the atom has a complex structure. In 1896, the French scientist Henri Becquerel discovered that certain minerals containing uranium continuously emit invisible rays that cause photographic plates to darken. Later, Marie and Pierre Curie studied the newly discovered phenomenon and found that uranium atoms decay spontaneously. The phenomenon, later observed in other elements, is called natural radioactivity.

Natural radioactivity, cathode rays, electrolysis, the periodic law of the properties of chemical elements change the existing ideas about the atom. Thanks to these discoveries, scientists have come to the conclusion that the atom is not the smallest indivisible particle. Data on the electron, the proton, the neutron were obtained. Models of the structure of the atom have also been created.

The first atomic model was created by Thomson (1904). It represents the atom as a sphere, on the surface of which are uniformly positively charged particles, and inside "float" electrons - negatively charged. The number of electrons is equal to the number of positive charges. The system is electroneutral. Later, Rutherford (1911) irradiated thin metal plates (copper, platinum) with α-particles and proved that the positive charge of an atom is not homogeneously distributed in a sphere, but is concentrated in a very small part of the atomic volume called the atomic nucleus. The electrons are outside the nucleus and orbit around it, analogous to the planets around the Sun. This model is called planetary. It has significant drawbacks and is contrary to classical physics, according to which, moving around the nucleus, the electron continuously radiates energy as it approaches the nucleus and at some point, should end up on its surface, i.e., be destroyed. This contradiction was overcome by Niels Bohr. Two postulates underlie his model:

1. Electrons move around the nucleus in well-defined stationary orbits without emitting energy;
2. Energy is emitted only when an electron transitions from an energy-rich orbit to another, energy-poorer orbit.

Bohr's theory is applicable only to the hydrogen atom, but not to atoms of more complex structure.

Atomic nucleus

Each atom consists of a nucleus and an electron shell. The atomic nucleus constitutes the central part of the atom. It is assumed to have a spherical shape. Compared to the atom, its size is extremely small - 10^{-14} to 10^{-15} m. However, almost the entire mass of the atom is concentrated in the nucleus. Therefore, the mass of the nucleus is almost equal to the mass of the corresponding atom. The nucleus of an atom is very stable and does not change in chemical reactions. The atomic nucleus is made of two types of particles: **protons** and **neutrons**.

A proton is a material particle whose mass is approximately equal to the mass of a hydrogen atom. It has a positive electric charge, which is taken to be unity. The proton is $_1$p denoted by the symbol p^+. The number of protons determines the number of positive charges of the atom, and also the sequence number (Z) of the element in the Periodic Table.

The neutron is a material particle with a mass equal to the mass of the hydrogen atom and therefore also to the mass of the proton. The neutron is electroneutral and is denoted by the 1 symbol o.

Protons and neutrons are also called nucleons (nucleons - nuclear particles). They are bound to the atomic nucleus by nuclear forces acting at a distance of about 10^{-15} m.

The total number of protons and neutrons is called the mass number (M), which is always an integer:

$$N + Z = A$$

If the Z-number (the number of protons) and the mass number (M) of an atom are known, the number of neutrons in its nucleus can be found.

$$A - Z = N$$

Fluorine has mass number 19, sequence number 9, neutron number 10.

Chemical element. Isotopes. Isobars

According to the atomic-molecular theory, **a chemical element is a collection of atoms of the same kind**. It later became clear that this formulation was inaccurate. The number of protons in the nuclei of the atoms is related to the sequence number of the element and its place in the Periodic Table, and hence the corresponding properties. **A chemical element is therefore a collection of atoms with the same number of protons in the atomic nucleus**. Chemical elements are constituents of simple substances and chemical compounds. So far, 118 chemical elements are known. This means that the number of protons in their nuclei ranges from 1 to 118, respectively.

Until the creation of the modern theory of the structure of the atom, it was believed that each chemical element consisted of one kind of atom. The end product of the natural decay of uranium is the isotope of lead with a mass number of 206, while lead derived from lead ores it has a mass number of 207. This is the first time that scientists have encountered the phenomenon of **isotopy**.

Because of the same number of protons, both types of lead have the same sequence number and the same place in the Periodic Table. The differences in mass numbers are due to the different number of neutrons. **Isotopes of a chemical element have the same number of protons in their nuclei but different numbers of neutrons.** All isotopes of a chemical element occupy the same place in the Periodic Table because they have the same number of protons in their nuclei.

An important characteristic of each isotope is the mass number M. Isotopes of a chemical element have different mass numbers due to the different number of neutrons in their nuclei.

Three isotopes of hydrogen are known. Two of them are found in nature and the third is artificially produced.

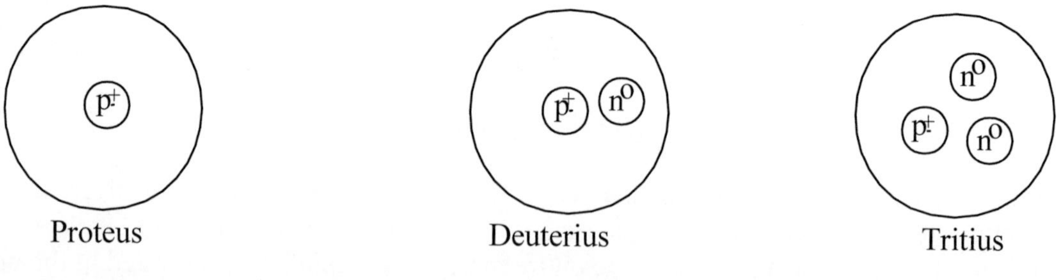

Proteus $^{1}_{1}H$ Deuterius $^{2}_{1}H$ Tritius $^{3}_{1}H$

Oxygen also has three isotopes with mass numbers 16, 17 and 18. Isotopes of a chemical element have the same chemical properties.

The presence of isotopes with different mass numbers affects the magnitude of the atomic mass of the chemical elements. The atomic mass of most chemical elements is not an integer. For example, the atomic mass of chlorine is 35.5 because it is a mixture of two isotopes with mass numbers 35 and 37. The content of the lighter isotope is 75% and the content of the heavier isotope is 25%. The atomic mass is calculated as follows:

$$35 \times 0.75 + 37 \times 0.25 = 35.5$$

Isobars are called atoms of different chemical elements with the same mass number. They exhibit different chemical properties, which are determined by the different number of protons and electrons of a given element. A well-known example of isobar elements is the triad Ar^{40}, K^{40} and Ca^{40}, the isobar pairs Cr^{54} and Fe^{54}, Zn^{70} and Ge^{70}. **The existence of isobars demonstrates unequivocally that the mass of an atom does not determine its chemical properties.**

Of the 118 known elements, 88 occur in nature. The remaining 30 are obtained artificially through nuclear reactions.

State of the electron in the electron shell of an atom

In the first quarter of the 20th century, Louis de Broglie hypothesized that **electrons have a dual nature. They have the properties of a particle and of a wave.** This assumption was dictated by the accumulated experimental data and evidence of the dual nature of light - diffraction and interference (wave properties of light), linear spectrum, photoelectric effect (particle properties of light). All material objects exhibit a dual nature according to de Broglie. The wave/particle ratio is such that as the mass of micro particles decreases, their wave properties strengthen and their corpuscular properties weaken. This means that the electron, in addition to having the properties of a particle, is also treated as a wave.

Based on this, the German physicist Heisenberg formulated the **uncertainty principle.** The principle states that one **cannot measure with precision both the coordinates and the velocity of an electron.**

The uncertainty principle assumes the notion of **regions of space around the nucleus of an atom, called electron clouds, in which the electron density is high – electrons can be found there with high probability (about 95%).**

The electron shell of each atom consists of a number of electrons equal to the number of protons in the atomic nucleus **Electrons are material particles whose mass is equal to 1/1837th of the mass of a proton.** An electron has a negative electric charge, which is denoted by -1. It is equal in magnitude to the charge of the proton. The electron is denoted by the symbol e or is e^-. In addition to mass and electric charge, the electron has spin. Spin is a mechanical characteristic of the electron. This can be visually represented as spin about its own axis. It is known that when an electrically charged particle spins, an electromagnetic field characterized by a certain directionality of the lines of force is generated around it. The electron can therefore be regarded as a small magnet. Graphically, the spin is expressed by an arrow, thus emphasizing the directionality of the field.

In the electron shell, the electrons are oriented relative to each other in two ways. When two electrons have the same field orientation, they have parallel spin and are graphically indicated by equally oriented arrows $\uparrow\uparrow$.

When two electrons have opposite field orientation, they have opposite spins and are graphically denoted by two oppositely directed arrows $\uparrow\downarrow$.

In the electron shell of atoms, electrons with opposite spins are grouped into electron pairs $\uparrow\downarrow$. Most of the electrons in the shell of atoms are in this state. Only a small number of electrons in the atom remain single. They have increased activity and, to some extent, determine the properties of the atom.

Each electron has a certain amount of energy, which is determined by its speed of motion and its distance from the nucleus. The energy of the electrons increases with distance from the nucleus. Depending on their amount of energy, electrons arrange themselves around the nucleus in a certain way called the electron shell structure. The electrons around the nucleus are in constant motion and their location at any given time cannot be fixed at one point. One can only speak of a certain part of the space around the atomic nucleus in which the electron can be found with the greatest probability. This space around the atomic nucleus is called the **electron cloud**. Each electron cloud is characterized by a certain size, density, shape, and also directionality in space. Electron clouds can be spherical in shape, differ in size and density distribution, others are shaped like a volumetric figure eight, differ in their orientation in space, and there are electron clouds with a more complex shape.

The shape, size, and spatial directionality of an electron cloud can be determined by a mathematical function called the atomic electron wave function (a solution to the Schrödinger equation). The state of the electron is characterized by four quantum numbers:

The principal quantum number (n) - it is related to the energy of the electron and the size of the electron cloud. The larger the energy, the larger the number n. In the same order, the sizes of the electron clouds increase. The principal quantum number takes integer values from 1 to $+\infty$. Each orbital in the atom has a definite principal quantum number. Orbitals that have the same principal quantum number form an electron layer.

The orbital (lateral) quantum number (l) takes integer values from 0 to (n - 1), indicating that the values of l depend on n. The orbital quantum number also accounts for differences in the shape of electron clouds. At l = 0 the orbitals are spherical in shape and have no spatial orientation:

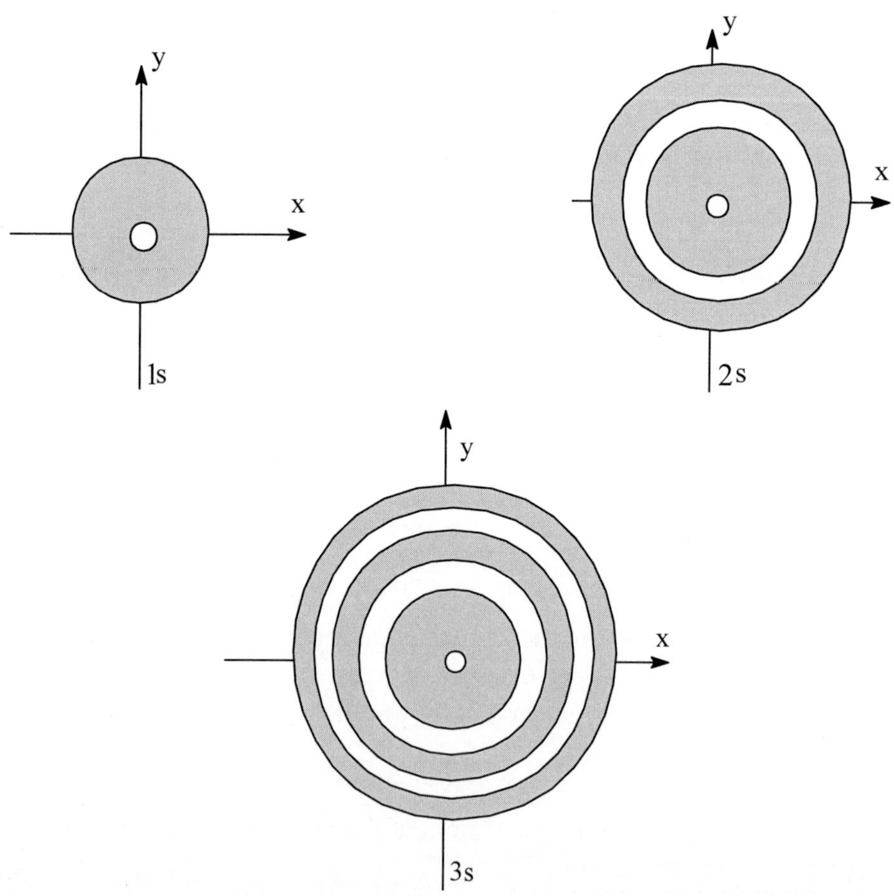

At l = 1 the p-orbital has the shape of a volume eight. They are distinguished by their size, by the distribution of the electron density in them and by their orientation in space.

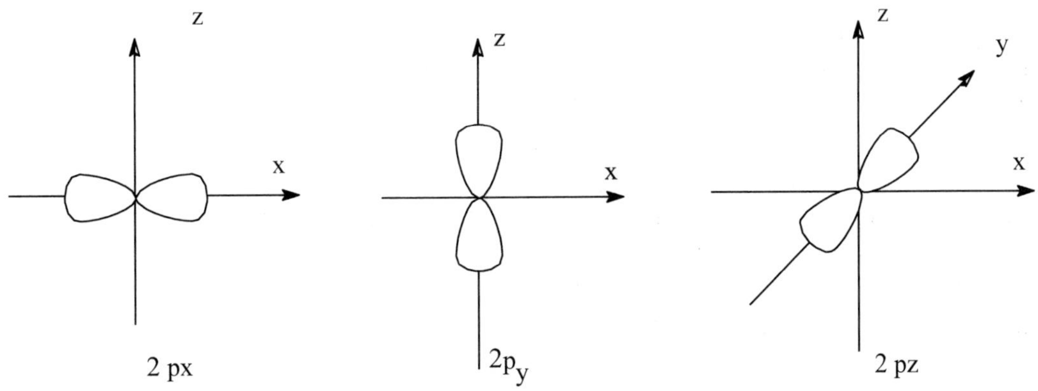

At l = 2 the shape is more complicated, but in most cases, it is a spatial clover. These orbitals are called d-orbitals.

The 1s state is characterized by a principal quantum number of 1 and an orbital of 0.

The state 2p corresponds to a principal quantum number of 2 and to an orbital of 1.

The state 4f - of n = 4 and l = 3, l = 2, l = 1, l = 0, etc.

Orbitals characterized by the same principal quantum number n and the same lateral quantum number l form a sublayer. The number of sublayers in a layer is equal to the layer number n. The orbital quantum number also determines the number of orbitals of a given species by the formula 2l + 1.

Thus, at l = 0 there is only one s-orbital, at l = 1 there are three orbitals - p_x, p_y, p_z; at l = 2 there are 5 orbitals, and so on.

Magnetic quantum number (m) values are related to orbital quantum number values. It takes the values of integers from - l to + l, including zero. This number is not related to the energy of the electron. It takes into account the shape and orientation of electron clouds. States which, for the same values of the principal and orbital quantum numbers, differ only in their magnetic numbers have the same energy. The number of orbitals on which these states lie is equal to 2l + 1.

A spin quantum number (m_s) is introduced for spin decay. It can take only two values

+ 1/2 and -1/2, since the electron rotates about its own axis conditionally clockwise and counterclockwise. Each orbital is expressed by a quantum cell and the electrons by oppositely directed arrows.

This means that the same electron cloud **characterized by n, l and m** can correspond to **two electrons** simultaneously. They will be located in the part of space defined by the given cloud and will differ only in their spins. Therefore, the state of an electron in the electron shell of an atom can be characterized by a combination of four quantum numbers that are not repeated for other electrons belonging to the same atom. This is the essence of **Pauli's principle, which** is defined as follows: *in an atomic system, no two electrons are in the same quantum state, or no two electrons have four identical quantum numbers. They will differ at least in spin quantum number.*

Structure of the electronic shell

In the space around the atomic nucleus, electrons are arranged in a strictly defined way, denoted as the structure of the electron shell. The number of electrons is fixed and constant for each atom. It is equal to the number of protons in its nucleus. The atom is electroneutral. The ordinal number expresses not only the number of protons but also the number of electrons. Hydrogen has an ordinal number of one, which means that it has one proton and one electron. Oxygen's ordinal number is eight; it has eight protons and eight electrons.

The structure of the electronic shell is defined by the four quantum numbers introduced. The basic grouping of electrons is done according to the principal quantum number. All states with the same principal quantum number belong to the same electron layer, have approximately the same energy and are located at approximately the same distance from the atomic nucleus. The layers are conventionally represented by concentric circles and denoted by Arabic numerals or capital letters of the Latin alphabet.

(((((((⊕)))))))

7 6 5 4 3 2 1 K L M N O P Q

The numerical or letter designation begins at the innermost layer (closest to the nucleus) of electrons and continues to the periphery of the atom. Some atoms have a single electron layer, such as hydrogen and helium. The elements from lithium to neon have two electron layers, whereas the others have three or more electron layers. The maximum number of electrons in an electron layer is calculated by the formula $2n^2$ with n denoting the layer number.

When the electron layer is last, regardless of its number (except for the first layer), it can contain, at most, eight electrons. The eight-electron grouping in the last electron layer is stable.

The second grouping of the quantum state is based on the orbital quantum number. The group of states with the same principal and the same orbital quantum number belong to a given sublayer. The number of sublayers in each electron layer is strictly defined and is equal to the layer number.

Layer	Number of sublayers	Designation of sublayers
K (1)	1	1s
L (2)	2	2s2p
M (3)	3	3s3p3d
N (4)	4	4s4p4d4f

The orbital quantum number determines the **number of orbitals in a sublayer** according to the formula 2l + 1. The maximum number of electrons that can occupy a sublayer is determined by the relation 2(2l + 1). Thus on the s-sublayer (l = 0) the maximum number of electrons is two, on the p-sublayer (l = 1) - six electrons, d-sublayer (l = 2) - ten electrons, and so on. A maximum of two electrons can be located on each orbital.

The fundamental law in the construction of electronic layers is the **law of minimum energy**. This means that the electrons occupy successively, in ascending order of energy, the quantum states 1s, 2s, 2p, 3s, 3p... Thus, in the hydrogen atom (z = 1), the single electron occupies the lowest energy state. Its energy diagram is:

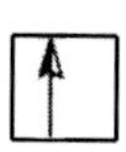 - 1s.

Helium (z = 2) has two electrons in the electron shell. They occupy the 1s-orbital with antiparallel spins and thus form the stable two-electron configuration characteristic of the first electron layer - 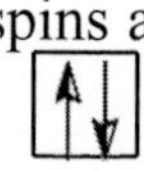 . In the case of helium, the construction of the first electron layer is complete.

The second electron layer begins its construction at lithium. Electrons from the second layer are arranged in two sublayers, 2s and 2p. The 2s-sublayer is built up at lithium and beryllium:

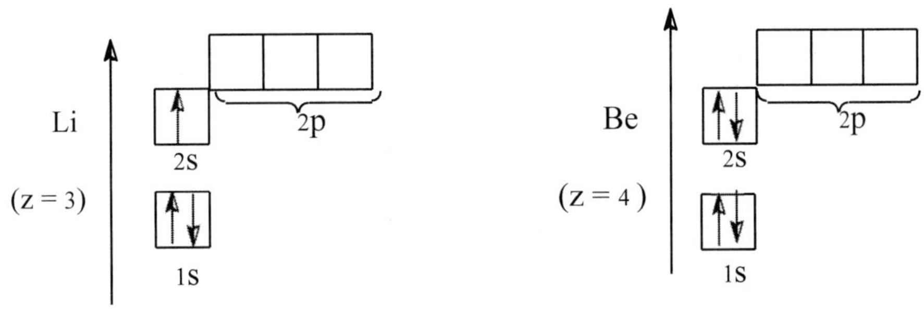

There is some specificity in the construction of the 2p-sublayer. It finds expression in **Hund's rule**, which states that orbitals of equal energy are initially occupied by single electrons, after which electron pairing begins. Accordingly, the construction of the 2p-sublayer in the elements boron (z = 5), carbon (z = 6), and nitrogen (z = 7) proceed as follows:

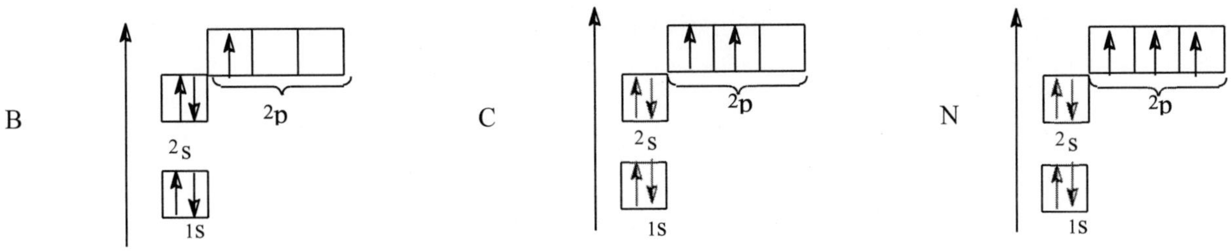

The pairing of electron spins on the p-orbitals starts at oxygen (z = 8) and ends at the neon (z = 10):

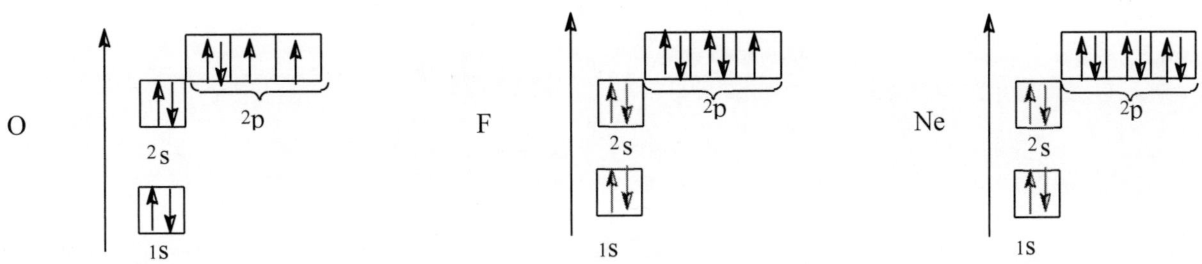

The constructed axisymmetric electron configuration of the second electron layer in the neon is stable.

The third electron layer contains from one to eighteen electrons. All elements where the third electron layer is built up have first and second electron layers built up. The electrons of the third layer are arranged in three sublayers of increasing energy - 3s, 3p and 3d. The 3s- and 3p- sublayers are filled as shown:

Na - $1s^2\ 2s^2\ 2p^6\ 3s^1$; (z = 11)   Mg - $1s^2\ 2s^2\ 2p^6\ 3s^2$ (z = 12)

Al - $1s^2\ 2s^2\ 2p^6\ 3s^2\ 3p^1$; (z = 13)   Si - $1s^2\ 2s^2\ 2p^6\ 3s^2\ 3p^2$; (z = 14)   P - $1s^2\ 2s^2\ 2p^6\ 3s^2\ 3p$; (z = 15)

S - $1s^2\ 2s^2\ 2p^6\ 3s^2\ 3p^4$; (z = 16)   Cl - $1s^2\ 2s^2\ 2p^6\ 3s^2\ 3p^4$; (z = 17)   Ar - $1s^2\ 2s^2\ 2p^6\ 3s^2\ 3p^6$ (z = 18)

After filling the 3d-sublayer at argon (Ag) in order of increasing energy, the occupation of the 3d-sublayer should begin. However, at the next element (potassium, z = 19) the construction of the 4s-sublayer begins. This apparent anomaly in the construction of electron layers and sublayers is also observed elsewhere in the Periodic Table. Studies show that the energy of the 4s-orbital is lower than that of the 3d-orbital. Thus, the ground state electronic configuration of the atoms of these elements is $4s^1$ for potassium and $4s^2$ for calcium. It is explained by **Kleczkowski's rule**. *The energy of the quantum states increases in the order of the increase of the sum of the principal and orbital quantum numbers. For two groups of states that have the same sums, the one that has less principal quantum number has the lower energy*. It follows from the rule that all states that belong to the same sublayer have the same energy. In accordance with Kleczkowski's rule, the energy of the states of the different sublayers of the electron shell increases in order:

$$\mathbf{1s < 2s < 2p < 3s < 3p < 4s < 3d < 4p < 5s < 4d < ....}$$

Schematically, the energy growth of the individual states united in the electronic shell of the atom can be represented as follows:

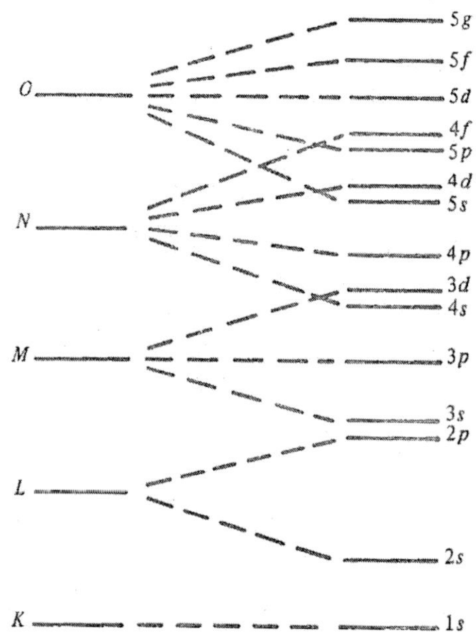

The 4s-orbital is energetically poorer than the 3d-orbital and will therefore be occupied by electrons before it.

In scandia, the filling of the 3d-orbital begins and the ground-state electronic configurations of the ten 3d-elements are represented as follows:

Sc - $1s^2\ 2s^2\ 2p^6\ 3s^2\ 3p^6\ 4s^2\ 3d^1$; Ti - $1s^2\ 2s^2\ 2p^6\ 3s^2\ 3p^6\ 4s^2\ 3d^2$;

V - $1s^2\ 2s^2\ 2p^6\ 3s^2\ 3p^6\ 4s^2\ 3d^3$; Cr - $1s^2\ 2s^2\ 2p^6\ 3s^2\ 3p^6\ 4s^1\ 3d^5$;

Mn - $1s^2\ 2s^2\ 2p^6\ 3s^2\ 3p^6\ 4s^2\ 3d^5$; Fe - $1s^2\ 2s^2\ 2p^6\ 3s^2\ 3p^6\ 4s^2\ 3d^6$

Co - $1s^2\ 2s^2\ 2p^6\ 3s^2\ 3p^6\ 4s^2\ 3d$; Ni - $1s^2\ 2s^2\ 2p^6\ 3s^2\ 3p^6\ 4s^2\ 3d^8$;

Cu - $1s^2\ 2s^2\ 2p^6\ 3s^2\ 3p^6\ 4s^1\ 3d^{10}$; Zn - $1s^2\ 2s^2\ 2p^6\ 3s^2\ 3p^6\ 4s^2\ 3d^{10}$.

The chemical properties of elements are not always uniquely determined by the electronic configuration of the atoms in the ground, unexcited state. When atoms enter into chemical interactions, their electronic configurations change, processes of valence excitation of atoms, ionization, etc. occur. For example, judging from the electronic structure of the metal copper (serial No. 29) in the ground state $1s^2\ 2s^2\ 2p^6\ 3s^2\ 3p^6\ 4s^1\ 3d^{10}$ one would expect that copper would be univalent in its co-units, since the 3d orbital is fully built and only a single 4s electron is present. This means copper should exhibit analogous properties to potassium. However, the weak mutual shielding of the 3d- and 4s-electrons determines a higher ionization energy of copper compared to potassium (ordinate No 19), and in general copper is mainly in the state of a divalent element.

The fifth period begins with the construction of the 5s-floor at Rb (serial No. 37 - IA group) and Sr (serial No. 38 - IIA group). The construction of the 4d- and 5p-atom orbitals follows, i.e., the analogy with period IV is complete. This determines the identical structure of the two periods (periods IV and V).

At Y (sequence No. 39) the construction of the 4d-orbital begins and the ground state electronic configurations of the ten 4d-elements are represented as follows

Y - $1s^22s^22p^63s^23p^64s^23d^{10}4p^65s^24d^1$; Zr - $1s^22s^22p^63s^23p^64s^23d^{10}4p^65s^24d^2$;

Nb - $1s^22s^22p^63s^23p^64s^23d^{10}4p^65s^14d^4$; Mo - $1s^22s^22p^63s^23p^64s^23d^{10}4p^65s^14d^5$;

Tc - $1s^2\ 2s^2\ 2p^6\ 3s^2\ 3p^6\ 4s^2\ 3d^{10}\ 4p^6\ 5s^2\ 4d^5$; Ru - $1s^2\ 2s^2\ 2p^6\ 3s^2\ 3p^6\ 4s^2\ 3d^{10}\ 4p^6\ 5s^1\ 4d^7$;

Rd - $1s^2\ 2s^2\ 2p^6\ 3s^2\ 3p^6\ 4s^2\ 3d^{10}\ 4p^6\ 5s^1\ 4d^8$; Pd - $1s^2\ 2s^2\ 2p^6\ 3s^2\ 3p^6\ 4s^2\ 3d^{10}\ 4p^6\ 5s^0\ 4d^{10}$;

Ag - $1s^2\ 2s^2\ 2p^6\ 3s^2\ 3p^6\ 4s^2\ 3d^{10}\ 4p^6\ 5s^1\ 4d^{10}$; Cd - $1s^2\ 2s^2\ 2p^6\ 3s^2\ 3p^6\ 4s^2\ 3d^{10}\ 4p^6\ 5s^2$

Similar to the previous periods, the sixth period begins with the filling of the 6s orbitals at Cs and Ba, whose electronic structures are $6s^1$ and $6s^2$, respectively.

At the next element La (serial No. 57) a new series of transition elements - 5d-elements - begins. Here a feature appears which results from the combination of two effects – shielding and nuclear charge increase. In caesium (Cs) and barium (Ba) the energy of the 4f-orbitals is practically equal to the energy of the hydrogen atom, since they are strongly shielded by the energy of the inner electrons (we are talking about the third electron layer from the outside in). In the case of lanthanum (La), the nuclear charge increases by one unit and a d-electron appears, which is characterized by a smaller shielding coefficient. As a result, the energy of the 4f- orbitals is lowered and in cerium (Serial No. 58, Se) the 4f-orbitals lie lower than the 5d- and 6p-orbitals.

Thus begins the construction of the 4f-sublayer of the third outside-in electron layer and to form the lanthanide group, viz:

La - $1s^2\ 2s^2\ 2p^6\ 3s^2\ 3p^6\ 4s^2\ 3d^{10}\ 4p^6\ 5s^2\ 4d^{10}\ 5p^6\ 6s^2\ 5d^1$;
Ce - $1s^2\ 2s^2\ 2p^6\ 3s^2\ 3p^6\ 4s^2\ 3d^{10}\ 4p^6\ 5s^2\ 4d^{10}\ 5p^6\ 6s^2\ 5d^1\ 4f^1$;
Pr - $1s^2\ 2s^2\ 2p^6\ 3s^2\ 3p^6\ 4s^2\ 3d^{10}\ 4p^6\ 5s^2\ 4d^{10}\ 5p^6\ 6s^2\ 4f^3$;
Nd - $1s^2\ 2s^2\ 2p^6\ 3s^2\ 3p^6\ 4s^2\ 3d^{10}\ 4p^6\ 5s^2\ 4d^{10}\ 5p^6\ 6s^2\ 4f^4$;
Pm - $1s^2\ 2s^2\ 2p^6\ 3s^2\ 3p^6\ 4s^2\ 3d^{10}\ 4p^6\ 5s^2\ 4d^{10}\ 5p^6\ 6s^2\ 4f^5$;
Sm - $1s^2\ 2s^2\ 2p^6\ 3s^2\ 3p^6\ 4s^2\ 3d^{10}\ 4p^6\ 5s^2\ 4d^{10}\ 5p^6\ 6s^2\ 4f^6$;
Eu - $1s^2\ 2s^2\ 2p^6\ 3s^2\ 3p^6\ 4s^2\ 3d^{10}\ 4p^6\ 5s^2\ 4d^{10}\ 5p^6\ 6s^2\ 4f^7$;
Gd - $1s^2\ 2s^2\ 2p^6\ 3s^2\ 3p^6\ 4s^2\ 3d^{10}\ 4p^6\ 5s^2\ 4d^{10}\ 5p^6\ 6s^2\ 5d^1$;
Tb - $1s^2\ 2s^2\ 2p^6\ 3s^2\ 3p^6\ 4s^2\ 3d^{10}\ 4p^6\ 5s^2\ 4d^{10}\ 5p^6\ 6s^2\ 4f^9$;
Dy - $1s^2\ 2s^2\ 2p^6\ 3s^2\ 3p^6\ 4s^2\ 3d^{10}\ 4p^6\ 5s^2\ 4d^{10}\ 5p^6\ 6s^2\ 4f^{10}$;
Ho - $1s^2\ 2s^2\ 2p^6\ 3s^2\ 3p^6\ 4s^2\ 3d^{10}\ 4p^6\ 5s^2\ 4d^{10}\ 5p^6\ 6s^2\ 4f^{11}$;
Er - $1s^2\ 2s^2\ 2p^6\ 3s^2\ 3p^6\ 4s^2\ 3d^{10}\ 4p^6\ 5s^2\ 4d^{10}\ 5p^6\ 6s^2\ 4f^{12}$;
Tm - $1s^2\ 2s^2\ 2p^6\ 3s^2\ 3p^6\ 4s^2\ 3d^{10}\ 4p^6\ 5s^2\ 4d^{10}\ 5p^6\ 6s^2\ 4f^{13}$;
Yb - $1s^2\ 2s^2\ 2p^6\ 3s^2\ 3p^6\ 4s^2\ 3d^{10}\ 4p^6\ 5s^2\ 4d^{10}\ 5p^6\ 6s^2\ 4f^{14}$;
Lu - $1s^2\ 2s^2\ 2p^6\ 3s^2\ 3p^6\ 4s^2\ 3d^{10}\ 4p^6\ 5s^2\ 4d^{10}\ 5p^6\ 6s^2\ 5d^1\ 4f^{14}$;

The elements from Ce to Lu, where the 4f-sublayer is built up, are called **rare earth elements** or **lanthanides**.

The elements from Hf to Hg, together with La, at which the filling of the 5d-sublayer takes place are the next third group of d-transition elements.

Hf - $1s^2 2s^2 2p^6 3s^2 3p^6 4s^2 3d^{10} 4p^6 5s^2 4d^{10} 5p^6 4f^{14} 6s^2 5d^2$;
Ta - $1s^2 2s^2 2p^6 3s^2 3p^6 4s^2 3d^{10} 4p^6 5s^2 4d^{10} 5p^6 4f^{14} 6s^2 5d^3$;
W - $1s^2 2s^2 2p^6 3s^2 3p^6 4s^2 3d^{10} 4p^6 5s^2 4d^{10} 5p^6 4f^{14} 6s^2 5d^4$;
Re - $1s^2 2s^2 2p^6 3s^2 3p^6 4s^2 3d^{10} 4p^6 5s^2 4d^{10} 5p^6 4f^{14} 6s^2 5d^5$;
Os - $1s^2 2s^2 2p^6 3s^2 3p^6 4s^2 3d^{10} 4p^6 5s^2 4d^{10} 5p^6 4f^{14} 6s^2 5d^6$;
Ir- $1s^2 2s^2 2p^6 3s^2 3p^6 4s^2 3d^{10} 4p^6 5s^2 4d^{10} 5p^6 4f^{14} 6s^2 5d^7$;
Pt - $1s^2 2s^2 2p^6 3s^2 3p^6 4s^2 3d^{10} 4p^6 5s^2 4d^{10} 5p^6 4f^{14} 6s^1 5d^9$;
Au - $1s^2 2s^2 2p^6 3s^2 3p^6 4s^2 3d^{10} 4p^6 5s^2 4d^{10} 5p^6 4f^{14} 6s^1 5d^{10}$;
Hg - $1s^2 2s^2 2p^6 3s^2 3p^6 4s^2 3d^{10} 4p^6 5s^2 4d^{10} 5p^6 4f^{14} 6s^2 5d^{10}$.

From TI (serial No. 81) to Rn (serial No. 86) the 6p-floor is built, which completes the sixth period. It includes 32 elements.

The seventh period is incomplete. It includes the elements from Fr (Frantius, Serial No. 87 through element with Serial No. 118). Francium and radium contain one and two electrons, respectively, on the 7s-orbital. The structure of the outermost electron layer of actinium (Ac, sequence no. 89) is analogous to that of La - $7s^2 6d^1$. Therefore, only one oxidation state, +3, is known for Ac. The elements from Ac (serial No. 89) to Lr (Laurensium, serial No. 103) are called ***actinides***. In these elements the 5f-sublayer is built up. The electronic configurations of their ground state atoms are as follows:

Ac - $1s^2 2s^2 2p^6 3s^2 3p^6 4s^2 3d^{10} 4p^6 5s^2 4d^{10} 5p^6 4f^{14} 6s^2 5d^{10} 6p^6 7s^2 6d^1$;
Th - $1s^2 2s^2 2p^6 3s^2 3p^6 4s^2 3d^{10} 4p^6 5s^2 4d^{10} 5p^6 4f^{14} 6s^2 5d^{10} 6p^6 7s^2 6d^2$;
Pa - $1s^2 2s^2 2p^6 3s^2 3p^6 4s^2 3d^{10} 4p^6 5s^2 4d^{10} 5p^6 4f^{14} 6s^2 5d^{10} 6p^6 7s^2 6d^1 5f^2$;
U - $1s^2 2s^2 2p^6 3s^2 3p^6 4s^2 3d^{10} 4p^6 5s^2 4d^{10} 5p^6 4f^{14} 6s^2 5d^{10} 6p^6 7s^2 6d^1 5f^3$;
Np - $1s^2 2s^2 2p^6 3s^2 3p^6 4s^2 3d^{10} 4p^6 5s^2 4d^{10} 5p^6 4f^{14} 6s^2 5d^{10} 6p^6 7s^2 5f^5$;
Pu - $1s^2 2s^2 2p^6 3s^2 3p^6 4s^2 3d^{10} 4p^6 5s^2 4d^{10} 5p^6 4f^{14} 6s^2 5d^{10} 6p^6 7s^2 5f^6$;
Am - $1s^2 2s^2 2p^6 3s^2 3p^6 4s^2 3d^{10} 4p^6 5s^2 4d^{10} 5p^6 4f^{14} 6s^2 5d^{10} 6p^6 7s^2 5f^7$;
Cm - $1s^2 2s^2 2p^6 3s^2 3p^6 4s^2 3d^{10} 4p^6 5s^2 4d^{10} 5p^6 4f^{14} 6s^2 5d^{10} 6p^6 7s^2 6d^1 5f^7$;
Bk - $1s^2 2s^2 2p^6 3s^2 3p^6 4s^2 3d^{10} 4p^6 5s^2 4d^{10} 5p^6 4f^{14} 6s^2 5d^{10} 6p^6 7s^2 6d^1 5f^8$;
or $1s^2 2s^2 2p^6 3s^2 3p^6 4s^2 3d^{10} 4p^6 5s^2 4d^{10} 5p^6 4f^{14} 6s^2 5d^{10} 6p^6 7s^2 5f^9$;
Cf - $1s^2 2s^2 2p^6 3s^2 3p^6 4s^2 3d^{10} 4p^6 5s^2 4d^{10} 5p^6 4f^{14} 6s^2 5d^{10} 6p^6 7s^2 5f^{10}$;
Es - $1s^2 2s^2 2p^6 3s^2 3p^6 4s^2 3d^{10} 4p^6 5s^2 4d^{10} 5p^6 4f^{14} 6s^2 5d^{10} 6p^6 7s^2 5f^{11}$;
Fm - $1s^2 2s^2 2p^6 3s^2 3p^6 4s^2 3d^{10} 4p^6 5s^2 4d^{10} 5p^6 4f^{14} 6s^2 5d^{10} 6p^6 7s^2 5f^{12}$;
Md - $1s^2 2s^2 2p^6 3s^2 3p^6 4s^2 3d^{10} 4p^6 5s^2 4d^{10} 5p^6 4f^{14} 6s^2 5d^{10} 6p^6 7s^2 5f^{13}$;
No - $1s^2 2s^2 2p^6 3s^2 3p^6 4s^2 3d^{10} 4p^6 5s^2 4d^{10} 5p^6 4f^{14} 6s^2 5d^{10} 6p^6 7s^2 5f^{14}$;
Lr - $1s^2 2s^2 2p^6 3s^2 3p^6 4s^2 3d^{10} 4p^6 5s^2 4d^{10} 5p^6 4f^{14} 6s^2 5d^{10} 6p^6 7s^2 6d^1 5f^{14}$;

There are some differences between the electronic configurations of the 4f- and 5f- elements. The thorium atom (Th) contains no f-electrons, and the atoms of Pa and U each have one d-electron. This is due to the fact that the difference between the energies of the 5f- and 6d- orbitals is negligibly small. This fact finds expression in the degrees of oxidation of either group of elements.

Ground and excited state of atoms

The electronic configuration of an atom that is characterized by the lowest energy is called the **ground state** configuration. If the atom absorbs energy (in terms of chemical interactions) it can transition to an **excited state**. In this, an electron that is in the lower energy orbital transitions to the higher energy orbital. If the electron has been paired on the lower energy orbital, the excitation is accompanied by a spin reversal. For the carbon atom, the excitation proceeds as follows:

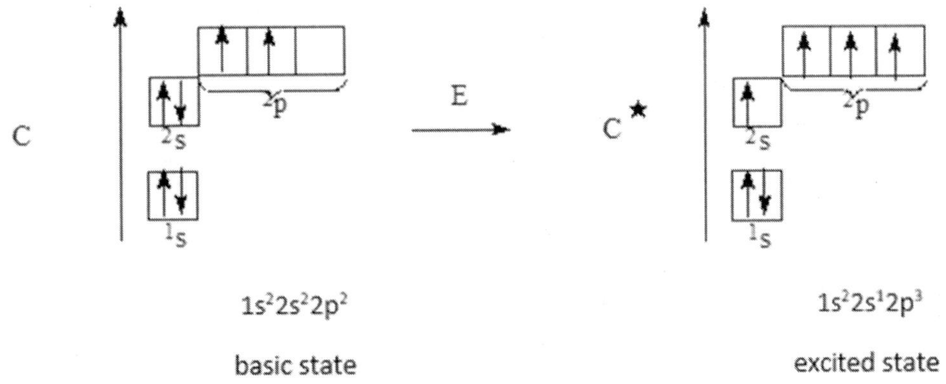

The electronic transition in carbon is from the 2s to the 2p-state, where the unpaired electrons of the atom go from two to four. This electronic transition explains the tetravalent nature of the carbon atom in its compounds.

The excited state of the atom is unstable. It is stabilised by giving up the absorbed energy, at which point the excited electron returns to its ground state.

Sample problems and solutions

Task 1

Which of the following electronic formulas is correct: $1s^2\ 2s^2\ 2p^6\ 3s^2\ 3p^1$, $1s^2\ 2s^2\ 2p^4\ 3s^1\ 3p^3$, $1s^2\ 2s^2\ 2p^5\ 3s^2\ 3p^2$? Explain why. What is the chemical element corresponding to the correct formula and what type is it (**s-, p- d-** or **f-**)? What is the chemical bond in its simple substance? Represent with quantum cells its valence electron layer in ground and excited states. What compound corresponds to this chemical element: hydroxide or oxoacid? Explain the properties of this compound and of the element's simple substance. Support your answers with chemical equations.

Exemplary solution

Of the three electron formulas presented, the first $1s^2\ 2s^2\ 2p^6\ 3s^2\ 3p^1$ is correct because the sequence of building the electron shell of the atom has been followed. In the next two electron formulas, this sequence is broken. Without completing the construction of the second electron layer, the construction of the third electron layer has begun, and before the first sublayer of the third electron layer is filled, the construction of the p-sublayer of the third electron layer has begun ($1s^2\ 2s^2\ 2p^4\ 3s^1\ 3p^3$ or $1s^2\ 2s^2\ 2p^5\ 3s^2\ 3p^2$).

The chemical element corresponding to the correct electronic formula is aluminium, a third period element, group IIIA, sequence number 13. This means that there are 13 p^+ in the nucleus of the atom. Since the p-sublayer of the third electron layer is built up, aluminium is a p-element. Aluminium is a metal, therefore the chemical bond in the simple substance is metallic. This bond also determines the properties of the elements with a metallic bond: ductility, ductility, thermal and electrical conductivity.

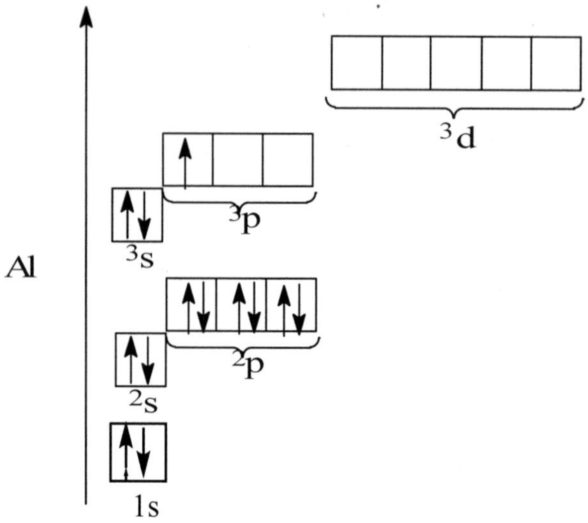

basic state

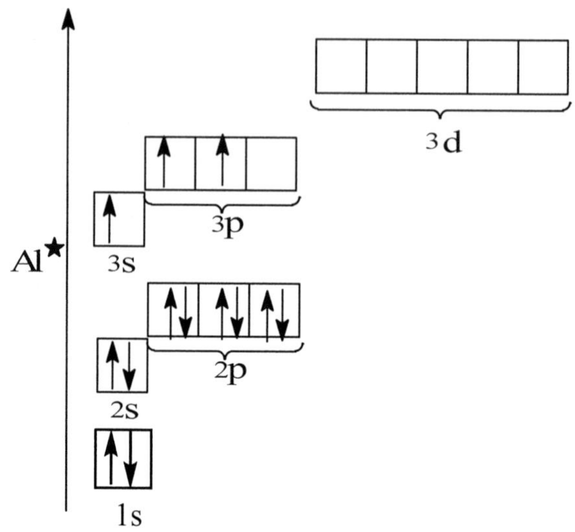

excited state

In the ground state the elements of group IIIA of the periodic system have one single (unpaired) p-electron in their valence electron layer. In the excited 3s state, the electron pair breaks, one of the electrons reverses its spin and switches to the nearby free-energy 3p-orbital. Thus, in its valence electron layer, aluminium in the excited state has three single electrons and this determines its constant third valence. It is known that the elements of groups IA, IIA and IIIA of the Perovician system exhibit a constant valence equal to the group number.

The three single electrons in the valence electron layer of aluminium have different energies - one 3s- and two 3p-. Their atomic orbitals are involved in the hybridization process, in this case sp^2. The three hybrid orbitals are arranged in a plane shaped like an equilateral triangle. With their bulky parts, the hybrid orbitals are directed towards the vertices of the triangle and their axes make angles of 120°. The aluminum compounds have a planar structure.

$$Al\ (Z = 13) - 1s^2\ 2s^2\ 2p^6\ 3s^2\ 3p^1$$

Aluminium is resistant to air, moisture and oxidizing oxoacids due to the rapid formation of a protective layer of Al_2O_3, which coats the surface of the metal. Once the protective layer is removed, its chemical activity increases.

Reduction activity is exhibited by p-elements with a smaller number of p-electrons in their outer electron layer and a larger atomic radius. These are the group IIIA elements and the elements to the left of the B/At line in the Periodic Table. Their simple substances are metals, which are chemically less active than those of the s-elements. Most of their hydrogen compounds are polymeric in nature.

Aluminium does not interact **directly with hydrogen**. Aluminium reacts with **water** to form a white pithy precipitate of Aluminium trihydroxide. Hydrogen is released.

$$2\,Al + 6\,H_2O \longrightarrow 2\,Al(OH)_3 + 3\,H_2\uparrow$$

$$\overset{0}{Al} - 3\bar{e} \longrightarrow Al^{+3} \quad \Big|\ 3\ \Big|\ \ 6\ \ \Big|\ 2$$

$$2\,H^+ + 2x1\bar{e} \longrightarrow 2\,\overset{0}{H} \longrightarrow H_2 \quad \Big|\ 2\ \Big|\ \ \ \ \ \Big|\ 3$$

Aluminium reacts with **acids** and hydrogen is released.

$$2Al + 6HCl \longrightarrow 2AlCl_3 + 3H_2\uparrow$$

$$\begin{array}{lll|c|c|c} Al^0 - 3\bar{e} & \longrightarrow & Al^{+3} & 3 & \multirow{2}{*}{6} & 2 \\ 2H^+ + 2\times 1\bar{e} & \longrightarrow 2H^0 \longrightarrow & H_2 & 2 & & 3 \end{array}$$

The process can also be represented in the following way with the formation of an aqua complex:

$$2Al + \underbrace{6H^+ + 6Cl^-} + 12H_2O \longrightarrow 2[Al(H_2O)_6]^{+3} + 6Cl^-$$

hexaquaaluminum trichloride

Due to its amphoteric nature, aluminium also reacts with **bases**. Concentrated solutions of alkaline bases readily dissolve the metal and a hexahydroxyaluminate anion is produced:

$$Al(OH)_3 + \underbrace{3Na^+ + 3\bar{O}H} \longrightarrow 3Na^+ + [Al(OH)_6]^{-3}$$

$$2Al + \underbrace{6Na^+ + 6\bar{O}H} + 6H_2O \longrightarrow 6Na^+ + 2[Al(OH)_6]^{-3} + 3H_2\uparrow$$

If aluminium is fused with solid sodium base, sodium metaaluminate is obtained:

$$2Al + 2NaOH + 2H_2O \xrightarrow{t^o} 2NaAlO_2 + 3H_2\uparrow$$

$$\begin{array}{lll|c|c|c} Al^0 - 3\bar{e} & \longrightarrow & Al^{+3} & 3 & \multirow{2}{*}{6} & 2 \\ 2H^+ + 2\times 1\bar{e} & \longrightarrow 2H^0 \longrightarrow & H_2 & 2 & & 3 \end{array}$$

The water required for the reaction is not added because it is present in the solid sodium base due to its hygroscopicity.

Aluminium trihydroxide is a pythium amorphous substance. It is obtained by mixing solutions containing aluminium ions with hydroxide ions:

$$Al^{+3} + 3\bar{O}H \longrightarrow Al(OH)_3\downarrow$$

Aluminium trihydroxide has amphoteric properties – it interacts with both acids and bases:

$$2Al(OH)_3 + \underbrace{6H^+ + 3SO_4^{-2}}_{3H_2SO_4} \longrightarrow \underbrace{2Al^{+3} + 3SO_4^{-2}} + 6H_2O$$

When fused with alkaline bases, aluminium trihydroxide yields metaaluminates:

$$Al(OH)_3 + NaOH \xrightarrow{t^o} NaAlO_2 + 2H_2O$$

If concentrated solutions of alkali hydroxides are used, aluminium trihydroxide dissolves to form hydroxocomplexes:

$$Al(OH)_3 + \underbrace{3Na^{+} + 3\bar{O}H} \longrightarrow 3\,Na^{+} + [Al(OH)_6]^{-3}$$

2. Periodic Law and Periodic System

Origin and development of the periodic law and the periodic system. Relationship between the structure of the atom and the structure of the Periodic System. Regularities in the construction of periods and groups. The place of hydrogen in the Periodic System. Variation of properties of chemical elements on the Periodic Table. Atomic and ionic radii. Ionization energy. Electronic affinity. Electronegativity. Redox properties of atoms and ions. Example problems and solutions.

Origin and development of the periodic law and the Periodic Table

Like every great idea and scientific generalization, the periodic law has a long history. In the second half of the 17th century, an intensive study of the chemical elements began, resulting in the discovery of many new elements. While by the middle of the 18th century 15 elements were known, by the end of the century their number had become 30, and by the 1860s 63 elements had been discovered and described. As the number of elements discovered increased, the question of their arrangement and classification arose.

Already Lavoisier divided the elements into metals and non-metals.

In 1829, Diebbeiner developed the idea of triads by grouping the elements known up to that time.

In 1862 Shankurtois proposed a spatial scheme for the arrangement of the chemical elements. It was in the form of a cylinder, with the elements drawn in ascending order of their atomic masses along a propeller line at an angle of 45°. The scheme was unfortunate, since elements very different in their properties fell into the same group.

Newlands created the law of octaves (1862 - 1866), according to which the elements can be arranged in rows of eight with their properties repeated periodically after each eight elements.

In the period 1864 to 1869, Lothar Mayer compiled a table of the chemical elements by dividing them into six vertical groups according to their valence and established a regularity in the variation of their atomic masses.

These researchers have discovered elements of the nature of the periodic law, but none of them have clearly defined it or attempted scientific predictions and conclusions.

In 1869 Mendeleev formulated the periodic law: the **properties of chemical elements and their compounds are periodically dependent on their atomic weights.** In Mendeleev's published table of the chemical elements there are blanks left out. Applying the laws of periodic variation to the properties of the chemical elements, he predicted the properties of three undiscovered elements, which he called ekabor, ekaaluminum, and ecasilicon. These elements were soon discovered - germanium, scandium, gallium.

Mendeleev arranged the chemical elements in ascending order based on atomic masses, believing this to be their most characteristic property, yet he violated this principle. The elements tellurium (Te) and iodine (I), potassium (K) and argon (Ar), cobalt (Co) and nickel (Ni) - Tellurium, potassium and cobalt have greater atomic mass than iodine, argon and nickel respectively. Obviously if the ordering of the elements by atomic mass is followed then nickel and cobalt must change places, iodine must be placed in the sixth group, tellurium in the seventh group and the patterns in both the sixth and seventh groups broken.

In the 1890s, Relay, Ramsay and Travers discovered the noble gas group. The displacement of argon and potassium would lead to even greater disturbances of the regularities in the properties of the noble gases and alkali metals. Mendeleev himself violated the law he formulated and put the properties of the elements first, not the atomic mass.

Obviously, there must be another criterion by which to order the elements in the Periodic Table. This criterion is the ordinal number of atoms, which corresponds to the **number of protons in the atomic nucleus**.

The development of the periodic law is characterized by two stages - chemical and physical.

The chemical stage is characterized by the fact that the atomic mass of the elements is the basis of classification. It follows that Mendeleev did not explain the reason for the periodic change of the properties of the elements. The discovery of this cause could be made on the basis of the structure of the atom, whose structure received a theoretical explanation from quantum mechanics.

The physical stage is related to the emergence of quantum mechanics. At this stage, theory and practice were enriched with new information about the structure of the atom, laws were found that allowed to explain the structure of the Periodic Table of the elements and to reveal the physical meaning of periodicity.

An important role in explaining periodicity has been played by establishing the **physical meaning of the ordinal number**. Thanks to Rutherford's research, the ordinal number of an element acquired a concrete meaning. The ordinal number was identified with the charge of the nucleus. Mosley experimentally confirmed this identity:

Sequence number of the element = positive charge of the nucleus = number of protons in the nucleus = number of electrons in the electron shell.

A deeper knowledge of the structure of the atom makes it possible to formulate the periodic law as follows:

The properties of the chemical elements and their compounds are periodically dependent on the sequence numbers of the elements, and more precisely on the electronic structure of the atom.

This formulation of the law reflects the idea that the **periodic recurrence of the properties of the elements is related to the structure of the electron shell of the atom**. The appearance of electrons with different quantum numbers determines new properties of the elements. Repeatability of such properties is caused by **periodically repeating identical electron configurations** of atoms. This is the physical meaning of periodicity.

Relationship between atomic structure and structure of the Periodic Table

About 100 variations have been proposed for the Periodic Table, but two are established and used: the **compact form** (Mendeleev) and the **long-period form**. Both are built on the same principle - the **principle of electronic analogy**.

The first three periods are small and the rest are large. Each group consists of a **main (A group) and a secondary (B group) group.**

A period is a horizontal order in the Periodic Table. In the ground state of the atoms of the elements of each period, the electrons are arranged on the same number of electron layers whose number corresponds to the **period number and the principal quantum number (n).**

First period n = 1;
Second period n = 2;
Third period n = 3;
Fourth period n = 4;
Fifth period n = 5;
Sixth period n = 6;
Seventh period n = 7.

A group is a vertical order in the Periodic Table. It arranges elements **with analogous structures to the outermost electron layer**. The main groups contain elements with a stable configuration of the penultimate electron layer and with the same structure of the last layer. **The number of electrons in it determines the group number, i.e., same number of electrons in the last layer, same structure of the last layer.**

In the **secondary groups are found elements with an unstable configuration of the penultimate electron layer and a uniform structure of the last layer**. This defines an analogy in chemical properties. The electronic analogy is complete for the elements of the "B" groups. The d-orbitals are taken up consistently. **The group number is determined by the sum of the valence d-electrons and those located at the next s-level.**

The groups of lanthanides and actinides have an analogous structure to the last two electron layers. This determines the great similarity in their chemical properties. They **differ only in the configuration of the third outward electron layer**.

The long-period version of the Periodic Table corresponds to modern ideas about the structure of the atom. It also consists of seven periods arranged horizontally and 18 groups arranged vertically, consisting of analogous elements. Each group consists of elements whose atoms have the same outer and penultimate layers, i.e., elements that have similar electronic configurations are grouped together. Since there cannot be more than eight electrons in the last electron layer of the atoms, there are eight main groups.

The groups are arranged as follows:

IA	IIA	IIIB	IVB	VB	VIB	VIIB	VIIIB
s^1	s^2	d^1	d^2	d^3	d^4	d^5	$d^{6,7,8}$
IB	IIB	IIIA	IVA	VA	VIA	VIIA	VIIIA
d^9	d^{10}	p^1	p^2	p^3	p^4	p^5	p^6

The Periodic Table is currently **viewed as a classification of the atoms of the elements by the structure of their electron shell**. The following types of element classifications exist:

According to the construction of the electronic shell, the elements are s-, p-, d- and f-.

The s- and p-elements are of group A. The number of s- and p-elements is 44.

In the case of **s-elements,** the penultimate layers are stable and there are only s-electrons in the last electron layer (IA and IIA groups).

In the case of **p-elements,** the penultimate layers are stable and the last layer is composed of s- and p-electrons (from group IIIA to VIIIA).

In the case of **d-elements,** the s-electrons are in the outermost electron layer and the d- sublayer (from IIIB to ...IB, IIB) is built up at these electrons. They are B group elements.

Their number is 32.

In the case of **f-elements,** the s-electrons are in the outermost electron layer, contain (or do not contain) one d-electron in the penultimate layer, and the f-sublayer is built up in the third outward layer. Their number is 28 (14 lanthanides - 4 f and 14 actinides - 5 f).

If there are **four to seven electrons** in the outer layer, the elements are classified as **non- metals**. To this group belong hydrogen and boron (with 1 and 3 electrons). Their number is 22, they are all **p-elements** except hydrogen. The noble gases contain eight electrons in the outer electron layer except helium (2 electrons). Elements that have **one to three electrons** in the outer electron layer **are metals.**

Regularities in the construction of periods and groups

The periodic system consists of seven periods.

First period includes two elements. For n = 1, l = 0, m = 0 there is one s-orbital on which two electrons can be located and hence two elements, hydrogen (H) and helium (He).

Second period. For n = 2, l = 0, 1 there are s- and p-orbitals. Total number of electrons - eight, elements - eight. From lithium to neon there are two electron sublayers, the number of electrons increasing from one to eight, viz:

$$2s^1, 2s^2, 2p^1, 2p^2, 2p^3\ 2p^4\ 2p^5\ 2p^6$$

Third period n = 3, l = 0, 1, 2. Three electron sublayers: 3s, 3p, 3d, but the construction of the 3d-sublayer starts after the 4s-sublayer is built. In this period are the elements from sodium (Na) to argon (Ar).

The first, second and third periods are small and the rest are large.

Fourth period begins with the construction of a new electron layer, n = 4, l = 0, 1, 2, 3. There are four electron sublayers - 4 s, 4 p, 4 d, 4 f. There are 18 elements located in this period from potassium (K) to krypton (Kr). For potassium and calcium, the 4 s-sublayer is built up, followed by ten d-elements from scandium (Sc) to zinc (Zn) and finally six p-elements from gallium (Ga) to krypton (Kr).

In the same way, the **fifth, sixth and the unfinished seventh** period are being built. In the sixth and seventh periods there are 4 f - and 5 f-elements, which are 14 each.

Each period begins with an alkali metal and ends with a noble element, an inert gas.

There are eight main groups. The elements of a group have the same electron configuration on the outermost layer, but differ in the number of electron layers.

Group IA - the electronic configuration of the outermost electron layer is ns^1, where n is the principal quantum number and the period number. Group IIA is ns^2, and from IIIA to VIIIA is $ns^2\ np^1$ to $ns^2\ np^6$. Group VIIIA is the noble gases, and Group VIIB in Period IV is the triad of iron (Fe), cobalt (Co), nickel (Ni), and in Periods V and VI the platinum group elements (Pt).

Hydrogen's place in the Periodic Table

The hydrogen atom has a single electron in its electron layer and in this it resembles the alkali metals. It can therefore be placed in group IA. In the free state it is a gas, while the elements of group IA are solids – metals. Compared to the helium atom, the hydrogen atom has one less electron and is therefore adjacent to helium. The atoms of the halogen elements have one less electron than the atoms of their neighbouring noble gases. These similarities also allow hydrogen to be placed above the Group VIIA elements. Despite these similarities, hydrogen cannot be assigned to either group IA or group VIIA.

Alter the properties of chemical elements under the Periodic system

The periodic recurrence of identical electronic configurations leads to a regular variation of properties in groups and periods.

Elements, depending on their properties, are divided into elements with **metallic and non-metallic properties**. The simple substances of the chemical elements are called metals, respectively non-metals. The oxides and hydroxides of metals have basic properties (Ca, CaO, $Ca(OH)_2$) and those of nonmetals have acidic properties (N_2, N_2O_3, NO_2, HNO_3). There are also amphoteric oxides and hydroxides (Zn, ZnO, $Zn(OH)_2$).

At the beginning of each period are the elements with the smallest number of electrons, and at the end - those with the largest number of electrons in the last electron layer. The elements at the beginning of the periods exhibit metallic properties, their oxides and hydroxides - basic. The elements at the end of the periods exhibit non-metallic properties and their oxides and hydroxides have acidic properties.

Consequently, as the number of electrons in the last electron layer of atoms increases, the properties of the elements change from metallic to non-metallic, and those of their corresponding oxides and hydroxides from basic to acidic. This change takes place gradually. This indicates that in the middle of each period there must be elements whose oxides and hydroxides exhibit both types of properties. For example, in the middle of the third period is the element aluminium (Al). In physical properties, aluminum is a metal, and Al_2O_3 and $Al(OH)_3$ are amphoteric compounds (dissolve in acids and bases, as does aluminum itself).

In each period, as the number of electrons in the last electron layer of the atoms of the main group elements increases, the metallic properties weaken and the non-metallic properties strengthen. When moving from one period to another, which corresponds to the beginning of the construction of a new electron layer, the properties of the elements change by leaps and bounds.

Variation of properties is also observed in each group. As the ordinal number of the element in the main groups (I to VII) increases, the metallic properties strengthen and the non-metallic properties weaken.

E.g., in group VA: valence layer ns^2 p^3 the properties change as follows: Nitrogen N - non-metallic properties;

Phosphorus P - non-metallic properties;

Arsenic As - non-metallic, but also some metallic properties; Antimony Sb - non-metallic, but also some metallic properties;

Bismuth Bi - metal properties.

The elements with the most pronounced metallic properties are located in the **lower left part of the periodic table (caesium, francium, barium, radium)**, and those with the most pronounced non-metallic properties are located in the **upper right part of the periodic table (fluorine, chlorine, oxygen).**

Atomic and ionic radii

It is customary to consider atoms and ions as spherical particles of a certain radius.

In the **groups,** as the sequence number increases, the radius of the atoms increases as the energy levels and the total number of electrons increase. This regularity is true for the main groups, i.e., the s- and p-elements. For the "B" groups in the case of the d-elements, this regularity is broken due to lanthanide contraction (shrinkage), since the f-electrons are strongly attracted by the increased charge on the nucleus and practically the radius of the atom does not change.

Each **period** starts with the element with the largest radius. As the sequence number increases, the radius decreases at first and increases at the end of the period. Thus, the atoms of the elements in the middle of the period have a minimum radius value. If this minimum is ignored, the radii of the atoms of the elements can be thought of as decreasing across periods (from left to right). In general, in a given period, the radii of the atoms decrease from left to right. This is explained by the fact that in a given period the number of energy layers does not change, but the number of electrons and the charge on the nucleus increases. As a result of the increase in charge, the electrons are attracted more strongly and the radius of the atoms decreases.

Ionization energy

The energy required to remove electrons from an atom is called ionization energy. It is measured in electron volts (eV). It depends on the structure of the electron shell of the atom. The atomic radius is the determining quantity. The ionization energy depends on the nuclear charge, the particular orbital occupied by the electrons, the electron-electron interaction.

In the main groups, as the sequence number increases, the radius of the atom increases, the attraction of electrons from the nucleus weakens, the ionization energy decreases.

In the secondary groups with increasing order number the radius practically does not change and the ionization energy increases.

The ionization energy increases in periods as the sequence number increases.

Electronic affinity

The energy released when an atom accepts an electron is called electron affinity and is a measure of the tendency of atoms to become negatively charged ions. It is measured in eV.

In a given group, the electronic affinity decreases as the ordinal number of the element increases, and in the period, it increases as the ordinal number increases. Halogen elements have the highest electronic affinity.

Eletronegativity

Electronegativity is a measure of the affinity of atoms for electrons of other atoms in the composition of chemical compounds. The most strongly electronegative element is fluorine, and the most electropositive are the alkali metals. The electronegativity depends on the valence of the element, the type of compound, etc. The electronegativity decreases in groups from top to bottom and in periods from right to left.

Redox properties of atoms and ions

Oxidation properties are determined by the ability of atoms and ions to attach electrons.

The electron affinity energy is used as a measure.

Reduction properties are determined by the ability of atoms and ions to give off electrons. A quantitative characteristic is the ionization energy.

In periods of increasing sequence number, oxidative properties are strengthened and reductive properties are weakened.

In the groups, as the sequence number increases, the oxidation properties weaken and the reduction properties strengthen.

Strong reducers are atoms of alkali metals and hydrogen, and oxidizers are atoms of halogen elements.

The redox properties of the ions change by periods and groups as for atoms. Cations are characterized by oxidation properties, while anions are more often characterized by reduction properties:

Ionic oxidants	Ion reducers
$Ag^{+} + e^{-} \longrightarrow Ag^{0}$	$Cl^{-} - e^{-} \longrightarrow Cl^{0}$
$Cu^{+2} + 2e^{-} \longrightarrow Cu^{0}$	$S^{-2} - 2e^{-} \longrightarrow S^{0}$

If an element occurs in a different valence, the redox properties depend on the radius of the ion.

Sample problems and solutions

Task 1

The chemical elements of simple substances A and B are of the same period. Simple substance A is an active metal that is stored under petroleum, and B is a yellow-green gas. Identify A and B. Represent the possible reaction between them. With which of the following substances would A and B interact: H_2, O_2, H_2O, HI?

Exemplary solution

The presented characteristics of the elements A and B give grounds to decide with certainty that A is an alkaline element (Na), which due to its high reactivity is stored under oil. Sodium is in group IA of the periodic table, period III, sequence number 11. There are 11 protons in the nucleus of the atom and the electron shell is made of 11 electrons. Its electronic configuration is $1s^2\ 2s^2\ 2p^6\ 3s^1$. This means that there is one electron in the s-sublayer of the third electron layer, which determines the reducing properties of sodium and its high reactivity. It interacts with non-metals, water, acids and amphoteric hydroxides.

Element B, a yellow-green gas, is chlorine (Cl_2). It is found in the third period, group VIIA, sequence number 17. The element's sequence number matches the number of protons in its nucleus and the number of electrons in its electron shell. Its electronic configuration is $1s^2\ 2s^2\ 2p^6\ 3s^2\ 3p^5$. There is a single electron in the valence electron layer. This determines its oxidizing ability - by taking up one electron in the course of chemical interaction, chlorine acquires the completed electron configuration of its neighboring inert element argon (Ar). Chlorine interacts with active metals, non-metals, water, alkaline bases.

Sodium reacts vigorously with chlorine to produce sodium chloride, a salt widely used in everyday life.

$$2\,Na + Cl_2 \longrightarrow 2\,NaCl$$

$$\begin{array}{lcl|c|c|c}
\overset{0}{Na} - e^- & \longrightarrow & Na^+ & 1 & & 2 \\
 & & & & 2 & \\
Cl_2 \longrightarrow 2\,\overset{0}{Cl} + 2\times 1\,e^- & \longrightarrow & 2\,Cl^- & 2 & & 1
\end{array}$$

The sodium atom readily gives up its only valence electron and becomes a sodium ion (exhibits a reducing action). The chlorine atom accepts one electron and becomes a chloride anion (exhibits an oxidizing action).

The sodium atom reacts with hydrogen to produce sodium hydride:

$$2\,Na + H_2 \longrightarrow 2\,\overset{+}{Na}\overset{-}{H}$$

$$\begin{array}{lcl|c|c|c}
\overset{0}{Na} - 1\bar{e} & \longrightarrow & \overset{+1}{Na} & 1 & & 2 \\
 & & & & 2 & \\
H_2 \longrightarrow 2\,\overset{0}{H} + 2\times 1\bar{e} & \longrightarrow & 2\,\overset{-}{H} & 2 & & 1
\end{array}$$

Only in hydrides hydrogen transforms in hydride anion.

Sodium forms oxide under special conditions - under cooling and oxygen deficiency, and in excess of oxygen - peroxide.

$$4\,Na + O_2 \longrightarrow 2\,Na_2O$$

$$\begin{array}{lcl|c|c|c}
\overset{0}{Na} - e^- & \longrightarrow & Na^+ & 1 & & 4 \\
 & & & & 4 & \\
O_2 \longrightarrow 2\,\overset{0}{O} + 2\times 2\,e^- & \longrightarrow & 2\,\overset{-2}{O} & 4 & & 1
\end{array}$$

and in excess of oxygen - peroxide.

$$2\,Na + O_2 \longrightarrow Na_2O_2 \quad (Na^+\ \bar{O}{-}\bar{O}\ Na^+)$$

$$\begin{array}{lcl|c|c|c}
\overset{0}{Na} - e^- & \longrightarrow & Na^+ & 1 & & 2 \\
 & & & & 2 & \\
O_2 \longrightarrow 2\,\overset{0}{O} + 2\times 1\,e^- & \longrightarrow & 2\,\overset{-1}{O} & 2 & & 1
\end{array}$$

Metals that are ahead of hydrogen in the order of relative activity react with water, releasing hydrogen and producing hydroxides.

$$2\,Na + 2\,H_2O \longrightarrow 2\,NaOH + H_2\uparrow$$

$$\begin{array}{lll|c|c|c} & Na^{0} - 1\bar{e} & \longrightarrow Na^{+1} & 1 & & 2 \\ & & & & 2 & \\ 2\,H^{+} + 2x1\bar{e} & \longrightarrow 2\,H^{0} & \longrightarrow H_2 & 2 & & 1 \end{array}$$

The activity of alkaline elements is even greater against acids than against water because in an acidic environment the concentration of hydrogen ions is significantly higher than in pure water.

$$2\,Na + 2\,HI \longrightarrow 2\,NaI + H_2\uparrow$$

$$\begin{array}{lll|c|c|c} & Na^{0} - 1\bar{e} & \longrightarrow Na^{+1} & 1 & & 2 \\ & & & & 2 & \\ 2\,H^{+} + 2x1\bar{e} & \longrightarrow 2\,H^{0} & \longrightarrow H_2 & 2 & & 1 \end{array}$$

Halogens have high chemical activity. They react with almost all simple substances. The interaction with hydrogen is an oxidation-reduction process and leads to the formation of halogenated hydrocarbons. Chlorine in direct sunlight reacts with an explosion by a chain- radical mechanism:

$$H_2 + Cl_2 \longrightarrow 2\,HCl$$

$$\begin{array}{ll|c|c|c} 2\,H^{0} - 2x1\bar{e} & \longrightarrow 2\,H^{+} & 2 & & 1 \\ & & & 1 & \\ 2\,Cl^{0} + 2x1\bar{e} & \longrightarrow 2\,Cl^{-} & 2 & & 1 \end{array}$$

Halogenated hydrocarbons are molecules with a covalent polar bond. They have an unpleasant suffocating odor, and dissolve well in water. Their aqueous solutions exhibit acidic properties due to their dissociation to hydrogen ions and halide anions caused by dipole water molecules.

Chlorine reacts with water by the equation:

$$Cl_2 + H_2O \longrightarrow HCl + HOCl$$

$$\begin{array}{ll|c|c|c} Cl^{0} + e^{-} & \longrightarrow Cl^{-} & 1 & & 1 \\ & & & 1 & \\ Cl^{0} - e^{-} & \longrightarrow Cl^{+} & 1 & & 1 \end{array}$$

The mixture of two acids is produced – hydrochloric (HCl) and hypochlorous (HOCl) acids. The process is known as disproportionation.

Chlorine exhibits a stronger oxidizing action than iodine and displaces it from its compounds. A redox process takes place and elemental iodine is released according to Eq:

$$Cl_2 + 2HI \longrightarrow 2\,HCl + I_2$$

$$\begin{array}{rl|c|c|c}
 & 2\,Cl^{0} + 2x1e^{-} \longrightarrow 2\,Cl^{-} & 2 & & 1 \\
 & & & 2 & \\
2\,I^{-} - 2x1e^{-} \longrightarrow 2\,I^{0} \longrightarrow & I_2 & 2 & & 1
\end{array}$$

3. CHEMICAL BOND

CAUSES OF BONDING OF ATOMS. ELECTRONEGATIVITY. FORMATION OF COVALENT NON-POLAR AND COVALENT POLAR BOND. DELOCALIZED CHEMICAL BOND. COORDINATIVE BOND. NATURE OF IONIC BOND AND PROPERTIES OF IONIC BOND SUBSTANCES. DIRECTIONALITY OF CHEMICAL BONDS. TYPES OF MOLECULES DEPENDING ON THE TYPE OF COVALENT CHEMICAL BOND. EXAMPLE PROBLEMS AND SOLUTIONS.

Reasons for bonding atoms. Electronegativity

Without knowing the nature of the interaction between atoms, one cannot understand the reason for the existence of different chemical compounds, the mechanism of their formation, their composition, structure and reactivity.

Atoms are hardly found in the free state. Under ordinary conditions inert gases exist as free atoms. This is due to their stable outer electron layer. For example, helium (He) has two electrons in its only electron layer, while the other inert elements have eight electrons. The atoms of the other elements join with each other or with atoms of other elements by chemical bonds to form simple substances or chemical compounds. The molecules of most substances have been found to be stable under ordinary conditions, e.g., the molecules of the substances hydrogen, oxygen, nitrogen, water, hydrogen chloride, ammonia, etc. The molecules of hydrogen can be broken down into atoms by heating the gas to over 3000° C. Experimentally, it has been found that the process is carried out with energy absorption:

$$H_2 \longrightarrow 2\,H^0 \; - \; 436 \text{ kJ/mol}$$

The reverse process, the bonding of hydrogen atoms into molecules, occurs with the release of energy:

$$H^0 + H^0 \longrightarrow H_2 + 436 \text{ kJ/mol}$$

Water molecules decompose with a constant electric current or at high temperature.

This shows that the molecules of substances are more stable than free atoms. The reason for this is that molecules have less energy than free atoms. When atoms bond, besides the decrease in energy, there is also an interaction between their electrons from the last electron layer.

The first **chemical bond theory** was developed by **Kossel**. According to him, the atoms of the elements are converted into ions by giving up and receiving electrons, whereby each atom acquires the electron configuration of the nearest noble gas. The ions interact **electrostatically with each other**, according to Coulomb's law, and so chemical bonds are formed. According to this theory, all compounds should have **an ionic structure** and **the chemical bond should be ionic only.** Thus, the formation of a chemical bond between identical atoms - hydrogen, chlorine, oxygen, nitrogen, etc. - cannot be explained.

Lewis's theory (1916) has a broader scope and includes as a special case Kossel's theory. According to this theory, in the formation of molecules, **atoms attain a stable axisymmetric electron configuration at the last electron layer** by the formation of one or more **common electron pairs**.

It was later found that many molecules are formed **without reaching the axisymmetric electronic structure of their constituent atoms.** It is assumed that the main reason for the formation of molecules is the **reduction of the energy of the atoms when they bond**. This bonding is widespread in various simple and complex substances. The name **covalent bond** was proposed by Langmuir in 1919.

To represent the covalent bond in chemical formulas, Lewis proposed to put as many dots around the chemical sign of each atom as there are **electrons in the last electron layer**. The electrons that make up the common electron pair are denoted by dots placed between the signs of the two atoms. For example, the formation of a fluorine molecule is expressed by the following diagram:

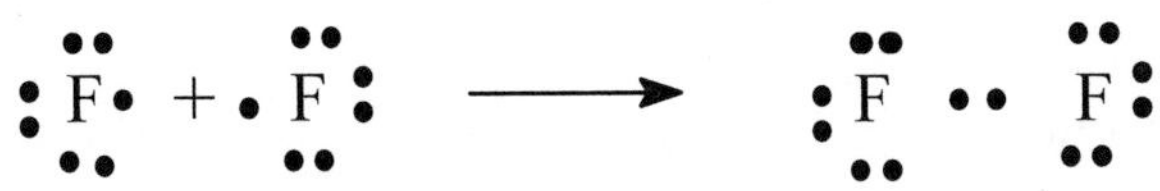

The outer electron layer of the fluorine atom contains seven electrons. In an effort to build a **stable outer axisymmetric electron layer similar to that** of its nearest inert gas, the neonium, the fluorine atom bonds into a molecule with another fluorine atom by creating a **common electron pair**. In this way, each of the two fluorine atoms in the fluorine molecule acquires a stable outer eight-electron layer.

Formation of covalent non-polar and covalent polar bond

Two hydrogen atoms join together to form a hydrogen molecule by releasing a certain amount of energy and their electrons **pair up.** In the formation of the molecule, the two atoms converge. An interaction occurs between **their single electrons**, which have antiparallel spins, and they bond into an electron pair that belongs to both **atoms at the same time** and is therefore called **a common electron pair**. Due to it, the electron shell of each atom in the hydrogen molecule is as in the helium atom. In the molecule formed, the two hydrogen atoms are linked by a strong chemical bond made by the common electron pair. Such a bond is called a **covalent bond**.

$$H\bullet + \bullet H \longrightarrow H{\bullet\bullet}H \quad (H{-}H)$$

It is customary to denote a common electron pair by two dots or a dash placed between the chemical symbols of the elements. When the arrangement of electrons in molecules is expressed by dots, the **formulas are called electronic**, and if dashes are used, **structural**. The electronic and structural formulas show in what **sequence the atoms in the molecule are bonded** and also their **valency**.

The atoms that make up a molecule may possess not only shared, but also **unshared electron pairs**. Unshared electron pairs belong to only one of the atoms in the molecule and **are not involved in chemical bonding**. For example, in the chlorine molecule, each chlorine atom possesses one shared and three unshared electron pairs:

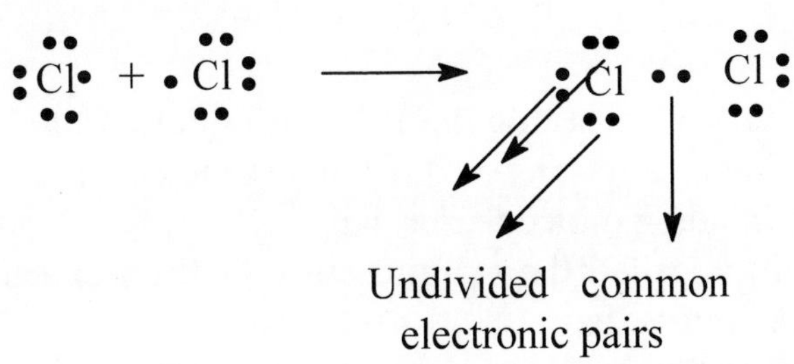

More than one common electron pair can form between two atoms. For example, in the oxygen molecule, the two oxygen atoms are bonded by **two common electron pairs:**

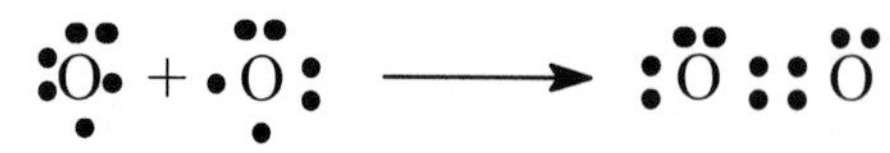

In the nitrogen molecule, the chemical bond between the two nitrogen atoms is made by **three common electron pairs:**

A covalent chemical bond that is made by one common electron pair is called simple, and by two or three common electron pairs is called complex. The bond between the atoms in the molecules of **hydrogen, fluorine, chlorine is simple** and in the molecules of **oxygen and nitrogen is complex**.

According to modern ideas, when a common electron pair is formed between two atoms, their atomic orbitals partially overlap. For example, in the formation of the hydrogen molecule, the 1s-atom orbitals of the two hydrogen atoms partially overlap

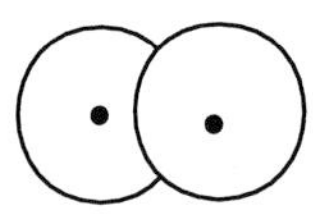

H—H

A greater electron density is created between the nuclei of the two atoms and the two electrons become **common to both nuclei**. These electrons are now in a new state that can be characterized by a **molecular wave function called a molecular orbital.** To each molecular orbital corresponds a well-defined energy and electron cloud that indicates the distribution of electron density in the molecule. The **number of molecular orbitals that form when atoms bond is equal to the number of atomic orbitals that overlap**. For example, in the formation of the hydrogen molecule, two molecular orbitals are formed from the two 1s-atom orbitals of the hydrogen atoms. The molecular orbital that has an energy less than the energy of the 1s atomic orbital is called the **binding molecular orbital**, and the molecular orbital that has an **energy greater than the 1s orbital is called the antibinding molecular orbital.**

In a molecular orbital, as in an atomic orbital, there cannot be two electrons in the same state, i.e., there can be **two electrons with opposite spins (Pauli's principle).** In the construction of the electron shell of molecules, electrons occupy **those free molecular orbitals that have minimum energy (minimum energy principle). Therefore, in the hydrogen molecule, the two electrons occupy the bonding molecular orbital.** When the electrons occupy the bonding molecular orbital in the formation of molecules energy is released. Therefore, the energy of the hydrogen molecule is less than the sum of the energies of the two atoms:

$$E_{H_2} < E_H + E_H$$

The hydrogen molecule is therefore **more stable than the individual atoms**. A stable formation of atoms is formed only if the number of electrons on the bonding molecular orbitals is greater than the number of electrons on the antibonding molecular orbitals.

The state with minimum energy is called the **ground state of the molecule.** If the molecule accepts energy, it can go into an **excited state.**

The covalent bond **is non-polar and polar** depending on the electron density distribution in the cloud corresponding to the molecular orbital. The electron density distribution in this cloud affects the strength of the chemical bond formed. **A non-polar covalent bond** is formed between atoms of the **same chemical element and is characterized by a uniform distribution of electron cloud**

density around the nuclei of the bonding atoms. Such is the bond in diatomic molecules of hydrogen, oxygen, nitrogen, fluorine, chlorine, etc.

An ordered covalent bond is realized between atoms of different elements and is characterized by a **non-uniform distribution of the electron cloud density around the two nuclei of the bonding atoms. For example, in molecules of water, ammonia, hydrogen chloride, etc., the bonds formed are polar covalent.** The polarity of the covalent bond depends on the properties of the atoms forming the molecule. The more the two atoms differ in nature, the more polar the covalent bond.

The ability of atoms in molecules to attract to themselves common electron pairs is called electronegativity. In order of their electronegativity values, some of the chemical elements are listed in the following sequence:

$$F > O . N > Cl > S > P > C > H > Al > Mg > Ca > Na > K$$

The elements that come **before hydrogen in this order are called electronegative because** their atoms pull the electron pair towards them when they bond with hydrogen. Elements that are **after hydrogen are called electropositive because they have no affinity for electrons.**

When a covalent bond is formed between two atoms of different electronegativity, the common electron pair is pulled towards the **atom of greater electronegativity. In the molecules of hydrogen chloride, water, ammonia, the common electron pairs are drawn closer to the atoms of chlorine, oxygen and nitrogen, respectively, because their electronegativity is greater than that of hydrogen.** In these cases, the electron pair is common to the two bonding atoms, but is located **asymmetrically between the two nuclei - closer to the nucleus of the more electronegative element.** It is usually said to be partially pulled towards it. Partial pulling of the common electron pair towards the **atom of greater electronegativity results in two electric poles in the molecule, a negative one (δ–) around the atom of greater electronegativity and a positive one (δ+) around the other atom.**

$$\underset{\delta+}{H} \longrightarrow \underset{\delta-}{Cl}$$

These charges are significantly smaller in magnitude than the charge of an electron, so they are called **partial charges**. The greater the difference in electronegativity of the bonding atoms, the more polar the chemical bond:

$$CH_4 \quad NH_3 \quad H_2O \quad HF$$

$$\xrightarrow{\hspace{6cm}}$$

The polarity of the connection increases

Molecules with a covalent bond only decay to atoms at high temperature. This requires a large amount of energy. The covalent bond is therefore strong.

A delocalized chemical bond. Coordinative bond

In many cases, the common electron pair is not localized between just two atoms, but can be found with significant probability around more atomic nuclei. Such relocalization of the electron pair leads practically to an electron cloud that binds more atoms. For example, the carbon atom in the carbonate anion $(CO_3)^{-2}$ is bonded to three oxygen atoms that are equivalent. The carbon atom is in sp^2 -hybridization and is bonded by three σ-bonds to the three oxygen atoms. The fourth electron of the carbon atom is not involved in the hybridization, together with electrons from the oxygen atoms, forms a delocalized π-bond, which is realized by a common electron cloud spanning all four nuclei:

The dashed line indicates the delocalized π-connection.

The delocalized π-bond is commonly found in organic compounds. In the benzene molecule, the six carbon atoms are bonded in a ring and are in a sp^2 -hybrid state. They are linked to each other and to the six hydrogen atoms by σ-bonds through their three equivalent sp^2 -hybrid orbitals. At each carbon atom, one unhybridized 2p-orbital perpendicular to the σ- skeleton remains. All of the non-hybridized 2p-orbitals are parallel to each other, allowing each of them to overlap with two adjacent 2p-orbitals since the distance is the same in both directions. This produces a single six-centre π-molecule orbital of lower energy spanning the nuclei of the six carbon atoms, ensuring an even distribution of electron density around each.

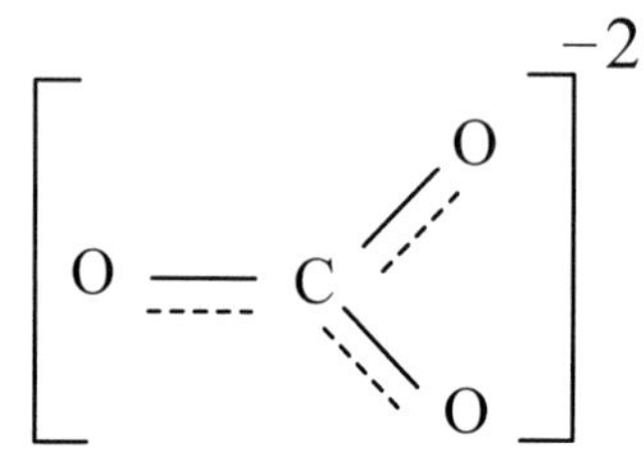

In agreement with Pauli's principle, the six π-electrons occupy the three low-energy bonding molecular orbitals of benzene.

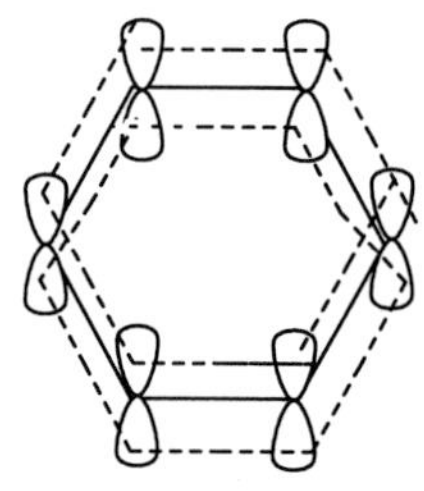

In some molecules, there are atoms with unshared electron pairs not involved in chemical bonds in the molecule. These electron pairs may be involved in forming covalent bonds with atoms from other molecules that have unoccupied orbitals in their outermost electron layer. The atom that provides the finished electron pair is called **the donor**, and the one that accepts it is called the **acceptor**. The bond is known as a **donor-acceptor** or **coordinative bond**. It is a variety of the covalent bond. Once formed, it is indistinguishable from the covalent bond formed by electron pairs involving one electron per atom. For example, in the formation of the ammonium ion (NH^{+}), one of the bonds is formed by the donor-acceptor mechanism:

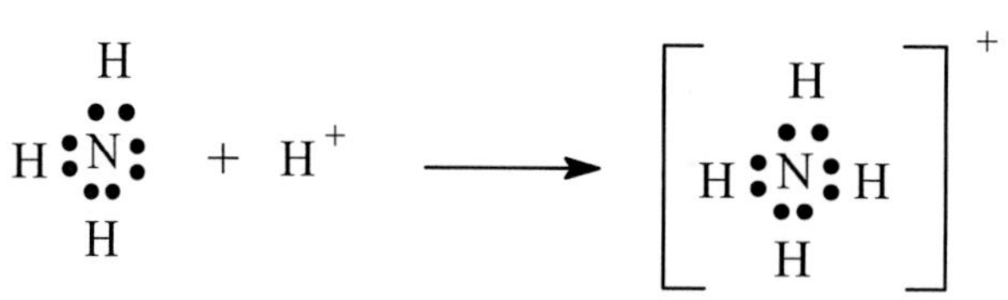

Nature of the ionic bond and properties of ionic bond substances

When the atoms involved in the bond formation have a large difference in electronegativity, the electron pair can be localized around only one nucleus. This gives rise to the ionic bond, which can be seen as a borderline case of the polar covalent bond.

The valence electrons in the ionic bond pass to the atom with the greater electronegativity. Two ions are formed, a positive and a negative. For example, sodium combines violently with chlorine to form the compound sodium chloride. Sodium is an electropositive element and chlorine is electronegative. The difference in electronegativity of these elements is very large. Under the conditions of the chemical reaction, each chlorine atom takes one electron from a corresponding sodium atom. As a result of these changes, the sodium atom becomes a sodium ion (cation), which has the electron configuration of its preceding inert gas, neon (Ne), and the chlorine atom becomes a chloride ion (anion) with the electron configuration of its following inert gas, argon (Ar). Two types of ions are produced, positive and negative, which bond by electrostatic forces to form the compound sodium chloride. The occurrence of an ionic bond is expressed by equations in which the transition of electrons is shown schematically. For example, the formation of sodium chloride is expressed by the following equation:

$$2\,Na + Cl_2 \longrightarrow 2\,\overset{+}{Na}\overset{-}{Cl}$$

$$\begin{array}{lll|c|c|c} & Na^{0} - e^{-} & \longrightarrow Na^{+} & 1 & & 2 \\ & & & & 2 & \\ Cl_2 \longrightarrow & 2\,Cl^{0} + 2 \times 1\,e^{-} & \longrightarrow 2\,Cl^{-} & 2 & & 1 \end{array}$$

Energy is released when the formed ions bond, i.e., the resulting compound is more stable because it is poorer in energy.

A chemical bond that takes place between oppositely charged ions is called an ionic bond. An ionic bond is formed when the ionization energy of the atom giving off electrons is small and the electron affinity of the atom receiving electrons is large.

Chemical compounds that are made up of ions are called ionic compounds. These are almost all salts, most of the metal oxides, and some acids and bases.

In ionic compounds, each ion attracts a relatively large number of oppositely charged ions. As a result, the ions arrange themselves properly relative to each other and form a crystal in which all the ions are bound together by electrostatic forces and there are no individual molecules. The chemical formula of ionic compounds shows **the simplest relationship between the ions in the crystal**. The formulas of ionic compounds **do not express molecules.** The ionic bond is very strong and determines the most characteristic properties of substances with an ionic crystal lattice:

- high melting temperature;
- fragility.

Directivity of chemical bonds

Unlike the ionic bond**, the covalent bond is directed in space**. **Directionality** is one of the characteristic features of chemical bonds. Schematically, the formation of a hydrogen molecule can be represented as follows:

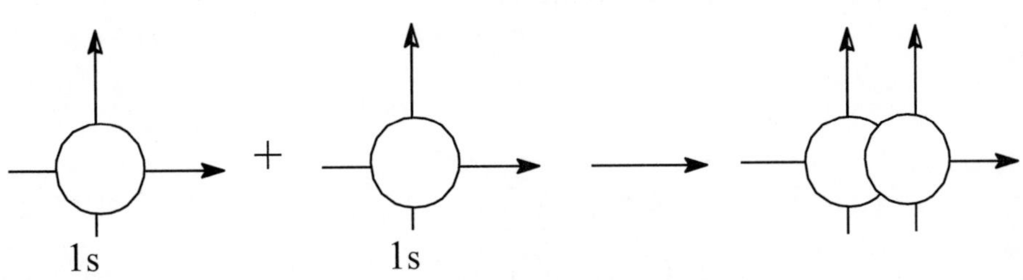

s-Electron clouds, which have a spherical shape, are not directed in space. When the two spherical electron clouds of the 1s-electrons of the hydrogen atoms overlap, a new, larger electron cloud is produced that covers both nuclei.

Depending on the overlap of the electron clouds (s-, p-, d-), and on their orientation relative to the internuclear axis of the bonded atoms, the bonds are sigma- (σ-), pi- (π-) and delta- (δ-).

The sigma- (σ-) bond is formed as the atomic orbitals overlap along the axis connecting the two atomic nuclei. The maximum electron density is found between the atomic nuclei. A sigma- (σ-) bond is formed when the:

s- and s-orbital

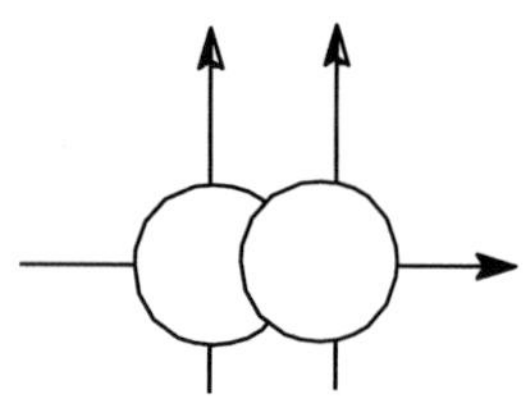

H–H

σ

s- and p-orbital

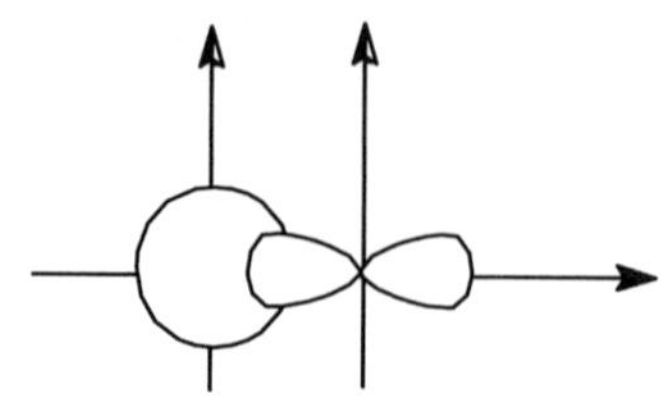

H–Cl

σ

H–F

p- and p-orbital

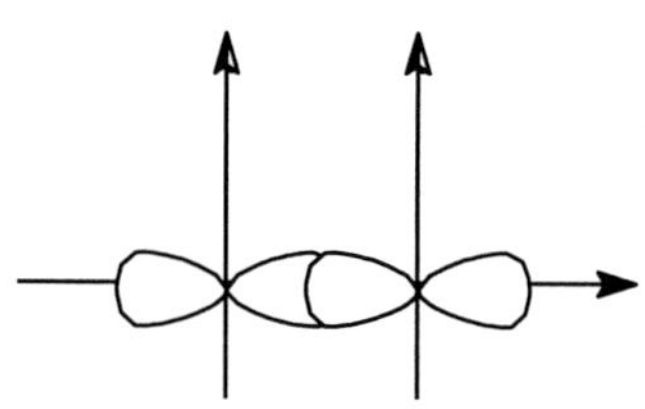

d- and d-orbital

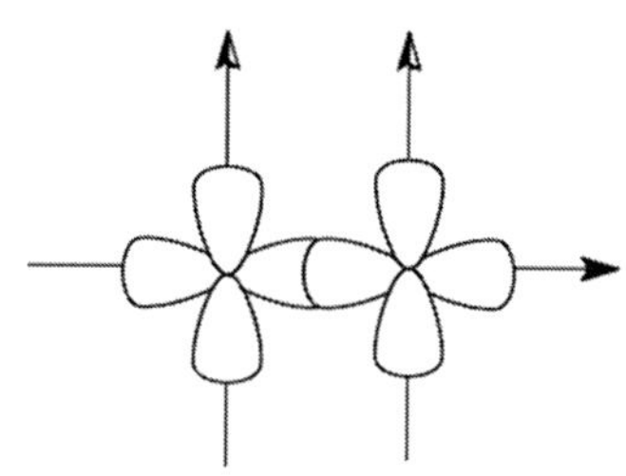

F–F

σ

Cl–Cl

σ-link

When the axis connecting the two kernels coincides with the overlap axis, a σ- connection is realized.

When the axis connecting the two nuclei is perpendicular to the overlap axis, a pi- (π-) connection is realized. It is formed after a σ-bond has been created between the two atoms. The π-bond is weaker than the σ-bond. π-Bonds can form electrons from p- and d-orbitals. These bonds are made by two regions of overlapping atomic orbitals on either side of the axis joining the two nuclei:

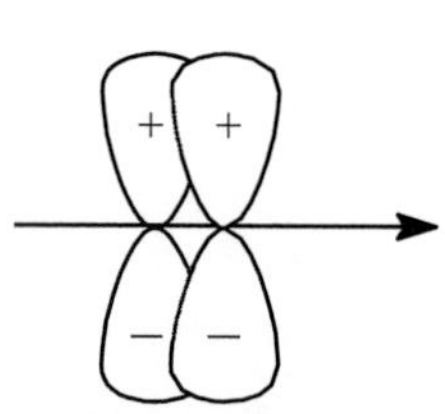

p-p
π-link

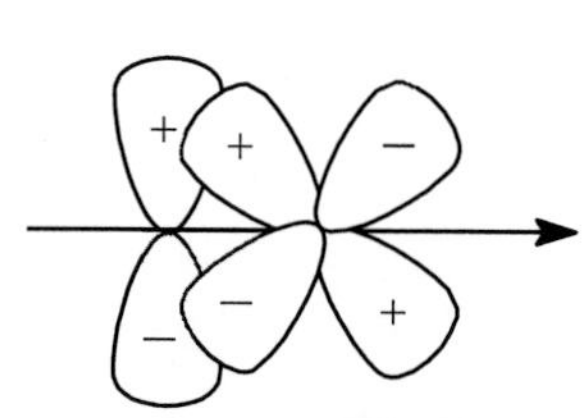

p-d
π-link

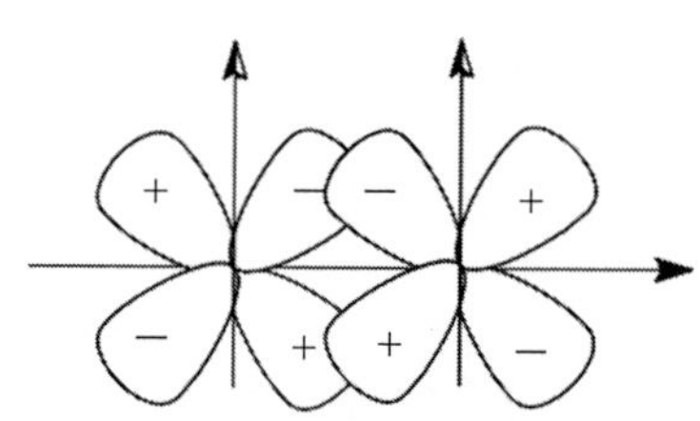

d-d
π-link

The δ-Couplings occur when the four sectors of the d-electron clouds located in two parallel planes overlap simultaneously.

In the hydrogen fluoride (HF) molecule, one s- and one 2p-atom orbitals overlap and a σ-bond is formed.

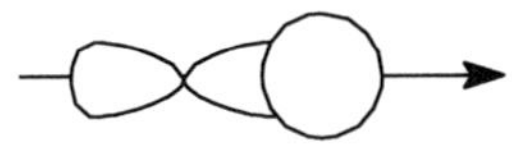

HF; p ---- s σ-link

When an atom is involved in the formation of chemical bonds with several atoms of another chemical element, symmetrical orientation of the bonds in space is often observed. Experimentally, it has been found that the ammonia molecule (NH_3) has a pyramidal structure and the methane molecule (CH_4) is a regular tetrahedron. **The carbon atom is located in the center of the tetrahedron and the four hydrogen atoms are at the vertices of the tetrahedron.**

In the boron trichloride (BCl_3) molecule, the three bonds are in the **same plane and make an angle of 120° with each other. The beryllium dichloride molecule ($BeCl_2$) is linear with the two bonds making an angle of 180° between them. In all these cases, s- and p- electrons from the parent atoms are involved in the formation of the bonds.** It is known that in the source atoms the p-electron clouds conclude angles of 90° between themselves. Clearly, in these cases the orientation of the valence bonds in space does not coincide with the orientation of the electron clouds in the free atoms.

At the moment of chemical interaction and bond formation, the states of the electrons in the source atoms change – they **hybridize**. New, hybrid states are produced. The **total energy does not change but is redistributed**. The **electron density** distribution**, the shape and the orientation of the hybrid electron clouds change significantly.**

Possible cases of hybridization between s- and p-states are as follows:

Hybridization of **one s- and one 2p-state** yields **two sp-hybrid** states. This type of hybridization is called **digonal**. The valence electrons of beryllium in beryllium dichloride ($BeCl_2$), of carbon in the ethene molecule, of hydrogen cyanide, etc., are in such a state.

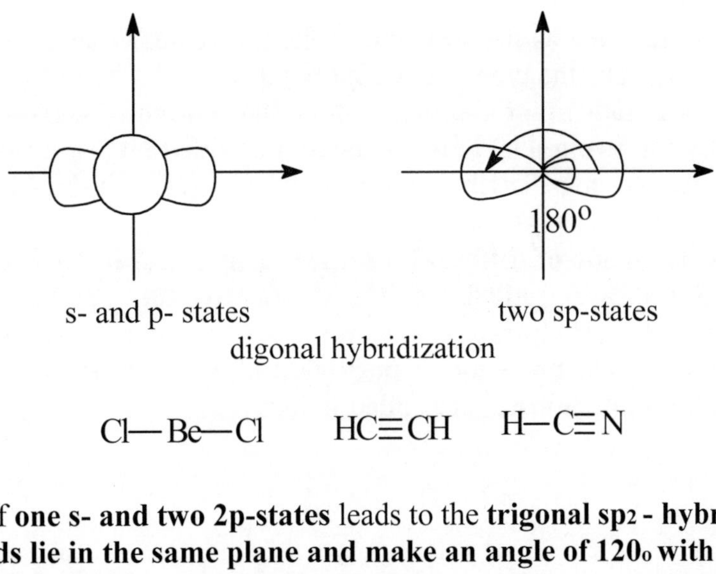

digonal hybridization

Cl—Be—Cl HC≡CH H—C≡N

The interaction of **one s- and two 2p-states** leads to the **trigonal sp^2 - hybridization.** In this case, **the three bonds lie in the same plane and make an angle of 120° with each other.** Such is the hybridization of boron (B) in boron trichloride (BCl_3), of carbon in the molecule of ethene, butadiene, benzene, etc.

sp^2 -hybridization.

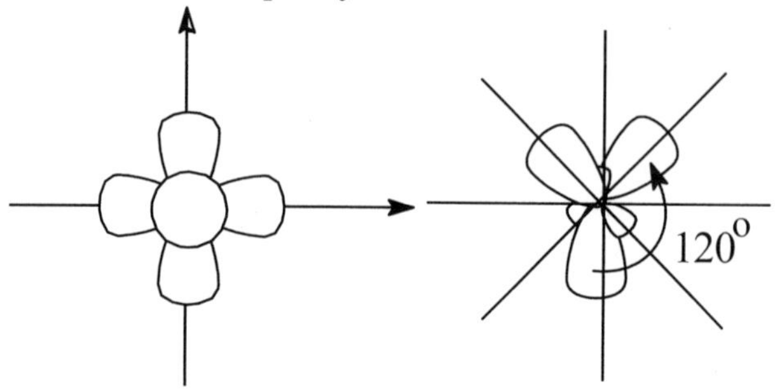

(s + p + p)-orbital three sp^2 -orbital

The tetragonal sp^3 -hybridization occurs between one s- and three 2p-states. The bonds formed are directed to the **vertices of a regular tetrahedron.** This is the case for the valence electrons of the **nitrogen atom in the ammonium ion, of the carbon in the methane molecule and alkanes, etc.**

Types of molecules depending on the covalent bond

Most of the covalently bonded substances exist in the form of molecules that include varying numbers of atoms.

The simplest are diatomic molecules. They can be made of **identical atoms** that are bonded by a **non-polar covalent bond (H_2, O_2 , N_2 , Cl_2 , etc.)** or of **different atoms** that are bonded by a **covalent polar bond** (HCl, HBr, HI, CO, NO, etc.). There are molecules in which the number of atoms is 3, 4 and can reach thousands. Usually, these molecules are made of different atoms, therefore the bonds between them are covalent polar.

In a molecule, the atoms are arranged in space in a specific way relative to each other. This spatial arrangement of atoms is denoted as the structure of the molecule. What types of molecules exist depending on their structure and the type of covalent bond?

In diatomic molecules made up of identical atoms, the common electron pairs are attracted with equal strength by the two nuclei and are located equidistant from them. Therefore, in a molecule the electric charges are distributed symmetrically (H_2, N_2, O_2, Cl_2). Molecules with a symmetrical arrangement of charges are called non-polar.

In diatomic molecules made of different atoms, the chemical bond is covalent polar. The common electron pairs are pulled (shifted) toward the atoms with the greater electronegativity, so the electric charges in the molecules are distributed asymmetrically. As a result, two poles form in each molecule, a positive and a negative pole. Molecules with an asymmetrical arrangement of charges are called polar.

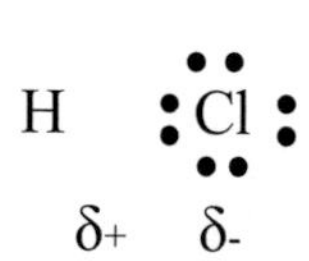

In triatomic molecules with a covalent polar bond, the atoms are arranged in two ways - angularly or linearly. For example, in the polar water molecule, the two bonds are at an angle of 104.5°.

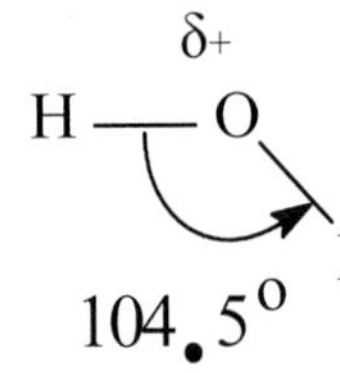

polar water molecule

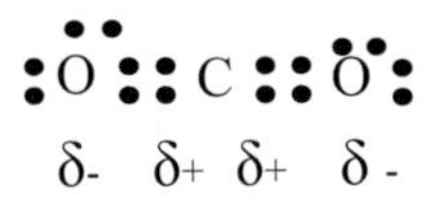

In the **carbon dioxide** molecule, **the bonds are polar, but the molecule is non-polar because it has a linear structure. Its positive and negative charges are symmetrically arranged in space.**

Therefore, depending on the spatial structure, triatomic molecules with a covalent polar bond can be polar and non-polar.

Polyatomic molecules can also be polar or non-polar. This depends on the spatial arrangement of the atoms bonded by a polar covalent bond.

According to **the covalent bond theory,** when **metal** atoms bond, their orbitals overlap forming molecular orbitals whose number is equal to the number of atomic orbitals. In a metal crystal, each metal atom overlaps its orbitals with the orbitals of its neighboring atoms. In this way, a large number of molecular orbitals having very close energies arise in a crystal on a given type of atomic orbitals. Neighboring orbitals in the crystal differ insignificantly in energy, then **all energy states are admissible for the electron.** The model allows the whole crystal to be viewed as one huge molecule consisting of n atoms. The general properties of metals, **thermal and electrical conductivity,** are due to the structure.

Sample problems and solutions

Task 1

According to the nature of the chemical bonds and the structure of the molecules, which of the following substances should dissolve in water and which are very slightly or practically insoluble: KNO_3, HCl (gas), $C_{15}H_{31}COOH$, NH_4Cl, C_8H_{18}, NH_3, H_2, C_2H_5OH, N_2? Which

of the gaseous substances in this sequence obey Henry's law and which do not? Compare the osmotic pressures of the solutions obtained from the solid and liquid solutes in the same sequence if taken at the same molar concentration at T = const. Justify each answer.

Exemplary solution

The substances listed in the problem statement can be divided into the following groups:

With **ionic bonding** - KNO_3, NH_4Cl - strong electrolytes. Dissociate completely in aqueous solution.

When the atoms involved in bond formation differ in their electronegativity, the electron pair is localized around only one nucleus. This gives rise to the ionic bond, which may be regarded as a borderline case of the polar covalent bond. When the ions are bonded together, energy is released, i.e., the resulting compound is more stable because it is poorer in energy.

An ionic bond is formed when the ionization energy of the electron-emitting atom is small and the electron affinity of the electron-receiving atom is large.

Chemical compounds that are made up of ions are called ionic compounds. These are almost all salts, most of the metal oxides, and some acids and bases.

$$KNO_3 \longrightarrow K^+ + NO_3^-$$

$$NH_4Cl \longrightarrow NH_4^+ + Cl^-$$

With a **covalent polar bond** - HCl (gas), NH_3 and C_2H_5OH. When a covalent bond is formed between two atoms of different electronegativity, the common electron pair is pulled toward the atom of greater electronegativity. In the molecules of hydrogen chloride, ammonia, and ethanol, the common electron pairs are pulled closer to the atoms of chlorine, nitrogen, and oxygen, respectively, because their electronegativity is greater than that of hydrogen. The electron pair is common to the two bonding atoms, but is located asymmetrically between the two nuclei - closer to the nucleus of the more electronegative element. These compounds are water soluble.

With a **weakly polar covalent bond** - C_8H_{18} and $C_{15}H_{31}COOH$. They are practically insoluble in water. This property is due to the weakly polar carbon-hydrogen bonds in the octane molecule and the long hydrocarbon moiety (one methyl and fourteen methylene groups) linked to the carboxyl group in hexadecanoic (palmitic) acid.

With a **non-polar covalent bond** - H_2, N_2. In diatomic molecules composed of identical atoms, the common electron pairs are attracted with equal strength by the two nuclei and are spaced equidistant from them. Therefore, in a molecule the electric charges are distributed symmetrically (H_2, N_2, O_2, Cl_2). Molecules with a symmetrical arrangement of charges are called non-polar.

The gaseous substances are HCl (gas), NH_3, H_2 and N_2. Henry's law is valid for those gaseous substances that do not interact with water. While hydrogen chloride and ammonia interact with water. The aqueous solution of hydrogen chloride gas is called hydrochloric acid and the aqueous solution of ammonia is a weak and perishable ammonium hydroxide. Henry's law does not apply to them.

The osmotic pressures of KNO_3 and HCl solutions at the same concentration and at T = const. are twice as high as expected. The reason for this is the dissociation of these compounds.

$$KNO_3 \longrightarrow K^+ + NO_3^- \qquad \pi \approx 2$$

$$HCl \longrightarrow H^+ + Cl^- \qquad \pi \approx 2$$

Ammonium chloride is a salt of strong acid and weak base. It dissociates and hydrolyzes. Hydrolysis is a reversible interaction of salt ions and water ions in which the ratio between hydrogen and hydroxide ions changes.

$$NH_4Cl \rightleftharpoons NH_4^+ + Cl^-$$

$$\underbrace{NH_4^+ + Cl^-} + \underbrace{H^+ + OH^-}_{\substack{\uparrow\downarrow \\ H_2O}} \rightleftharpoons NH_4OH + \underbrace{H^+ + Cl^-}$$

The ammonium ions from the salt bind with the hydroxide ions from the water in the weak and perishable ammonium hydroxide. Hydrogen ions remain in excess in the solution and this determines the acid character of the solution. $1 < \pi < 2$.

Ammonia interacts with water. The proton from water by the mechanism of donor-acceptor interaction binds to the uncharged electron pair and forms ammonium ion. It, with the hydroxide ion from water, forms the weak and perishable ammonium hydroxide. NH_4OH dissociates reversibly and the aqueous solution has a basic character.

$1 < \pi < 2$.

4. CHEMICAL ELEMENTS

ISOTOPES. TYPES OF CHEMICAL ELEMENTS.
S-ELEMENTS, P-ELEMENTS, D-ELEMENTS, F-ELEMENTS. ELECTRONIC AFFINITY. IONIZATION ENERGY. ELECTRONEGATIVITY. DEGREE OF OXIDATION. ELEMENTS OF METALLIC CHEMICAL CHARACTER. ELEMENTS OF NON-METALLIC CHEMICAL CHARACTER. CHEMICAL PROPERTIES OF METALS. INTERACTION OF METALS WITH HYDROGEN. INTERACTION OF METALS WITH OXYGEN. INTERACTION OF METALS WITH NON-METALS. INTERACTION OF METALS WITH WATER. INTERACTION OF METALS WITH ACIDS. INTERACTION OF METALS WITH SALTS. INTERACTION OF METALS WITH BASES. CHEMICAL PROPERTIES OF NON- METALS. INTERACTION OF NON-METALS WITH HYDROGEN. INTERACTION OF NON-METALS WITH OXYGEN. INTERACTION WITH OXYGEN-FREE SALTS OF NON-METALS, WHICH HAVE WEAK OXIDIZING PROPERTIES. INTERACTION WITH NON-METALS. INTERACTION WITH METALS.

According to empirical notions, a chemical element is defined as an indecomposable constituent of simple and complex substances. In terms of atomic-molecular theory, a chemical element is defined as a collection of atoms with the same chemical properties.

On the basis of the modern theory of the structure of the atom, the term **chemical element refers to** a collection of **atoms with the same number of protons in their atomic nuclei**. The most significant feature of atoms is the **number of protons in their nuclei**, which determines the nuclear charge, the number of electrons in the electron shell, the sequence number and the place of each element in the Periodic Table.

Isotopes

From this point of view, **isotopes are varieties of atoms of the same chemical element that have the same number of protons in their nuclei but different mass numbers due to the different number of neutrons.** In nature, most chemical elements exist as a mixture of isotopes, so their atomic masses are usually not whole numbers.

It is customary to write the number of protons (sequence number) as a subscript before the sign of the chemical element, and the mass number as a superscript. For example, the chemical element hydrogen is a mixture of three isotopes:

- **Protium**, which has a proton in its nucleus - ${}^{1}_{1}H$;
- **Deuterium**, which has one proton and one neutron in its nucleus - ${}^{2}_{1}H$;
- **Tritium**, which has one proton and two neutrons in its nucleus - ${}^{3}_{1}H$.

The chemical element helium (He) is a mixture of four isotopes:

$${}^{3}_{2}He\ ;\ {}^{4}_{2}He;\ {}^{5}_{2}He\ ;\ {}^{6}_{2}He$$

and the chemical element oxygen - a mixture of three isotopes:

$${}^{16}_{8}O, \quad {}^{17}_{8}O, \quad {}^{18}_{8}O$$

The number of known isotopes exceeds 1300. There are also "pure" elements that consist of only one atomic species - e.g., sodium (Na), beryllium (Be), fluorine (F_2).

Atoms are chemically indivisible particles, so, **in chemical interactions, the number of protons in their nuclei is conserved.** Under the conditions of chemical reactions, the chemical elements are conserved, but the structure of the outer electron layers of the atom changes. A complete characterization of a chemical element includes:

- **sequence number and the place of the element in the Periodic Table;**
- **structure of the electronic shell of its atoms;**
- **its valence in the ground and excited states, as well as its characteristic degrees of oxidation;**
- **the activity of its atoms under conditions of chemical interaction;**
- **the type of simple substance or simple substances in the presence of allotropic species;**
- **the properties of oxides, hydroxides or oxoacids of elements.**

Types of chemical elements

According to modern ideas about the structure of the atom, it is made of a **nucleus and an electron shell**. The nucleus **is made up of protons and neutrons** with the **number of protons determining the element's sequence number, its place in the Periodic Table and which chemical element it is. The structure of the electron shell determines most of the characteristics of a chemical element. The electron shell of atoms is built up in a regular way by successively filling the s-, p-, d-, and f-sublayers with electrons. The following general rules are followed in filling the allowed shell states with electrons:**

- **electrons occupy the lowest energy levels (Kleczkowski's rule);**
- **the Pauli principle;**
- **Hund's rule.**

The electronic configuration of the atoms of the chemical elements and the values of the nuclear charges and atomic radii determines the chemical properties of the elements and their compounds.

The chemical character of elements is determined by the ability of their atoms to give up or accept electrons, i.e., by their reducing and oxidizing power. This ability depends on the structure of their electron shells and the size of their atomic radius. Depending on this they are conventionally divided into s-, p-, d- and f-elements.

Based on these facts, the following conclusions can be drawn:

Elements with reducing properties (s-, p-, d- and f-elements) form simple metal substances. The s-elements form basic oxides and hydroxides. The p-, d- and f-elements are characterized by basic oxides and hydroxides in their lower oxidation states (+2, +3) and amphoteric in their intermediate oxidation states.

The chemical properties of metals and non-metals differ mainly in their activity towards hydrogen and oxygen. Non-metals exhibit greater chemical activity towards hydrogen and metals, while metals exhibit greater chemical activity towards oxygen and non-metals. Many of the reactions take place under ordinary conditions for the typical metals (group IA) and the typical nonmetals (group VIIA) or under heating for the other metals and nonmetals. Therefore, each metal or non-metal must be considered as a state of matter which, depending on conditions, may exhibit varying degrees of metallic or dual character as well as non-metallic or dual character.

Chemical elements are divided into s-, p-, d- and f-elements.

s- Elements

Elements in which the **s-sublayer of the outer electron layer** of atoms is built up and characterized by an electron configuration ns^{1-2}, are called **s-elements**. They are located in the left part of the Periodic Table (groups I A and II A). Their second outward electron layer, conventionally called the penultimate layer, has the structure of an inert gas. For example, sodium (Na) has the following electronic configuration - $1s^2\ 2s^2\ 2p^6\ 3s^1$, with its penultimate electron layer having the configuration of the neon (Ne) -. In chemical interactions, the **s- element** atoms give up the weakest bound s-electrons from their outer layer and form stable positively charged ions of the form:

$$E^{+1};\ E^{+2}$$

These elements have **low values of ionization energy, electron affinity and electronegativity.** Their activity **increases in each group as their sequence numbers increase.**

In s-elements, valence electrons occupy the **s-AO.**

p- Elements

The elements located in the right part of the Periodic Table, which **make up the p- sublayer of the outer electron layer** and are characterized by an electronic configuration $ns^2\ np^{1-6}$ are called **p-elements.**

They exhibit a variable degree of oxidation, varying in sign and numerical value. The maximum degree of oxidation of a p-element is equal to the sum of the total number of valence s- and p-electrons.

For example, the maximum oxidation state of sulfur is six, since the sum of the total number of valence s- and p-electrons is six ($3\ s^2\ 3p^4$).

In a given main group, the atoms of the first p-element exhibit certain features compared to the other elements of the group.

The atoms of the **p-elements of the third, fourth, etc. periods,** in addition to their valence s- and p-electrons, can also form bonds with the **d-AO** of the penultimate layer (in the formation of complex compounds).

These **differences significantly affect the degree of oxidation of p- elements, the properties of simple substances and of chemical compounds.** In chemical reactions, atoms of the **p-elements with a larger number of electrons in the outer electron layer exhibit predominantly oxidative action** relative to atoms with reducing properties. These **p-elements have higher values of ionization energy, electron affinity and electronegativity.** Atoms of **p-elements with a lower number of p-electrons in the outer layer are chemically less active and give up their valence electrons less easily**. They are weaker reducers than the s-elements. **The p-elements of group VIIIA** have a stable outer electron layer and are therefore chemically weakly active. They are characterized by **high ionization energy values and low electron affinity values. These are so-called inert elements.**

d- Elements

The elements that occupy the secondary groups of the Periodic Table from Group IB to Group VIIIB are called **d-elements**. They are found only in the major periods. In the electron shell of their atoms, **the d-sublayer of the second electron layer is built from the outside in**. In most cases, they have an electronic configuration of $ns^2\ (n-1)d^{1-10}$.

They are characterized by a higher value of ionization energy, exhibit a variable and always positive degree of oxidation. For example, chromium (Cr) is characterised by +2, +3 and +6 oxidation states.

In d- elements, valence electrons **occupy s- and d-AO.**

f- Elements

Lanthanides and actinides are f-elements. In the electron shell of their atoms **the f- sublayer of the third outward-inward electron layer is built up**.

f- Elements of the lanthanide group are characterized by a variable and always positive degree of oxidation (most often +3).

f- Elements of the actinide group exhibit a greater number of oxidation grades than the lanthanides. This is explained by the close energy of the 5f-, 6d- and 7s-sublayers in the shells of their atoms. Therefore, the following conclusion can be drawn:

1. The ionization energy increases with increasing sequence number of elements by period, and decreases in the main groups.

2. Period by period, as the ordinal numbers increase, the electronic affinity usually increases, while in the main groups it decreases.

3. In general, as the sequence number increases by period, the electronegativity increases and by group decreases.

4. Period by period, as the sequence number of the elements increases, their higher oxidation state increases from +1 to +7. In the case of the f-elements, the valence electrons occupy - **s-, d- and very rarely f-AO.**

For chemical interactions between atoms of chemical elements, so-called **valence electrons** are important. Valence electrons are electrons with which an atom of a chemical element **can participate in the creation of a chemical bond**.

Electronic affinity

Electron affinity is determined by the energy released when an electron attaches to the electron shell of a free atom:

$$E + \bar{e} \longrightarrow \bar{E}$$

Ionisation energy

The redox activity of chemical elements, i.e., their ability to give up or accept electrons. Quantitative measures of this ability are the quantities of **ionization energy and electron affinity. Ionization energy is the energy required to detach the weakest bound electrons from a free atom in the ground state.**

$$E \longrightarrow E^{+} + \bar{e}$$

It can be determined experimentally and is marked in electron volts.

Electronegativity

Electronegativity characterizes the ability of atoms of one chemical element to attract electron density when covalently bonded to atoms of another chemical element.

Degree of oxidation

Degree of oxidation refers to the **basic charge of the atoms of the** chemical elements in a compound**, assuming that all bonds are ionic.** Elements of greater electronegativity conditionally accept electrons and exhibit a negative oxidation state, and elements of lesser electronegativity conditionally give up electrons and exhibit a positive oxidation state. The degrees of oxidation of the same chemical element may vary in value and sign in different compounds depending on the elements with which the bonds are realized.

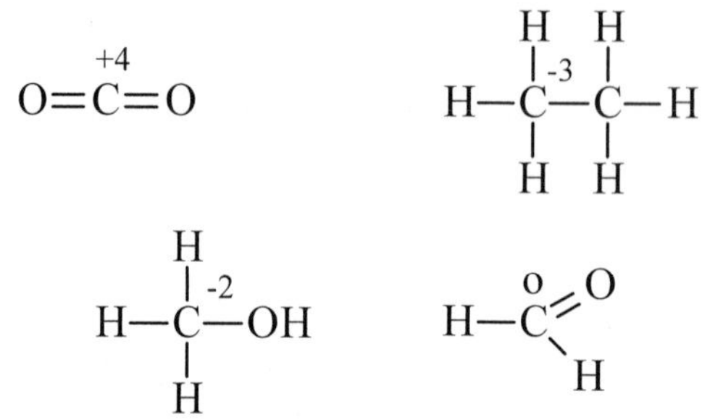

Elements of metallic chemical character

Elements of metallic chemical character are those which exhibit reducing properties, forming simple metallic substances and basic or amphoteric oxides and hydroxides.

The s-Elements, with the exception of helium (He), most readily give up electrons from their outermost electron layer and accordingly exhibit the highest reduction activity. They are characterized by the lowest values of ionization energy, electron affinity and electronegativity. Their simple substances are chemically active metals. Their hydrogen compounds are solids, and their oxides and hydroxides have strongly marked basic properties. The compounds corresponding to them are of ionic-crystalline structure. The elements of group IA have the most pronounced metallic character. Reducing properties are also exhibited by d-elements when they participate in chemical reactions with only s-electrons from the outer electron layer. For example, iron (Fe) in its outer electron layer has the following configuration $4s^23d^6$. Compounds of divalent iron are largely ionic in nature. Iron manifests itself as a reducing agent. Similarly, the f-elements exhibit their reducing properties.

d- and f-Elements in their lower oxidation state form basic oxides and hydroxides, and in the intermediate – amphoteric. Their oxides and hydroxides exhibit basic or amphoteric character.

Acid oxides and oxoacids form only the d-elements in their higher oxidation states - e.g., Mn_2O_7 (dimanganese heptaoxide) and the corresponding acid $HMnO_4$ (permanganic acid).

Elements of a non-metallic chemical nature

Chemical elements that exhibit oxidizing properties, form simple substances non- metals, acid oxides and oxoacids, are elements of non-metallic character. Non-metallic chemical character is exhibited by p-elements with a larger number of electrons in their outer electron layer (4 to 7) and with a smaller atomic radius, to the right of the B/At line. These p- elements are characterized by high values of ionization energy, electron affinity and electronegativity. The compounds in the form in which the non-metallic elements exist are usually of molecular structure – carbon, hydrogen, oxygen – and between themselves they form covalent compounds with different bond polarities. Depending on the nature of the other chemical element, these elements can exhibit both oxidizing and reducing ability when chemically interacting. For example, sulfur is a p-element with an electronic configuration of the valence electron layer of $3s^23p^4$. When it interacts with hydrogen, it exhibits its oxidizing ability,

$$S + H_2 \longrightarrow H_2S$$

$$S^0 + 2\bar{e} \longrightarrow S^{-2} \quad | \ 2 \ | \quad | \ 1$$

$$\phantom{S^0 + 2\bar{e}} \quad | \ 2 \ |$$

$$H_2 \longrightarrow 2H^0 - 2 \times 1\bar{e} \longrightarrow 2H^+ \quad | \ 2 \ | \quad | \ 1$$

,

and in interaction with oxygen – its reducing ability.

$$S + O_2 \longrightarrow SO_2$$

$$\begin{array}{lll|c|c|c}
 & S^0 - 4\bar{e} & \longrightarrow S^{+4} & 4 & & 1 \\
 & & & & 4 & \\
O_2 \longrightarrow & 2O^0 + 2\times 2\bar{e} & \longrightarrow 2O^{-2} & 4 & & 1
\end{array}$$

With hydrogen, oxygen forms two compounds: water and hydrogen peroxide (H_2O and H_2O_2). With the exception of water, the hydrides of the elements of group VIA are gases with pungent, unpleasant odors. They are toxic. Their stability decreases in the order:

$$H_2O > H_2S > H_2Se > H_2Te$$

Their aqueous solutions are acidic in nature. In the major groups, the non-metallic character weakens with an increasing ordinal number. The most pronounced non-metallic character is found in the p-elements of group VIIA. Elements of group VIIIA, which are characterized by complete electronic configurations of the outermost layer, are distinguished by very low chemical activity and are therefore called inert elements. The atoms of inert elements show high ionisation energy values and low electron affinity values.

Chemical properties of metals

A metallic crystal lattice is characteristic of metals. In a metallic crystal lattice, the valence electrons do not form common electron pairs, but are common to all the atoms and ions that make up the lattice. The electrons are in continuous motion, like particles in a gas phase, and so by analogy they are called free electrons or "electron gas". There are electrical interactions between the positive ions and the "electron gas", which is also responsible for the strength of the crystal lattice. The chemical bond in a metallic crystal lattice is called **a metallic bond**. The nature of the bonds is explained by the so-called zone theory. According to this theory, in the metallic crystal lattice the interacting particles have completely lost their individuality. The valence atomic orbitals of adjacent closely spaced metal atoms overlap and interact with each other, forming one or more partially electron-filled energy zones (s-, p-, d-, etc.). The electron zones are completely delocalized, relatively free and very mobile. The metallic bond is very strong, unsaturated and undirected. It can also be seen as an extreme degree of delocalization of the covalent bond between identical atoms. Metals usually crystallize in a cubic or hexagonal crystal lattice. The similar structure of the metals determines their similar aggregate state under ordinary conditions - solid (except for Hg), and their common physical properties. The metals are characterized by metallic lustre, high density, malleability and ductility, thermal and electrical conductivity.

Metals are exclusively referred to as reducers. Depending on the nature of the partner in the chemical process, some of the non-metals also exhibit reducing properties to a lesser extent. Metals interact with hydrogen, oxygen and other non-metals, with water, with salts of metals which, in the order of relative activity, stand to the right of the metal concerned.

K, Ca, Na, Mg, Al, Zn, Cr, Fe, Ni, Sn, Pb, H, Cu, Hg, Ag, Au

$$\xrightarrow{\hspace{4cm}}$$

decreasing activity

order of relative activity of metals

Metal atoms give up electrons from their outer electron layer and positively charged ions (cations) are formed.

$$\underset{\text{metal}}{M} - ne^- \longrightarrow M^{n+}$$

In these processes, the degree of oxidation of the atoms of the elements with metallic character increases. The metals therefore act as reducing agents towards hydrogen, oxygen, other non-metals, the hydrogen ions of water and the acids and metal ions of aqueous solutions of the salts concerned.

The main types of chemical reactions involving metals can be expressed by the following examples:

Interaction of metals with hydrogen

The compounds of metals with hydrogen are called hydrides. They are solids, readily dissolve in water and have a markedly high reactivity. In these reactions, hydrogen appears as an oxidizing agent.

$$2\,Na + H_2 \longrightarrow \overset{+1}{Na}\overset{-1}{H}$$

$$\begin{array}{lll|c|c|c} & \overset{o}{Na} - 1\bar{e} & \longrightarrow \overset{+1}{Na} & 1 & \multirow{2}{*}{2} & 2 \\ H_2 \longrightarrow & 2\,\overset{o}{H} + 2\text{x}1\bar{e} & \longrightarrow 2\,\bar{H} & 2 & & 1 \end{array}$$

$$Ca + H_2 \longrightarrow \overset{+2}{Ca}\overset{-1}{H_2}$$

$$\begin{array}{lll|c|c|c} & \overset{o}{Ca} - 2\bar{e} & \longrightarrow \overset{+2}{Ca} & 2 & \multirow{2}{*}{2} & 1 \\ H_2 \longrightarrow & 2\,\overset{o}{H} + 2\text{x}1\bar{e} & \longrightarrow 2\,\bar{H} & 2 & & 1 \end{array}$$

Interaction of metals with oxygen

This reaction is more characteristic of metals than their interaction with hydrogen. The activity of metals towards oxygen increases as their metallic properties increase. For example, the activity of group IA metals increases from lithium to cesium. These interactions produce **oxides**.

$$2\,Mg + O_2 \longrightarrow \overset{+2}{Mg}\overset{-2}{O}$$

$$\begin{array}{lll|c|c|c} & \overset{o}{Mg} - 2\bar{e} & \longrightarrow \overset{+2}{Mg} & 2 & \multirow{2}{*}{4} & 2 \\ O_2 \longrightarrow & 2O^{o} + 4e^- & \longrightarrow 2\,O^{-2} & 4 & & 1 \end{array}$$

$$4\,Al + 3\,O_2 \longrightarrow 2\,Al_2O_3$$

$$\begin{array}{lll|c|c|c} & Al^{0} - 3\bar{e} & \longrightarrow Al^{+3} & 3 & \multirow{2}{*}{12} & 4 \\ O_2 \longrightarrow & 2\,O^{0} + 4\bar{e} & \longrightarrow 2\,O^{-2} & 4 & & 3 \end{array}$$

Interaction of metals with non-metals

These processes produce salts.

$$Fe + S \longrightarrow \overset{+2}{Fe}\overset{-2}{S}$$

$$\begin{array}{ll|c|c|c} Fe^{0} - 2\bar{e} & \longrightarrow Fe^{+2} & 2 & \multirow{2}{*}{2} & 1 \\ S^{0} + 2\bar{e} & \longrightarrow S^{-2} & 2 & & 1 \end{array}$$

Interaction of metals with water

Metals that are ahead of hydrogen in the order of relative activity react with water, releasing hydrogen and producing **hydroxides**.

$$Ca + \underbrace{2\,H^{+} + 2\,\bar{O}H}_{\uparrow\downarrow\; 2\,H_2O} \longrightarrow \underbrace{Ca^{+2} + 2\,\bar{O}H} + H_2\uparrow$$

$$\begin{array}{lll|c|c|c} & Ca^{0} - 2\bar{e} & \longrightarrow Ca^{+2} & 2 & \multirow{2}{*}{2} & 1 \\ 2\,H^{+} + 2\bar{e} & \longrightarrow 2\,H^{0} & \longrightarrow H_2\uparrow & 2 & & 1 \end{array}$$

Interaction of metals with acids

With respect to acids, metals are treated differently. Those metals, whose atoms are stronger reducers than hydrogen, react with dilute acids reducing the proton from the acids, and metals whose atoms are weaker reducers than hydrogen (located after it in the Order of Relative Activity of the Metals (ORAM)) react with concentrated acids, which have a strong oxidizing effect.

$$2\,Al + \underbrace{6\,H^{+} + 3\,SO_4^{-2}} \longrightarrow \underbrace{2\,Al^{+3} + 3\,SO_4^{-2}} + 3\,H_2\uparrow$$

$$\begin{array}{lll|c|c|c} & Al^{0} - 3\bar{e} & \longrightarrow Al^{+3} & 3 & \multirow{2}{*}{6} & 2 \\ 2\,H^{+} + 2\bar{e} & \longrightarrow 2\,H^{0} & \longrightarrow H_2\uparrow & 2 & & 3 \end{array}$$

Some metals of oxidizing oxoacids (Al, Cr, Fe) form an oxide layer on the surface that protects them from further changes. The process is known as **passivation.**

Interaction of metals with salts

Metals that are before hydrogen in the order of relative activity can interact with salts of metals that are after hydrogen in the order of relative activity.

$$\overset{0}{Fe} + \underbrace{Cu^{+2} + SO_4^{-2}} \longrightarrow \underbrace{Fe^{+2} + SO_4^{-2}} + \overset{0}{Cu}$$

$$\begin{array}{lcl|c|c|c} \overset{0}{Fe} - 2\bar{e} & \longrightarrow & \overset{+2}{Fe} & 2 & \multirow{2}{*}{2} & 1 \\ \overset{+2}{Cu} + 2\bar{e} & \longrightarrow & Cu^{0} & 2 & & 1 \end{array}$$

Interaction of metals with bases

Some of the less active metals can react with acids as well as bases:

$$Zn^{0} + 2NaOH + H_2O \longrightarrow Na_2[Zn(OH)_4] + H_2\uparrow$$

disodium tetrahydroxyzincate

$$\begin{array}{lcl|c|c|c} Zn^{0} - 2\bar{e} & \longrightarrow & Zn^{+2} & 2 & \multirow{2}{*}{2} & 1 \\ 2H^{+} + 2\times 1\bar{e} \longrightarrow 2\overset{0}{H} & \longrightarrow & H_2 & 2 & & 1 \end{array}$$

Depending on the activity of the metals, reactions can take place under both ordinary and heated conditions. For example, the simple substances of many of the d-elements do not react under ordinary conditions, but their activity increases as the temperature increases.

Chemical properties of non-metals

In chemical reactions involving non-metals, atoms of elements of non-metallic character exhibit oxidizing action against atoms of elements of metallic character, ions of elements of non-metallic character being weaker oxidizers. With the atoms of other elements of non-metallic character forming compounds with varying degrees of polarity of bonds in the molecules. Non- metals interact with hydrogen, oxygen (there are non-metals that cannot react directly with oxygen), with each other, with metals, with oxygen-free salts of elements of non-metallic character that are weaker oxidants.

Interaction of non-metals with hydrogen

When interacting with hydrogen, non-metals act as oxidizing agents, accepting electrons and lowering their degree of oxidation:

$$N_2 + 3H_2 \longrightarrow 2NH_3$$

$$\begin{array}{lll|c|c|c} N_2 \longrightarrow & 2\overset{0}{N} + 2 \times 3\bar{e} \longrightarrow & 2N^{-3} & 6 & \multirow{2}{*}{6} & 1 \\ H_2 \longrightarrow & 2\overset{0}{H} - 2 \times 1\bar{e} \longrightarrow & 2H^{+} & 2 & & 3 \end{array}$$

$$O_2 + 2H_2 \longrightarrow 2H_2O$$

$$\begin{array}{lll|c|c|c} O_2 \longrightarrow & 2\overset{0}{O} + 2 \times 2\bar{e} \longrightarrow & 2O^{-2} & 4 & \multirow{2}{*}{4} & 1 \\ H_2 \longrightarrow & 2\overset{0}{H} - 2 \times 1\bar{e} \longrightarrow & 2H^{+} & 2 & & 2 \end{array}$$

$$\overset{0}{Cl_2} + \overset{0}{H_2} \longrightarrow 2HCl$$

$$\begin{array}{lll|c|c|c} Cl_2 \longrightarrow & 2\overset{0}{Cl} + 2 \times 1\bar{e} \longrightarrow & 2Cl^{-} & 2 & \multirow{2}{*}{2} & 1 \\ H_2 \longrightarrow & 2\overset{0}{H} - 2 \times 1\bar{e} \longrightarrow & 2H^{+} & 2 & & 1 \end{array}$$

Hydrogen compounds of the elements of non-metallic character have a molecular structure. Reaction with hydrogen is more characteristic of non-metals since hydrogen has an electronic configuration similar to metals.

Interaction of non-metals with oxygen

Fluorine and inert gases do not react with oxygen. Reaction with oxygen is more characteristic of metals, so the interaction of non-metals with oxygen takes place at high temperatures.

$$S + O_2 \xrightarrow{t^0} \overset{+4}{S}\overset{-2}{O_2}$$

$$\begin{array}{lll|c|c|c} & S^0 - 4\bar{e} \longrightarrow & S^{+4} & 4 & \multirow{2}{*}{4} & 1 \\ O_2 \longrightarrow & 2O^0 + 2 \times 2\bar{e} \longrightarrow & 2O^{-2} & 4 & & 1 \end{array}$$

$$4P + 3O_2 \xrightarrow{t^0} 2P_2O_3$$

$$\begin{array}{lll|c|c|c} & P^0 - 3\bar{e} \longrightarrow & P^{+3} & 3 & \multirow{2}{*}{12} & 4 \\ O_2 \longrightarrow & 2O^0 + 2 \times 2\bar{e} \longrightarrow & 2O^{-2} & 4 & & 3 \end{array}$$

Interaction with oxygen-free salts of non-metals, which have weak oxidizing properties

Interaction with non-metals

$$F_2 + \underbrace{2\overset{+}{K} + 2\overset{-}{Cl}} \longrightarrow \underbrace{2\overset{+}{K} + 2\overset{-}{F}} + Cl_2$$

$$F_2 \longrightarrow 2\overset{o}{F} + 2 \times 1\bar{e} \longrightarrow 2F^- \quad \left| 2 \right| 2 \left| 1 \right.$$

$$2\overset{-}{Cl} - 2 \times 1\bar{e} \longrightarrow 2\overset{o}{Cl} \longrightarrow Cl_2 \quad \left| 2 \right| \quad \left| 1 \right.$$

$$2P + 3Cl_2 \longrightarrow 2PCl_3$$

$$\overset{o}{P} - 3e^- \longrightarrow P^{+3} \quad \left| 3 \right| 6 \left| 2 \right.$$

$$Cl_2 \longrightarrow 2\overset{o}{Cl} + 2 \times 1e^- \longrightarrow 2\overset{-}{Cl} \quad \left| 2 \right| \quad \left| 3 \right.$$

Interaction with metals

$$N_2 + 3Mg \longrightarrow Mg_3N_2$$

magnesium nitride

$$\overset{o}{Mg} - 2\bar{e} \longrightarrow Mg^{+2} \quad \left| 2 \right| 6 \left| 3 \right.$$

$$N_2 \longrightarrow 2\overset{o}{N} + 2 \times 3e^- \longrightarrow 2N^{-3} \quad \left| 6 \right| \quad \left| 1 \right.$$

$$2P + 3Mg \longrightarrow Mg_3P_2$$

magnesium phosphide

$$\overset{o}{Mg} - 2\bar{e} \longrightarrow Mg^{+2} \quad \left| 2 \right| 6 \left| 3 \right.$$

$$\overset{o}{P} + 3e^- \longrightarrow P^{-3} \quad \left| 3 \right| \quad \left| 2 \right.$$

$$3S + 2Al \longrightarrow Al_2S_3$$

aluminium sulfide

$$\overset{o}{Al} - 3\bar{e} \longrightarrow Al^{+3} \quad \left| 3 \right| 6 \left| 2 \right.$$

$$\overset{o}{S} + 2e^- \longrightarrow S^{-2} \quad \left| 2 \right| \quad \left| 3 \right.$$

5. Chemical Properties of the Elements by Group in the Periodic Table

Group IA of the Periodic Table

Group IA includes the alkali metal elements lithium (Li), sodium (Na), potassium (K), rubidium (Rb), caesium (Cs) and francium (Fr).

Elements of group IA **interact actively with hydrogen**. Lithium reacts most easily. The resulting compounds are called hydrides and are solids with an ionic crystal lattice:

$$2Li + H_2 \longrightarrow 2LiH$$

$$\begin{array}{lll|l|l|l} & \overset{0}{Li} - e^- & \longrightarrow Li^+ & 1 & & 2 \\ & & & & 2 & \\ H_2 \longrightarrow & 2\overset{0}{H} + 2x1e^- & \longrightarrow 2H^- & 2 & & 1 \end{array}$$

Due to their strong reducing properties, alkali metals are reducers even towards hydrogen, which only in this interaction manifests itself as an oxidant and is in the oxidation state (-1).

Alkali metals **interact actively with oxygen**, and the activity is enhanced from Li to Cs.

$$4Li + O_2 \longrightarrow 2Li_2O$$

$$\begin{array}{lll|l|l|l} & \overset{0}{Li} - e^- & \longrightarrow Li^+ & 1 & & 4 \\ & & & & 4 & \\ O_2 \longrightarrow & 2\overset{0}{O} + 2x2e^- & \longrightarrow 2\overset{-2}{O} & 4 & & 1 \end{array}$$

Only lithium forms only Li_2O. Sodium and potassium form oxides under special conditions - under cooling and oxygen deficiency, and peroxides in excess of oxygen.

$$2K + O_2 \longrightarrow K_2O_2$$

$$\begin{array}{lll|l|l|l} & \overset{0}{K} - e^- & \longrightarrow K^+ & 1 & & 2 \\ & & & & 2 & \\ O_2 \longrightarrow & 2\overset{0}{O} + 2x1e^- & \longrightarrow 2\overset{-1}{O} & 2 & & 1 \end{array}$$

The remaining alkali elements self-ignite and burn to peroxides (E_2O_2 and superoxides (EO_2).

Alkali metals react very actively with **halogen elements**. The activity decreases from fluorine (F_2) to iodine (I_2) and from caesium (Cs) to lithium (Li).

$$2Na + Cl_2 \longrightarrow 2NaCl$$

$$\begin{array}{lll|l|l|l} & \overset{0}{Na} - e^- & \longrightarrow Na^+ & 1 & & 2 \\ & & & & 2 & \\ Cl_2 \longrightarrow & 2\overset{0}{Cl} + 2x1e^- & \longrightarrow 2Cl^- & 2 & & 1 \end{array}$$

Mixing the alkali elements with sulphur results in **sulfides** with an ionocrystalline lattice, well soluble in water:

$$2\,Na + S \longrightarrow Na_2S$$

$$\begin{array}{l|c|c|c} \overset{0}{Na} - e^{-} \longrightarrow Na^{+} & 1 & & 2 \\ & & 2 & \\ \overset{0}{S} + 2e^{-} \longrightarrow S^{-2} & 2 & & 1 \end{array}$$

In relation to **nitrogen,** the activity of alkaline elements decreases. Only lithium reacts under ordinary conditions, the other elements at very high temperatures. The compounds are called **nitrides**.

$$6\,Li + N_2 \longrightarrow 2\,Li_3N$$

$$\begin{array}{l|c|c|c} \overset{0}{Li} - e^{-} \longrightarrow Li^{+} & 1 & & 6 \\ & & 6 & \\ N_2 \longrightarrow 2\,\overset{0}{N} + 2 \times 3e^{-} \longrightarrow 2\,N^{-3} & 6 & & 1 \end{array}$$

Alkaline elements react very actively with **water**. The activity is enhanced from lithium to cesium.

$$2K + 2\,H_2O \longrightarrow 2\,KOH + H_2\uparrow$$

$$\begin{array}{l|c|c|c} \overset{0}{K} - e^{-} \longrightarrow K^{+} & 1 & & 2 \\ & & 2 & \\ 2\,H^{+} + 2 \times 1e^{-} \longrightarrow 2\,\overset{0}{H} \longrightarrow H_2 & 2 & & 1 \end{array}$$

The reaction is highly exothermic; therefore, the alkaline elements are stored under oil and should **never be touched with hands!**

The activity of alkaline elements is even greater against **acids** than against water because in an acidic environment, the concentration of hydrogen ions is significantly higher than in pure water.

$$2\,Na + 2\,HCl \longrightarrow 2\,NaCl + H_2\uparrow;$$

$$\begin{array}{l|c|c|c} \overset{0}{Na} - e^{-} \longrightarrow Na^{+} & 1 & & 2 \\ & & 2 & \\ 2\,H^{+} + 2 \times 1e^{-} \longrightarrow 2\,\overset{0}{H} \longrightarrow H_2 & 2 & & 1 \end{array}$$

The alkali elements are produced by electrolysis of a melt of their chlorides. The alkali elements or their ions can be identified using an alcohol lamp.

The lithium tints the flame **carmine red,**
The sodium – in **yellow,**
Potassium – in **violet**

Sample problems and solutions

Task 1.

Make a characterization of the chemical element **Na**. Express with equations the transitions indicated on the diagram explaining the processes involved.

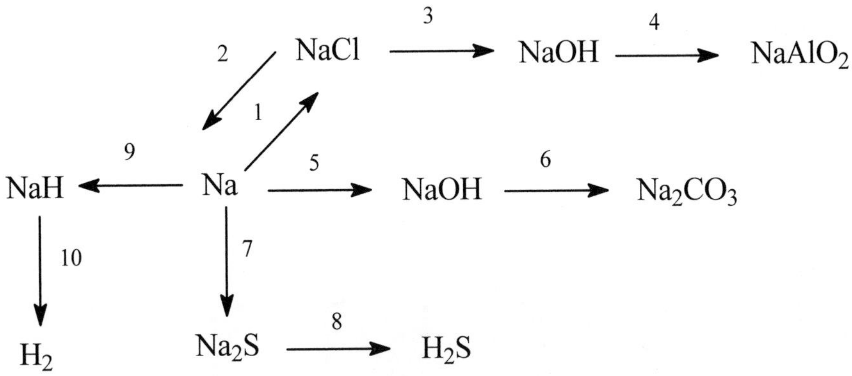

Exemplary solution

Sodium is a chemical element with the sequence number 11. It contains 11 protons in the atomic nucleus and 11 electrons in the electron shell, distributed in layers and sublayers according to the principles and rules of electron shell construction. The electronic formula of sodium is $1s^2\ 2s^2\ 2p^6\ 3s^1$.

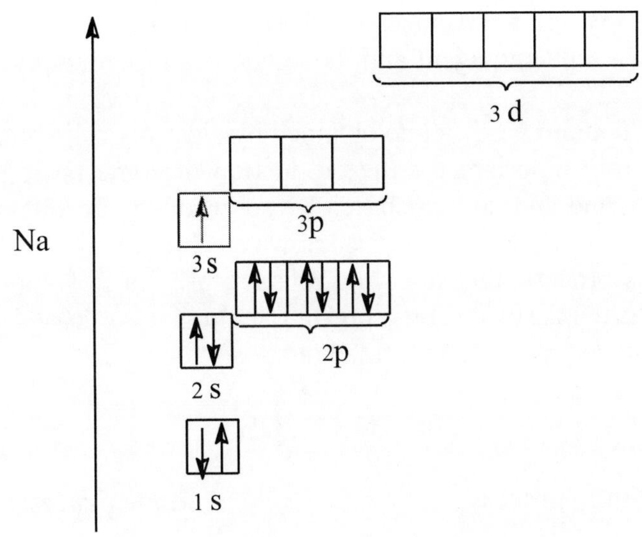

Sodium is in group I A, period III of the Periodic Table. An alkali metal with high chemical activity.

Sodium interacts very actively with halogen elements. With chlorine it forms sodium chloride (1). The process is redox.

$$2Na + Cl_2 \longrightarrow 2NaCl \quad (1)$$

$$\overset{o}{Na} - e^- \longrightarrow Na^+ \quad \left|\ 1\ \right|\ 2\ \left|\ 2\right.$$

$$Cl_2 \longrightarrow 2\,Cl^o + 2\times1\,e^- \longrightarrow 2\,Cl^- \quad \left|\ 2\ \right|\ \ \left|\ 1\right.$$

Sodium can be obtained from sodium chloride by melt electrolysis. The redox processes that take place in melts or in solutions of electrolytes under the action of an electric current are called electrolytic, and the phenomenon is called electrolysis. Sodium cations and chloride anions move in the sodium chloride melt. When a constant electric current is passed through the melt, the chaotically moving ions are directed to the electrodes. The cations at the cathode accept electrons, i.e., are reduced and converted to neutral sodium atoms. Negative ions are directed to the anode. There the chloride anions give up electrons, i.e., are oxidized and converted to neutral chlorine atoms.

The processes can be expressed by the following scheme:

Available ions in the melt:

$$NaCl \longrightarrow \overset{+}{Na} + \overset{-}{Cl}$$

$$\overset{+}{Na} + \bar{e} \longrightarrow \overset{o}{Na}$$

(cathode process)
(2)

$$2\,\overset{-}{Cl} - 2 \times 1\,\bar{e} \longrightarrow 2\,\overset{o}{Cl} \longrightarrow Cl_2\uparrow$$

(anode process)

These processes are redox and are carried out under the action of an electric current. When it flows, the cathode acts as a reducing agent and the anode is an oxidizing agent. Electrolysis takes place in the melt.

From sodium chloride, sodium base can be obtained by electrolysis of an aqueous solution.

Electrolysis processes are important for the production of some basic products for inorganic synthesis - hydrogen, chlorine and sodium base. These products are obtained by electrolysis of aqueous sodium chloride solution. The process is known as chloralkali electrolysis.

Prior to the electrolysis process, the following ions are present in solution: sodium, hydrogen, chloride and hydroxide. During the electrolysis process, hydrogen is released at the cathode, chlorine is released at the anode and sodium base is formed in the solution.

Na^+, H^+, Cl^-, OH^-

cathode process:

$$2\,\overset{+}{H} + 2 \times 1\,e^- \longrightarrow 2\,\overset{o}{H} \longrightarrow H_2\uparrow$$

anode process:

$$2\,\overset{-}{Cl} - 2 \times 1\,\bar{e} \longrightarrow 2\,\overset{o}{Cl} \longrightarrow Cl_2\uparrow$$

Ions in solution (after electrolysis):

$$\underbrace{\overset{+}{Na} + \overset{-}{OH}}$$

(3)

At the cathode, hydrogen is emitted, not sodium, because hydrogen stands after sodium in the Order of Relative Activity of the Metals (ORAM). At the anode, chloride anions are electroneutralized, not hydroxide, because chloride ions come before hydroxide ions in the electrochemical order of the anions.

The solution produces sodium base, which could react with the chlorine that is released. Measures are taken, therefore, in the construction of the electrolysis bath to separate these substances.

$$2Al + 2NaOH + 2H_2O \xrightarrow{t^o} 2NaAlO_2 + 3H_2\uparrow$$

$$\begin{array}{lll|c|c|c} & Al^{0} - 3\bar{e} & \longrightarrow Al^{+3} & 3 & \multirow{2}{*}{6} & 2 \\ 2H^{+} + 2\times 1\bar{e} \longrightarrow & 2H^{0} & \longrightarrow H_2\uparrow & 2 & & 3 \end{array}$$

(4)

The water required for the reaction is not added because it is present in the solid sodium base due to its hygroscopicity.

Sodium interacts vigorously with water. Sodium base is obtained (5).

$$2Na + 2H{-}OH \longrightarrow 2NaOH + H_2\uparrow$$

$$\begin{array}{lll|c|c|c} & Na^{0} - e^{-} & \longrightarrow Na^{+} & 1 & \multirow{2}{*}{2} & 2 \\ 2H^{+} + 2\times 1\bar{e} \longrightarrow & 2H^{0} & \longrightarrow H_2 & 2 & & 1 \end{array}$$

The sodium base can interact with the acid oxide CO_2 or with its aqueous solution H_2CO_3. Disodium carbonate (6) is obtained.

$$2NaOH + CO_2 \longrightarrow Na_2CO_3 + H_2O$$

$$2NaOH + H_2CO_3 \rightleftharpoons Na_2CO_3 + 2H_2O$$

Mixing the alkali elements with sulfur results in sulfides with an ionocrystalline lattice, well soluble in water.

$$2Na + S \longrightarrow Na_2S$$

$$\begin{array}{ll|c|c|c} Na^{0} - e^{-} & \longrightarrow Na^{+} & 1 & \multirow{2}{*}{2} & 2 \\ S^{0} + 2e^{-} & \longrightarrow S^{-2} & 2 & & 1 \end{array}$$

(7)

The interaction of disodium sulfide with hydrochloric acid results in hydrogen sulfide. The stronger hydrochloric acid displaces the weaker hydrosulfuric acid from its salts.

$$Na_2S + 2HCl \longrightarrow 2NaCl + H_2S\uparrow$$

(8)

Elements of group IA interact actively with hydrogen. The resulting compounds are called hydrides and are solids with an ionic crystal lattice.

$$2Na + H_2 \longrightarrow 2NaH$$

$$\begin{array}{lll|c|c|c} & \overset{0}{Na} - e^- & \longrightarrow Na^+ & 1 & & 2 \\ & & & & 2 & \\ H_2 \longrightarrow & 2H^0 + 2\times1e^- & \longrightarrow 2H^- & 2 & & 1 \end{array}$$

(9)

The hydrides of the alkali elements have strong reducing properties. As a result, they readily react with oxygen to form oxides, and can also take oxygen away from oxides of other elements.

$$2NaH + MoO \longrightarrow Na_2O + Mo + H_2$$

(10)

$$\begin{array}{lll|c|c|c} & Mo^{+2} + 2e^- & \longrightarrow \overset{0}{Mo} & 2 & & 1 \\ & & & & 2 & \\ 2H^- - 2\times1e^- & \longrightarrow 2H^0 & \longrightarrow H_2\uparrow & 2 & & 1 \end{array}$$

Group IIA of the Periodic Table

Group IIA includes s-elements with configuration ns^2 in the last electron layer: beryllium (Be), magnesium (Mg), calcium (Ca), strontium (Sr), barium (Ba) and radium (Ra). They have metallic properties, but less pronounced than those of group **IA**. Barium (Ba) has strong metallic properties and beryllium (Be) has more specific properties due to the presence of only two electrons, $1s^2$ in the penultimate electron layer. The elements calcium (Ca), strontium (Sr), and barium (Ba) are called **alkaline earth elements** because their hydroxides are strong bases like the hydroxides of the alkali elements ("**alkalo-**"), and their oxides are highly fusible like those of aluminum and the heavy metals referred to by the term "**earths.**" **Group IIA** elements are active metals, but not as active as the alkali ones. Barium exhibits the strongest metallic properties and magnesium the weakest. Beryllium has amphoteric properties more similar to those of aluminum.

With **hydrogen,** the elements form solid hydrides as the activity decreases from magnesium (Mg) to barium (Ba). Beryllium (Be) does not react under ordinary conditions.

$$Ca + H_2 \longrightarrow CaH_2$$

$$\begin{array}{lll|c|c|c} & \overset{0}{Ca} - 2e^- & \longrightarrow Ca^{+2} & 2 & & 1 \\ & & & & 2 & \\ H_2 \longrightarrow & 2H^0 + 2\times1e^- & \longrightarrow 2H^- & 2 & & 1 \end{array}$$

The hydrides of the alkaline earth elements, analogous to the hydrides of the alkali elements, have pronounced reducing properties. As a result, they readily react with oxygen to form oxides, and they can also take oxygen away from oxides of other elements:

$$CaH_2 + O_2 \longrightarrow CaO + H_2O$$

$$\begin{array}{rl|c|c|c}
 & 2\overset{-}{H} - 2x2e^- \longrightarrow 2H^+ & 4 & & 1 \\
 & & & 4 & \\
O_2 \longrightarrow & 2\overset{0}{O} + 2x2e^- \longrightarrow 2\overset{-2}{O} & 4 & & 1
\end{array}$$

$$CaH_2 + MoO \longrightarrow CaO + Mo + H_2$$

$$\begin{array}{l|c|c|c}
2\overset{-}{H} - 2x1\bar{e} \longrightarrow 2\overset{0}{H} \longrightarrow H_2 & 2 & & 1 \\
 & & 2 & \\
\overset{+2}{Mo} + 2e^- \longrightarrow \overset{0}{Mo} & 2 & & 1
\end{array}$$

They react actively with **oxygen**, but less strongly than alkali metals.

$$2\,Ca + O_2 \longrightarrow 2\,CaO$$

$$\begin{array}{rl|c|c|c}
 & \overset{0}{Ca} - 2\bar{e} \longrightarrow Ca^{+2} & 2 & & 2 \\
 & & & 4 & \\
O_2 \longrightarrow & 2\overset{0}{O} + 2x2\bar{e} \longrightarrow 2\,O^{-2} & 4 & & 1
\end{array}$$

At higher temperatures, the active metals form peroxides:

$$Ba + O_2 \longrightarrow BaO_2$$

$$\begin{array}{rl|c|c|c}
 & \overset{0}{Ba} - 2\bar{e} \longrightarrow Ba^{+2} & 2 & & 1 \\
 & & & 2 & \\
O_2 \longrightarrow & 2\overset{0}{O} + 2x1\bar{e} \longrightarrow 2\,\overset{-1}{O} & 2 & & 1
\end{array}$$

Non-metals are readily reacted with as the activity increases with increasing sequence number of the element in the group. For example, beryllium reacts with **chlorine** on heating, while barium self-ignites.

$$Ba + Cl_2 \longrightarrow BaCl_2$$

$$\begin{array}{rl|c|c|c}
 & \overset{0}{Ba} - 2\bar{e} \longrightarrow Ba^{+2} & 2 & & 1 \\
 & & & 2 & \\
Cl_2 \longrightarrow & 2\overset{0}{Cl} + 2x1\bar{e} \longrightarrow 2\,Cl^- & 2 & & 1
\end{array}$$

All **Group IIA** elements react with **sulfur. Sulfides** are obtained:

$$Mg + S \longrightarrow MgS$$

$$\begin{array}{lcl|c|c|c} \overset{0}{Mg} - 2e^- & \longrightarrow & Mg^{+2} & 2 & \multirow{2}{*}{2} & 1 \\ \overset{0}{S} + 2e^- & \longrightarrow & S^{-2} & 2 & & 1 \end{array}$$

With **nitrogen** they interact directly at elevated temperature and form **nitrides**. Beryllium and magnesium react most readily.

$$3\,Mg + N_2 \longrightarrow Mg_3N_2$$

$$\begin{array}{lcl|c|c|c} \overset{0}{Mg} - 2e^- & \longrightarrow & Mg^{+2} & 2 & \multirow{2}{*}{6} & 3 \\ N_2 \longrightarrow 2\,\overset{0}{N} + 2 \times 3\,e^- & \longrightarrow & 2\,N^{-3} & 6 & & 1 \end{array}$$

With **carbon,** beryllium and magnesium form **carbides**:

$$2\,Be + C \longrightarrow Be_2C$$

$$\begin{array}{lcl|c|c|c} \overset{0}{Be} - 2e^- & \longrightarrow & Be^{+2} & 2 & \multirow{2}{*}{4} & 2 \\ \overset{0}{C} + 4\,e^- & \longrightarrow & C^{-4} & 4 & & 1 \end{array}$$

Calcium carbide (CaC_2), important for practice, is obtained from calcium oxide (quicklime) and coke (C):

$$CaO + 3\,C \longrightarrow CaC_2 + CO$$

$$\begin{array}{lcl|c|c|c} \overset{0}{C} - 2e^- & \longrightarrow & C^{+2} & 2 & \multirow{2}{*}{2} & 1 \\ 2\,\overset{0}{C} + 2 \times 1\,e^- & \longrightarrow & 2\,C^{-1} & 2 & & 1 \end{array}$$

The process is called disproportionation: atoms of the same chemical element simultaneously accept and give up electrons with one atom being a redox species and the other an oxidant.

Calcium carbide is used to produce acetylene:

$$CaC_2 + 2H_2O \longrightarrow H{-}C{\equiv}C{-}H + Ca(OH)_2$$

Beryllium does not react with **water** because it forms a protective oxide cover, and magnesium reacts when heated. The other elements of the group react with water under ordinary conditions to form hydroxides. The activity increases with increasing sequence number of the element. Barium is the most active towards water.

$$Ba + 2H_2O \longrightarrow Ba(OH)_2 + H_2\uparrow$$

$$\begin{array}{lll|l|l|l} & \overset{0}{Ba} - 2e^- \longrightarrow & Ba^{+2} & 2 & & 1 \\ & & & & 2 & \\ 2H^+ + 2 \times 1e^- \longrightarrow & 2\overset{0}{H} \longrightarrow & H_2 & 2 & & 1 \end{array}$$

With **dilute acids** the elements of group **IIA** interact and the activity increases towards barium.

$$Mg + 2HCl \longrightarrow MgCl_2 + H_2\uparrow$$

$$\begin{array}{lll|l|l|l} & \overset{0}{Mg} - 2e^- \longrightarrow & Mg^{+2} & 2 & & 1 \\ & & & & 2 & \\ 2H^+ + 2 \times 1e^- \longrightarrow & 2\overset{0}{H} \longrightarrow & H_2 & 2 & & 1 \end{array}$$

With magnesium deficiency in the human body, serious pathological changes are induced:
- memory disorder;
- serious rhythm disturbances in cardiac activity;
- muscle cramps.

Beryllium has amphoteric properties and interacts with both acids and bases:

$$Be + \underbrace{2H^+ + 2Cl^-} + 2H_2O \longrightarrow \underbrace{[Be(H_2O)_4]^{+2} + 2Cl^-} + H_2\uparrow$$

$$\begin{array}{lll|l|l|l} & \overset{0}{Be} - 2e^- \longrightarrow & Be^{+2} & 2 & & 1 \\ & & & & 2 & \\ 2H^+ + 2 \times 1e^- \longrightarrow & 2\overset{0}{H} \longrightarrow & H_2 & 2 & & 1 \end{array}$$

$$Be + \underbrace{2Na^+ + 2OH^-} + 2H_2O \longrightarrow Na_2[Be(OH)_4] + H_2\uparrow$$

disodium berylate

$$\begin{array}{lll|l|l|l} & \overset{0}{Be} - 2e^- \longrightarrow & Be^{+2} & 2 & & 1 \\ & & & & 2 & \\ 2H^+ + 2 \times 1e^- \longrightarrow & 2\overset{0}{H} \longrightarrow & H_2 & 2 & & 1 \end{array}$$

The alkaline earth elements are obtained by electrolysis of melts and mainly of chlorides.
The volatile salts of the alkaline earth elements color the flame of a spirit lamp characteristically: **Magnesium** - in **white; Calcium** - in **tile red**; **Strontium - bright red**; **Barium** - in **green.**

Sample problems and solutions

Task 1.

Explain and express with chemical equations the following chemical transformations:

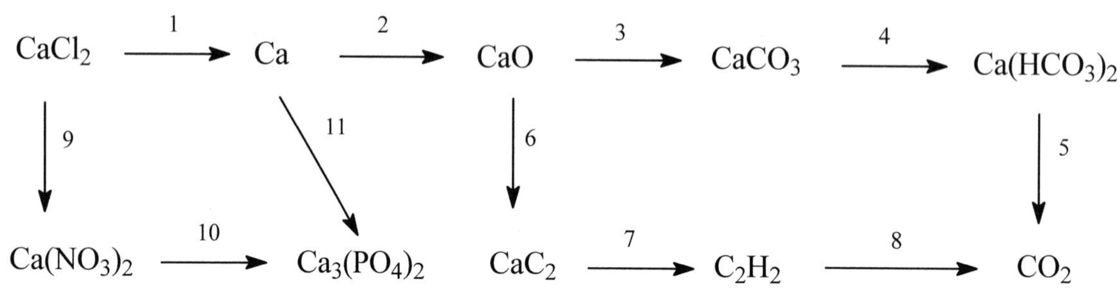

Exemplary solution

Alkali and alkaline earth metals are produced by electrolysis of molten salts. Passing a direct electric current through the calcium dichloride melt, calcium cations are directed to the cathode and chloride anions to the anode. At the cathode, the calcium cations accept electrons and are electroneutralised, and at the anode the chloride anions give up electrons and are converted to neutral atoms.

$$CaCl_2 \longrightarrow Ca^{+2} + 2\,Cl^{-}$$

$$Ca^{+2} + 2\,e^{-} \xrightarrow{1} Ca^{0} \quad | \; 2 \; | \; 2 \; | \; 1$$

$$2\,Cl^{-} - 2 \times 1\,e^{-} \longrightarrow 2\,Cl^{0} \longrightarrow Cl_2 \quad | \; 2 \; | \; \; | \; 1$$

Calcium reacts with oxygen to form calcium oxide. The process is redox.

$$2\,Ca + O_2 \xrightarrow{2} 2\,CaO$$

$$Ca^{0} - 2\,e^{-} \longrightarrow Ca^{+2} \quad | \; 2 \; | \; 4 \; | \; 2$$

$$O_2 \longrightarrow 2\,O^{0} + 2 \times 2\,e^{-} \longrightarrow 2\,O^{-2} \quad | \; 4 \; | \; \; | \; 1$$

Calcium oxide (basic oxide) reacts with carbon dioxide (acid oxide) to produce calcium carbonate.

$$CaO + CO_2 \xrightarrow{3} CaCO_3$$

Calcium carbonate with carbonic acid (an aqueous solution of carbon dioxide) forms the water-soluble calcium dihydrogen carbonate.

$$CaCO_3 + H_2CO_3 \xrightarrow{4} Ca(HCO_3)_2$$

When heated, calcium dihydrogen carbonate decomposes to release water, carbon dioxide and calcium carbonate.

$$Ca(HCO_3)_2 \xrightarrow{5} CaCO_3 + H_2O + CO_2\uparrow$$

Calcium oxide reacts with coke to form calcium carbide.

$$CaO + C \xrightarrow{6} CaC_2 + CO$$

Calcium carbide reacts with water to form acetylene.

$$CaC_2 + 2\,H_2O \xrightarrow{7} H{-}C{\equiv}C{-}H + Ca(OH)_2$$

In acetylene burners, ethyne burns with the release of carbon dioxide and water. The process is hig hly exothermic and develops too high a temperature. It is used in construction for welding metal structures.

$$2\,H{-}C{\equiv}C{-}H + 5\,O_2 \xrightarrow{8} 4\,CO_2 + 2\,H_2O + Q$$

$$C^{-} - 5\,e^{-} \longrightarrow C^{+4} \quad \left|\,5\,\right| \quad 20 \quad \left|\,4\right.$$

$$O_2 \longrightarrow 2\,O^{0} + 2\times 2\,e^{-} \longrightarrow 2\,O^{-2} \quad \left|\,4\,\right| \quad \left|\,5\right.$$

From calcium dichloride, calcium dinitrate can be produced by reaction with silver nitrate.

$$CaCl_2 + 2\,AgNO_3 \longrightarrow 2\,AgCl + Ca(NO_3)_2$$

$$\underbrace{Ca^{+2} + 2\,Cl^{-}} + \underbrace{2\,Ag^{+} + 2\,NO_3^{-}} \xrightarrow{9} 2\,AgCl\downarrow + \underbrace{Ca^{+2} + 2\,NO_3^{-}}$$

$$2\,Ag^{+} + 2\,Cl^{-} \longrightarrow 2\,AgCl\downarrow$$

Calcium dinitrate when mixed with trisodium phosphate releases a precipitate of tricalcium diphosphate.

$$3\,Ca(NO_3)_2 + 2\,Na_3PO_4 \longrightarrow 3\,Ca_3(PO_4)_2 + 6\,NaNO_3$$

$$\underbrace{3\,Ca^{+2} + 6\,NO_3^{-}} + \underbrace{6\,Na^{+} + 2\,PO_4^{-3}} \xrightarrow{10} Ca_3(PO_4)_2\downarrow + \underbrace{6\,Na^{+} + 6\,NO_3^{-}}$$

$$3\,Ca^{+2} + 2\,PO_4^{-3} \longrightarrow Ca_3(PO_4)_2\downarrow$$

Calcium interacts with phosphoric acid. Tricalcium diphosphate is produced and hydrogen is released. The process is redox.

$$3\,Ca + 2\,H_3PO_4 \xrightarrow{11} Ca_3(PO_4)_2 + 3\,H_2$$

$$\begin{array}{lcl|c|c|c}
\overset{0}{Ca} - 2\,e^- & \longrightarrow & Ca^{+2} & 2 & & 1 \\
 & & & & 2 & \\
2\,H^+ + 2 \times 1\,e^- \longrightarrow 2\,\overset{0}{H} & \longrightarrow & H_2 & 2 & & 1
\end{array}$$

Group IIIA of the Periodic Table

The elements boron (B), aluminium (Al), gallium (Ga), indium (In), and thallium (Tl) are found in **Group IIIA**. They have three electrons in their valence electron layer, **$ns^2 np^1$** configuration. The penultimate electron layer is substantially different. The element boron has two electrons, aluminium has eight, and the other elements gallium, indium, and thallium have eighteen electrons each. They are fully electron analogues. The electronic configuration is the reason why the metallic properties of the group elements are less pronounced than those of the first and second groups. In boron, the non-metallic properties are more pronounced. In their compounds, the elements of the group exhibit an oxidation degree of +3, but lower degrees appear as the sequence number of the elements increases. For example, in the case of thallium (Tl), the most stable compounds are those in which its oxidation degree is +1. In the order of boron to thallium, the metal properties are enhanced. For example, B_2O_3 exhibits an acidic character, the oxides of aluminium and indium an amphoteric character, and thallium a basic character. In practical terms, the most important elements of **Group IIIA** are aluminium and boron.

$$Al\ (Z = 13) - 1s^2\,2s^2\,2p^6\,3s^2\,3p^1$$

Aluminium is resistant to air, moisture and oxidizing oxoacids due to the rapid formation of a protective layer of Al_2O_3, which coats the surface of the metal. Once the protective layer is removed, its chemical activity increases.

Reduction activity, but weaker than that of s-elements, is also exhibited by p-elements with a smaller number of p-electrons in their outer electron layer and with a larger atomic radius. These are the group IIIA elements and the elements to the left of the B/At line in the periodic table. Their simple substances are metals, which are chemically less active than those of the s-elements. Most of their hydrogen compounds are polymeric in nature.

Aluminium does not interact **directly with hydrogen**.

Aluminium reacts with **water** to form a white pithy precipitate of aluminium trihydroxide:

$$2\,Al + 6\,H_2O \longrightarrow 2\,Al(OH)_3 + 3\,H_2\uparrow$$

$$\begin{array}{lcl|c|c|c}
\overset{0}{Al} - 3\,e^- & \longrightarrow & Al^{+3} & 3 & & 2 \\
 & & & & 6 & \\
2\,H^+ + 2 \times 1\,e^- \longrightarrow 2\,\overset{0}{H} & \longrightarrow & H_2 & 2 & & 3
\end{array}$$

In an **oxygen atmosphere**, aluminium burns with a dazzling brilliance.

$$4\,Al + 3\,O_2 \longrightarrow Al_2O_3 + Q$$

$$\begin{array}{lcl|c|c|c}
\overset{0}{Al} - 3\,e^- & \longrightarrow & Al^{+3} & 3 & & 2 \\
 & & & & 6 & \\
O_2 \longrightarrow 2\,\overset{0}{O} + 2 \times 2\,e^- & \longrightarrow & 2\,O^{-2} & 2 & & 3
\end{array}$$

The high heat effect of the reaction indicates the high chemical activity of aluminium towards oxygen. The reducing properties of aluminium are used to extract some high-melting metals from their oxides. The method is called **aluminothermy**.

$$Cr_2O_3 + 2Al \longrightarrow 2Cr + Al_2O_3 + Q$$

$$\begin{array}{lll|c|c|c} 2\,\overset{o}{Al} - 2x3e^- & \longrightarrow & 2\,Al^{+3} & 6 & \multirow{2}{*}{6} & 1 \\ 2\,Cr^{+3} + 2\,x3e^- & \longrightarrow & 2Cr^{o} & 6 & & 1 \end{array}$$

$$Fe_2O_3 + 2Al \longrightarrow 2Fe + Al_2O_3 + Q$$

$$\begin{array}{lll|c|c|c} 2\,\overset{o}{Al} - 2x3e^- & \longrightarrow & 2\,Al^{+3} & 6 & \multirow{2}{*}{6} & 1 \\ 2\,Fe^{+3} + 2\,x3e^- & \longrightarrow & 2\,Fe^{o} & 6 & & 1 \end{array}$$

With **acids,** aluminium interacts and hydrogen is released:

$$2Al + 6\,HCl \longrightarrow 2\,AlCl_3 + 3H_2\uparrow$$

$$\begin{array}{lll|c|c|c} & \overset{o}{Al} - 3e^- \longrightarrow & Al^{+3} & 3 & \multirow{2}{*}{6} & 2 \\ 2H^+ + 2x1e^- \longrightarrow & 2\,\overset{o}{H} \longrightarrow & H_2 & 2 & & 3 \end{array}$$

The process can also be represented in the following way with the formation of an aqua complex:

$$2Al + \underbrace{6\,H^+ + 6Cl^-} + 12H_2O \longrightarrow 2[Al(H_2O)_6]^{+3} + 6\,Cl^-$$

hexaquaaluminum trichloride

Due to its amphoteric nature, aluminium also reacts with **bases**. Concentrated solutions of alkaline bases readily dissolve the metal and a hexahydroxyaluminate anion is produced:

$$2Al + \underbrace{6\,Na^+ + 6\bar{O}H} + 6\,H_2O \longrightarrow 6\,Na^+ + 2[Al(OH)_6]^{-3} + H_2\uparrow$$

If aluminium is fused with solid sodium base, sodium metaaluminate is obtained:

$$2Al + 2NaOH + 2\,H_2O \xrightarrow{t^o} 2\,NaAlO_2 + 3H_2\uparrow$$

$$\begin{array}{lll|c|c|c} & \overset{o}{Al} - 3e^- \longrightarrow & Al^{+3} & 3 & \multirow{2}{*}{6} & 2 \\ 2H^+ + 2x1e^- \longrightarrow & 2\,\overset{o}{H} \longrightarrow & H_2 & 2 & & 3 \end{array}$$

The water required for the reaction is not added because it is present in the solid sodium base due to its hygroscopicity.

The interaction with **non-metal halide elements** is direct. The resulting halides have a covalent bond.

Aluminium reacts with **carbon** directly at high temperature (~2000° C). This produces aluminium carbide, which when reacted with water releases methane (laboratory method for producing methane):

$$4\,Al + 3\,C \longrightarrow Al_4C_3$$

$$\begin{array}{lll|c|c|c} \overset{0}{Al} - 3e^- & \longrightarrow & Al^{+3} & 3 & & 4 \\ & & & & 12 & \\ C + 4e^- & \longrightarrow & C^{-4} & 4 & & 3 \end{array}$$

$$Al_4C_3 + 12\,H_2O \longrightarrow Al(OH)_3 + 3\,CH_4\uparrow$$

Aluminum forms **alloys** with **metals**.

Aluminum trioxide can be produced by the combustion of aluminum powder in an oxygen environment in addition to the decomposition of aluminum trihydroxide:

$$2\,Al(OH)_3 \xrightarrow{t^o} Al_2O_3 + 3\,H_2O$$

Aluminium trioxide does not dissolve in water. It is amphoteric in nature - reacts with both acids and bases:

$$Al_2O_3 + 6\,HCl \longrightarrow 2\,AlCl_3 + 3\,H_2O$$

or

$$Al_2O_3 + \underbrace{6\,H^+ + 6Cl^-} + 12\,H_2O \longrightarrow 2\,[Al(H_2O)_6]^{+3} + 6\,Cl^-$$

When fused with alkaline bases, alumina forms metaaluminates:

$$Al_2O_3 + 2NaOH \xrightarrow{t^o} 2\,NaAlO_2 + H_2O$$

In concentrated solutions of alkaline bases, alumina forms hydroxocomplex ions:

$$Al_2O_3 + \underbrace{6\,Na^+ + 6\overline{O}H} + 3\,H_2O \longrightarrow \underbrace{6\,Na^+ + 2[Al(OH)_6]^{-3}}$$

Aluminium trihydroxide is a pythium amorphous substance. It is obtained by mixing solutions containing aluminium ions with hydroxide ions:

$$Al^{+3} + 3\,\overline{O}H \longrightarrow Al(OH)_3\downarrow$$

Aluminium trihydroxide has amphoteric properties - it interacts with both acids and bases:

$$2\,Al(OH)_3 + \underbrace{6\,H^+ + 3\,SO_4^{-2}}_{3\,H_2SO_4} \longrightarrow \underbrace{2\,Al^{+3} + 3\,SO_4^{-2}} + 6\,H_2O$$

When fused with alkaline bases, aluminium trihydroxide yields metaaluminates:

$$Al(OH)_3 + NaOH \xrightarrow{t^o} NaAlO_2 + 2\,H_2O$$

If concentrated solutions of alkali hydroxides are used, aluminium trihydroxide dissolves to form hydroxocomplexes:

$$Al(OH)_3 + \underbrace{3Na^+ + 3\,\bar{O}H} \longrightarrow 3\,Na^+ + [Al(OH)_6]^{-3}$$

The direct interaction of **boron and oxygen** forms B_2O_3. It is an acidic oxide that forms **orthoboric acid** when dissolved in water:

$$B_2O_3 + 3\,H_2O \longrightarrow 2\,H_3BO_3$$

Sample problems and solutions

Task 1.

Just a substance reacts with HCl, conc. NaOH, $ZnSO_4$ and Fe_2O_3. Its chloride is hydrolyzed. Its hydroxide reacts with NaOH and HCl. Which is the simple substance? Write the equations of all the reactions listed and characterize the processes.

Exemplary solution

The simple substance is aluminium, an element of the third group of the Periodic Table, third period: Al (Z = 13) - $1s^2\,2s^2\,2p^6\,3s^2\,3p^1$.

The element exhibits amphoteric properties. Reacts with both acids and bases. With hydrochloric acid it forms aluminium trichloride.

$$2\,Al + 6\,HCl \longrightarrow 2\,AlCl_3 + 3\,H_2\uparrow$$

$$\begin{array}{lll|l|l|l} \overset{0}{Al} - 3e^- & \longrightarrow & Al^{+3} & 3 & & 2 \\ & & & & 6 & \\ 2H^+ + 2 \times 1\bar{e} \longrightarrow 2\overset{0}{H} & \longrightarrow & H_2 & 2 & & 3 \end{array}$$

The process can also be represented in the following way with the formation of an aqua complex:

$$2\,Al + \underbrace{6\,H^+ + 6\,\bar{Cl}} + 12\,H_2O \longrightarrow \underbrace{2\,[Al(H_2O)_6]^{+3} + 6\,\bar{Cl}}_{\text{hexaaquaaluminum trichloride}} + 3\,H_2\uparrow$$

Due to its amphoteric nature, aluminium also reacts with bases. Concentrated solutions of alkaline bases readily dissolve the metal and a hexahydroxyaluminate anion is produced.

$$2Al + \underbrace{6\,Na^{+} + 6\,\overline{O}H}\; + 6\,H_2O \longrightarrow \underbrace{6\,Na^{+} + 2[Al(OH)_6]^{3-}} + 3\,H_2\uparrow$$

At high temperature aluminium reacts with sodium base to produce sodium metaaluminate. The water required for the reaction is not added because it is present in the solid sodium base due to its hygroscopicity.

$$2Al + 2NaOH + 2\,H_2O \xrightarrow{t^{o}} 2\,NaAlO_2 + 3H_2\uparrow$$

$$\begin{array}{lcl|c|c|c} & Al^{o} - 3e^{-} \longrightarrow & Al^{+3} & 3 & & 2 \\ & & & & 6 & \\ 2H^{+} + 2\times 1\bar{e} \longrightarrow 2\,H^{o} \longrightarrow & & H_2 & 2 & & 3 \end{array}$$

Aluminium displaces zinc from its compounds due to its place in Order of Relative Activity of the Metals (ORAM).

$$2\,Al + 3\,ZnSO_4 \xrightarrow{t^{o}} Al_2(SO_4)_3 + 3\,Zn$$

$$\begin{array}{lcl|c|c|c} Al^{o} - 3\,e^{-} & \longrightarrow & Al^{+3} & 3 & & 2 \\ & & & & 6 & \\ Zn^{+2} + 2\,e^{-} & \longrightarrow & Zn^{o} & 2 & & 3 \end{array}$$

The reducing properties of aluminium are used to extract some high melting metals from their oxides. The method is called aluminothermy.

$$Fe_2O_3 + 2\,Al \longrightarrow 2\,Fe + Al_2O_3 + Q$$

$$\begin{array}{lcl|c|c|c} Al^{o} - 3\,e^{-} & \longrightarrow & Al^{+3} & 3 & & 1 \\ & & & & 3 & \\ Fe^{+3} + 3\,e^{-} & \longrightarrow & Fe^{o} & 3 & & 1 \end{array}$$

Aluminium trichloride is the salt of the weak electrolyte aluminium trihydroxide and the strong hydrochloric acid. The salt hydrolyzes.

$$\underbrace{Al^{3+} + 3\,\overline{Cl}} + \underbrace{3\,H^{+} + 3\,\overline{OH}}_{\substack{\updownarrow \\ 3\,H_2O}} \rightleftharpoons Al(OH)_3 + \underbrace{3\,H^{+} + 3\,\overline{Cl}}$$

The aluminium ions from the salt bind the hydroxide ions from the water in the weak electrolyte aluminium trihydroxide. In excess, hydrogen ions remain in the solution, which determine the acidic character of the solution, $pH < 7$, the litmus changes color to red.

Aluminium trihydroxide reacts with hydrochloric acid. The process is called neutralization.

$$2\,Al(OH)_3 + \underbrace{6\,H^+ + 6\,Cl^-} \rightleftharpoons \underbrace{2\,Al^{+3} + 6\,Cl^-} + 6\,H_2O$$

The process is reversible because the salt hydrolyzes.

Aluminium trihydroxide is a white, pithy precipitate. The addition of sodium base causes a gradual dissolution of the pithy precipitate and clarification of the solution. The complex compound trisodium hexahydroxyaluminate is obtained.

$$Al(OH)_3 + \underbrace{3\,Na^+ + 3\,OH^-} \longrightarrow \underbrace{3\,Na^+ + [Al(OH)_6]^{-3}}$$

When fused with alkaline bases, aluminium trihydroxide yields metaaluminates:

$$Al(OH)_3 + NaOH \xrightarrow{t^o} NaAlO_2 + 2\,H_2O$$

Group IVA of the Periodic Table

Group IVA includes the elements carbon (C), silicon (Si), germanium (Ge), tin (Sn) and lead (Pb). The electronic configuration of the last electron layer of the elements is $\mathbf{ns^2np^2}$. In the penultimate electron layer, carbon and silicon have two and eight electrons, respectively, which is why these elements differ in properties from the others, which have eighteen electrons in their penultimate layer and are complete electron analogues.

The relatively large number of electrons in the outer electron layer determines the non- metallic nature of the elements. In the transition from **carbon** to **lead**, the atomic radius increases and the force of attraction of the electrons from the nucleus decreases. This leads to a weakening of the non-metallic character, so that in **germanium** the metallic properties already appear, and in **tin** and **lead** - the metallic properties predominate. **Silicon** and **germanium** and their compounds have the properties of semiconductors.

The nonmetallic character of **Group IVA** elements is less pronounced compared to those of **Group VA, VIA** and **VIIA.**

The simple substances of the elements of the group under ordinary conditions are solid.

Only **carbon** and **silicon** have allotropic forms.

The elements appear as oxidizing (oxidation degree - 4) and reducing (oxidation degree +4 and +2). In the group, as the sequence number of the element increases, the metallic character becomes stronger and the characteristic oxidation degree is +2.

The elements form compounds with **hydrogen**:

$$CH_4 \quad SiH_4 \quad GeH_4 \quad SnH_4 \quad PbH_4$$
$$\longleftarrow$$

The durability of hydrides is increasing

Methane (CH_4) has strong covalent bonds and is stable, while silane (SiH_4) has weaker bonds and decomposes on heating.

Aqueous solutions of hydrides are **neutral in nature.**

Relative to **oxygen,** the elements act as reducing agents and form two types of oxides (EO and EO_2) with oxidation degrees +2 and +4. The higher oxides of **carbon** and **silicon,** CO_2 and SiO_2 have strong covalent bonds, are stable and have an **acidic character**. The lower carbon oxide, CO, is less stable because of its strong reducing properties. It is a **neutral oxide**. For tin and lead, the lower oxides, SnO and PbO, are more durable. Both the oxides and the corresponding hydroxides $Sn(OH)_2$ $Pb(OH)_2$ have amphoteric properties with the basic character being more pronounced for

$Pb(OH)_2$.

Carbon is only chemically active at a high temperature. Diamond is the most resistant, and charcoal and carbon black are the most active because of their large surface areas.

Carbon combines with **hydrogen** at high temperature (1200° C) to form methane:

$$C + 2H_2 \rightleftharpoons CH_4 + Q$$

$$\begin{array}{rcl|c|c|c}
C^{0} + 4e^{-} & \longrightarrow & C^{-4} & 4 & & 1 \\
 & & & & 4 & \\
H_2 \longrightarrow 2H^{0} - 2\times 1e^{-} & \longrightarrow & 2H^{+} & 2 & & 2
\end{array}$$

The straight reaction is exothermic and the equilibrium may shift to the right on cooling, but in this case the process rate tends to be zero. In the presence of catalyst Ni and elevated pressure, the equilibrium can shift to the right (Le Chatelier-Brown principle).

Carbon reacts with **metals** at high temperature to form carbides:

$$4Al + 3C \longrightarrow Al_4C_3$$

$$\begin{array}{rcl|c|c|c}
Al^{0} - 3e^{-} & \longrightarrow & Al^{+3} & 3 & & 4 \\
 & & & & 12 & \\
C + 4e^{-} & \longrightarrow & C^{-4} & 4 & & 3
\end{array}$$

Carbides are compounds of ionic character. Compounds containing carbide ion C^{-4} readily hydrolyze in water with evolution of methane:

$$Al_4C_3 + 12H_2O \longrightarrow Al(OH)_3 + 3CH_4\uparrow$$

Carbon reacts with **silicon** to produce silicon carbide (**carborundum**).

$$C + Si \longrightarrow SiC$$

$$\begin{array}{rcl|c|c|c}
C^{0} + 4e^{-} & \longrightarrow & C^{-4} & 4 & & 1 \\
 & & & & 4 & \\
Si^{0} - 4e^{-} & \longrightarrow & Si^{+4} & 4 & & 1
\end{array}$$

Carborundum has an atomic crystal lattice similar to diamond, has high hardness and resistance at high temperatures.

Relative to **oxygen,** carbon is treated as a reducing agent. At full combustion, CO_2 is produced and a large amount of heat is given off.

$$C + O_2 \longrightarrow CO_2 + Q$$

$$\begin{array}{rcl|c|c|c}
C^{0} - 4e^{-} & \longrightarrow & C^{+4} & 4 & & 1 \\
 & & & & 4 & \\
O_2 \longrightarrow 2O^{0} + 2\times 2e^{-} & \longrightarrow & 2O^{-2} & 4 & & 1
\end{array}$$

CO_2 can also be produced by the decomposition of carbonates:

$$CaCO_3 \rightleftharpoons CaO + CO_2\uparrow$$

or when dissolved in acids:

$$CaCO_3 + 2HCl \longrightarrow CaCl_2 + H_2O + CO_2\uparrow$$

Carbon dioxide dissolves in water, and reacts with it to form the weak and perishable dibasic carbonic acid, H_2CO_3. The acid dissociates into two degrees:

$$H_2CO_3 + H_2O \rightleftharpoons H_3O^+ + HCO_3^-$$

$$HCO_3^- + H_2O \rightleftharpoons H_3O^+ + CO_3^{-2}$$

Carbon dioxide does not support combustion. It is heavier than air and is used in firefighting equipment. Small amounts of CO_2, dissolved in the blood, excite the respiratory centre. Larger amounts are asphyxiating.

Due to its significant density, CO_2 accumulates in the lower layers of the atmosphere and impedes the free exchange of heat on Earth with space, the so-called "**greenhouse effect**" occurs, which leads to a slow increase in the Earth's temperature with unpredictable environmental consequences.

Incomplete combustion of carbon produces carbon monoxide (CO), which is a colourless gas, slightly soluble in water, and does not react with it. It is a **neutral oxide**. It is highly poisonous and particularly dangerous because it has no characteristic odor and its poisoning is imperceptible and irreversible. Industrially, carbon monoxide is produced when solid fuel is converted into gaseous fuel, so-called '**coal gasification'**. The process takes place during incomplete combustion of the coal, in which the following interactions take place in the oxygen-carbon mixture:

$$C + O_2 \longrightarrow CO_2 + Q$$

$$\begin{array}{rcl|c|c|c} & C^0 - 4e^- \longrightarrow & C^{+4} & 4 & & 1 \\ & & & & 4 & \\ O_2 \longrightarrow & 2O^0 + 2\times 2e^- \longrightarrow & 2O^{-2} & 4 & & 1 \end{array}$$

So-called "**water gas**" is produced by blowing heated water vapor over glowing coal:

$$C_{(s.)} + H_2O_{(g.)} \rightleftharpoons CO_{(g)} + H_{2(g)} - Q$$

$$\begin{array}{rcl|c|c|c} & C^0 - 2e^- \longrightarrow & C^{+2} & 2 & & 1 \\ & & & & 2 & \\ 2H^+ + 2\times 1e^- \longrightarrow & 2H^0 \longrightarrow & H_2\uparrow & 2 & & 1 \end{array}$$

$$2C + O_2 \longrightarrow 2CO + Q$$

$$\begin{array}{rcl|c|c|c} & C^0 - 2e^- \longrightarrow & C^{+2} & 2 & & 2 \\ & & & & 4 & \\ O_2 \longrightarrow & 2O^0 - 2\times 2e^- \longrightarrow & 2O^{-2} & 4 & & 1 \end{array}$$

Water gas has a higher calorific value due to the presence of two combustible components: CO (g) + H_2 (g.).

The main property of carbon monoxide is its reducing action. It takes oxygen and converts it into carbon dioxide. It is used in metallurgy to obtain metals from their oxides:

$$2CO + O_2 \longrightarrow 2CO_2 + Q$$

$$\begin{array}{lcl|c|c|c} & C^{+2} - 2\bar{e} & \longrightarrow C^{+4} & 2 & & 2 \\ & & & & 4 & \\ O_2 \longrightarrow & 2O^{0} + 2\times 2\bar{e} & \longrightarrow 2O^{-2} & 4 & & 1 \end{array}$$

$$C + CO_2 \longrightarrow 2CO + Q$$

$$\begin{array}{lcl|c|c|c} C^{0} - 2\bar{e} & \longrightarrow & C^{+2} & 2 & & 1 \\ & & & & 2 & \\ C^{+4} + 2\bar{e} & \longrightarrow & C^{+2} & 2 & & 1 \end{array}$$

$$Fe_2O_3 + 3CO \longrightarrow 2Fe + 3CO_2$$

$$\begin{array}{lcl|c|c|c} Fe^{+3} + 3\bar{e} & \longrightarrow & Fe^{0} & 3 & & 2 \\ & & & & 6 & \\ C^{+2} - 2\bar{e} & \longrightarrow & C^{+4} & 2 & & 3 \end{array}$$

Carbon monoxide also reacts with other oxidants, for example chlorine, to produce carbonyl chloride according to equilibrium:

$$CO + Cl_2 \rightleftharpoons COCl_2$$

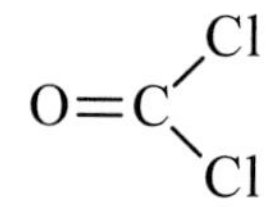

$$\begin{array}{lcl|c|c|c} & C^{+2} - 2\bar{e} & \longrightarrow C^{+4} & 2 & & 1 \\ & & & & 2 & \\ Cl_2 \longrightarrow & 2Cl^{0} + 2\times 1\bar{e} & \longrightarrow 2Cl^{-} & 2 & & 1 \end{array}$$

Carbonyl chloride, trivially called **phosgene**, is a warfare poison.

Carbon reacts with all **halogen elements.** The stability of halogen derivatives decreases from fluorine to iodine. Particularly important are **tetrachloromethane CCl_4** (a good organic solvent) and **tetrafluoromethane CF_4** (a raw material for making **polytetrafluoroethene - "Teflon"**).

At high temperature, carbon reacts with **sulfur. Sulfur carbon**, a widely used organic solvent, is produced.

$$C + 2S \longrightarrow CS_2$$

$$\begin{array}{lcl|c|c|c} C^{0} - 4\bar{e} & \longrightarrow & C^{+4} & 4 & & 1 \\ & & & & 4 & \\ S^{0} + 2\bar{e} & \longrightarrow & S^{-2} & 2 & & 2 \end{array}$$

Silicon is the second (after oxygen) most abundant chemical element on Earth. As quartz, silica is included in the composition of many minerals (silicates) and is a major element in the lithosphere. Industrially, silicon is produced by reducing silica sand with carbon:

$$SiO_2 + 2C \longrightarrow Si + 2CO$$

$$\begin{array}{llll|l|l|l} Si^{+4} & + 4e^- & \longrightarrow & Si^{0} & 4 & & 1 \\ & & & & & 4 & \\ C^{0} & - 2e^- & \longrightarrow & C^{+2} & 2 & & 2 \end{array}$$

In laboratory conditions – in the reduction of quartz sand with active metals:

$$SiO_2 + 2Mg \longrightarrow Si + 2MgO$$

$$\begin{array}{llll|l|l|l} Si^{+4} & + 4e^- & \longrightarrow & Si^{0} & 4 & & 1 \\ & & & & & 4 & \\ Mg^{0} & - 2e^- & \longrightarrow & Mg^{+2} & 2 & & 2 \end{array}$$

Silicon is a chemically inert element. At room temperature it reacts with fluorine, chlorine and alkaline bases:

$$Si + 2KOH + H_2O \longrightarrow K_2SiO_3 + 2H_2\uparrow$$

$$\begin{array}{llll|l|l|l} Si^{0} & - 4e^- & \longrightarrow & Si^{+4} & 4 & & 1 \\ & & & & & 4 & \\ 2H^{+} + 2 \times 1e^- & \longrightarrow 2H^{0} & \longrightarrow & H_2 & 2 & & 2 \end{array}$$

The silicon is dissolved in a mixture of **nitric and hydrofluoric acids** only and **hexafluorosilicic acid** is obtained:

$$Si + 4HNO_3 + 18HF \longrightarrow 3H_2SiF_6 + 4NO + 8H_2O$$

$$\begin{array}{llll|l|l|l} Si^{0} & - 4e^- & \longrightarrow & Si^{+4} & 4 & & 3 \\ & & & & & 12 & \\ N^{+5} & + 3e^- & \longrightarrow & N^{+2} & 3 & & 4 \end{array}$$

Interacts with active metals to form **silicides:**

$$2Ca + Si \longrightarrow Ca_2Si$$

$$\begin{array}{llll|l|l|l} Ca^{0} & - 2e^- & \longrightarrow & Ca^{+2} & 2 & & 2 \\ & & & & & 4 & \\ Si^{0} & + 4e^- & \longrightarrow & Si^{-4} & 4 & & 1 \end{array}$$

Hydrogen compounds of silicon (**silanes**) are produced when silicides react with acids:

$$Mg_2Si + 2H_2SO_4 \longrightarrow SiH_4 + 2MgSO_4$$

Compared to hydrocarbons, silanes are more reactive and therefore cannot form long silicon chains. Unlike methane, **silane reacts with water** to produce **metasilicic acid**:

$$SiH_4 + 3\,H_2O \longrightarrow H_2SiO_3 + 4\,H_2\uparrow$$

$$\overset{-4}{Si} - 8\,e^- \longrightarrow \overset{+4}{Si} \quad \left|\,8\,\right|\;8\;\left|\,1\right.$$

$$2\,H^+ + 2\times1\,\bar{e} \longrightarrow 2\,\overset{0}{H} \longrightarrow H_2 \quad \left|\,2\,\right|\;\;\;\left|\,4\right.$$

Silica SiO_2 (quartz) is stable and dissolves only in hydrofluoric acid:

$$SiO_2 + 4\,HF \longrightarrow SiF_4 + 2\,H_2O$$

When heated, the quartz melts and when cooled, a glassy mass is produced – quartz glass.

SiO_2 is the anhydride of orthosilicic acid $Si(OH)_4$ or H_4SiO_4. Its salts are stable only in strongly alkaline solutions. On acidification, it polymerizes rapidly and a product with a highly developed surface is obtained - $SiO_2 . xH_2O$, known as **silica gel** - a good adsorbent and drying agent.

At high temperature, silicon bonds with **carbon** to form carborundum - SiC.

Silicates are the most abundant and most important compounds of silicon. They make up the earth's crust, glass, ceramics, porcelain, cement, etc. One of the most important man- made silicates is glass. It is obtained by the co-fusion of soda, limestone and sand:

$$Na_2CO_3 + CaCO_3 + 6\,SiO_2 \longrightarrow \underset{\text{glass}}{Na_2O.\,CaO.\,6\,SiO_2} + 2CO_2\uparrow$$

The resulting colorless viscous melt solidifies into a transparent glassy mass.

Germanium, Tin and Lead

Lead and tin were some of the first metals used by man. The three elements are complete electronic analogues.

Germanium is a hard crumbly substance with a metallic lustre. It has the properties of a semiconductor. It is used in microelectronics.

Tin has three allotropic modifications - grey, white and rhombic (crumbly) tin.

Lead has only one modification - it is a grayish white metal, soft and malleable. One of the most dangerous to living organisms' chemical element. Once in the body, lead is not removed - it accumulates in the bone system and replaces calcium. Significant lead contamination is caused by internal combustion engines running on regular (leaded) gasoline.

The metals are produced from their most common ores (galena - **PbS**, cassiterite - **SnO_2**) by reduction processes:

$$SnO_2 + 2C \longrightarrow Sn + 2CO$$

$$Sn^{+4} + 4\,e^- \longrightarrow Sn^0 \quad \left|\,4\,\right|\;4\;\left|\,1\right.$$

$$C^0 - 2\,e^- \longrightarrow C^{+2} \quad \left|\,2\,\right|\;\;\;\left|\,2\right.$$

In lead, the ore is initially "roasted":

$$2PbS + 3O_2 \longrightarrow 2PbO + 2SO_2$$

$$\begin{array}{lll|c|c|c}
 & S^{-2} - 6e^- \longrightarrow & S^{+4} & 6 & & 2 \\
 & & & & 12 & \\
O_2 \longrightarrow 2O^{o} & + 2\times 2e^- \longrightarrow & 2O^{-2} & 4 & & 3
\end{array}$$

Oxide reduction follows:

$$PbO + C \longrightarrow Pb + CO$$

$$\begin{array}{ll|c|c|c}
Pb^{+2} + 2e^- \longrightarrow & Pb^{o} & 2 & & 1 \\
 & & & 2 & \\
C^{o} - 2e^- \longrightarrow & C^{+2} & 2 & & 1
\end{array}$$

$$PbO + CO \longrightarrow Pb + CO_2$$

$$\begin{array}{ll|c|c|c}
Pb^{+2} + 2e^- \longrightarrow & Pb^{o} & 2 & & 1 \\
 & & & 2 & \\
C^{+2} - 2e^- \longrightarrow & C^{+4} & 2 & & 1
\end{array}$$

$$2PbO + PbS \longrightarrow 3Pb + SO_2$$

$$\begin{array}{ll|c|c|c}
Pb^{+2} + 2e^- \longrightarrow & Pb^{o} & 2 & & 3 \\
 & & & 6 & \\
S^{-2} - 6e^- \longrightarrow & S^{+4} & 6 & & 1
\end{array}$$

They do not react directly with **hydrogen**, but hydrides of general formula EH_4 are known.

Lead dissolves readily in dilute nitric and acetic acids, but hydrochloric and sulfuric acids passivate it due to the formation of dense coatings of $PbCl_2$ and $PbSO_4$. Concentrated nitric acid also passivates lead due to the formation of a dense surface layer of PbO.

$$Pb + 4\text{ d. }HNO_3 \longrightarrow Pb(NO_3)_2 + 2NO_2 + 2H_2O$$

$$\begin{array}{ll|c|c|c}
Pb^{o} - 2e^- \longrightarrow & Pb^{+2} & 2 & & 1 \\
 & & & 2 & \\
N^{+5} + e^- \longrightarrow & N^{+4} & 1 & & 2
\end{array}$$

$$Pb + 2CH_3COOH \longrightarrow Pb(CH_3COO)_2 + H_2$$

$$\begin{array}{lll|c|c|c} & \overset{0}{Pb} - 2e^- & \longrightarrow Pb^{+2} & 2 & & 1 \\ & & & & 2 & \\ 2H^+ + 2 \times 1e^- & \longrightarrow 2\overset{0}{H} & \longrightarrow H_2 & 2 & & 1 \end{array}$$

Lead reacts with **soluble salts of metals** that are after it in the order of relative activity of the metals (ORAM).

$$Pb + CuSO_4 \longrightarrow PbSO_4 + Cu$$

$$\begin{array}{ll|c|c|c} \overset{0}{Pb} - 2e^- & \longrightarrow Pb^{+2} & 2 & & 1 \\ & & & 2 & \\ Cu^{+2} + 2e^- & \longrightarrow \overset{0}{Cu} & 2 & & 1 \end{array}$$

Tin and lead dissolve in **concentrated alkaline bases**:

$$Pb + 2NaOH \longrightarrow \underset{\text{disodium plumbate}}{Na_2PbO_2} + H_2\uparrow$$

$$\begin{array}{lll|c|c|c} & \overset{0}{Pb} - 2e^- & \longrightarrow Pb^{+2} & 2 & & 1 \\ & & & & 2 & \\ 2H^+ + 2 \times 1e^- & \longrightarrow 2\overset{0}{H} & \longrightarrow H_2 & 2 & & 1 \end{array}$$

In aqueous solutions of bases form complex compounds:

$$Pb + 2NaOH + 2H_2O \longrightarrow \underset{\text{disodium tetrahydroxyplumbate}}{Na_2[Pb(OH)_4]} + H_2\uparrow$$

$$\begin{array}{lll|c|c|c} & \overset{0}{Pb} - 2e^- & \longrightarrow Pb^{+2} & 2 & & 1 \\ & & & & 2 & \\ 2H^+ + 2 \times 1e^- & \longrightarrow 2\overset{0}{H} & \longrightarrow H_2 & 2 & & 1 \end{array}$$

Lead combines directly with **oxygen** to form PbO. Lead oxide can also be produced by thermal decomposition of salts of divalent lead. The oxide dissolves in acids, but not in hydrochloric and sulfuric acids.

Tin and lead react directly with the **halogen elements** to produce mostly tetrahalides - EX_4.

Of the sulfides, the most stable are those in which the elements are in the +2 oxidation state - tin sulfide (**SnS** - dark brown), lead sulfide (**PbS** - black). Lead sulfide is the least soluble compound of lead. It is obtained by precipitation of soluble salts of lead with hydrogen sulfide or soluble sulfides (alkali or ammonium):

$$Pb(CH_3COO)_2 + H_2S \longrightarrow PbS\downarrow + 2CH_3COOH$$

$$Pb(CH_3COO)_2 + Na_2S \longrightarrow PbS\downarrow + 2CH_3COONa$$

$$Pb(CH_3COO)_2 + (NH_4)_2S \longrightarrow PbS\downarrow + 2CH_3COONH_4$$

Sample problems and solutions

Task 1.

Make a characterization of the chemical element carbon. Express with equations the transitions indicated on the diagram explaining the processes involved:

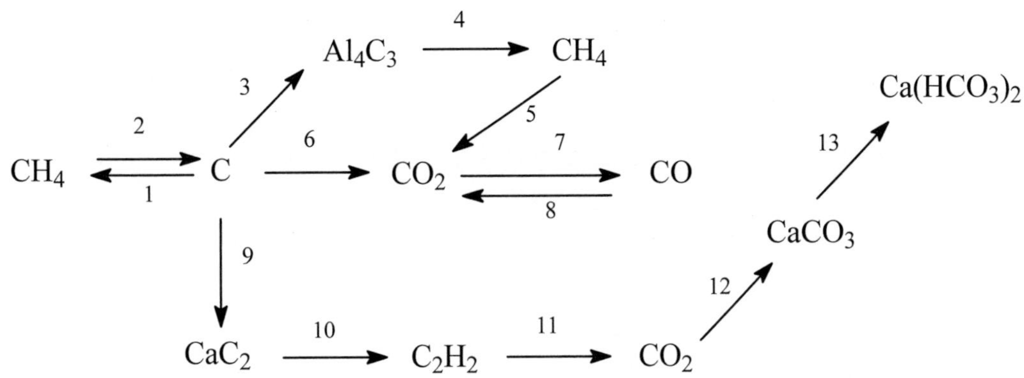

Exemplary solution

Carbon has the sequence number 6, group IV A, period II of the Periodic Table. It has six protons in its nucleus and six electrons in its electronic shell.

The electronic configuration of an atom that is characterized by the lowest energy is called the ground state configuration. If the atom absorbs energy (in terms of chemical interactions) it can transition to an excited state. In this, an electron that is in the lower energy orbital transitions to the higher energy orbital. If the electron has been paired on the lower energy orbital, the excitation is accompanied by a spin reversal. For the carbon atom, the excitation proceeds as follows:

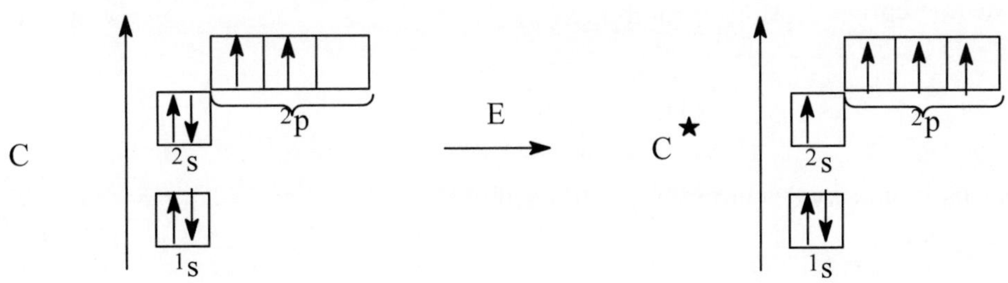

$1s^2\ 2s^2\ 2p^2$ basic state $\qquad$ $1s^2\ 2s^1\ 2p^3$ excited state

The electronic transition in carbon is from the 2s to the 2p-state, where the unpaired electrons of the atom go from two to four. This electronic transition explains the tetravalent nature of the carbon atom in its compounds.

The excited state of the atom is unstable. It is stabilised by giving up the absorbed energy, at which point the excited electron returns to its ground state.

Carbon is only chemically active at high temperature. Diamond is the most resistant, while charcoal and carbon black are the most active because of their large surface area.

Carbon combines with hydrogen at high temperature (1200° C) to form methane:

$$C + 2\,H_2 \overset{1}{\rightleftharpoons} CH_4 + Q$$

$$\begin{array}{lcl|c|c|c} & \overset{0}{C} + 4\,\bar{e} \longrightarrow & \overset{-4}{C} & 4 & \multirow{2}{*}{4} & 1 \\ H_2 \longrightarrow & 2\,\overset{0}{H} - 2 \times 1\,\bar{e} \longrightarrow & 2\,H^+ & 2 & & 2 \end{array}$$

The straight reaction is exothermic and the equilibrium may shift to the right on cooling, but in this case the process rate tends to zero. In the presence of catalyst Ni and elevated pressure, the equilibrium can shift to the right (Le Chatelier-Brown principle).

From the equilibrium process, it can be concluded that the decomposition of methane to carbon (2) and hydrogen is an endothermic process that takes place under reduced pressure.

$$CH_4 \overset{2}{\rightleftharpoons} C + 2\,H_2 - Q$$

$$\begin{array}{lcl|c|c|c} & \overset{-4}{C} - 4\,\bar{e} \longrightarrow & \overset{0}{C} & 4 & \multirow{2}{*}{4} & 1 \\ 2\,H^+ + 2 \times 1\,\bar{e} \longrightarrow & 2\,\overset{0}{H} \longrightarrow & H_2 & 2 & & 2 \end{array}$$

Carbon reacts with metals at high temperature to form carbides.

$$4\,Al + 3\,C \xrightarrow{3} Al_4C_3$$

$$\begin{array}{lcl|c|c|c} \overset{0}{Al} - 3\,e^- & \longrightarrow & Al^{+3} & 3 & \multirow{2}{*}{12} & 4 \\ C + 4\,\bar{e} & \longrightarrow & \overset{-4}{C} & 4 & & 3 \end{array}$$

Carbides are compounds of ionic character. Compounds containing carbide ion C^{-4} readily hydrolyze in water with evolution of methane.

$$Al_4C_3 + 12\,H_2O \xrightarrow{4} 4\,Al(OH)_3 + 3\,CH_4\uparrow$$

Methane burns in an oxygen atmosphere to carbon dioxide (5) and water with the release of a large amount of heat.

$$CH_4 + 2\,O_2 \xrightarrow{5} CO_2 + 2\,H_2O + Q$$

$$\begin{array}{lcl|c|c|c} & \overset{-4}{C} - 8\,\bar{e} \longrightarrow & C^{+4} & 8 & \multirow{2}{*}{8} & 1 \\ O_2 \longrightarrow & 2\,\overset{0}{O} + 2 \times 2\,\bar{e} \longrightarrow & 2\,\overset{-2}{O} & 4 & & 2 \end{array}$$

Relative to oxygen, carbon is treated as a reducing agent. At full combustion CO_2 is produced and a large amount of heat is given off.

$$C + O_2 \xrightarrow{6} CO_2 + Q$$

$$\begin{array}{lll|c|c|c} & C^{0} - 4e^{-} & \longrightarrow C^{+4} & 4 & \multirow{2}{*}{4} & 1 \\ O_2 \longrightarrow & 2O^{0} + 2\,x2e^{-} & \longrightarrow 2O^{-2} & 4 & & 1 \end{array}$$

Incomplete combustion of carbon or coal produces the highly toxic carbon monoxide.

$$C + CO_2 \xrightarrow{7} 2\,CO + Q$$

$$\begin{array}{ll|c|c|c} C^{0} - 2e^{-} & \longrightarrow C^{+2} & 2 & \multirow{2}{*}{2} & 1 \\ C^{+4} + 2e^{-} & \longrightarrow C^{+2} & 2 & & 1 \end{array}$$

The main property of carbon monoxide is its reducing action. It takes oxygen and converts it into carbon dioxide. It is used in metallurgy to obtain metals from their oxides.

$$Fe_2O_3 + 3\,CO \xrightarrow{8} 2\,Fe + 3\,CO_2$$

$$\begin{array}{ll|c|c|c} Fe^{+3} + 3e^{-} & \longrightarrow Fe^{0} & 3 & \multirow{2}{*}{6} & 2 \\ C^{+2} - 2e^{-} & \longrightarrow C^{+4} & 2 & & 3 \end{array}$$

Carbon reacts with calcium oxide at high temperature to produce calcium carbide and carbon monoxide.

$$CaO + 3\,C \xrightarrow{9} CaC_2 + CO$$

$$\begin{array}{ll|c|c|c} C^{0} - 2e^{-} & \longrightarrow C^{+2} & 2 & \multirow{2}{*}{2} & 1 \\ 2C^{0} + 2x1e^{-} & \longrightarrow 2C^{-1} & 2 & & 1 \end{array}$$

Calcium carbide when mixed with water reacts violently and releases acetylene (ethyne) and calcium dihydroxide. The reaction finds wide application in construction.

$$CaC_2 + 2\,H_2O \xrightarrow{10} HC{\equiv}CH + Ca(OH)_2$$

When ethyne is burned, a large amount of heat is released, as well as carbon dioxide and water.

$$2\ HC\equiv CH + 5\ O_2 \xrightarrow{11} 4\ CO_2 + 2\ H_2O$$

$$\begin{array}{lllll|c|c|c}
 & \overset{-}{C} - 5\ \bar{e} & \longrightarrow & C^{+4} & & 5 & & 4 \\
 & & & & & & 20 & \\
O_2 \longrightarrow & 2\ O^{0} + 2 \times 2\ \bar{e} & \longrightarrow & 2\ O^{-2} & & 4 & & 5
\end{array}$$

Carbon dioxide reacts with calcium dihydroxide to produce calcium carbonate.

$$Ca(OH)_2 + CO_2 \xrightarrow{12} CaCO_3 + H_2O$$

If calcium carbonate is treated with carbonic acid it is converted to the soluble calcium dihydrogen carbonate.

$$CaCO_3 + H_2CO_3 \xrightarrow{13} Ca(HCO_3)_2$$

VA group of the Periodic Table

The **VA group** includes the elements nitrogen (**N_2**), phosphorus (**P**), arsenic (**As**), antimony (**Sb**), and bismuth (**Bi**). The electronic configuration of their last electron layer is **ns^2np^3**. Nitrogen and phosphorus are distinguished from the other elements of the group because of the presence of two and eight electrons in their penultimate electron layer. On this basis, only arsenic, antimony, and bismuth, which have 18 electrons in their penultimate electron layer, are complete electron analogues. From nitrogen to bismuth, as the sequence number of the element increases, the number of electron layers as well as the atomic radius increases, and the ionization energy, electron affinity and electronegativity decrease, the metallic character is enhanced. Nitrogen and phosphorus are typical non-metals, but with weaker non-metallic properties than those of the halogen elements and of oxygen and sulfur. Antimony and bismuth are metals, but with weaker metallic propertics than the metals of the **IA** and **IIA** groups. Characteristic oxidation degrees for the elements of the **VA group** are -3 as oxidizers and +3 and +5 as reducers.

Relative to **hydrogen,** the elements of the group appear as stronger oxidants, in the oxidation state -3. In the order:

$$NH_3 \quad PH_3 \quad AsH_3 \quad SbH_3 \quad BiH_3$$
$$\xleftarrow{\hspace{6cm}}$$

their sustainability grows

The aqueous solutions of most of the hydrogen compounds have a basic character, unlike the hydrogen compounds of the elements of **the VIA** and **VIIA** groups.

In relation to **oxygen,** the elements of group **VA** appear as reducers. The characteristic degree of oxidation is +3 and +5 forming two types of oxides, E_2O_3 and E_2O_5.

The chemical bond in the compounds of nitrogen and phosphorus is covalent, that of arsenic and antimony is of considerable polarity, and in bismuth the bonds are ionic.

Nitrogen is a gas as the simple substance is a diatomic molecule - N_2, and the rest are solids. Phosphorus is soft with no metallic lustre. In arsenic and antimony, the hardness increases and antimony has a metallic lustre. In bismuth, the typical malleability of metals also appears.

Nitrogen is the most abundant element in the Earth's crust and atmosphere. It is the main constituent of air (78%). It is called lifeless because it does not support respiration and combustion. Typical oxidation states of nitrogen are -3, 0 (in N_2), +3, +5. It is a typical non-metal. Relative to oxygen it exhibits variable degrees of oxidation: +1, +2, +3, +4, +5.

At ordinary temperature, nitrogen is too inert and does not enter into chemical interactions. At high temperature it reacts with hydrogen, oxygen and metals.

With **hydrogen** forms ammonia. The reaction is reversible and exothermic.

$$N_2 + 3H_2 \longrightarrow 2NH_3 + Q$$

$$N_2 \longrightarrow 2N^0 + 2 \times 3\bar{e} \longrightarrow 2N^{-3} \quad \left|\, 6 \,\right| \; 6 \; \left|\, 1 \right.$$

$$H_2 \longrightarrow 2H^0 - 2 \times 1e^- \longrightarrow 2H^+ \quad \left|\, 2 \,\right| \qquad \left|\, 3 \right.$$

In accordance with the Le Chatelier-Brown principle, a straight reaction is favored at high pressure and low temperature.

Thermal effects of chemical interactions of nitrogen with **hydrogen, oxygen and metals** show that nitrogen has greater chemical activity towards **hydrogen and metals** than towards **oxygen**.

Laboratory **ammonia** is obtained:

- in thermal dissociation of ammonium chloride:

$$NH_4Cl \overset{t^o}{\rightleftharpoons} NH_3\uparrow + HCl\uparrow$$

,

- in the reaction of **ammonium salts with concentrated solutions of bases**:

$$NH_4Cl + NaOH \longrightarrow NaCl + NH_4OH$$

$$NH_4OH \rightleftharpoons H_2O + NH_3\uparrow$$

The chemical properties of ammonia are determined by the presence of a free electron pair at the nitrogen atom, which determines its tendency to attract hydrogen ions (protons) from water and acids, i.e., gives it the properties of a base.

Ammonia dissolves very well in **water** and reacts with it. Its aqueous solutions have the basic character of changing the color of litmus to blue and of phenolphthalein to crimson.

$$NH_3 + H_2O \rightleftharpoons NH_4^+ + \bar{O}H$$

With **acids,** ammonia forms salts called ammonium salts:

$$NH_3 + \underbrace{H^+ + \bar{Cl}} \longrightarrow \underbrace{NH_4^+ + \bar{Cl}}$$

Under ordinary conditions, ammonia does not react with **oxygen**, but when heated it forms an explosive mixture. In an oxygen environment and in the absence of a catalyst, ammonia oxidizes to nitrogen:

$$4NH_3 + 3O_2 \longrightarrow 2N_2 + 6H_2O$$

$$\begin{array}{l|c|c|c}
2N^{-3} - 2 \times 3\bar{e} \longrightarrow 2N^{0} \longrightarrow N_2 & 6 & & 2 \\
 & & 12 & \\
O_2 \longrightarrow 2O^{0} + 2 \times 2\bar{e} \longrightarrow 2O^{-2} & 4 & & 3
\end{array}$$

Of great practical importance is the catalytic oxidation of ammonia to NO, which is used to produce nitric acid:

$$4NH_3 + 5O_2 \longrightarrow 4NO + 6H_2O + Q$$

$$\begin{array}{l|c|c|c}
N^{-3} - 5\bar{e} \longrightarrow N^{+2} & 5 & & 4 \\
 & & 20 & \\
O_2 \longrightarrow 2O^{0} + 2 \times 2\bar{e} \longrightarrow 2O^{-2} & 4 & & 5
\end{array}$$

When reacting with **oxygen,** nitrogen forms oxides, in which the degree of oxidation is from +1 to +5.

Nitrogen interacts directly with **oxygen** at a very high temperature - under voltaic arc conditions:

$$N_2 + O_2 \rightleftharpoons 2NO - Q$$

$$\begin{array}{l|c|c|c}
N_2 \longrightarrow 2N^{0} - 2 \times 2\bar{e} \longrightarrow 2N^{+2} & 4 & & 1 \\
 & & 4 & \\
O_2 \longrightarrow 2O^{0} + 2 \times 2\bar{e} \longrightarrow 2O^{-2} & 4 & & 1
\end{array}$$

Similar processes take place in the atmosphere in lightning.

Diazote oxide (N_2O) is produced by the thermal decomposition of ammonium nitrate:

$$NH_4NO_3 \xrightarrow{t^{o}} N_2O + 2H_2O$$

$$\begin{array}{l|c|c|c}
N^{-3} - 4\bar{e} \longrightarrow N^{+1} & 4 & & 1 \\
 & & 4 & \\
N^{+5} + 4\bar{e} \longrightarrow N^{+1} & 4 & & 1
\end{array}$$

A colorless gas with a characteristic pleasant odor, the so-called "**paradise gas**". It has a narcotic effect and is therefore used in surgical practice. It does not dissolve in water and does not react with it, it is not an anhydride of known acids. It is a **neutral oxide**.

Nitric oxide (NO) is produced by oxidation processes of nitric acid with noble metals:

$$3\,Ag + 4\,HNO_3 \longrightarrow 3\,AgNO_3 + NO + 2\,H_2O$$

$$\begin{array}{lll|l|l|l} Ag^{0} - e^{-} & \longrightarrow & Ag^{+} & 1 & & 3 \\ & & & & 3 & \\ N^{+5} + 3\,e^{-} & \longrightarrow & N^{+2} & 3 & & 1 \end{array}$$

$$Au + \underbrace{HNO_3 + 3\,HCl}_{\text{(royal water)}} \longrightarrow AuCl_3 + NO + 2\,H_2O$$

$$\begin{array}{lll|l|l|l} Au^{0} - 3\,e^{-} & \longrightarrow & Au^{+3} & 3 & & 1 \\ & & & & 3 & \\ N^{+5} + 3\,e^{-} & \longrightarrow & N^{+2} & 3 & & 1 \end{array}$$

It dissolves slightly in water, but does not react with it; it is not an anhydride of any of the acids of nitrogen. In air, it spontaneously oxidizes to **nitrogen dioxide**:

$$2\,NO + O_2 \rightleftharpoons 2\,NO_2$$

$$\begin{array}{lll|l|l|l} N^{+2} - 2\,e^{-} & \longrightarrow & N^{+4} & 2 & & 2 \\ & & & & 4 & \\ O_2 \longrightarrow 2\,O^{0} + 2\times 2\,e^{-} & \longrightarrow & 2\,O^{-2} & 4 & & 1 \end{array}$$

Diazote trioxide (N_2O_3) is a typical acid oxide. It interacts with alkaline bases to form **nitrites** (salts of nitrous acid - **HNO_2**):

$$N_2O_3 + 2\,NaOH \longrightarrow 2\,NaNO_2 + H_2O$$

N_2O_3 e anhydride of nitrous acid.

Nitrogen dioxide NO_2 is a brown colored, highly poisonous gas. It is produced by the oxidation of **NO** by oxygen in the air. It interacts vigorously with water to form a mixture of nitric and nitrous acids. The process is disproportionation.

$$2\,NO_2 + H_2O \longrightarrow HNO_2 + HNO_3$$

$$\begin{array}{lll|l|l|l} N^{+4} + e^{-} & \longrightarrow & N^{+3} & 1 & & 1 \\ & & & & 1 & \\ N^{+4} - e^{-} & \longrightarrow & N^{+5} & 1 & & 1 \end{array}$$

It is therefore a mixed anhydride. In the presence of oxygen, it reacts with water to form nitric acid:

$$4\,NO_2 + O_2 + 2\,H_2O \longrightarrow 4\,HNO_3$$

$$N^{+4} - e^- \longrightarrow N^{+5} \quad | 1 | 4 | 4$$
$$O_2 \longrightarrow 2O^0 + 2 \times 2e^- \longrightarrow 2O^{-2} \quad | 4 | | 1$$

Nitric acid is a colorless liquid. It readily decomposes on heating or by direct sunlight:

$$4HNO_3 \longrightarrow 4\,NO_2 + 2\,H_2O + O_2$$

$$N^{+5} + e^- \longrightarrow N^{+4} \quad | 1 | 4 | 4$$
$$2O^{-2} - 2 \times 2e^- \longrightarrow 2O^0 \longrightarrow O_2 \quad | 4 | | 1$$

Nitric acid is a strong monobasic acid. In water it dissociates into hydrogen (hydroxonium) cations and nitrate anions:

$$HNO_3 + H_2O \longrightarrow H_3O^+ + NO_3^-$$

Due to its strong oxidative properties, it destroys plant and animal tissues, staining them yellow (**xanthoprotein reaction**). Its salts are called **nitrates**.

Nitric acid reacts with almost all metals (except noble metals). If the acid is concentrated at the beginning an oxide is produced. If the oxide does not dissolve in the acid (e.g., oxides of **Al, Fe, Cr**) the reaction stops. The phenomenon is called **passivation**.

Important: When dissolving metals in nitric acid no hydrogen is released!

With the active metals of the **IA** and **IIA** groups, the **HNO_3** concentrate releases **N_2O**, and the dilute acid, **NH_3**, which in a nitric acid environment is converted to **NH_4NO_3**, **ammonium** nitrate.

$$4Ca + \underbrace{10H^+ + 10NO_3^-} \longrightarrow \underbrace{4\,Ca^{+2} + 8\,NO_3^-} + N_2O + 5\,H_2O$$

$$Ca^0 - 2e^- \longrightarrow Ca^{+2} \quad | 2 | 4 | 2$$
$$N^{+5} + 4e^- \longrightarrow N^{+1} \quad | 4 | | 1$$

Similar is the action of dilute nitric acid on Zn and Sn:

$$4\,Zn + \underbrace{9H^+ + 9NO_3^-} \longrightarrow \underbrace{4\,Zn^{+2} + 8\,NO_3^-} + NH_3 + 3H_2O$$

$$NH_3 + \underbrace{H^+ + NO_3^-} \longrightarrow NH_4NO_3$$

$$\begin{array}{llll|c|c|c} N^{+5} & + 8\bar{e} & \longrightarrow & N^{-3} & 8 & \multirow{2}{*}{8} & 1 \\ Zn^{0} & - 2\bar{e} & \longrightarrow & Zn^{+2} & 2 & & 4 \end{array}$$

Concentrated **HNO3** oxidizes the metals Hg and Cu. A reddish-brown fume of nitrogen dioxide (**NO2**) is given off, a qualitative reaction for nitric acid and nitrate ions. As with concentrated sulfuric acid, the reaction proceeds in two stages:

$$Cu + 2HNO_3 \longrightarrow CuO + 2NO_2\uparrow + H_2O$$

$$\begin{array}{llll|c|c|c} Cu^{0} & - 2\bar{e} & \longrightarrow & Cu^{+2} & 2 & \multirow{2}{*}{2} & 1 \\ N^{+5} & + \bar{e} & \longrightarrow & N^{+4} & 1 & & 2 \end{array}$$

$$CuO + 2HNO_3 \longrightarrow Cu(NO_3)_2 + H_2O$$

In sum:

$$Cu + 4HNO_3 \longrightarrow Cu(NO_3)_2 + 2H_2O + 2NO_2\uparrow$$

Nitric acid also dissolves silver with the release of **NO**. Noble metals are dissolved only by "royal water", a mixture of **HNO3** and **HCl** (in a 1:3 ratio).

Concentrated **HNO3** oxidizes some nonmetals **(S, P, C)** to the corresponding oxides:

$$S + 4HNO_3 \longrightarrow SO_2\uparrow + 4NO_2\uparrow + 2H_2O$$

$$\begin{array}{llll|c|c|c} S^{0} & - 4\bar{e} & \longrightarrow & S^{+4} & 4 & \multirow{2}{*}{4} & 1 \\ N^{+5} & + \bar{e} & \longrightarrow & N^{+4} & 1 & & 4 \end{array}$$

Dilute 10% **HNO3** reacts with bases, basic oxides and some salts:

$$\underbrace{Na^{+} + \overline{O}H} + \underbrace{H^{+} + NO_3^{-}} \longrightarrow \underbrace{Na^{+} + NO_3^{-}} + H_2O$$

$$MgO + \underbrace{2H^{+} + 2NO_3^{-}} \longrightarrow \underbrace{Mg^{+2} + 2NO_3^{-}} + H_2O$$

$$\underbrace{2Na^{+} + S^{-2}} + \underbrace{2H^{+} + 2NO_3^{-}} \longrightarrow \underbrace{2Na^{+} + 2NO_3^{-}} + H_2S\uparrow$$

All nitrates are soluble in water and are strong electrolytes. When heated they decompose, giving off oxygen and the other reaction products depend on the metal cation of the salt:

- The nitrates of alkali metals decompose to nitrite:

$$2\,NaNO_3 \xrightarrow{t^o} 2\,NaNO_2 + O_2$$

$$\begin{array}{l|c|c|c}
2\,N^{+5} + 2\times 2\bar{e} \longrightarrow 2\,N^{+3} & 4 & & 1 \\
 & & 4 & \\
2\,O^{-2} - 2\times 2\,\bar{e} \longrightarrow 2\,O^{0} \longrightarrow O_2 & 4 & & 1
\end{array}$$

- The nitrates of the remaining metal ions decompose to release oxides:

$$2\,Cu(NO_3)_2 \xrightarrow{t^o} 2\,CuO + 4\,NO_2 + O_2$$

$$\begin{array}{l|c|c|c}
2\,N^{+5} + 2\times 1\bar{e} \longrightarrow 2\,N^{+4} & 2 & & 2 \\
 & & 4 & \\
2\,O^{-2} - 2\times 2\,\bar{e} \longrightarrow 2\,O^{0} \longrightarrow O_2 & 4 & & 1
\end{array}$$

$$2\,Mg(NO_3)_2 \xrightarrow{t^o} 2\,MgO + 4\,NO_2 + O_2$$

$$\begin{array}{l|c|c|c}
2\,N^{+5} + 2\times 1\bar{e} \longrightarrow 2\,N^{+4} & 2 & & 2 \\
 & & 4 & \\
2\,O^{-2} - 2\times 2\,\bar{e} \longrightarrow 2\,O^{0} \longrightarrow O_2 & 4 & & 1
\end{array}$$

The nitrates of the precious metals decompose with the release of free metal:

$$2\,AgNO_3 \xrightarrow{t^o} 2\,Ag + 2\,NO_2 + O_2$$

$$\begin{array}{l|c|c|c}
2\,Ag^{+} + 2\times 1\bar{e} \longrightarrow 2\,Ag^{0} & 2 & & 2 \\
2\,N^{+5} + 2\times 1\bar{e} \longrightarrow 2\,N^{+4} & 2 & 4 & 2 \\
2\,O^{-2} - 2\times 2\,\bar{e} \longrightarrow 2\,O^{0} \longrightarrow O_2 & 4 & & 1
\end{array}$$

The reaction for proving ammonium salts is the interaction with strong bases:

$$\underbrace{NH_4^{+} + Cl^{-}} + \underbrace{Na^{+} + \overline{OH}} \longrightarrow \underbrace{Na^{+} + Cl^{-}} + NH_4OH \rightleftharpoons H_2O + NH_3\uparrow$$

as well as thermal dissociation of solid ammonium salts:

$$NH_4Cl \rightleftharpoons NH_3\uparrow + HCl\uparrow$$

Phosphorus is an element of great importance for living nature.

Phosphorus is a typical non-metal, but with weaker non-metallic properties than nitrogen. The most typical oxidation state is +5, but it also exhibits +3 and -3.

Phosphorus is produced by reduction of tricalcium diphosphate with coal in the presence of silica according to equation:

$$2\,Ca_3(PO_4)_2 + 10\,C + 6\,SiO_2 \longrightarrow 6\,CaSiO_3 + 4\,P + 10\,CO$$

$$\begin{array}{lcl|c|c|c} 10\,C^{0} - 10 \times 2\,e^- & \longrightarrow & 10\,C^{+2} & 20 & \multirow{2}{*}{20} & 1 \\ 4\,P^{+5} + 4 \times 5\,e^- & \longrightarrow & 4\,P^{0} & 20 & & 1 \end{array}$$

Phosphorus is an active element.

- It forms compounds called **phosphides** with **metals**. These are of high hardness (**Na_3P, Ca_3P_2**). Some phosphides are semiconductors (of **Ga, In**).
- It forms two types of oxides with **oxygen**: **P_2O_3** and **P_2O_5**.
- The **halogen elements** produce PCl_3 and PCl_5.
- Phosphorus does not interact with water and mineral acids.

White phosphorus has a marked reactivity and is therefore stored under water.

With **hydrogen,** phosphorus forms a gaseous hydride called **phosphine**. Phosphine is produced indirectly:

$$Ca_3P_2 + 6\,HCl \longrightarrow 3\,CaCl_2 + 2\,PH_3\uparrow$$

Phosphine is a colourless gas with a garlic smell. It is highly poisonous. Its aqueous solutions have a basic character, but weaker compared to ammonia.

In an **oxygen** deficient atmosphere, phosphorus burns to **P_2O_3**, and in an oxygen atmosphere to **P_2O_5**. Diphosphorus pentoxide greedily absorbs water and is one of the most strongly hygroscopic substances. In laboratory practice it is used in desiccators to provide a dry, anhydrous environment. The process is highly exothermic and **orthophosphoric acid** is produced.

$$P_2O_5 + 3\,H_2O \longrightarrow 2\,H_3PO_4$$

Orthophosphoric acid is tribasic. It is an electrolyte of medium strength ($3\% < \alpha < 30\%$). It dissociates to three degrees and forms three types of salts:

$$H_3PO_4 \longrightarrow H^+ + H_2PO_4^- \qquad \text{(I)}$$

$$H_2PO_4^- \rightleftharpoons H^+ + HPO_4^{-2} \qquad \text{(II)}$$

$$HPO_4^{-2} \rightleftharpoons H^+ + PO_4^{-3} \qquad \text{(III)}$$

$$H_3PO_4 + NaOH \rightleftharpoons NaH_2PO_4 + H_2O$$

monosodium dihydrogen phosphate

$$H_3PO_4 + 2\,NaOH \rightleftharpoons Na_2HPO_4 + 2\,H_2O$$

disodium monohydrogen phosphate

$$H_3PO_4 + 2\,NaOH \rightleftharpoons Na_2HPO_4 + 2\,H_2O$$

trisodium phosphate

A large number of phosphates are slightly soluble white substances. On this basis, an important qualitative reaction for the demonstration of phosphate ions in aqueous solutions is the reaction with silver nitrate, which produces the light yellow precipitate of trisilver phosphate:

$$\underbrace{3\,Na^{+} + PO_4^{-3}} + \underbrace{3\,Ag^{+} + 3\,NO_3^{-}} \longrightarrow Ag_3PO_4\downarrow + \underbrace{3\,Na^{+} + 3\,NO_3^{-}}$$

Sample problems and solutions

Task 1.

Write the reactions that with an element of group VA to get an oxoacid. Choose a metal with which the concentrated acid reacts to give a colored salt. From the salt through the appropriate hydroxide, obtain the metal again. Characterize all the chemical reactions. What type of chemical bond do the simple substances and their compounds used have?

Sample solution

The element of interest is nitrogen, since nitric acid forms colored salts with some metals. The production of this acid is carried out by the following reactions:

(a) nitrous oxide production:

$$4\,NH_3 + 5\,O_2 \longrightarrow 4\,NO + 6\,H_2O$$

$$\begin{array}{lcl|c|c|c} & N^{-3} - 5\,e^{-} \longrightarrow & N^{+2} & 5 & & 4 \\ & & & & 20 & \\ O_2 \longrightarrow & 2\,O^{0} + 2\,x2\,e^{-} \longrightarrow & 2\,O^{-2} & 4 & & 5 \end{array}$$

The process is redox and occurs in the presence of a catalyst (Pt) and elevated temperature. For the production of nitric oxide, the reaction is also taken into account

$$N_2 + O_2 \rightleftharpoons 2\,NO - Q,$$

although it has no practical value.

(b) oxidation of nitric oxide to nitrogen dioxide

$$2\,NO + O_2 = 2\,NO_2$$

$$\begin{array}{lcl|c|c|c} & N^{+2} - 2e^{-} \longrightarrow & N^{+4} & 2 & & 2 \\ & & & & 4 & \\ O_2 \longrightarrow & 2\,O^{0} + 2\,x2e^{-} \longrightarrow & 2\,O^{-2} & 4 & & 1 \end{array}$$

(c) preparation of nitric acid

$$2\overset{+4}{N}O_2 + H_2O \longrightarrow H\overset{+5}{N}O_3 + H\overset{+3}{N}O_2$$

The reaction is a more special case of an oxidation-reduction process, since atoms of the same element (the nitrogen atom of the NO_2 molecule) are oxidants and reducers.

A characteristic property of nitric acid is its oxidizing ability. As a result, it can also interact with metals that are after hydrogen in the Order of Relative Activity of the Metals. In this case the color salt is produced with the metal copper. The reaction proceeds in two stages: first, the acid oxidizes the metal to an oxide, and then with the oxide forms a salt.

$$\overset{o}{Cu} + 2\,H\overset{+5}{N}O_3 \longrightarrow CuO + 2\,\overset{+4}{N}O_2 + H_2O$$

$$CuO + 2\,HNO_3 \longrightarrow Cu(NO_3)_2 + H_2O$$

Total:

$$\overset{o}{Cu} + 4\,HNO_3 \longrightarrow Cu(NO_3)_2 + 2\,NO_2 + 2\,H_2O$$

The conversion of salt to hydroxide is carried out with the help of an alkaline base. The process is expressed by molecular, full ionic and shortened ionic equations.

$$Cu(NO_3)_2 + 2\,NaOH \longrightarrow Cu(OH)_2\downarrow + 2\,NaNO_3$$

Complete ionic equation:

$$\underbrace{Cu^{+2} + 2NO_3^-} + \underbrace{2\,Na^+ + 2\overline{O}H} \longrightarrow Cu(OH)_2\downarrow + \underbrace{2\,Na^+ + 2NO_3^-}$$

Abbreviated ionic equation:

$$Cu^{+2} + 2\overline{O}H \longrightarrow Cu(OH)_2\downarrow$$

The reaction is ion exchange.

Two approaches are possible to obtain the metal. One is to convert the hydroxide to an oxide and the other to a salt convenient for electrolysis (e.g., $CuSO_4$).

For the correct solution to the problem, the presentation of one of the possible approaches is sufficient. We will consider the first case:

$$Cu(OH)_2 \xrightarrow{t^o} CuO + H_2O$$

When heated, copper dihydroxide readily converts to oxide:

The oxide is reduced to elemental copper:

$$\overset{+2}{Cu}O + \overset{0}{C} \longrightarrow \overset{0}{Cu} + \overset{+2}{C}O$$

$$\begin{array}{lll|c|c|c} Cu^{+2} + 2e^- & \longrightarrow & Cu^{0} & 2 & \multirow{2}{*}{2} & 1 \\ C^{0} - 2e^- & \longrightarrow & C^{+2} & 2 & & 1 \end{array}$$

The simple substances and binary compounds involved in the processes under consideration are: nitrogen, ammonia, oxygen, nitrogen monoxide, nitrogen dioxide, water, copper oxide, carbon and copper.

In the simple substances nitrogen, oxygen and carbon, the chemical bond is covalent non-polar.

In the compounds nitrogen monoxide, nitrogen dioxide and water, the bond is covalent polar, i.e., the common electron pair is pulled towards the atom of the more electronegative element.

In copper oxide, the bond is ionic-covalent.

In copper, the connection is metal. It is characterized by a high degree of delocalization of valence electrons.

VIA group of the Periodic Table

The VIA group includes the elements oxygen (O_2), sulfur (S), selenium (Se), tellurium (Te), and polonium (Po). The electronic configuration of the last electron layer is $\mathbf{ns^2np^4}$ and defines an oxidation state of -2 as typical for these elements.

The oxygen atom differs from the other elements of the group because they have free **d-orbitals**, which allows the realization of two excited states with four and six valence electrons.

Unlike **oxygen** and **sulfur**, **selenium**, **tellurium** and **polonium** each have 18 electrons in the penultimate electron layer and are complete electron analogues.

On this basis only oxygen exhibits an oxidation state of -2 with the only exception at OF_2 - +2. The other elements exhibit a variable oxidation state of -2, +4, +6.

Group VIA elements are typical non-metals as the non-metallic character weakens from **oxygen** to **polonium**, but compared to the halogen elements they have less pronounced non- metallic properties. **Polonium** is a radioactive and perishable element.

All the elements of group VIA form more than one simple substance - for example, oxygen (O_2) and ozone (O_3), sulfur - monoclinic and rhombic, etc. The phenomenon is called **allotropy**. Oxygen and ozone are gases, and the allotropic forms of sulfur are solids. They have a non-metallic chemical character.

The simple substances of **selenium** and **tellurium** are hard but have a metallic luster. At ordinary temperature they have a weak electrical conductivity which increases on heating, i.e., they exhibit semiconductor properties.

Oxygen is the most abundant element on Earth - 47% by mass (air, water, rocks, minerals, living organisms, etc.).

Atmospheric air is a mixture of gases: nitrogen (78%), oxygen (20%), carbon dioxide (0.03%), water vapor (2-3%).

Air is liquefied at low temperatures (-190° C) and has a variety of uses, including the production of oxygen.

Industrial oxygen is obtained from air. Laboratory oxygen is produced by the decomposition of bertolite salt (potassium chlorate):

$$2\,KClO_3 \xrightarrow[MnO_2]{t^o} 2\,KCl + 3\,O_2$$

$$\begin{array}{lll|c|c|c} & Cl^{+5} + 6\,e^- & \longrightarrow Cl^- & 6 & & 2 \\ & & & & 12 & \\ 2\,O^{-2} - 2 \times 2\,e^- \longrightarrow & 2\,O^0 & \longrightarrow O_2 & 4 & & 3 \end{array}$$

and electrolysis of water or aqueous solutions of bases.

The following process takes place at the anode A^+:

$$4\,\bar{O}H - 4\,e^- \longrightarrow 2\,H_2O + O_2\uparrow$$

Oxygen is one of the most active chemical elements. With the exception of inert gases, halogen elements, and precious metals (gold, platinum, etc.), it also interacts directly with all other elements at ordinary temperatures or when heated. The reaction is most often exothermic; it is called combustion. Oxides are formed.

Oxygen is a vital element associated with respiration, decay, combustion, corrosion. In medicine it is applied to breathing difficulties, and in engineering to high temperatures, explosives.

When electrical sparks are passed into an oxygen or air environment, a characteristic odor - that of **ozone - is produced**. Similar is the action of special devices called ozonators. Oxygen and ozone are bound in equilibrium:

$$3\,O_2 \rightleftharpoons 2O_3$$

Oxides are compounds of oxygen with chemical elements that are more electropositive than it. These are all elements except fluorine. With it, oxygen forms fluoride - OF_2.

Oxides are produced either by direct interaction of the elements with oxygen or by thermal decomposition of hydroxides or salts of oxygen-containing acids:

$$S + O_2 \longrightarrow SO_2$$

$$\begin{array}{ll|c|c|c} & S^o - 4\,\bar{e} \longrightarrow S^{+4} & 4 & & 1 \\ & & & 4 & \\ O_2 \longrightarrow 2\,O^o & + 2 \times 2\,\bar{e} \longrightarrow 2\,O^{-2} & 4 & & 1 \end{array}$$

$$2\,Al(OH)_3 \xrightarrow{t^o} Al_2O_3 + 3\,H_2O$$

$$2\,AgNO_3 \xrightarrow{t^o} 2\,Ag + 2\,NO_2 + O_2$$

$$\begin{array}{lll|c|c|c}
2\,Ag^{+} + 2 \times 1e^{-} & \longrightarrow & 2\,Ag^{0} & 2 & & 2 \\
2N^{+5} + 2 \times 1e^{-} & \longrightarrow & 2\,N^{+4} & 2 & 4 & 2 \\
2O^{-2} - 2 \times 2\,e^{-} & \longrightarrow & 2O^{0} \longrightarrow O_2 & 4 & & 1
\end{array}$$

The classification of oxides is based on:

- the nature of the chemical bond (from typically ionic to typically covalent);
- based on their relationships to water, acids and bases: basic, acidic, amphoteric and neutral.

Oxygen forms compounds with other elements called **peroxides**. Peroxides of the elements of the IA and IIA groups are produced by direct interaction in an oxygen environment:

$$2K + O_2 \longrightarrow K_2O_2$$

$$Ba + O_2 \longrightarrow BaO_2$$

Oxygen gives two compounds with **hydrogen - water and hydrogen peroxide**.

Water is the most important compound between the two elements and its importance to life on Earth is enormous.

Due to its polarity, water is a good dissociating solvent. It interacts with solutes and hydrates them. Water is a natural medium in which a variety of reactions take place between inorganic substances. In many cases it acts as a catalyst. Its electrolytic dissociation characterizes it as an amphoteric electrolyte. It is an **ampholyte**.

Hydrogen peroxide is produced when barium peroxide reacts with sulfuric acid:

$$BaO_2 + H_2SO_4 \longrightarrow H_2O_2 + BaSO_4\downarrow$$

It is perishable and breaks down according to the scheme:

$$2\,H_2O_2 \longrightarrow 2\,H_2O + O_2$$

$$\begin{array}{lll|c|c|c}
2O^{-} + 2 \times 1\,e^{-} & \longrightarrow & 2O^{-2} & 2 & \multirow{2}{*}{2} & 1 \\
2O^{-} - 2 \times 1\,e^{-} & \longrightarrow & 2O^{0} \longrightarrow O_2 & 2 & & 1
\end{array}$$

Many catalysts accelerate the process - metals, oxides, etc. This makes it necessary to stabilize the commercial product. This is achieved with barbituric acid, which acts as a **negative catalyst (inhibitor)**.

Hydrogen peroxide is mainly used as an oxidant and disinfectant. a 3% solution of H_2O_2 is called "oxygenated water". A 6 - 9 % solution is used in cosmetics.

Sulfur has been known to mankind since antiquity. It occurs in the free state (native sulfur) and in the form of compounds - sulfates and sulfides. It dissolves well in carbon sulfide and organic solvents. It does not dissolve in water.

Native sulfur is extracted using special technology;

Where there are no deposits of native sulfur (for example in Bulgaria), the following scheme applies:

$$2H_2S + SO_2 \longrightarrow 3S + 2H_2O$$

$$\begin{array}{llll|l|l|l} S^{-2} & - 2e^- & \longrightarrow & S^0 & 2 & \multirow{2}{*}{4} & 2 \\ S^{+4} & + 4e^- & \longrightarrow & S^0 & 4 & & 1 \end{array}$$

This enables the recovery of some waste gases.
Sulfur can also be produced by incomplete combustion of hydrogen sulfide:

$$2H_2S + O_2 \longrightarrow 2S + 2H_2O$$

$$\begin{array}{llll|l|l|l} & S^{-2} - 2e^- & \longrightarrow & S^0 & 2 & 4 & 2 \\ O_2 \longrightarrow 2O^0 & + 2 \times 2e^- & \longrightarrow & 2O^{-2} & 4 & & 1 \end{array}$$

Sulfur interacts with almost all elements except precious metals, inert gases, and iodine, but its activity toward metals is much greater than toward nonmetals.

With the **active metals** (IA and IIA groups), shearing yields the corresponding sulfides (M_2S, MS).

In glowing sulfur vapor the **copper** burns to **copper sulfide**.

$$Cu + S \longrightarrow CuS$$

$$\begin{array}{llll|l|l|l} S^0 & + 2e^- & \longrightarrow & S^{-2} & 2 & 2 & 1 \\ Cu^0 & - 2e^- & \longrightarrow & Cu^{+2} & 2 & & 1 \end{array}$$

$$H_2S + Cu \longrightarrow CuS + H_2\uparrow$$

$$\begin{array}{llll|l|l|l} & Cu^0 - 2e^- & \longrightarrow & Cu^{+2} & 2 & 2 & 1 \\ 2H^+ & + 2 \times 1e^- & \longrightarrow 2H^0 \longrightarrow & H_2 & 2 & & 1 \end{array}$$

Silver and mercury react with sulfur at room temperature. Black colored sulfides are produced:

$$Hg + S \longrightarrow HgS$$

$$\begin{array}{llll|l|l|l} S^0 & + 2e^- & \longrightarrow & S^{-2} & 2 & 2 & 1 \\ Hg & - 2e^- & \longrightarrow & Hg^{+2} & 2 & & 1 \end{array}$$

This is an important reaction for removing scattered mercury on premises!

$$2Ag + S \longrightarrow Ag_2S$$

$$\begin{array}{lcl|c|c|c} S^{0} + 2e^{-} & \longrightarrow & S^{-2} & 2 & \multirow{2}{*}{2} & 1 \\ Ag^{0} - \bar{e} & \longrightarrow & Ag^{+} & 1 & & 2 \end{array}$$

Silver objects blacken over time from the sulfide that forms.
Iron reacts with sulfur on moderate heating:

$$Fe + S \longrightarrow FeS$$

$$\begin{array}{lcl|c|c|c} S^{0} + 2e^{-} & \longrightarrow & S^{-2} & 2 & \multirow{2}{*}{2} & 1 \\ Fe^{0} - 2\bar{e} & \longrightarrow & Fe^{+2} & 2 & & 1 \end{array}$$

With the exception of the sulfides of the elements of groups IA and IIA and ammonium sulfide, $(NH_4)_2S$, all the other sulfides are very slightly soluble compounds, and among them are some of the most insoluble precipitates in general, HgS, Ag_2S, etc. They do not dissolve even in acids, unless the acid also possesses an oxidizing action - nitric acid or "royal water."

$$CuS + 4HNO_3 \longrightarrow Cu(NO_3)_2 + 2NO_2 + S + 2H_2O$$

$$\begin{array}{lcl|c|c|c} S^{-2} - 2e^{-} & \longrightarrow & S^{0} & 2 & \multirow{2}{*}{2} & 1 \\ N^{+5} + e^{-} & \longrightarrow & N^{+4} & 1 & & 2 \end{array}$$

Similarly, sulfur is oxidized by nitric acid to oxide:

$$S + 4HNO_3 \longrightarrow SO_2 + 4NO_2 + 2H_2O$$

$$\begin{array}{lcl|c|c|c} S^{0} - 4e^{-} & \longrightarrow & S^{+4} & 4 & \multirow{2}{*}{4} & 1 \\ N^{+5} + e^{-} & \longrightarrow & N^{+4} & 1 & & 4 \end{array}$$

Sulfur finds application in industry, in vulcanization of rubber, etc. Rubber with a high sulfur content is called **ebonite** and has valuable qualities as an insulator.

In practice, the most important are **sulfur dioxide** and **sulfur trioxide**. Sulfur dioxide is produced:

- in direct combustion of sulfur in an air environment:

$$S + O_2 \longrightarrow SO_2$$

$$\begin{array}{lcl|c|c|c} S^{0} - 4e^{-} & \longrightarrow & S^{+4} & 4 & \multirow{2}{*}{4} & 1 \\ O_2 \longrightarrow 2O^{0} + 2 \times 2\bar{e} & \longrightarrow & 2O^{-2} & 4 & & 1 \end{array}$$

- when "frying" sulfides:

Particularly important is the "sintering" process of the iron sulfide, **pyrite (FeS_2)**, which is the basis of the first stage of sulfuric acid production:

$$FeS_2 + O_2 \xrightarrow{t^o} Fe_2O_3 + SO_2$$

$$\begin{array}{lll|c|c|c} Fe^{+2} - e^- & \longrightarrow & Fe^{+3} & 1 & & 4 \\ & & & & 4 & \\ O_2 \longrightarrow 2O^o + 2x2e^- & \longrightarrow & 2O^{-2} & 4 & & 1 \end{array}$$

$$4\,FeS_2 + 11O_2 \xrightarrow{t^o} 2\,Fe_2O_3 + 8\,SO_2$$

Although the sulfur atom also changes its degree of oxidation in the above redox process, the material balance in the equation is achieved without including the sulfur atom and its changes in oxidation degrees.

- in the oxidation of metals with sulfuric acid. The reaction is carried out in two stages:
  Stage I - redox process:

$$Cu + H_2SO_4 \longrightarrow CuO + SO_2 + H_2O$$

$$\begin{array}{lll|c|c|c} Cu^o - 2\,e^- & \longrightarrow & Cu^{+2} & 2 & & 1 \\ & & & & 2 & \\ S^{+6} + 2\,e^- & \longrightarrow & S^{+4} & 2 & & 1 \end{array}$$

Stage II - exchange reaction:

$$CuO + H_2SO_4 \longrightarrow CuSO_4 + H_2O$$

Total:

$$Cu + 2H_2SO_4 \longrightarrow CuSO_4 + SO_2 + 2\,H_2O$$

$$\begin{array}{lll|c|c|c} Cu^o - 2\,e^- & \longrightarrow & Cu^{+2} & 2 & & 1 \\ & & & & 2 & \\ S^{+6} + 2\,e^- & \longrightarrow & S^{+4} & 2 & & 1 \end{array}$$

Sulfur dioxide is a colorless gas with a pungent smell. It dissolves in and reacts with water. It produces **sulfurous acid, H_2SO_3**, which is a perishable dibasic acid:

$$SO_2 + H_2O \rightleftharpoons H_2SO_3$$

Sulfurous acid is dibasic and dissociates in two steps:

$$H_2SO_3 + H_2O \rightleftharpoons H_3O^+ + HSO_3^-$$

$$HSO_3^- + H_2O \rightleftharpoons H_3O^+ + SO_3^{-2}$$

Sulfurous acid is neutralized with bases and its salts are called **sulfites**, in incomplete neutralization - **hydrogensulfites**.

Sulfur dioxide is a typical acid oxide and reacts with basic oxides and bases. Sulfur dioxide is oxidized by oxygen to sulfur trioxide:

$$2\,SO_2 + O_2 \longrightarrow 2\,SO_3$$

$$S^{+4} - 2e^- \longrightarrow S^{+6} \quad \left|\, 2 \,\right| \; \left|\, 4 \,\right| \; \left|\, 2 \,\right.$$

$$O_2 \longrightarrow 2\,O^0 + 2 \times 2e^- \longrightarrow 2O^{-2} \quad \left|\, 4 \,\right| \; \left| \; \right| \; \left|\, 1 \,\right.$$

The reaction is carried out under heating and in the presence of catalysts - V_2O_5 or Pt (heterogeneous catalysis) or NO (homogeneous catalysis).

Sulfur trioxide is an important product in the production of sulfuric acid - an acidic oxide, highly hygroscopic, sulfuric acid anhydride:

$$SO_3 + H_2O \longrightarrow H_2SO_4$$

Sulfur trioxide reacts with basic oxides and bases:

$$SO_3 + CaO \longrightarrow CaSO_4$$

$$SO_3 + 2NaOH \longrightarrow Na_2SO_4 + H_2O$$

Sulfuric acid is produced in several steps:

- Production of sulfur dioxide (**SO_2**) by burning sulfur or "frying" sulfides (pyrite - FeS_2);
- Oxidation of sulfur dioxide to sulfur trioxide.
- absorption of sulfur trioxide from water to sulfuric acid:

$$SO_3 + H_2O \longrightarrow H_2SO_4$$

The properties of **dilute** and **concentrated sulfuric acid** differ. Regardless of its concentration, sulfuric acid is a **strong dibasic** acid:

$$H_2SO_4 + H_2O \longrightarrow H_3O^+ + HSO_4^-$$

$$HSO_4^- + H_2O \longrightarrow H_3O^+ + SO_4^{-2}$$

It forms two types of salts – **hydrogen sulfates** and **sulfates**. Dissociation by the second degree is significant. Sulfuric acid forms mostly normal salts - **sulfates**.

Dilute sulfuric acid reacts with active metals, basic oxides and hydroxides - neutralization.

Concentrated sulfuric acid is **highly hygroscopic** and therefore hazardous in handling. In addition to the typical properties of an acid (analogous to dilute acid), it also has an oxidizing effect (see reaction with copper).

A qualitative reaction for sulfuric acid and soluble sulfates is the interaction with barium dichloride solution:

$$Ba^{+2} + SO_4^{-2} \longrightarrow BaSO_4\downarrow$$

Sulfuric acid has wide application in practice - for the production of fertilizers, the preparation of other acids, the production of dyes, oil refining, for batteries, etc.

Hydrogen sulfide is the most important hydride of sulfur. It is produced industrially by the reaction of sulfur with hydrogen:

$$S + H_2 \longrightarrow H_2S$$

$$S^{0} + 2e^{-} \longrightarrow S^{-2} \quad | \; 2 \; | \; 2 \; | \; 1$$

$$H_2 \longrightarrow 2H^{0} - 2 \times 1e^{-} \longrightarrow 2H^{+} \quad | \; 2 \; | \; \; | \; 1$$

Laboratory **hydrogen sulfide** is produced from metal sulfides and mineral acids:

$$FeS + 2HCl \longrightarrow H_2S + FeCl_2$$

$$Na_2S + 2HCl \longrightarrow H_2S + 2NaCl$$

Hydrogen sulfide occurs in volcanic gases and in some thermal waters. It has a pungent odor of rotten eggs and is poisonous.

Hydrogen sulfide has a reducing effect. In an air environment it burns according to the equation:

$$2H_2S + 3O_2 \longrightarrow 2H_2O + 2SO_2$$

$$S^{-2} - 6e^{-} \longrightarrow S^{+4} \quad | \; 6 \; | \; 12 \; | \; 2$$

$$O_2 \longrightarrow 2O^{0} + 2 \times 2e^{-} \longrightarrow 2O^{-2} \quad | \; 4 \; | \; \; | \; 3$$

Hydrogen sulfide dissolves well in water. Its aqueous solution is a weak dibasic acid which yields two types of salts, **hydrogen sulfides** and **sulfides**:

$$H_2S + H_2O \rightleftharpoons H_3O^{+} + HS^{-}$$

$$HS^{-} + H_2O \rightleftharpoons H_3O^{+} + S^{-2}$$

The sulfide ion is a very strong reducing agent. Even from atmospheric oxygen, elemental sulfur is released from its solutions:

$$H_2S + O_2 \longrightarrow S\downarrow + 2H_2O$$

$$S^{-2} - 2e^{-} \longrightarrow S^{0} \quad | \; 2 \; | \; 4 \; | \; 2$$

$$O_2 \longrightarrow 2O^{0} + 2 \times 2e^{-} \longrightarrow 2O^{-2} \quad | \; 4 \; | \; \; | \; 1$$

The sulfides of almost all elements are slightly soluble compounds. When hydrogen sulfide (gas) or even aqueous solutions of hydrogen sulfide are passed through, all **soluble** and **insoluble** compounds of the elements (with the exception of the elements of groups IA and IIA and of the ammonium ion) are converted into colored precipitates of sulfides.

The hydrides of the elements of group VI A, with the exception of water, are gases with a pungent odor. They are poisonous. Their durability decreases in the order:

$$H_2O > H_2S > H_2Se > H_2Te$$

Their aqueous solutions are acidic in nature.

Group VIA elements form two types of **oxides**, dioxides (**EO_2**) and trioxides (**EO_3**).

Sulfur dioxide is a gas, and all other dioxides and trioxides are solids with a molecular structure. They are typical acid oxides and are anhydrides of the corresponding acids.

In the order

$$S \longrightarrow Te$$

the strength of the acids decreases for both the lower (+4) and the higher oxidation degree (+6).

Sample problems and solutions

Task 1.

The aqueous solution of solid **A** changes the red litmus to blue. When **A** interacts with acids, gas **B** with an unpleasant odor and highly toxic action is released. Incomplete combustion of **B** produces the chemical element **C.** From **B** you know two oxoacids. Derive them and describe the properties of the acid in which **B** is in its higher oxidation state.

Sample solution

The alkaline nature of the aqueous solution of substance **A** suggests that it is probably a salt of a strong base and a weak acid. The evolved gas and the other data in the condition indicate that **A** is an alkali sulfide, **B** is hydrogen sulfide and **C** is the element sulfur.

An ion exchange reaction takes place with acids:

$$Na_2S + 2HCl \longrightarrow H_2S + 2NaCl$$

molecular equation

$$\underbrace{2Na^+ + S^{-2}} + \underbrace{2H^+ + 2Cl^-} \longrightarrow \underbrace{2Na^+ + 2Cl^-} + H_2S$$

complete ionic equation

$$2H^+ + S^{-2} \longrightarrow H_2S$$

abbreviated ionic equation.

The reaction proceeds to completion as hydrogen sulfide is a gas and a weak electrolyte.

Hydrogen sulfide is easily oxidised. Ignited in the air, it burns. In incomplete combustion, the process is represented by the equation:

$$2\,H_2S + O_2 \longrightarrow 2\,S^{0} + 2\,H_2O^{-2}$$

$$S^{-2} - 2e^{-} \longrightarrow S \quad | \; 2 \; | \; 4 \; | \; 2$$

$$O_2 \longrightarrow 2O^{0} + 2 \times 2e^{-} \longrightarrow 2O^{-2} \quad | \; 4 \; | \; \; | \; 1$$

The process is redox.

The process of giving up electrons is called oxidation and the process of taking up electrons is called reduction. The substance that gives off electrons is oxidized and is a reductant, and the substance that accepts electrons is reduced and is an oxidizer.

Two sulfuric acids: sulfurous (H_2SO_3) and sulfuric (H_2SO_4), are described in a high school chemistry and environmental science course. These can be obtained by dissolving SO_2 and SO_3 in water.

$$S + O_2 \longrightarrow SO_2$$

$$S^{0} - 4e^{-} \longrightarrow S^{+4} \quad | \; 4 \; | \; 4 \; | \; 1$$

$$O_2 \longrightarrow 2O^{0} + 2 \times 2e^{-} \longrightarrow 2O^{-2} \quad | \; 4 \; | \; \; | \; 1$$

When heated, sulfur interacts vigorously with oxygen - it burns. The process is redox.

$$SO_2 + H_2O \rightleftharpoons H_2SO_3$$

It is a reaction of synthesis.

When heated in an oxygen atmosphere, sulfur dioxide is oxidized to sulfur trioxide:

$$2\,SO_2 + O_2 \rightleftharpoons 2\,SO_3$$

$$SO_3 + H_2O \longrightarrow H_2SO_4$$

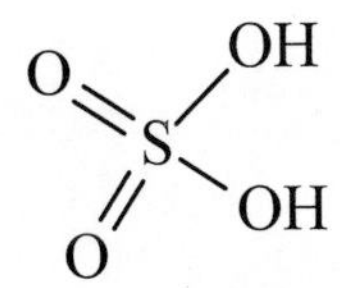

Sulfuric acid is a strong dibasic acid. It dissociates in two degrees.

Dilute sulfuric acid differs in properties from concentrated oxidizing oxoacid. Dilute sulfuric acid exhibits properties like all acids. It interacts with metals that are ahead of hydrogen in the Order of Relative Activity of the Metals, with basic and amphoteric oxides and hydroxides, and with salts of weaker acids:

$$Zn + H_2SO_4 \longrightarrow ZnSO_4 + H_2$$

$$Zn^{0} - 2e^{-} \longrightarrow Zn^{+2} \quad | \; 2 \; | \; 2 \; | \; 1$$

$$2\,H^{+} + 2 \times 1e^{-} \longrightarrow 2\,H^{0} \longrightarrow H_2 \quad | \; 2 \; | \; \; | \; 1$$

The process is redox

$$ZnO + H_2SO_4 \longrightarrow ZnSO_4 + H_2O$$

$$H_2SO_4 + 2NaOH \longrightarrow Na_2SO_4 + 2\,H_2O$$

molecular equation

The process is neutralization.

$$\underbrace{2\,H^+ + SO_4^{-2}} + \underbrace{2\,Na^+ + 2\bar{O}H} \longrightarrow \underbrace{2\,Na^+ + SO_4^{-2}} + H_2O$$

complete ionic equation

$$H^+ + \bar{O}H \longrightarrow H_2O$$

abbreviated ionic equation

The process is ion exchange, in particular neutralization, and proceeds to completion due to the production of the weak electrolyte water.

$$H_2SO_4 + Na_2CO_3 \longrightarrow Na_2SO_4 + H_2CO_3$$

$$H_2CO_3 \rightleftharpoons CO_2 + H_2O$$

$$\underbrace{2\,H^+ + SO_4^{-2}} + \underbrace{2\,Na^+ + CO_3^{-2}} \longrightarrow \underbrace{2\,Na^+ + SO_4^{-2}} + H_2CO_3 \rightleftharpoons H_2O + CO_2\uparrow$$

$$2H^+ + CO_3^{-2} \longrightarrow H_2CO_3 \rightleftharpoons CO_2 + H_2O$$

The process proceeds to completion due to the production of the weak and unstable electrolyte carbonic acid.

Concentrated sulfuric acid exhibits an oxidizing action, which is related to the tendency of its molecules to decompose.

$$H_2SO_4 \longrightarrow SO_2 + H_2O + O$$

$$\begin{array}{lll|c|c|c} S^{+6} + 2\bar{e} & \longrightarrow & S^{+4} & 2 & & 1 \\ & & & & 2 & \\ O^{-2} - 2e^- & \longrightarrow & O^0 & 2 & & 1 \end{array}$$

For this reason, it can also react with some metals that come after hydrogen in the order of relative metal activity:

$$Cu + 2\,H_2SO_4 \longrightarrow CuSO_4 + SO_2 + 2\,H_2O$$

$$\begin{array}{lcl|c|c|c}
Cu^{0} & \xrightarrow{-2e^-} & Cu^{+2} & 2 & & 1 \\
 & & & & 2 & \\
S^{+6} & \xrightarrow{+2e^-} & S^{+4} & 2 & & 1
\end{array}$$

The reaction proceeds in two stages, as the metal is first oxidized to copper oxide and this in the next stage reacts with the acid.

Due to its oxidizing action, concentrated sulfuric acid passivates some metals - **iron, aluminum, chromium.** It is highly hygroscopic, taking hydrogen and oxygen from their compounds in the form of water. This explains its charring action on organic compounds.

Group VIIA of the Periodic Table

The group includes the elements fluorine (F_2), chlorine (Cl_2), bromine (Br_2), iodine (I_2), and astatine (At). They are called 'halogenated', which means 'brineable'. Astatine is unstable, is radioactive, and is therefore unavailable and not well studied. For convenience, it is customary to denote halogen elements by the symbol **-X** in summary reactions and descriptions.

These are p-elements with electronic configuration $\mathbf{ns^2np^5}$, and therefore have significant electronic affinity, readily attach an electron and acquire the electronic configuration of the inert element standing behind them. They are typical nonmetals with pronounced oxidative properties. A special place among the halogen elements is occupied by **fluorine**, which, unlike the others, has no d-sublayer. It appears only in the -1oxidation state and is an oxidant even with respect to oxygen. The other halogen elements exhibit a variable degree of oxidation - besides the most characteristic -1, also from +1 to +7. In the order from fluorine to iodine, the atomic radius increases and the electron affinity and associated non-metallic and oxidizing properties of the elements decrease. Halogen elements are chemically very active, and therefore occur only in the form of compounds.

Fluorine and **chlorine** are gases, **bromine** is the only liquid non-metal, it has a high vapor pressure. **Iodine** is a solid with a weak metallic lustre, it evaporates very easily, the phenomenon is called **sublimation**.

When in the gaseous state, halogen elements are diatomic molecules and have a very suffocating odor. When inhaled they cause severe poisoning, damaging the respiratory system.

Their solubility in water is small. Aqueous solutions of **chlorine** and **bromine** exhibit an oxidizing action and are called '**chlorine water**' and '**bromine water**'. They are used as safer reagents (oxidizing agents) than gaseous simple substances. **Fluorine** does not dissolve in water because it decomposes it. **Bromine** and **iodine** dissolve readily in many organic solvents. An alcoholic solution of iodine (10 %) is called '**iodine tincture**'.

Free halogens have high chemical activity. They react with almost all simple substances. Although bromine and iodine are not as active as fluorine and chlorine, their chemical activity is too great.

The interaction with **hydrogen is an oxidation-reduction process** and results in the production of halogenated hydrocarbons. **Fluorine** reacts with an explosion under ordinary conditions, and **chlorine** in direct sunlight also reacts with an explosion by a chain-radical mechanism:

$$H_2 + X_2 \longrightarrow 2\,HX$$

$$\begin{array}{l|c|c|c}
H_2 \longrightarrow 2\,H^{0} - 2 \times 1e^- \longrightarrow 2\,H^+ & 2 & & 1 \\
 & & 2 & \\
X_2 \longrightarrow 2\,X^{0} + 2 \times 1e^- \longrightarrow 2\,X^- & 2 & & 1
\end{array}$$

In contrast to metal hydrides, hydrogen halides are molecules with a covalent polar bond. They all have an unpleasant suffocating odor, and dissolve well in water. Their acqueous solutions exhibit acidic properties due to their dissociation to hydrogen ions and halide anions caused by the dipole molecules of water.

Halogen elements react actively with all **metals**. Salts called **halides** are produced.

$$2\,Na + Cl_2 \longrightarrow 2NaCl$$

$$\begin{array}{rcl|c|c|c} \overset{0}{Na} - 1e^- & \longrightarrow & Na^+ & 1 & & 2 \\ & & & & 2 & \\ Cl_2 \longrightarrow 2\overset{0}{Cl} + 2 \times 1e^- & \longrightarrow & 2\,Cl^- & 2 & & 1 \end{array}$$

If iron filings are sprinkled into a cylinder of chlorine, a spectacular reaction is observed, the so-called "**fire rain**":

$$2Fe + 3Cl_2 \longrightarrow 2\,FeCl_3$$

$$\begin{array}{rcl|c|c|c} \overset{0}{Fe} - 3e^- & \longrightarrow & Fe^{+3} & 3 & & 2 \\ & & & & 6 & \\ Cl_2 \longrightarrow 2\,\overset{0}{Cl} + 2x1e^- & \longrightarrow & 2Cl^- & 2 & & 3 \end{array}$$

Metal halides have crystal lattice. With the exception of silver halides (AgX), they dissolve in water and are strong electrolytes.

The activity towards metals is so great that it can occur in the interaction between metal and chlorine (bromine) water. For example, zinc decolors bromine water:

$$Zn + Br_2 \longrightarrow ZnBr_2$$

$$\begin{array}{rcl|c|c|c} \overset{0}{Zn} - 2e^- & \longrightarrow & Zn^{+2} & 2 & & 1 \\ & & & & 2 & \\ Br_2 \longrightarrow 2\,\overset{0}{Br} + 2x1e^- & \longrightarrow & 2\,Br^- & 2 & & 1 \end{array}$$

With **nonmetals,** halogens exhibit great chemical activity. **Fluorine** reacts with all non-metals - S, P, C, Si, even with bromine and iodine, always in the -1oxidation state. The other halogen elements also interact, but not as actively as fluorine.

$$2P + 3Cl_2 \longrightarrow 2PCl_3$$

$$\begin{array}{rcl|c|c|c} \overset{0}{P} - 3e^- & \longrightarrow & P^{+3} & 3 & & 2 \\ & & & & 6 & \\ Cl_2 \longrightarrow 2\overset{0}{Cl} + 2x1e^- & \longrightarrow & 2Cl^- & 2 & & 3 \end{array}$$

,

and in excess of chlorine

$$PCl_3 + Cl_2 \rightleftharpoons PCl_5$$

$$\begin{array}{lll|c|c|c} & P^{+3} - 2\,e^- & \longrightarrow P^{+5} & 2 & & 1 \\ & & & & 2 & \\ Cl_2 \longrightarrow & 2\,Cl^{0} + 2 x 1 e^- & \longrightarrow 2\,Cl^- & 2 & & 1 \end{array}$$

Fluorine reacts with **water** by decomposing it:

$$2\,F_2 + 2H_2O \longrightarrow 4\,HF + O_2$$

$$\begin{array}{lll|c|c|c} 2\,O^{-2} - 2 x 2\,e^- & \longrightarrow 2\,O^{0} & \longrightarrow O_2 & 4 & & 1 \\ & & & & 4 & \\ F_2 \longrightarrow & 2\,F^{0} + 2 x 1 e^- & \longrightarrow 2\,F^- & 2 & & 2 \end{array}$$

Chlorine is not as active, but reacts according to the equation:

$$Cl_2 + H_2O \longrightarrow HCl + HOCl$$

$$\begin{array}{ll|c|c|c} Cl_2 \longrightarrow 2\,Cl^{0} + 2 x 1 e^- & \longrightarrow 2\,Cl^- & 2 & & 1 \\ & & & 2 & \\ Cl_2 \longrightarrow 2\,Cl^{0} - 2 x 1 e^- & \longrightarrow 2\,Cl^+ & 2 & & 1 \end{array}$$

The process is known as **disproportionation**.

Hypochlorous acid - HOCl has a strong oxidizing and bleaching effect, as it releases atomic (nascent) oxygen in sunlight:

$$HOCl \longrightarrow HCl + \overset{0}{O}$$

$$\begin{array}{ll|c|c|c} \overset{-2}{O} - 2\,e^- & \longrightarrow \overset{0}{O} & 2 & & 1 \\ & & & 2 & \\ Cl^+ + 2\,e^- & \longrightarrow Cl^- & 2 & & 1 \end{array}$$

Halogens react actively with aqueous solutions of **bases** by disproportionation to the corresponding **halides** and **hypohalenites**. Fluorine does not react with bases because it interacts vigorously with water.

$$Cl_2 + 2\,NaOH \longrightarrow NaCl + NaOCl + H_2O$$

$$\begin{array}{ll|c|c|c} Cl_2 \longrightarrow 2\,Cl^{0} + 2 x 1 e^- & \longrightarrow 2\,Cl^- & 2 & & 1 \\ & & & 2 & \\ Cl_2 \longrightarrow 2\,Cl^{0} - 2 x 1 e^- & \longrightarrow 2\,Cl^+ & 2 & & 1 \end{array}$$

When saturating an aqueous solution of sodium base with chlorine, "bleach" is produced. Its bleaching action is due to the reactions:

$$NaOCl + H_2O \longrightarrow NaOH + HOCl$$

$$HOCl \longrightarrow HCl + \overset{0}{O}$$

$$\begin{array}{lll|c|c|c} \overset{-2}{O} - 2\bar{e} & \longrightarrow & \overset{0}{O} & 2 & \multirow{2}{*}{2} & 1 \\ Cl^{+} + 2\bar{e} & \longrightarrow & Cl^{-} & 2 & & 1 \end{array}$$

If dry calcium base is treated with chlorine, "**chlor lime**" is produced, which has a strong oxidizing effect and is used as a disinfectant.

$$Ca(OH)_2 + Cl_2 \longrightarrow CaOCl_2 + H_2O$$

Chlor lime is a white powdery substance with a pungent chlorine smell. In moist air and under the action of carbon dioxide it decomposes according to the following scheme:

$$2CaOCl_2 + CO_2 + H_2O \longrightarrow CaCO_3\downarrow + CaCl_2 + 2\,HOCl$$

A qualitative reaction for the detection of halogens and soluble halides is interaction with silver nitrate:

$$\underbrace{Ag^{+} + NO_3^{-}} + \underbrace{K^{+} + Cl^{-}} \longrightarrow AgCl\downarrow + \underbrace{K^{+} + NO_3^{-}}$$

$$\underbrace{Ag^{+} + NO_3^{-}} + \underbrace{K^{+} + Br^{-}} \longrightarrow AgBr\downarrow + \underbrace{K^{+} + NO_3^{-}}$$

$$\underbrace{Ag^{+} + NO_3^{-}} + \underbrace{K^{+} + I^{-}} \longrightarrow AgI\downarrow + \underbrace{K^{+} + NO_3^{-}}$$

The iodine colors the starch an intense blue.

Fluorine is produced by electrolysis of potassium fluoride (KF) and **chlorine** by chloralkali electrolysis of aqueous sodium chloride solution.

Laboratory chlorine is obtained from hydrochloric acid and manganese dioxide:

$$MnO_2 + 4\,HCl \longrightarrow MnCl_2 + Cl_2 + 2\,H_2O$$

$$\begin{array}{lll|c|c|c} Mn^{+4} + 2\bar{e} & \longrightarrow & Mn^{+2} & 2 & \multirow{2}{*}{2} & 1 \\ 2\,Cl^{-} - 2 \times 1\,e^{-} \longrightarrow 2\,\overset{0}{Cl} & \longrightarrow & Cl_2 & 2 & & 1 \end{array}$$

Bromine and **iodine** are most commonly obtained from aqueous solutions of salts under the action of an oxidizing agent, such as chlorine water:

$$2NaBr + Cl_2 \longrightarrow 2NaCl + Br_2$$

$$\begin{array}{l|c|c|c} Cl_2 \longrightarrow 2Cl^{0} + 2 \times 1\,e^{-} \longrightarrow 2Cl^{-} & 2 & & 1 \\ & & 2 & \\ 2Br^{-} - 2 \times 1\,e^{-} \longrightarrow 2Br^{0} \longrightarrow Br_2 & 2 & & 1 \end{array}$$

$$2NaI + Cl_2 \longrightarrow 2NaCl + I_2$$

$$\begin{array}{l|c|c|c} Cl_2 \longrightarrow 2Cl^{0} + 2 \times 1\,e^{-} \longrightarrow 2Cl^{-} & 2 & & 1 \\ & & 2 & \\ 2I^{-} - 2 \times 1\,e^{-} \longrightarrow 2I^{0} \longrightarrow I_2 & 2 & & 1 \end{array}$$

Halogenated hydrocarbons are derived from halides (solid state) and sulfuric acid:

$$2NaCl + H_2SO_4 \longrightarrow Na_2SO_4 + 2HCl$$

Halogenated hydrocarbons (hydrogen halides) dissolve in water and are strong acids because, with the exception of hydrogen fluoride, they completely dissociate into ions (hydroxonium cations and halide anions). They are strong acids. They color the litmus red.

Of the hydrochloric acids, hydrochloric acid is the most widely used. At maximum saturation of water with hydrogen chloride, concentrated hydrochloric acid (37 %) is obtained. It is called fuming because, when left in the air, it gives off white smoke with a pungent suffocating odor – hydrogen chloride.

$$H_2 + Cl_2 \longrightarrow 2HCl$$

$$\begin{array}{l|c|c|c} Cl_2 \longrightarrow 2Cl^{0} + 2 \times 1\,e^{-} \longrightarrow 2Cl^{-} & 2 & & 1 \\ & & 2 & \\ H_2 \longrightarrow 2H^{0} - 2 \times 1\,e^{-} \longrightarrow 2H^{+} & 2 & & 1 \end{array}$$

A mixture of 3-parts hydrochloric acid and 1-part nitric acid is called "**royal water**" because it has a very strong oxidizing effect and dissolves precious metals – gold, platinum, etc. The dissolving action of royal water is due to the chlorine released when the acids are mixed:

$$3HCl + HNO_3 \longrightarrow NOCl + Cl_2 + 2H_2O$$

$$\begin{array}{l|c|c|c} 2Cl^{-} - 2 \times 1\,e^{-} \longrightarrow 2Cl^{0} \longrightarrow Cl_2 & 2 & & 1 \\ & & 2 & \\ N^{+5} + 2e^{-} \longrightarrow N^{+3} & 2 & & 1 \end{array}$$

$$Au + 3\,HCl + HNO_3 \longrightarrow AuCl_3 + NO + 2\,H_2O$$

$$\begin{array}{lcl|c|c|c} Au^{0} - 3\,e^{-} & \longrightarrow & Au^{+3} & 3 & \multirow{2}{*}{3} & 1 \\ N^{+5} + 3\,e^{-} & \longrightarrow & N^{+2} & 3 & & 1 \end{array}$$

In the order HF, HCl, HBr, HI, the strongest is iodohydric acid and the weakest is hydrofluoric acid. Nevertheless, hydrofluoric acid corrodes glass and is used in the glass industry for etching glassware.

HI, HBr and HCl are completely dissociated in water, i.e., they are strong acids, interact with bases (neutralization), with alkali and alkaline earth metals (IA and IIA groups). With metals after hydrogen in the order of relative activity, they generally do not interact. In the presence of oxygen (oxidizing agent) they can oxidize semi-noble metals (Cu, Hg, etc.):

$$2\,Cu + 4HCl + O_2 \longrightarrow CuCl_2 + 2H_2O$$

$$\begin{array}{lcl|c|c|c} Cu^{0} - 2\,e^{-} & \longrightarrow & Cu^{+2} & 2 & \multirow{2}{*}{4} & 2 \\ O_2 \longrightarrow 2\,O^{0} + 2\times 2\,e^{-} & \longrightarrow & 2\,O^{-2} & 4 & & 1 \end{array}$$

The reaction proceeds in two stages:

$$2\,Cu + O_2 \longrightarrow 2\,CuO$$

$$\begin{array}{lcl|c|c|c} Cu^{0} - 2\,e^{-} & \longrightarrow & Cu^{+2} & 2 & \multirow{2}{*}{4} & 2 \\ O_2 \longrightarrow 2\,O^{0} + 2\times 2\,e^{-} & \longrightarrow & 2\,O^{-2} & 4 & & 1 \end{array}$$

$$CuO + 2\,HCl \longrightarrow CuCl_2 + H_2O$$

Hydrofluoric acids interact with basic oxides.
The reaction with **silver nitrate** is qualitative for halide ions:

$$Ag^{+} + X^{-} \longrightarrow AgX\downarrow$$

Oxides and **oxoacids** form only **chlorine, bromine** and **iodine. Fluorine** forms fluorides.

$$Cl_2O_7 + H_2O \longrightarrow 2\,HClO_4$$

Chlorine, bromine and **iodine** have corresponding **perchloric ($HClO_4$), perbromic ($HBrO_4$)** and **periodic (HIO_4)** acids.

In order:

The strength of the acids and their resistance grows. →

$HOCl$ $HClO_2$ $HClO_3$ $HClO_4$

←

The growing oxidizing action that is due to the increasing instability of acids - decompose with the release of oxygen.

Perchloric acid **$HClO_4$** is the strongest of all known acids.

Of the salts of **$HClO_3$**, potassium chlorate **$KClO_3$,** the so-called **"Bertholite salt"**, is of importance. From this, oxygen is produced under laboratory conditions by heating or in the presence of a catalyst **MnO_2**:

$$2\,KClO_3 \longrightarrow KCl + 3\,O_2$$

$$\begin{array}{lll|c|c|c} & Cl^{+5} + 6\,e^{-} \longrightarrow & Cl^{-} & 6 & & 2 \\ & & & & 12 & \\ 2\,O^{-2} - 2\times2\,e^{-} \longrightarrow & 2\,O^{0} \longrightarrow & O_2 & 4 & & 3 \end{array}$$

Bertholite salt is an explosive, especially when in contact with organic substances.

Sample problems and solutions

Task 1.

Explain the possible valence states of the chemical elements fluorine and chlorine based on their electronic configurations expressed in terms of quantum cells. Express with equations the transitions indicated on the diagram by explaining the processes involved:

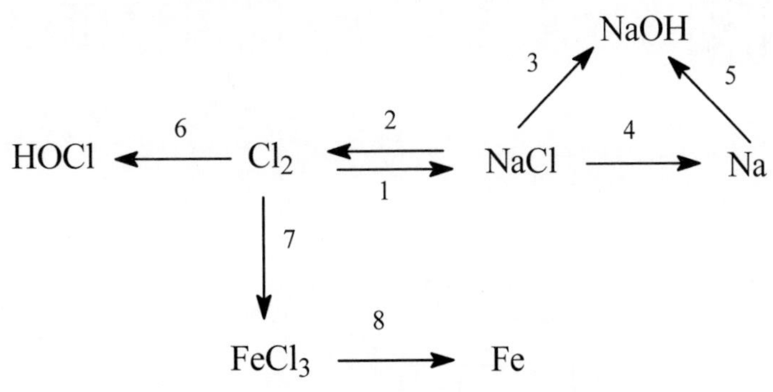

Exemplary solution

Fluorine is an element of period II, group VII A of the Periodic Table with sequence number 9. This means that it has 9 protons in its atomic nucleus and 9 electrons in its electron shell, distributed as follows: $1s^2\ 2s^2\ 2p^5$.

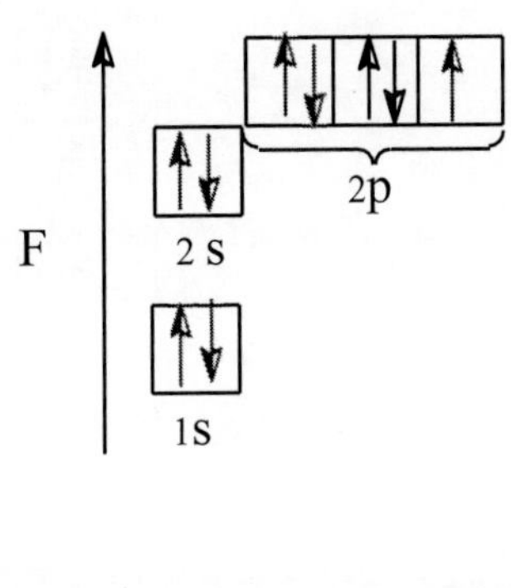

In the electron shell in the 2p-sublayer, the fluorine atom has a single electron. It lacks one electron to reach the completed electron octet of the inert element neon (Ne). Fluorine has only one valence state. Fluorine occupies a special place among the halogen elements because, unlike the others, it has no d-sublayer in its electronic shell. It appears only in the -1 oxidation state and is an oxidant even with respect to oxygen.

The remaining halogen elements exhibit variable degrees of oxidation - in addition to the most characteristic -1, also from +1 to +7. In the order from fluorine to iodine, the atomic radius increases and the electron affinity and associated non-metallic and oxidation properties of the elements decrease. The halogen elements are chemically very active, and therefore occur only in the form of compounds.

Chlorine is an element of period III, group VII A of the Periodic Table with sequence number 17. This means that it has 17 protons in its atomic nucleus and 17 electrons in its electron shell, distributed as follows: $1s^2\ 2s^2\ 2p^6\ 3s^2\ 3p^5$. In its ground state, the chlorine atom lacks one electron to acquire the electron configuration of the inert element argon (Ar). Third period elements have three sublayers, s-, p- d-, in their final electron layer. The presence of unoccupied 3d-orbitals allows the atom to accept energy during a chemical interaction, whereby an electron pair from the 3p-layer is broken, one electron reverses its spin and occupies a vacant 3d-orbital. This produces the four valence states of the chlorine atom: +1, +3, +5, and +7.

Cl

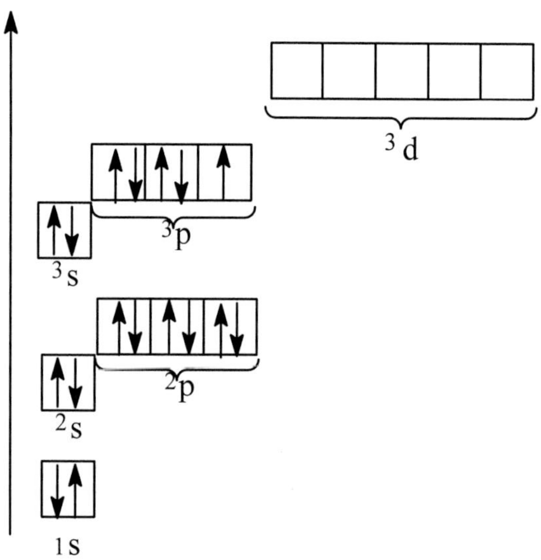

Cl★

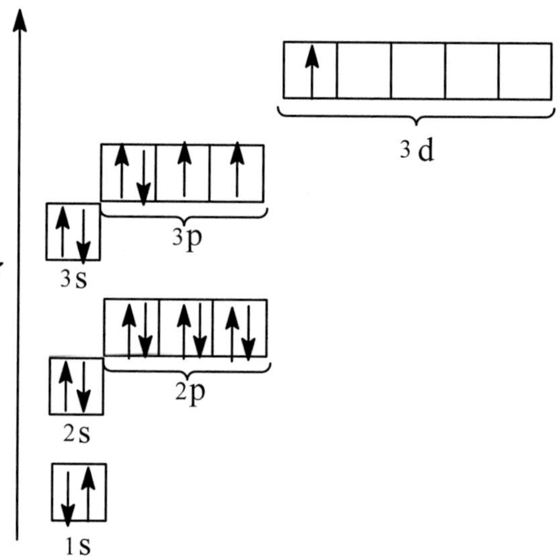

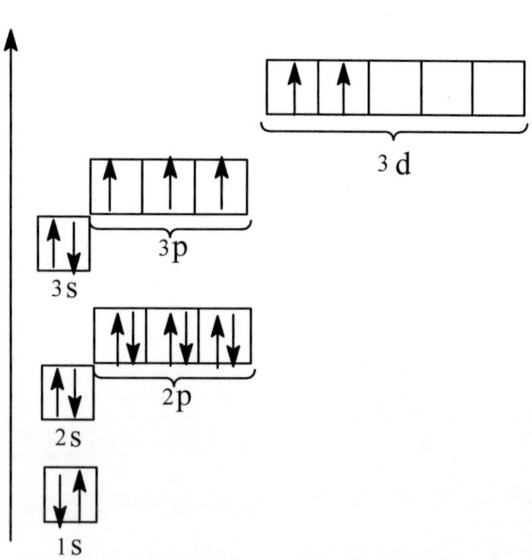

Cl★★

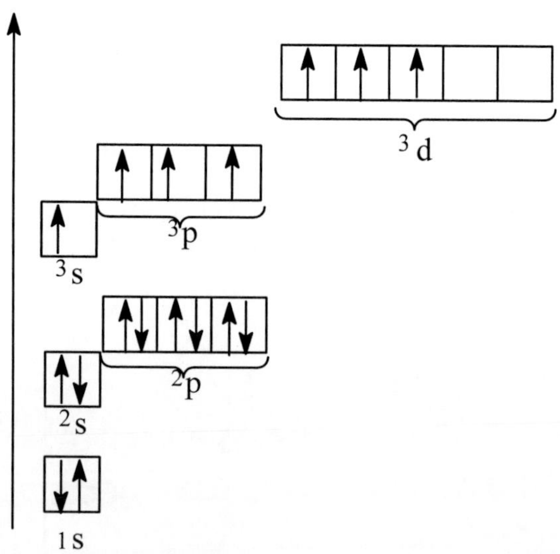

Cl★★★

The different valence states of chlorine can be illustrated by the following examples: HCl - the chlorine atom in hydrochloric acid is in oxidation state -1, HOCl - the oxidation state of the chlorine atom in **hypochlorous acid** is +1, $HClO_2$ - the chlorine atom in **chlorous acid** is in oxidation state +3, $HClO_3$ - the chlorine atom in **chloric acid** is in oxidation state +5 and $HClO_4$ - the chlorine atom in **perchloric acid** is in oxidation state +7.

Halogen elements react actively with all metals. Salts called halides are produced. Sodium with chlorine forms sodium chloride.

$$2\,Na + Cl_2 \xrightarrow{1} 2\,NaCl$$

$$\begin{array}{lll|c|c|c}
 & Na - e^- & \longrightarrow Na^+ & 1 & & 2 \\
 & & & & 2 & \\
Cl_2 \longrightarrow & 2\,Cl^0 + 2 \times 1\,e^- & \longrightarrow 2\,Cl^- & 2 & & 1
\end{array}$$

In sodium chloride melt electrolysis, the metal sodium (4) is electroneutralized at the cathode and the non-metal chlorine (2) at the anode.

In the electrolysis of an aqueous solution of sodium chloride, four types of ions are formed in solution. Passing a direct electric current through the sodium chloride solution, hydrogen and sodium cations are directed to the cathode and chloride and hydroxide anions are directed to the anode. At the cathode, the hydrogen cations are electroneutralised as they are located further to the right of the sodium cations in the Order of Relative Activity of the Metals (ORAM). Hydrogen is released at the cathode. At the anode the chloride anions are neutralised. At the anode, chlorine is released. Sodium cations and hydroxide anions remain in the electrolysis bath, i.e., sodium base (3).

$$2\,Na + 2\,H\text{—}OH \xrightarrow{5} 2\,NaOH + H_2$$

$$\begin{array}{ll|c|c|c}
Na - e^- & \longrightarrow Na^+ & 1 & & 2 \\
 & & & 2 & \\
2\,H^+ + 2 \times 1\,e^- \longrightarrow 2\,H^0 & \longrightarrow H_2 & 2 & & 1
\end{array}$$

The reaction of chlorine with water produces a mixture of two acids: hydrochloric and hypochlorous.

$$Cl_2 + H_2O \xrightarrow{6} HCl + HOCl$$

$$\begin{array}{ll|c|c|c}
Cl^0 + e^- & \longrightarrow Cl^- & 1 & & 1 \\
 & & & 1 & \\
Cl^0 - e^- & \longrightarrow Cl^+ & 1 & & 1
\end{array}$$

The process is known as disproportionation.

If iron filings are sprinkled into a cylinder of chlorine, a spectacular reaction is observed, the so-called "fire rain".

$$2\,Fe + 3\,Cl_2 \xrightarrow{7} 2\,FeCl_3$$

$$\begin{array}{lllll} Fe^{0} - 3\,e^{-} \longrightarrow Fe^{+3} & 3 & & 2 \\ & & 6 & \\ Cl_2 \longrightarrow 2\,Cl^{0} + 2 \times 1\,\bar{e} \longrightarrow 2\,Cl^{-} & 2 & & 3 \end{array}$$

Chlorine can be obtained by electrolysis of a melt or aqueous solution of ferric trichloride. At the cathode the ferric ions are electroneutralised and iron (8) is released, at the anode the chloride anions are neutralised and chlorine is released.

Group VIIIB of the Periodic Table

The eighth group of the Periodic Table differs from all others in that it does not contain elements of the minor periods; nor does it contain subgroups of elements sharply different in electronic structure and in atomic properties. It is composed of **triads of** the elements of the major periods.

Proceeding from the presence of no more than two electrons in the outer layers of all the atoms under consideration, it may be expected that the tendency to join new electrons would not be characteristic of them. The elements of group VIIIB are therefore metallic in character.

In the totality of their properties, iron (Fe), cobalt (Co) and nickel (Ni) are quite similar and different from the other elements of the group. Therefore, iron, cobalt and nickel are regarded as members of the **iron family**.

The similarity of most of the properties of the elements in the series ruthenium (Ru), rhodium (Rh), palladium (Pd) and osmium (Os), iridium (Ir), platinum (Pt) makes it possible to unite them in the family of **platinum metals**.

Of the iron family, we will consider only the most common element.

Iron has two electrons in its outer electron layer, and the penultimate electron is not built up. There are 14 electrons. In chemical interactions, it participates with the electrons from the outer and penultimatc electron layers. The most characteristic oxidation degrees of the element are +2 and +3.

Iron is a highly fusible metal and easily submitted to processing - forging, drawing, rolling.

The most important feedstocks for iron are its mineral ores: hematite $\mathbf{Fe_2O_3}$, magnetite $\mathbf{Fe_3O_4}$, siderite $\mathbf{FeCO_3}$, limonite $\mathbf{2Fe_2O_3 . 3H_2 O}$, and pyrite $\mathbf{FeS_2}$.

The preparation of iron is carried out in blast furnaces, where at high temperature the oxides are reduced from the carbon:

$$C + O_2 \longrightarrow CO_2$$

$$\begin{array}{lllll} C^{0} - 2\,e^{-} \longrightarrow C^{+2} & 2 & & 2 \\ & & 4 & \\ O_2 \longrightarrow 2\,O^{0} + 2 \times 2\,e^{-} \longrightarrow 2\,O^{-2} & 4 & & 1 \end{array}$$

$$CO_2 + C \rightleftharpoons 2\,CO$$

$$\begin{array}{lllll} C^{+4} + 2\,e^{-} \longrightarrow C^{+2} & 2 & & 1 \\ & & 2 & \\ C^{0} - 2\,e^{-} \longrightarrow C^{+2} & 2 & & 1 \end{array}$$

$$Fe_2O_3 + 3\,C \longrightarrow 2\,Fe + 3\,CO$$

$$\begin{array}{lcl|c|c|c} Fe^{+3} + 3\,\bar{e} & \longrightarrow & Fe^{o} & 3 & \multirow{2}{*}{6} & 2 \\ C^{o} - 2\,e^{-} & \longrightarrow & C^{+2} & 2 & & 3 \end{array}$$

$$Fe_2O_3 + 3\,CO \longrightarrow 2\,Fe + 3\,CO_2$$

$$\begin{array}{lcl|c|c|c} Fe^{+3} + 3\,\bar{e} & \longrightarrow & Fe^{o} & 3 & \multirow{2}{*}{6} & 2 \\ C^{+2} - 2\,e^{-} & \longrightarrow & C^{+4} & 2 & & 3 \end{array}$$

At high temperature, the iron absorbs some of the carbon and cast iron or steel is produced (temperature control of carbon content is applied).

In its chemical relations, iron is a reducing agent. It gives up its valence electrons and oxidizes to Fe^{+2} or Fe^{+3}.

Under ordinary conditions of dry air, iron does not change. At a temperature of about 200° C it is covered with a thin cover of oxide which protects it from further oxidation:

$$2Fe + O_2 \longrightarrow 2\,FeO$$

$$\begin{array}{lcl|c|c|c} Fe^{o} - 2\,\bar{e} & \longrightarrow & Fe^{+2} & 2 & \multirow{2}{*}{4} & 2 \\ O_2 \longrightarrow 2O^{o} + 2 \times 2\,\bar{e} & \longrightarrow & 2\,O^{-2} & 4 & & 1 \end{array}$$

In an oxygen environment, iron burns violently and a large amount of heat and light is given off. Fe is formed Fe_3O_4, which is a mixed oxide FeO. Fe_2O_3.

$$3\,Fe + 2\,O_2 \longrightarrow Fe_3O_4 + Q$$

$$\begin{array}{lcl|c|c|c} Fe^{o} - 2\,\bar{e} & \longrightarrow & Fe^{+2} & 2 & \multirow{2}{*}{4} & 2 \\ O_2 \longrightarrow 2O^{o} + 2 \times 2\,\bar{e} & \longrightarrow & 2\,O^{-2} & 4 & & 1 \end{array}$$

$$\begin{array}{lcl|c|c|c} Fe^{o} - 3\,\bar{e} & \longrightarrow & Fe^{+3} & 3 & \multirow{2}{*}{12} & 4 \\ O_2 \longrightarrow 2O^{o} + 2 \times 2\,\bar{e} & \longrightarrow & 2\,O^{-2} & 4 & & 3 \end{array}$$

In an environment of moist air, iron oxidizes and becomes coated with a loose oxide layer called **"rust"** - **Fe_2O_3. xH_2O**. The layer easily crumbles and exposes new surfaces of the metal. The phenomenon is called corrosion.

Under ordinary conditions, **iron** does not react with **water**, but oxidizes to FeO or to Fe_3O_4 as the temperature rises, depending on conditions:

$$3Fe + 4H_2O \xrightarrow{t < 570^{o}C} Fe_3O_4 + 4H_2\uparrow$$

$$Fe + H_2O \xrightarrow{t > 570^{o}C} FeO + H_2\uparrow$$

Iron reacts with various hydrochloric and sulfuric acids. The corresponding salts are formed and hydrogen is evolved:

$$Fe + HCl \longrightarrow FeCl_2 + H_2\uparrow$$

$$\begin{array}{lll|l|l|l} & Fe^{o} - 2\bar{e} \longrightarrow & Fe^{+2} & 2 & \multirow{2}{*}{2} & 1 \\ 2H^{+} + 2x1\,\bar{e} \longrightarrow & 2H^{o} \longrightarrow & H_2 & 2 & & 1 \end{array}$$

$$Fe + H_2SO_4 \longrightarrow FeSO_4 + H_2\uparrow$$

$$\begin{array}{lll|l|l|l} & Fe^{o} - 2\bar{e} \longrightarrow & Fe^{+2} & 2 & \multirow{2}{*}{2} & 1 \\ 2H^{+} + 2x1\,\bar{e} \longrightarrow & 2H^{o} \longrightarrow & H_2 & 2 & & 1 \end{array}$$

Dilute nitric acid oxidizes iron to Fe^{+3} with the evolution of nitric oxide NO:

$$Fe + 4HNO_3 \longrightarrow Fe(NO_3)_3 + NO + 2H_2O$$

$$\begin{array}{ll|l|l|l} \overset{o}{Fe} - 3\bar{e} \longrightarrow & Fe^{+3} & 3 & \multirow{2}{*}{3} & 1 \\ N^{+5} + 3\bar{e} \longrightarrow & N^{+2} & 3 & & 1 \end{array}$$

Concentrated sulfuric and nitric acids passivate iron.

Iron reacts actively with **non- metals with oxidizing action**- halogen elements, sulfur, nitrogen,

$$2Fe + 3Cl_2 \longrightarrow 2FeCl_3$$

$$\begin{array}{lll|l|l|l} & \overset{o}{Fe} - 3\bar{e} \longrightarrow & Fe^{+3} & 3 & \multirow{2}{*}{6} & 2 \\ Cl_2 \longrightarrow & 2\overset{o}{Cl} + 2x1\,\bar{e} \longrightarrow & 2Cl^{-} & 2 & & 3 \end{array}$$

At high temperature, iron can dissolve in concentrated bases. It exhibits weak amphoteric properties:

$$Fe + \underbrace{6K^{+} + 6\bar{O}H} + 6H_2O \longrightarrow 2K_3[Fe(OH)_6] + 3H_2\uparrow$$

$$\begin{array}{lll|c|c|c}
Fe^{0} - 3\bar{e} & \longrightarrow & Fe^{+3} & 3 & \multirow{2}{*}{6} & 2 \\
2H^{+} + 2 \times 1\,e^{-} \longrightarrow 2H^{0} & \longrightarrow & H_2 & 2 & & 3
\end{array}$$

$$K_3[Fe(OH)_6] \longrightarrow 3K^{+} + [Fe(OH)_6]^{-3}$$

Iron oxide is a black, water-insoluble compound. It dissolves in acids (basic oxide):

$$FeO + 2HCl \longrightarrow FeCl_2 + H_2O$$

Its hydroxide is produced by alkalinizing solutions of divalent iron Fe^{+2}.

$$\underbrace{Fe^{+2} + 2C\bar{l}} + \underbrace{2K^{+} + 2\bar{O}H} \longrightarrow Fe(OH)_2\downarrow + \underbrace{2K^{+} + 2C\bar{l}}$$

Iron dihydroxide is a white, water-insoluble substance. It dissolves in acids. In air, it spontaneously oxidizes to the trivalent iron hydroxide:

$$4Fe(OH)_2\downarrow + 2H_2O + O_2 \longrightarrow 4Fe(OH)_3\downarrow$$

$$\begin{array}{lll|c|c|c}
Fe^{+2} - \bar{e} & \longrightarrow & Fe^{+3} & 1 & \multirow{2}{*}{4} & 4 \\
O_2 \longrightarrow 2O^{0} + 2 \times 2\bar{e} & \longrightarrow & 2\bar{O}^{2} & 4 & & 1
\end{array}$$

Ferric trioxide Fe_2O_3 is a reddish-brown substance insoluble in water. It dissolves in acids. It is the anhydride of ferric trihydroxide $Fe(OH)_3$, a rusty brown precipitate that exhibits amphoteric properties.

$$Fe(OH)_3 + \underbrace{3H^{+} + 3C\bar{l}} \rightleftharpoons \underbrace{Fe^{+3} + 3C\bar{l}} + H_2O$$

$$Fe(OH)_3 + 3K^{+} + 3\bar{O}H \rightleftharpoons 3K^{+} + [Fe(OH)_6]^{-3}$$

hexahydroxyferrate of potassium

Iron is a metal with significant applications in modern engineering. A major problem in the use of iron is the fight against corrosion. Iron is an important bioelement. It is a constituent of hemoglobin. Iron deficiency in the body causes anemia. Nowadays, iron-containing preparations have been developed, which are successfully applied in the fight against iron deficiency anemia.

Sample problems and solutions

Task 1.

Make a characterization of the chemical element iron. Express with equations the transitions indicated on the diagram explaining the processes involved:

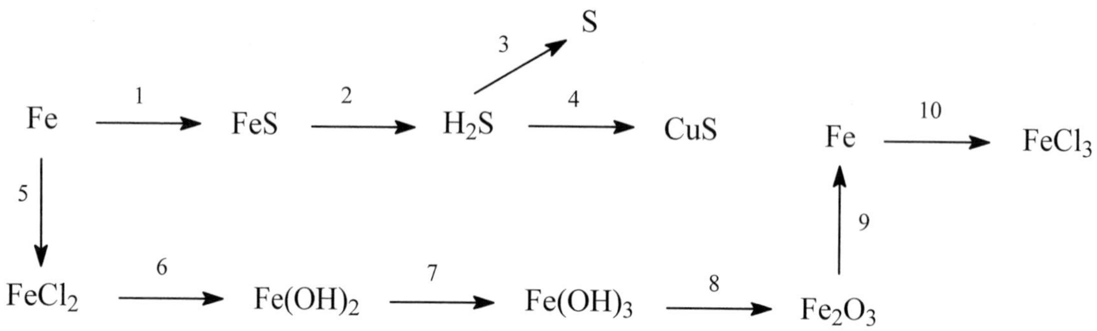

Exemplary solution

Iron is an element of group VIIIB of the Periodic Table with the sequence number 26, which means that it contains 26 protons in its atomic nucleus and 26 electrons in its electron shell, arranged as follows in the different layers and sublayers $1s^2\ 2s^2\ 2p^6\ 3s^2\ 3p^6\ 4s^2\ 3d^6$. Using quantum cells, the electronic structure of an iron:

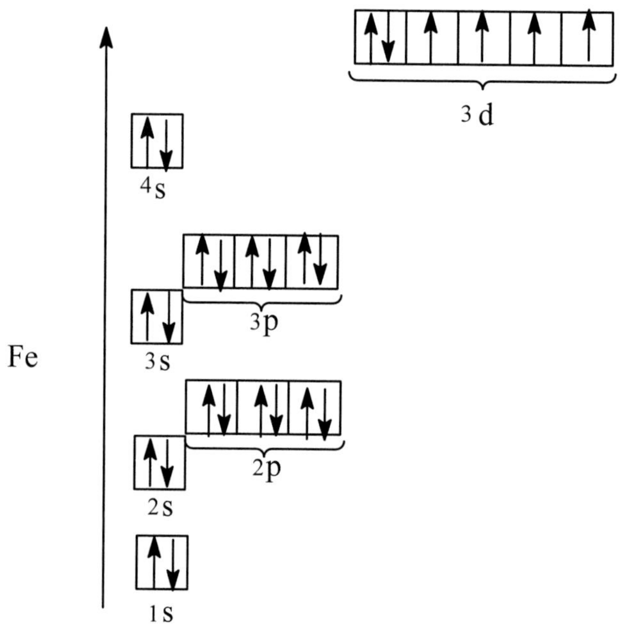

The eighth group of the Periodic System differs from all others in that it does not contain elements of the minor periods; nor does it contain subgroups of elements sharply different in electronic structure and in atomic properties. It is composed of triads of the elements of the major periods.

In the totality of their properties, iron (Fe), cobalt (Co) and nickel (Ni) are quite similar and different from the other elements of the group. Therefore, iron, cobalt and nickel are regarded as members of the iron family.

Proceeding from the presence of no more than two electrons in the outer 4s-sublayer of the fourth electron layer, it can be expected that the tendency to join new electrons is not characteristic of the iron atom and triad members. Therefore, the elements of group VIIIB have a metallic character.

Iron has two electrons in its outer electron layer and the penultimate electron is not built up. There are 14 electrons. In chemical interactions, it participates with the electrons from the outer (4s-sublayer) and the penultimate electron layer. The most characteristic oxidation degrees of the element are +2 and +3.

Iron is a highly fusible metal and easily submitted to processing - forging, drawing, rolling.

The most important feedstocks for iron are its mineral ores: hematite $\mathbf{Fe_2O_3}$, magnetite $\mathbf{Fe_3O_4}$, siderite $\mathbf{FeCO_3}$, limonite $2Fe_2O_3 .3H_2O$ $\mathbf{[(FeO(OH).n(H_2O))]}$, brown limonite $\mathbf{Fe_2O_3 .2Fe(OH)_3}$, and pyrite $\mathbf{FeS_2}$.

Iron reacts with sulfur on moderate heating. The process is redox. The iron atom is the reducing agent and the sulfur atom is the oxidizing agent.

$$Fe + S \xrightarrow{1} FeS$$

$$\begin{array}{lcl|c|c|c} Fe^{0} - 2e^{-} & \longrightarrow & Fe^{+2} & 2 & & 1 \\ & & & & 2 & \\ S^{0} + 2e^{-} & \longrightarrow & S^{-2} & 2 & & 1 \end{array}$$

The stronger hydrochloric acid displaces the weaker hydrogen sulfide acid from its salts as hydrogen sulfide is released. The reaction is substitutive.

$$FeS + 2\,HCl \xrightarrow{2} FeCl_2 + H_2S$$

Where there are no deposits of native sulfur, e.g., in Bulgaria, sulfur is obtained according to the following scheme:

$$2\,H_2S + SO_2 \xrightarrow{3} 3\,S + 2\,H_2O$$

$$\begin{array}{lcl|c|c|c} S^{-2} - 2e^{-} & \longrightarrow & S^{0} & 2 & & 2 \\ & & & & 4 & \\ S^{+4} + 4e^{-} & \longrightarrow & S^{0} & 4 & & 1 \end{array}$$

This enables the recovery of some waste gases.

Sulfur can also be produced by incomplete combustion of hydrogen sulfide.

$$2\,H_2S + O_2 \xrightarrow{3} 2\,S + 2\,H_2O$$

$$\begin{array}{lcl|c|c|c} S^{-2} - 2e^{-} & \longrightarrow & S^{0} & 2 & & 2 \\ & & & & 4 & \\ O_2 \longrightarrow 2\,O^{0} + 2 \times 2\bar{e} & \longrightarrow & 2\,O^{-2} & 4 & & 1 \end{array}$$

Hydrogen sulfide reacts with the metal copper and converts it to copper sulfide. The process is redox.

$$H_2S + Cu \xrightarrow{4} CuS + H_2\uparrow$$

$$\begin{array}{lcl|c|c|c} Cu^{0} - 2e^{-} & \longrightarrow & Cu^{+2} & 2 & & 1 \\ & & & & 2 & \\ 2\,H^{+} + 2 \times 1\bar{e} & \longrightarrow & 2\,H^{0} \longrightarrow H_2 & 2 & & 1 \end{array}$$

Iron reacts with hydrochloric acid to produce ferric dichloride.

$$Fe + 2\,HCl \xrightarrow{5} FeCl_2 + H_2\uparrow$$

$$\begin{array}{lll|c|c|c} Fe^{0} - 2\,e^{-} & \longrightarrow & Fe^{+2} & 2 & & 1 \\ & & & & 2 & \\ 2\,H^{+} + 2\times 1\,\bar{e} & \longrightarrow 2\,H^{0} \longrightarrow & H_2 & 2 & & 1 \end{array}$$

With sodium base, ferric dichloride is converted to a white precipitate of ferric dihydroxide, which rapidly turns green from the presence of hydrated ferric ions.

$$FeCl_2 + 2\,NaOH \longrightarrow Fe(OH)_2\downarrow + 2\,NaCl$$

$$\underbrace{Fe^{+2} + 2\,Cl^{-}} + \underbrace{2Na^{+} + 2OH^{-}} \longrightarrow Fe(OH)_2\downarrow + \underbrace{2Na^{+} + 2\,Cl^{-}}$$

$$Fe^{+2} + 2\,OH^{-} \longrightarrow Fe(OH)_2\downarrow$$

Iron dihydroxide is a white, water-insoluble substance. It dissolves in acids. In air, it spontaneously oxidizes to the trivalent iron hydroxide:

$$4\,Fe(OH)_2 + 2\,H_2O + O_2 \xrightarrow{7} 4\,Fe(OH)_3$$

$$\begin{array}{lll|c|c|c} & Fe^{+2} - e^{-} & \longrightarrow Fe^{+3} & 1 & & 4 \\ & & & & 4 & \\ O_2 \longrightarrow & 2\,O^{0} + 2\times 2\bar{e} & \longrightarrow 2\,O^{-2} & 4 & & 1 \end{array}$$

Ferric trioxide Fe_2O_3 is a reddish-brown substance insoluble in water. It dissolves in acids. It is the anhydride of iron trihydroxide $Fe(OH)_3$, a rusty brown precipitate with amphoteric properties.

$$2\,Fe(OH)_3 \xrightarrow{8} Fe_2O_3 + 3\,H_2O$$

The preparation of iron is carried out in blast furnaces, where at high temperature the oxides are reduced from the carbon.

$$Fe_2O_3 + 3\,C \xrightarrow{9} 2\,Fe + 3\,CO$$

$$\begin{array}{lll|c|c|c} Fe^{+3} + 3\,e^{-} & \longrightarrow & Fe^{0} & 3 & & 2 \\ & & & & 6 & \\ C^{0} - 2\,e^{-} & \longrightarrow & C^{+2} & 2 & & 3 \end{array}$$

Iron interacts with chlorine as well as with non-oxidizing metals.

$$2\,Fe + 3\,Cl_2 \xrightarrow{10} 2\,FeCl_3$$

$$\begin{array}{lcl|c|c|c}
 & Fe^{o} - 3\,e^{-} \longrightarrow & Fe^{+3} & 3 & & 2 \\
 & & & & 6 & \\
Cl_2 \longrightarrow & 2\,Cl^{o} + 2\,x\,1\,\bar{e} \longrightarrow & 2\,Cl^{-} & 2 & & 3
\end{array}$$

6. CHEMICAL PROCESSES

TYPES OF CHEMICAL PROCESSES. MOLECULAR, IONIC, RADICAL PROCESSES. REDOX AND NON-ELECTRON TRANSFER PROCESSES. EXCHANGE, ADDITION AND DECOMPOSITION REACTIONS. REACTIONS IN HOMOGENEOUS AND HETEROGENEOUS MEDIA. EXOTHERMIC AND ENDOTHERMIC REACTIONS. SYNTHESIS, DECOMPOSITION, SUBSTITUTION, ADDITION, HYDROLYSIS AND NEUTRALIZATION REACTIONS. THERMOCHEMISTRY. HESS'S LAW

There are many processes in nature - physical, chemical, biological, etc. Only a few of them are chemical processes. In these processes, one substance is transformed into another, different in composition, structure and properties from the original one. The necessary condition for a process to be chemical is that it involves the breaking of existing chemical bonds and the formation of new ones. For example, the radioactive transformation of some atoms into others is not a chemical process but a nuclear, i.e., physical process. The processes of evaporation of water and liquefaction of water vapor are physical, since no qualitatively new compound is formed, but only the aggregate state is changed.

This means that the chemical process is not a conversion at all, but a conversion in which the chemical elements are preserved.

There are a number of exceptions to the definition of chemical processes. The transformation of one polymorphic form into another, for example, monoclinic sulfur into rhombic sulfur, or of white phosphorus into red, which is in fact the conversion of one simple substance into another, is also regarded as a chemical rather than a physical process. These examples show that there is no sharp boundary between chemical and physical processes.

Chemical processes proceed according to the basic laws of nature - the laws of conservation of mass and energy of substances. They enable reactions to be expressed in terms of chemical equations that contain quantitative and qualitative information about the processes involved. A chemical equation is a conventional notation for chemical reactions that uses accepted chemical symbols and formulas. To construct chemical equations, it is necessary to know the formulas of the starting substances and the products of the reaction, as well as their quantities. The two parts are connected by arrows that indicate the direction of the process.

$$S + O_2 \longrightarrow SO_2$$

$$\begin{array}{rcl|c|c|c} & S^0 - 4e^- \longrightarrow & S^{+4} & 4 & & 1 \\ & & & & 4 & \\ O_2 \longrightarrow & 2O^0 + 2\times 2\bar{e} \longrightarrow & 2O^{-2} & 4 & & 1 \end{array}$$

The equation does not imply that the reaction ends after the interaction of one sulfur atom and one oxygen molecule. It shows what the quantitative ratio between the interacting substances in an elementary chemical act is.

Chemical equations contain qualitative and quantitative information about the processes they describe. For example, sulfur dioxide (SO_2) is a qualitatively new compound with different properties from those of its constituent elements sulfur and oxygen. The quantitative information it contains is that one atom of sulfur and one molecule of oxygen produces one molecule of sulfur dioxide. The number of oxygen atoms on the left-hand side of the equation is equal to their number on the right-hand side. The same is true for sulfur atoms. This shows that the equation is composed in agreement with the law of conservation of mass of substances. In most cases, to obey this law, it is necessary to make the equation equal. In doing so, the number of any type of atoms on one side of the equation must be equal to the number of the same type of atoms on the other side. The equations written below satisfy this condition:

$$2H_2 + O_2 \longrightarrow 2H_2O$$

$$\begin{array}{llll|c|c|c}
H_2 \longrightarrow & 2H^0 & - 2 \times 1\bar{e} & \longrightarrow 2H^+ & 2 & \multirow{2}{*}{4} & 2 \\
O_2 \longrightarrow & 2O^0 & + 2 \times 2\bar{e} & \longrightarrow 2O^{-2} & 4 & & 1
\end{array}$$

$$N_2 + 3H_2 \longrightarrow 2NH_3$$

$$\begin{array}{llll|c|c|c}
H_2 \longrightarrow & 2H^0 & - 2 \times 1\bar{e} & \longrightarrow 2H^+ & 2 & \multirow{2}{*}{6} & 3 \\
N_2 \longrightarrow & 2N^0 & + 2 \times 3\bar{e} & \longrightarrow 2N^{-3} & 6 & & 1
\end{array}$$

$$2HgO \longrightarrow 2Hg + O_2$$

$$\begin{array}{lll|c|c|c}
Hg^{+2} + 2\bar{e} & & \longrightarrow Hg^0 & 2 & \multirow{2}{*}{4} & 2 \\
2O^{-2} - 2 \times 2\bar{e} & \longrightarrow 2O^0 & \longrightarrow O_2 & 4 & & 1
\end{array}$$

They are therefore subject to the law of conservation of mass of substances, i.e., they are correctly constituted.

Chemical equations are gross equations. They give only the material balance of the system, but do not give detailed information about the mechanism of reactions.

Types of chemical processes

The classification of chemical processes can be made on the basis of various features distinguishing one chemical reaction from another: on the basis of the type of reacting particles, from the point of view of atomic-molecular theory, **homogeneity** of the system, heat effect of the reaction, etc. This separation of chemical processes is not absolutely strict. The same chemical reaction may be assigned to different groups depending on the leading feature in the system.

According to the type of interacting particles, chemical processes are **molecular, ionic and radical.**

In **molecular processes,** the interaction takes place between molecules. It involves the breaking of chemical bonds in the parent molecules and the creation of new chemical bonds in the product molecules. Molecular processes proceed slowly.

$$CH_3COOH + CH_3OH \rightleftharpoons CH_3COOCH_3 + H_2O$$

$$CH_4 + 2O_2 \longrightarrow CO_2 + 2H_2O$$

$$\begin{array}{llll|c|c|c}
 & C^{-4} & - 8\bar{e} & \longrightarrow C^{+4} & 8 & \multirow{2}{*}{8} & 1 \\
O_2 \longrightarrow & 2O^0 & + 2 \times 2\bar{e} & \longrightarrow 2O^{-2} & 4 & & 2
\end{array}$$

In **ionic processes** the interaction takes place between ions. Most often they take place in solutions in which the substances are dissociated. Slightly soluble substances and weak electrolytes are written in undissociated form in the ionic equations. In most cases, reactions between ions in solution proceed **instantaneously**. This is due to the fact that they do not break existing bonds, or more accurately, only the solvate shells of the ions are broken. For example, when solutions of sodium chloride and silver nitrate are mixed, a white precipitate of silver chloride is immediately formed.

$$\underbrace{Na^{+} + Cl^{-}} + \underbrace{Ag^{+} + NO_3^{-}} \longrightarrow AgCl\downarrow + \underbrace{Na^{+} + NO_3^{-}}$$

The process is abbreviated as follows:

$$Ag^{+} + Cl^{-} \longrightarrow AgCl\downarrow$$

Ionic processes are also the following:

$$\underbrace{Na^{+} + OH^{-}} + \underbrace{H^{+} + Cl^{-}} \longrightarrow \underbrace{Na^{+} + Cl^{-}} + H_2O$$

$$\underbrace{2Na^{+} + CO_3^{-2}} + \underbrace{2H^{+} + 2\,Cl^{-}} \longrightarrow \underbrace{2\,Na^{+} + 2\,Cl^{-}} + H_2O + CO_2\uparrow$$

$$\underbrace{Cu^{2+} + SO_4^{-2}} + \underbrace{2\,Na^{+} + 2OH^{-}} \longrightarrow \underbrace{2\,Na^{+} + SO_4^{-2}} + Cu(OH)_2\downarrow$$

If the interaction also involves complex ions that are destroyed, the process may proceed more slowly, for example, the reaction between potassium permanganate and oxalic acid:

$$2\,K^{+} + 2\,MnO_4^{-} + 16\,H^{+} + 5\,C_2O_4^{-2} \longrightarrow 2\,Mn^{+2} + 2\,K^{+} + 10\,CO_2 + 8\,H_2O$$

In organic chemistry, chemical processes are known to proceed by an ionic mechanism. Such examples occur in substitution reactions (alkyl halides - alcohols), electrophilic addition in alkenes, nucleophilic addition in carbonyl compounds, etc.

Radical reactions take place between radicals - particles that contain a single electron. Radicals, like free atoms, are chemically very reactive. A typical example of a radical reaction is the halogenation of alkanes by UV light:

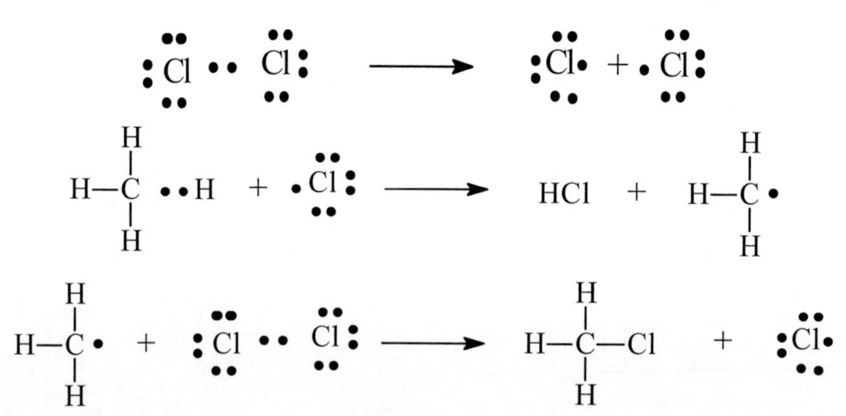

The reaction continues until the free radicals are exhausted. In addition to alkyl halides, alkanes, for example, can also be produced:

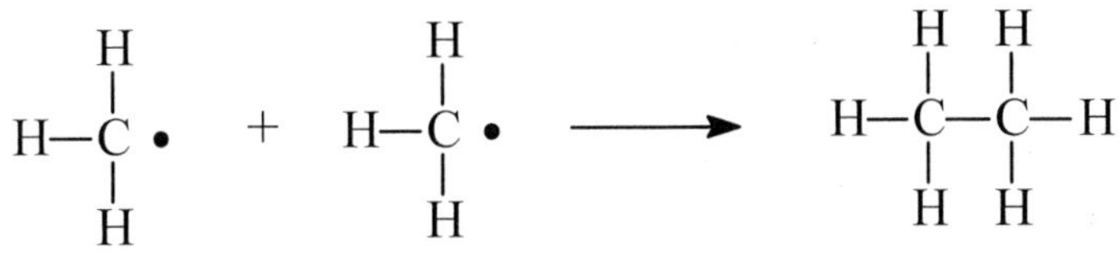

Some polymerization processes also proceed by a radical mechanism.

According to the presence of **an electronic transition,** the reactions are with (redox) and without an electronic transition.

When involved in redox processes, substances are divided into **oxidants and reducers**. This division is conditional because any oxidant can end up as a reductant in another process as long as it interacts with a stronger oxidant than itself. In the course of the process, **the oxidant** is reduced and lowers its positive valence, and the **reductant** - vice versa. For example, the equation of the oxidation reaction of oxalic acid goes through the following steps:

1. **Write down the formulae** of the reactants and the reaction products

$$KMnO_4 + H_2C_2O_4 + H_2SO_4 \longrightarrow MnSO_4 + K_2SO_4 + CO_2 + H_2O$$

2. **Identify** those elements that change valence upon interaction, i.e., the **oxidant and the reductant.** The electronic equations are written which give a quantitative description of the mutually exchanged electrons between the oxidant and the reductant:

$$\overset{+7}{Mn} + 5\bar{e} \longrightarrow \overset{+2}{Mn} \quad \Big| \; 5 \; \Big| \; \; \Big| \; 2$$
$$\quad \quad \quad \quad \quad \quad \quad \quad \quad \quad \quad \; 10$$
$$\overset{-2}{C_2O_4} - 2\bar{e} \longrightarrow 2\,CO_2 \quad \Big| \; 2 \; \Big| \; \; \Big| \; 5$$

3. **An electron balance** of the process is made and the least common multiple is found since the number of electrons given up is equal to the number of electrons received (2, 5).

4. **Determine the coefficients** in front of the oxidizer and reducer by calculating that 10 electrons will be accepted by two seven-valent manganese ions and given up by five divalent oxalate ions.

$$2\,KMnO_4 + 5\,H_2C_2O_4 + H_2SO_4 \longrightarrow 2\,MnSO_4 + K_2SO_4 + 10\,CO_2 + H_2O$$

5. **The remaining coefficients** are **calculated** and recorded:

$$2\,KMnO_4 + 5\,H_2C_2O_4 + 3\,H_2SO_4 \longrightarrow 2\,MnSO_4 + K_2SO_4 + 10\,CO_2 + 8\,H_2O$$

Finally, the number of oxygen atoms is checked.

$$2\,SO_2 + O_2 \longrightarrow 2\,SO_3$$

$$\overset{+4}{S} - 2\bar{e} \longrightarrow \overset{+6}{S} \quad \Big| \; 2 \; \Big| \; \; \Big| \; 2$$
$$\quad \quad \quad \quad \quad \quad \quad \quad \quad \quad \quad \; 4$$
$$O_2 \longrightarrow 2\,O^0 + 2 \times 2\bar{e} \longrightarrow 2\,\overset{-2}{O} \quad \Big| \; 4 \; \Big| \; \; \Big| \; 1$$

Reactions without electron transfer - all chemical processes between organic and inorganic substances in which there is no oxidant and reductant in the reaction system, i.e., no changes in valence.

Reactions without electronic exchange can be:

Exchange - these cover all processes in which atomic groupings of at least two reactants are mutually exchanged, and in the reaction products at least one of the substances is a precipitate, gas or weak electrolyte.

$$AgNO_3 + NaCl \longrightarrow AgCl\downarrow + NaNO_3$$

$$NH_4Cl + NaOH \longrightarrow NaCl + H_2O + NH_3\uparrow$$

$$CuSO_4 + 2\,NaOH \longrightarrow Cu(OH)_2\downarrow + Na_2SO_4$$

Adjunctive reactions:

$$CO_2 + H_2O \rightleftharpoons H_2CO_3$$

$$CaO + CO_2 \longrightarrow CaCO_3$$

Decomposition reactions:

$$CaCO_3 \xrightarrow{t^o} CaO + CO_2$$

$$2\,Fe(OH)_3 \longrightarrow Fe_2O_3 + 3\,H_2O$$

According to the homogeneity or heterogeneity of the system, the processes are homogeneous and heterogeneous.

Homogeneous processes occur in homogeneous systems. A system is homogeneous when its properties and composition at each point are the same or change smoothly. For example, air is a homogeneous system. The changes (temperature, pressure, etc.) that occur in the composition and properties of this gas mixture occur very smoothly, gradually as one passes from one point to another.

The following processes are homogeneous:

$$\underset{g}{N_2} + \underset{g}{3\,H_2} \rightleftharpoons \underset{g}{2\,NH_3}$$

$$\underset{liq}{CH_3COOH} + \underset{liq}{C_2H_5OH} \rightleftharpoons \underset{liq}{CH_3COOC_2H_5} + \underset{liq}{H_2O}$$

A characteristic feature of reactions occurring in homogeneous systems is that the interaction takes place throughout the volume of the reaction system. These reactions take place in the gas phase or in solution.

In **heterogeneous systems,** there is an abrupt jump in the properties or composition of the system upon crossing some boundary, called a phase boundary. A heterogeneous system is, for example, the **ice-water-water vapor** system. Although the composition is the same, the properties of this system (pressure, density, hardness, etc.) change abruptly when the ice-liquid, ice-vapor, liquid-vapor boundaries are crossed. The oil-water system is also heterogeneous. The individual **homogeneous parts** of a heterogeneous system are called **phases**. The distinct phases are ice, water, and vapor in the former case and oil and water in the latter. There are two liquid phases in the oil-water system. Heterogeneous processes take place between reactants in different phases. In these, chemical, physical and physicochemical phases are distinguished.

For example, the interaction between

$$\underset{\text{(solid)}}{CaO} + \underset{\text{(gas)}}{CO_2} \longrightarrow \underset{\text{(solid)}}{CaCO_3}$$

gas with a solid requires the gas molecules to reach the surface of the solid, which is accomplished by diffusion; adsorption of the gas onto the surface of the solid; chemical interaction between the gas and the solid; desorption (separation of the adsorbed substance from the surface); and reverse diffusion of the reaction products. This multi-step process affects the rate of the process and, in addition to the familiar factors of temperature, concentration and pressure, the important conditions for the process are the size of the surface, the agitation of the reaction mixture, etc.

Examples of **heterogeneous processes**:

$$\underset{\text{(solid)}}{CaCO_3} + \underset{\text{(liq)}}{2\,HCl} \longrightarrow CaCl_2 + \underset{\text{(gas)}}{CO_2} + \underset{\text{(liq)}}{H_2O}$$

$$\underset{\text{(solid)}}{CaCO_3} \xrightarrow{t^o} \underset{\text{(solid)}}{CaO} + \underset{\text{(gas)}}{CO_2}$$

According to the **thermal effect** that accompanies the reactions, they are exothermic and endothermic.

Exothermic processes are processes which are accompanied by the **release of heat.**
The energy of the system decreases.

$$N_2 + 3\,H_2 \rightleftharpoons 2\,NH_3 + Q$$

$$2\,SO_2 + O_2 \longrightarrow 2SO_3 + Q$$

Endothermic processes are processes that involve the **absorption of heat, in which the system increases its energy:**

$$N_2 + O_2 \longrightarrow 2\,NO - Q$$

$$2HgO \rightleftharpoons 2\,Hg + O_2 - Q$$

In many cases, the thermal effect is the sum of the thermal effects of several successive processes, some of them even physical. This is, for example, the case for dissolution.

Chemical processes can also be classified by a number of other attributes:

Synthesis reactions - chemical processes in which two or more starting substances yield a qualitatively new product.

$$2\,H_2 + O_2 \xrightarrow{Pt} 2\,H_2O$$

$$N_2 + 3\,H_2 \underset{}{\overset{Fe}{\rightleftharpoons}} 2\,NH_3$$

Chemical decomposition reactions:

$$CaCO_3 \xrightarrow{t^o} CaO + CO_2$$

$$NH_4Cl \xrightarrow{t^o} NH_3 + HCl$$

Substitution reactions:

$$CH_4 + Cl_2 \longrightarrow CH_3Cl + HCl$$

Addition reactions - usually occur in the presence of **multiple bonds:**

$$H_2C{=}CH_2 + Br_2 \longrightarrow CH_2(Br){-}CH_2(Br)$$

ethene 1,2-dibromoethane

$$HC{\equiv}CH + H_2O \xrightarrow{Hg^{+2},\ H^+} \left[H_2C{=}CH(OH)\right] \rightleftharpoons CH_3CHO$$

ethyne ethenol ethanal

Hydrolysis processes:

$$\underbrace{2Na^+ + CO_3^{-2}} + \underbrace{2\,H^+ + 2\,\bar{O}H} \rightleftharpoons \underbrace{2Na^+ + 2\bar{O}H} + H_2CO_3$$

$$2\,H^+ + 2\,\bar{O}H \rightleftharpoons 2\,H_2O \qquad H_2CO_3 \rightleftharpoons H_2O + CO_2\uparrow$$

Neutralization processes:

$$\underbrace{Na^+ + \bar{O}H} + \underbrace{H^+ + Cl^-} \longrightarrow \underbrace{Na^+ + Cl^-} + H_2O$$

The same chemical reaction can be assigned to different types of chemical processes depending on which characteristic trait is being reported, e.g., the reaction

$$2\,SO_2 + O_2 \longrightarrow 2SO_3 + Q$$

is an exothermic, homogeneous, molecular, redox, synthesis reaction.

The knowledge of the thermal effects of chemical processes is very important, since the conditions for the realization of the processes, the design of the reaction apparatus, the energy consumption, etc. depend on them. Some chemical processes are carried out in order to use the energy that is released during their course.

Thermochemistry. Hess's Law

The branch of chemistry that studies the thermal effects of chemical processes is called **thermochemistry.**

Thermal changes in chemical processes are recorded using **thermochemical equations.** In them, along with the formulas of the reactants and products, the **thermal effects** are also given, which are denoted by Q. In the thermochemical equations, the **aggregate state** of the starting substances and products is also always denoted. The processes of change of aggregate state are associated with energetic changes:

$$\underset{(gas)}{H_2} + \underset{(gas)}{\frac{1}{2}O_2} \longrightarrow \underset{(gas)}{H_2O} + 242{\cdot}5\ kJ$$

$$\underset{(gas)}{H_2} + \underset{(gas)}{\frac{1}{2}O_2} \longrightarrow \underset{(liq)}{H_2O} + 285{\cdot}8\ kJ$$

The heat effect which accompanies the production of a mole of a compound from the corresponding simple substances is called the **heat of formation** of that compound.

The heats of formation of the compounds are different in sign and value. The compounds resulting from **endothermic processes** are **more energy rich** than the starting ones and are therefore **less stable** than them. The greater the heat of formation of a compound produced by **an exothermic process**, the more stable it is.

In thermochemistry, two basic laws are known, initially established experimentally. They are a consequence of the first thermodynamic principle and the law of conservation of energy.

The Lavoisier-Laplace law states that the heat of formation of a compound is equal in value and opposite in sign to the heat of decomposition of the same compound:

$$\underset{(gas)}{CO} + \underset{(gas)}{H_2O} \longrightarrow \underset{(gas)}{CO_2} + \underset{(gas)}{H_2} + 41{\cdot}55\ kJ$$

$$\underset{(gas)}{CO_2} + \underset{(gas)}{H_2} \longrightarrow \underset{(gas)}{CO} + \underset{(gas)}{H_2O} - 41{\cdot}55\ kJ$$

If these heats were not equal it would mean that an extra amount of heat is produced in one of the reactions without expending any other kind of energy. This contradicts the law of conservation of energy and is not possible.

According to Hess's law, the variation of the heat effect of a reaction under given conditions depends only on the **initial and final states of** the system, but does not depend on the intermediate pathway(s), i.e., if the starting substances are given different pathways can produce the same final product, **the heat effect is always the same.** This law makes it possible to calculate the thermal effects of reactions that do not proceed to completion or for which experimental data are lacking.

For example, the heat of formation of carbon monoxide (CO) is difficult to measure directly because it is difficult to stop the oxidation process of carbon to carbon monoxide. However, the heat of formation of carbon dioxide (CO_2) can be easily determined when graphite (C) and carbon monoxide (CO) are used as starting materials.

$$\underset{\text{(graphite)}}{C} + \underset{\text{(gas)}}{O_2} \longrightarrow \underset{\text{(gas)}}{CO_2} + 393{\cdot}5\ \text{kJ}$$

$$\underset{\text{(gas)}}{CO} + \underset{\text{(gas)}}{\frac{1}{2}O_2} \longrightarrow \underset{\text{(gas)}}{CO_2} + 283\ \text{kJ}$$

The energetic scheme of these processes can be represented as follows:

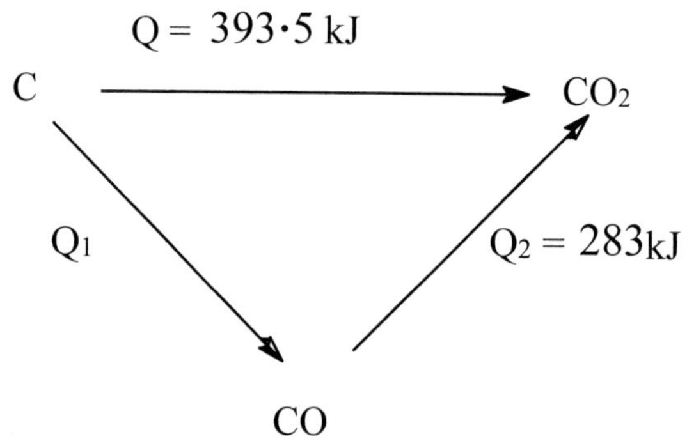

From it follows the energy equation:

$Q = Q_1 + Q_2$; 393.5 kJ $= Q_1 + 283$ kJ; $Q_1 = 110.5$ kJ

Many of the problems related to the determination of the thermal effect of chemical processes are solved in a similar way. Thc general rule is that the heat of reaction is equal to the difference of the heat of formation of the products and the heat of formation of the starting substances. Mathematically, the expression has the following form:

$Q_{reaction} = \Sigma Q_{products} - \Sigma Q_{source}$ substances

This form of Hermann Hess's law is most commonly used in thermochemical equations. Here the heat of formation of simple substances is assumed to be zero. An illustration of Hess's law is the following example:

$$CO + H_2O \longrightarrow CO_2 + H_2 + Q$$

The thermal effect of the process is determined by considering the heat of formation of CO_2 - 393.5 kJ/mol, CO - 110.5 kJ/mol and H_2O - 242.5 kJ/mol. According to Hess's law

$$Q = 393.5 - (242.5 + 110.5) = 40.5\ \text{kJ/mol}.$$

This means that the process is exothermic, involving the release of 40.5 kJ/mol of heat.

Sample problems and solutions

Task 1.

Express with chemical equations the processes occurring in the transition:

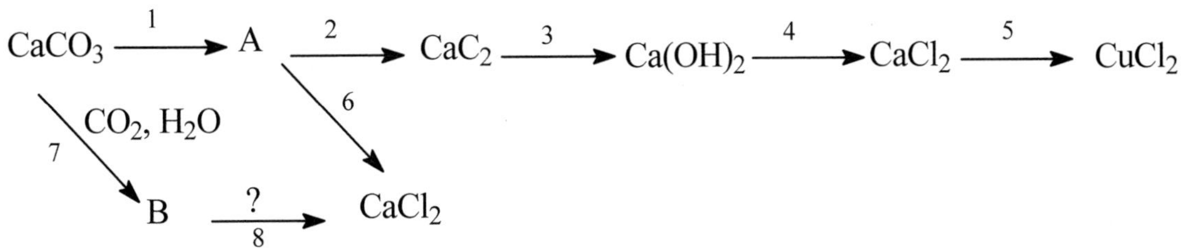

What products will be emitted at the anode and cathode during electrolysis of melt and aqueous $CuCl_2$ solution?

Exemplary solution

When heated to about 900° C, $CaCO_3$ decomposes into calcium oxide (CaO) and carbon dioxide (CO_2).

$$CaCO_3 \xrightarrow{1} \underset{(A)}{CaO} + CO_2\uparrow$$

In the presence of carbon, and on heating, calcium oxide is converted to calcium carbide. Carbon monoxide is released.

$$CaO + 3C \xrightarrow{2} CaC_2 + CO$$

$$\begin{array}{lll|c|c|c} C^{0} - 2e^{-} & \longrightarrow & C^{+2} & 2 & & 1 \\ & & & & 2 & \\ C^{0} + e^{-} & \longrightarrow & C^{-} & 1 & & 2 \end{array}$$

The process is redox - disproportionation:

A carbon atom gives up two electrons and goes into a second positive oxidation state. Two other carbon atoms take one electron and go into the first negative oxidation state. In this case, calcium carbide, the carbon atoms are in the sp-hybrid state.

$$CaC_2 + 2H_2O \xrightarrow{3} Ca(OH)_2 + HC{\equiv}CH$$

The interaction of calcium carbide with water is used to produce **ethyne** (acetylene) in industrial and laboratory practice.

The reaction between calcium dihydroxide and hydrochloric acid is called neutralization and produces salt and water. The process is irreversible since calcium dichloride is a salt of a strong base and a strong acid.

$$Ca(OH)_2 + 2\ HCl \xrightarrow{4} CaCl_2 + 2\ H_2O$$

$$\underbrace{Ca^{+2} + 2\ \bar{O}H} + \underbrace{2\ H^+ + 2\ \bar{Cl}} \longrightarrow \underbrace{Ca^{+2} + 2\ \bar{Cl}} + 2\ H_2O$$

$$2\ H^+ + 2\ \bar{O}H \rightleftharpoons 2\ H_2O$$

The conversion of calcium dichloride to copper dichloride occurs by an ion exchange reaction between calcium dichloride and copper sulfate. Ion-exchange reactions proceed as a precipitate, gas or weak electrolyte is released. They are expressed by molecular, full ionic and reduced ionic equations.

$$CaCl_2 + CuSO_4 \xrightarrow{5} CuCl_2 + CaSO_4\downarrow$$

$$\underbrace{Ca^{+2} + 2Cl^-} + \underbrace{Cu^{+2} + SO_4^{-2}} \longrightarrow CaSO_4\downarrow + \underbrace{Cu^{+2} + 2\bar{Cl}}$$

$$Ca^{+2} + SO_4^{-2} \longrightarrow CaSO_4\downarrow$$

The reaction between ions of group IIA elements with sulfuric acid or its soluble sulfates is used in analytical practice to prove sulfate ions.

$$CaO + 2HCl \xrightarrow{6} CaCl_2 + H_2O$$

The above equation represents the interaction between a basic oxide and an acid. Salt and water are obtained.

Limestone rocks, under the action of rainwater containing carbon dioxide, are converted into soluble calcium dihydrogen carbonate. As the process is reversible the dihydrogen carbonate slowly decomposes and becomes calcium carbonate. Deposition of the calcium carbonate takes place and so the stalactites grow. The process is slow and takes tens and hundreds of years. The length of the stalactites in some cases reaches several meters. A stalactite (from the Greek for 'dripping one') is a formation of calcium carbonate hanging from the ceiling of caves in a place where water is dripping. Stalagmite (from the Greek for 'dripping') is a sedimentary formation layered on the floor of the caves in a place where water drips. The term was introduced by Danish explorer Ole Worm in 1655. Stalactites and stalagmites have formed on this basis over the millennia in some of the most beautiful caves on Earth. The reaction is reversible.

$$CaCO_3 + CO_2 + H_2O \overset{7}{\rightleftharpoons} Ca(HCO_3)_2$$

When calcium dihydrogen carbonate is treated with hydrochloric acid, a so- called "noisy sample" takes place – carbon dioxide is violently released.

$$Ca(HCO_3)_2 + 2\ HCl \xrightarrow{8} CaCl_2 + 2CO_2\uparrow + 2\ H_2O$$

Electrolysis of a copper dichloride melt releases elemental copper at the cathode and chlorine at the anode.

$$CuCl_2 \xrightarrow{t^o} \overset{+2}{Cu} + 2\,\overline{Cl}$$

At high temperature, the ionic crystalline compounds fuse and divalent ions are produced. If a constant electric current is passed through the copper dichloride melt, copper cations are directed to the cathode and chloride anions to the anode. The following processes take place:

$$\overset{+2}{Cu} + 2e^- \longrightarrow \overset{o}{Cu}$$

$$2\,\overline{Cl} - 2 \times 1\,\bar{e} \longrightarrow 2\,\overset{o}{Cl} \longrightarrow Cl_2$$

Similar processes are carried out in electrolyte solutions. When a direct electric current is passed through an aqueous solution of copper dichloride, elemental copper is released at the cathode and chlorine at the anode. Hydrogen and copper ions are directed to the cathode and hydroxide and chloride ions to the anode. The copper ions are electroneutralised at the cathode because they are after the hydrogen in the Order of Relative Activity of Metals (ORAM). The chloride ions are electroneutralised at the anode. They are before the hydroxide ions in the order of activity of the anions.

$$CuCl_2 \longrightarrow \overset{+2}{Cu} + 2\,\overline{Cl}$$

$$H_2O \rightleftharpoons H^+ + \overline{O}H$$

7. Redox Processes

OXIDATION AND REDUCTION. NATURE OF REDOX PROCESSES. REACTIONS BETWEEN METALS AND SOLUTIONS OF ELECTROLYTES CONTAINING METAL IONS. OXIDATION AND REDUCTION PROPERTIES OF METALS AND NON-METALS. ORDER OF RELATIVE ACTIVITY OF METALS.

Electrolysis. Application

Oxidation and reduction

Historically, the concept of **redox processes** arose in the study of reactions involving the element oxygen. When elements combine with oxygen, they are oxidized and the process is oxidation. When copper reacts with oxygen, copper oxide is produced.

$$2\,Cu + O_2 \longrightarrow 2\,CuO$$

$$\begin{array}{rl|c|c|c}
 & Cu^{0} - 2\bar{e} \longrightarrow Cu^{+2} & 2 & \multirow{2}{*}{4} & 2 \\
O_2 \longrightarrow & 2\,O^{0} + 2 \times 2\bar{e} \longrightarrow 2O^{-2} & 4 & & 1
\end{array}$$

Copper oxidizes.

The oxygen atom is strongly electronegative; when it bonds with other atoms, it accepts their valence electrons and becomes a negative ion, and the corresponding atoms that have given up electrons become positive ions. The process of oxidation of copper proceeds as presented above.

When oxygen is removed from an oxide, it is reduced, and the process is reduction.

The element that takes away the oxygen is a reductant. When copper oxide reacts with hydrogen, copper and water are produced:

$$CuO + H_2 \longrightarrow Cu + H_2O$$

Copper oxide is reduced and hydrogen is oxidized. Common to both substances, the process is redox. Hydrogen is a reductant as it takes away oxygen and copper oxide is an oxidant.

The process of reduction of copper oxide with hydrogen is written as follows:

$$\begin{array}{rl|c|c|c}
 & Cu^{+2} + 2\bar{e} \longrightarrow Cu^{0} & 2 & \multirow{2}{*}{2} & 1 \\
H_2 \longrightarrow & 2\,H^{0} - 2 \times 1\bar{e} \longrightarrow 2H^{+} & 2 & & 1
\end{array}$$

The essence of the process is that the hydrogen atoms give electrons to the copper ion.

There are many processes in which, like processes involving oxygen, the most significant feature is the transition of electrons from one atom (or ion) to another atom (or ion).

Nature of redox processes

Processes characterized by the transfer of electrons from one type of atoms, ions, or molecules to other atoms, ions, or molecules are called **redox processes.** Redox processes are:

$$Cu + Cl_2 \longrightarrow CuCl_2$$

$$\begin{array}{lcl|c|c|c} Cu^{0} - 2e^{-} & \longrightarrow & Cu^{+2} & 2 & \multirow{2}{*}{2} & 1 \\ Cl_2 \longrightarrow 2\,Cl^{0} + 2 \times 1\,\bar{e} & \longrightarrow & 2\,Cl^{-} & 2 & & 1 \end{array}$$

$$CuO + C \longrightarrow Cu + CO$$

$$\begin{array}{lcl|c|c|c} Cu^{+2} + 2\bar{e} & \longrightarrow & Cu^{0} & 2 & \multirow{2}{*}{2} & 1 \\ C^{0} - 2\bar{e} & \longrightarrow & C^{+2} & 2 & & 1 \end{array}$$

Atoms or ions which, like oxygen atoms, take up electrons in chemical processes are called **oxidants**. In the processes under consideration the oxidizing agents are chlorine and the copper ion of copper oxide, respectively.

$$Cl_2 \longrightarrow 2\,Cl^{0} + 2 \times 1\,\bar{e} \longrightarrow 2\,Cl^{-}$$

$$\overset{+2}{Cu} + 2e^{-} \longrightarrow Cu^{0}$$

Atoms or ions that give off electrons in chemical processes are called **reducers**. In the above examples, the reducers are copper and carbon atoms, respectively:

$$\overset{0}{Cu} - 2e^{-} \longrightarrow \overset{+2}{Cu}$$

$$\overset{0}{C} - 2e^{-} \longrightarrow \overset{+2}{C}$$

The oxidation and reduction processes always occur simultaneously. In them, one of the substances is an **oxidizer** (accepts electrons) and the other is a **reducer** (gives off electrons). These processes are therefore called **redox processes**.

Atoms of elements with a metallic character have a small number of electrons in their last electron layer. Therefore, in redox processes they give up electrons, i.e., they are **reducers**. The strongest reducers are the elements of group IA of the Periodic Table – the alkali elements. In the groups of the Periodic System, as the sequence number of the elements increases, so does their reducing power because, as the atomic radius increases and the valence electrons move away from the nucleus, the atoms more readily give them up. In addition to metals, hydrogen and carbon also exhibit reducing properties. These properties are used in metallurgy.

It is more typical for the atoms of the non-metallic elements in redox processes to accept electrons, i.e., to act as **oxidants**. Fluorine and the other halogen elements, as well as oxygen, are typical oxidants.

Some non-metals, for example sulfur, can both **accept** and **give up** electrons in redox processes. Depending on the conditions, they exhibit the properties of **oxidizing or reducing agents**. For example, when sulfur burns in an oxygen atmosphere, it manifests itself as a reducing agent, and when it interacts with hydrogen, it is an oxidizing agent.

$$S + O_2 \longrightarrow SO_2$$

$$\begin{array}{lcl|c|c|c} & S^{0} - 4e^{-} & \longrightarrow S^{+4} & 4 & & 1 \\ & & & & 4 & \\ O_2 \longrightarrow & 2\,O^{0} + 2 \times 2\bar{e} & \longrightarrow 2O^{-2} & 4 & & 1 \end{array}$$

$$S + H_2 \longrightarrow H_2S$$

,

$$\begin{array}{lcl|c|c|c} & S^{0} + 2\,e^{-} & \longrightarrow S^{-2} & 2 & & 1 \\ & & & & 2 & \\ H_2 \longrightarrow & 2\overset{0}{H} - 2 \times 1\bar{e} & \longrightarrow 2\,H^{+} & 2 & & 1 \end{array}$$

The giving up and **taking up** of electrons is associated **with a change in the valence of the corresponding atom**. Oxidizers accept electrons, which lowers the positive valence (lowers the oxidation state). Reducers give up electrons, which increases the positive valence or oxidation state.

$$Ca + Cl_2 \longrightarrow CaCl_2$$

$$\begin{array}{lcl|c|c|c} & Ca - 2\bar{e} & \longrightarrow Ca^{+2} & 2 & & 1 \\ & & & & 2 & \\ Cl_2 \longrightarrow & 2\,Cl^{0} + 2 \times 1\bar{e} & \longrightarrow 2Cl^{-} & 2 & & 1 \end{array}$$

$$2\,NaBr + Cl_2 \longrightarrow 2\,NaCl + Br_2$$

$$\begin{array}{lcl|c|c|c} 2\,Br^{-} - & 2 \times 1\bar{e} \longrightarrow 2\,Br^{0} & \longrightarrow Br_2 & 2 & & 1 \\ & & & & 2 & \\ Cl_2 \longrightarrow & 2\,Cl^{0} + 2 \times 1\bar{e} & \longrightarrow 2Cl^{-} & 2 & & 1 \end{array}$$

$$Ca + S \longrightarrow CaS$$

$$\begin{array}{cl|c|c|c} Ca^{0} - 2\bar{e} & \longrightarrow Ca^{+2} & 2 & & 1 \\ & & & 2 & \\ S^{0} + 2\,e^{-} & \longrightarrow S^{-2} & 2 & & 1 \end{array}$$

$$2\,HgO \longrightarrow 2\,Hg + O_2$$

$$\begin{array}{lll|c|c|c} & Hg^{+2} + 2\,e^{-} \longrightarrow & Hg^{0} & 2 & & 2 \\ & & & & 4 & \\ 2\,O^{-2} - 2 \times 2\,\bar{e} \longrightarrow & 2\,O^{0} \longrightarrow & O_2 & 4 & & 1 \end{array}$$

Processes are redox, because they are associated with an electronic transition and a change in valence.

Oxidation-reduction processes take place between substances of different composition and structure. Therefore, the processes are widely used in practice - in metallurgy, in electrolysis, to obtain various substances.

Reactions between metals and electrolyte solutions, which contain metal ions

If a copper plate is introduced into a solution containing silver ions (for example, a solution of silver nitrate), it is noticed that the copper plate becomes covered with a layer of shiny silver, and with this the solution takes on a blue color. The changes indicate that the following redox process has taken place on the surface of the copper plate:

$$Cu^{0} + \underbrace{2\,Ag^{+} + 2\,NO_3^{-}} \longrightarrow 2\,Ag^{0} + \underbrace{Cu^{+2} + 2\,NO_3^{-}}$$

$$\begin{array}{ll|c|c|c} Cu^{0} - 2\,\bar{e} \longrightarrow & Cu^{+2} & 2 & & 1 \\ & & & 2 & \\ Ag^{+} + e^{-} \longrightarrow & Ag^{0} & 1 & & 2 \end{array}$$

Copper displaces silver from its compounds, i.e., it is a **more active element** than silver. **The reverse process does not take place** - if a silver plate is placed in a solution containing copper ions, **no reaction takes place**.

If a zinc plate is introduced into a solution containing copper ions, for example a copper sulfate solution, copper is found to be released on the zinc plate and the blue color of the solution fades. This means that an oxidation-reduction process is taking place:

$$Zn^{0} + \underbrace{\overset{+2}{Cu} + \overset{-2}{SO_4}} \longrightarrow \underbrace{\overset{+2}{Zn} + \overset{-2}{SO_4}} + Cu^{0}$$

$$\begin{array}{ll|c|c|c} Zn^{0} - 2\,\bar{e} \longrightarrow & Zn^{+2} & 2 & & 1 \\ & & & 2 & \\ Cu^{+2} + 2\,e^{-} \longrightarrow & Cu^{0} & 2 & & 1 \end{array}$$

From the full ionic equation of the process, it can be concluded that zinc is a more active element than copper because it displaces it from its compounds, i.e., zinc atoms give electrons to copper ions. Zinc is a stronger reductant than copper.

Since zinc displaces copper from its compounds, and copper displaces silver, therefore zinc displaces silver from its compounds.

$$\overset{o}{Zn} + \underbrace{2\,\overset{+}{Ag} + 2\,\overset{-}{NO_3}} \longrightarrow 2\,\overset{o}{Ag} + \underbrace{\overset{+2}{Zn} + 2\,\overset{-}{NO_3}}$$

$$\begin{array}{lll|c|c|c} \overset{o}{Zn} - 2\bar{e} & \longrightarrow & Zn^{+2} & 2 & & 1 \\ & & & & 2 & \\ Ag^{+} + e^{-} & \longrightarrow & Ag^{o} & 1 & & 2 \end{array}$$

Order of relative activity of metals

It can be seen from the examples that if the three elements zinc, copper and silver are arranged in ascending order of their activity (of their reducing power), the following order is obtained:

$$\underset{\text{activity increases}}{\underrightarrow{Ag \quad Cu \quad Zn}}$$

If the metals are arranged in descending order of their reducing activity, the **order of the relative activity of the metals is obtained**:

Li, K, Ca, Na, Mg, Al, Zn, Fe, Ni, Sn, Pb, H, Cu, Hg, Ag, Au

Each element in the top row can displace from their compounds, the elements to the right of it, as they are less active than it. The strongest reductant is lithium, and the strongest oxidant is the trivalent ions of gold. In this sense, the displacement of copper by zinc is possible because the electrode potential of Zn/Zn^{+2} is lower than that of Cu/Cu^{+2}.

Each metal in the order of relative activity can only interact with the ions of the metals after it.

Again, from the order of relative activity it is seen that all elements before hydrogen have a lower value of electrode potential. Therefore, all elements to the left of hydrogen can displace it from the acids.

The interaction of metals with acids is an oxidation-reduction process, since it involves the transfer of electrons from the metal atoms to hydrogen ions. For example, when zinc is dissolved in hydrochloric acid, the process takes place:

$$Zn + \underbrace{2\,\overset{+}{H} + 2\,\overset{-}{Cl}} \longrightarrow \underbrace{\overset{+2}{Zn} + 2\,\overset{-}{Cl}} + H_2$$

$$\begin{array}{lllll|c|c|c} & \overset{o}{Zn} - 2\bar{e} & & \longrightarrow & Zn^{+2} & 2 & & 1 \\ & & & & & & 2 & \\ 2\,H^{+} + 2 \times 1\bar{e} & \longrightarrow & 2\,\overset{o}{H} & \longrightarrow & H_2 & 2 & & 1 \end{array}$$

resulting in zinc dichloride ($ZnCl_2$).

This means that the process is possible

$$M + nH^{+} \longrightarrow M^{n+} + nH^{o}$$

where M is a metal atom, n is the number of hydrogen ions, respectively the number of positive charges of the metal ion. In this case, the metal atoms have a reducing action with respect to the hydrogen ions.

The metals standing to the left in order of relative activity interact with solutions of salts of the metals standing after them (provided the metals do not interact with water).

The order studied can serve to determine the course of reactions between metals and salts. For example, reactions between

$$Zn + NaCl \not\rightarrow$$
$$Mg + KNO_3 \not\rightarrow$$
$$Fe + CaCl_2 \not\rightarrow$$

do not proceed until the reactions

$$Sn + CuSO_4 \longrightarrow SnSO_4 + Cu$$

$$Hg + 2AgNO_3 \longrightarrow Hg(NO_3)_2 + 2Ag$$

flow as tin is before copper and mercury is before silver.

Elements after hydrogen in the order of relative activity cannot displace it from its compounds because they have higher electrode potential values.

Electrolysis

The redox processes discussed so far occur spontaneously, i.e., without external influence. There are chemical processes which can only take place under external influence, for example under the action of an electric current.

The redox processes that take place in melts or in solutions of electrolytes under the action of an electric current are called electrolytic, and the phenomenon is called electrolysis.

Metals are conductors of the first kind, in them the flow of current is carried by electrons, in which no chemical changes are induced.

Solutions and melts of electrolytes are referred to as conductors of the second genus. The flow of current through a solution or melt (direct electric current) causes the differently charged ions to move simultaneously in two opposite directions. The positive ions are directed towards the negative electrode (cathode), and are therefore called cations. Negative ions are directed towards the positive electrode (anode) and are called anions. This results in changes in the state of the electrolytes.

Salts, for example NaCl, Na_2SO_4, KNO_3, have an ionic crystal lattice. When they melt, the crystal lattice breaks down and ions move into the melt. For example, sodium cations and chloride anions move in the sodium chloride melt.

When a **constant electric current** is passed through **the melt,** the chaotically moving ions are directed to the electrodes. The cations at the cathode accept electrons, i.e., they are reduced and converted to neutral sodium atoms. Negative ions are directed to the anode. There, the chloride anions give up electrons, i.e., are oxidized and converted to neutral chlorine atoms.

The processes can be expressed by the following scheme:

Available ions in the melt:

$$NaCl \longrightarrow \overset{+}{Na} + \overset{-}{Cl}$$

cathode process:

$$\overset{+}{Na} + \overset{-}{e} \longrightarrow \overset{o}{Na}$$

anode process:

$$2\,\bar{Cl} \; - \; 2 \times 1\,\bar{e} \longrightarrow 2\,\overset{o}{Cl} \longrightarrow Cl_2$$

These processes are redox and are carried out under the action of an electric current. When it flows, the cathode acts as a reducing agent and the anode is an oxidizing agent. Electrolysis takes place in the melt.

Similar processes are carried out in electrolyte solutions. When a **direct electric current** is passed through **an aqueous solution** of copper dichloride, elemental copper is released at the cathode and chlorine at the anode.

Available ions in solution:

$$CuCl_2 \longrightarrow \overset{+2}{Cu} + 2\,\bar{Cl}$$

$$H_2O \rightleftharpoons H^+ + \bar{O}H$$

cathode process:

$$\overset{+2}{Cu} + 2\,\bar{e} \longrightarrow \overset{o}{Cu}$$

anode process:

$$2\,\bar{Cl} \; - \; 2 \times 1\,\bar{e} \longrightarrow 2\,\overset{o}{Cl} \longrightarrow Cl_2$$

When there are more than two types of ions in solution, the electrolysis will electrify those ions further to the right in the Order of Relative Activity of the Metals at the cathode, and the anions further to the left at the anode.

For example, in the electrolysis of an aqueous solution of disodium sulfate, hydrogen and sodium cations are directed to the cathode, but hydrogen ions will be electroneutralized:

$$2\,H^+ + 2 \times 1\,\bar{e} \longrightarrow 2\,\overset{o}{H} \longrightarrow H_2$$

Hydroxide and sulfate anions are directed to the anode, but hydroxide anions are electroneutralized:

$$HO^- - e^- \longrightarrow HO$$

$$4\,HO \longrightarrow 2\,H_2O + O_2$$

Like metals, anions can also be arranged in order according to their ability to give off electrons, i.e., to oxidise:

$$S^{-2},\ I^-,\ Br^-,\ Cl^-,\ \bar{O}H,\ NO_3^-,\ SO_4^{-2}\ ,\ldots\ldots\ldots\ldots$$

Any anion in this row is more easily oxidized than any anion to the right of it (after it). The processes that take place at the electrodes depend too often on the type of electrode. Typically, electrodes are made of materials that do not react (interact) with the products produced by electrolysis. Electrodes of platinum or carbon are used.

Electrolysis processes are the basis of many industrial processes, which are called electrochemical. The electrolysis of molten salts is used to produce group IA and IIA metals.

Electrolysis processes are important for the production of some basic products for inorganic synthesis - hydrogen, chlorine and sodium base. These products are obtained by electrolysis of aqueous sodium chloride solution. The process is known as **chloralkali electrolysis**.

Prior to the electrolysis process, the following ions are present in solution: sodium, hydrogen, chloride and hydroxide. During the electrolysis process, hydrogen is released at the cathode, chlorine is released at the anode and sodium base is formed in the solution.

Ions in solution (before electrolysis):

$$Na^+,\ H^+,\ Cl^-,\ \overline{O}H$$

cathode process: **anode process:**

$$2\,H^+ + 2 \times 1\,e^- \longrightarrow 2\,\overset{0}{H} \longrightarrow H_2\ ; \quad 2\,Cl^- - 2 \times 1\,e^- \longrightarrow 2\,\overset{0}{Cl} \longrightarrow Cl_2$$

Ions in solution (after electrolysis):

$$\underbrace{\overset{+}{Na} + \overset{-}{OH}}$$

At the cathode, hydrogen is emitted, not sodium, because hydrogen stands after sodium in the Order of Relative Activity of the Metals. At the anode, chloride anions are electroneutralized, not hydroxide, because chloride ions come before hydroxide ions in the electrochemical order of the anions.

The solution produces a sodium base, which could react with the chlorine that is released. Measures are therefore taken in the construction of the electrolysis bath to separate these substances.

Electrolysis is also used to purify some metals. For example, copper is purified by electrolysis. In this case, a solution of copper sulfate and sulfuric acid is placed in the electrolysis bath. The copper to be purified is used as the anode and pure copper as the cathode. During electrolysis, the technical copper is dissolved and deposited on the cathode. The following processes take place:

cathode processes:

$$\overset{+2}{Cu} + 2\,e^- \longrightarrow \overset{0}{Cu}$$

anode processes:

$$\overset{0}{Cu} - 2\,e^- \longrightarrow \overset{+2}{Cu}$$

Substances that impurify copper remain in solution and precipitate out.

Electrolysis is also used to produce metal coatings that have a protective role. Such are the coatings of copper, zinc, chromium, nickel. Gold and silver coatings are used for special purposes.

An example of metal plating is the electrolysis of a silver nitrate solution. In this electrolysis process, silver is released at the cathode. If a metal object is placed on the cathode, it will be coated with a layer of silver.

cathode process

$$\overset{+}{Ag} + e^- \longrightarrow Ag^0$$

The process is called silvering. In order to not quickly deplete the silver ions from the solution, the anode is made of silver. In the electrochemical process, the anode is dissolved.

anode process

$$Ag^{0} - e^{-} \longrightarrow Ag^{+}$$

The resulting silver cations pass into solution and maintain a constant concentration of silver nitrate.

Sample problems and solutions

Task 1.

Express the given chemical transformations with chemical equations and explain the processes:

$$Cu \xrightarrow{1} CuSO_4 \xrightarrow{2} Cu(OH)_2 \xrightarrow{3} CuCl_2 \xrightarrow{4} ZnCl_2 \xrightarrow{5}$$

$$\xrightarrow{5} ZnS \xrightarrow{6} ZnO \xrightarrow{7} Zn \xrightarrow{8} Na_2[Zn(OH)_4]$$

Exemplary solution

Copper is located after hydrogen in the Order of Relative Activity of Metals and interacts only with concentrated oxidizing oxoacids (H_2SO_4, HNO_3). The process is carried out in two stages: in the first stage copper oxide is produced, and the oxide with a new molecule of sulfuric acid leads to the production of copper sulfate.

$$Cu + H_2SO_4 \longrightarrow CuO + SO_2 + H_2O$$

$$CuO + H_2SO_4 \longrightarrow CuSO_4 + H_2O$$

....

$$Cu + 2H_2SO_4 \longrightarrow CuSO_4 + SO_2 + 2H_2O$$

$$\begin{array}{lc|c|c|c} Cu^{0} - 2e^{-} \longrightarrow Cu^{+2} & & 2 & \multirow{2}{*}{2} & 1 \\ S^{+6} + 2e^{-} \longrightarrow S^{+4} & & 2 & & 1 \end{array}$$

Processes characterized by the transfer of electrons from one type of atoms, ions, or molecules to other atoms, ions, or molecules are called **redox processes.**

The copper sulfate is treated with sodium base to give a sky-blue precipitate of copper dihydroxide. The process is expressed by molecular, full ion and truncated ion equations:

$$CuSO_4 + 2NaOH \longrightarrow Cu(OH)_2 + Na_2SO_4$$

$$\underbrace{Cu^{+2} + SO_4^{-2}} + \underbrace{2Na^+ + 2\bar{O}H} \longrightarrow Cu(OH)_2\downarrow + \underbrace{2Na^+ + SO_4^{-2}}$$

$$Cu^{+2} + 2\bar{O}H \longrightarrow Cu(OH)_2\downarrow$$

Copper dichloride is produced by the process of neutralisation - copper dihydroxide reacts with hydrochloric acid:

$$Cu(OH)_2 + 2HCl \rightleftharpoons CuCl_2 + H_2O$$

$$Cu(OH)_2 + \underbrace{2H^+ + 2\bar{Cl}} \rightleftharpoons \underbrace{Cu^{+2} + 2\bar{Cl}} + H_2O$$

Ion exchange refers to the processes of interaction between salt solutions that produce a gas, precipitate or weak electrolyte. A similar example is the reaction to produce zinc sulfide.

$$CuCl_2 + Zn \longrightarrow ZnCl_2 + Cu$$

$$\begin{array}{lll|c|c|c} Cu^{+2} + 2\bar{e} & \longrightarrow & Cu^{0} & 2 & & 1 \\ & & & & 2 & \\ Zn^{0} - 2\bar{e} & \longrightarrow & Zn^{+2} & 2 & & 1 \end{array}$$

$$ZnCl_2 + Na_2S \longrightarrow ZnS\downarrow + 2NaCl$$

$$\underbrace{Zn^{+2} + 2\bar{Cl}} + \underbrace{2Na^+ + S^{-2}} \longrightarrow ZnS\downarrow + \underbrace{2Na^+ + 2\bar{Cl}}$$

$$Zn^{+2} + S^{-2} \longrightarrow ZnS\downarrow$$

In an oxygen atmosphere, zinc sulfide is converted to zinc oxide upon heating. Sulfur dioxide is also released.

$$2\,ZnS + 3\,O_2 \longrightarrow 2\,ZnO + 2\,SO_2$$

$$\begin{array}{lcl|c|c|c}
O_2 \longrightarrow 2\overset{0}{O} + 2 \times 2e^- & \longrightarrow & 2\overset{-2}{O} & 4 & & 3 \\
 & & & & 12 & \\
\overset{-2}{S} - 6e^- & \longrightarrow & S^{+4} & 6 & & 2
\end{array}$$

Zinc oxide with carbon in an oxidation-reduction process is converted into elemental zinc.

$$ZnO + C \longrightarrow Zn + CO$$

$$\begin{array}{lcl|c|c|c}
Zn^{+2} + 2e^- & \longrightarrow & \overset{0}{Zn} & 2 & & 1 \\
 & & & & 2 & \\
C^{0} - 2e^- & \longrightarrow & C^{+2} & 2 & & 1
\end{array}$$

Zinc metal exhibits amphoteric properties - it interacts with both acids and bases.

$$Zn + 2\,NaOH + 2\,H{-}OH \longrightarrow Na_2[Zn(OH)_4] + H_2$$

$$\begin{array}{lcl|c|c|c}
Zn^{0} - 2e^- & \longrightarrow & Zn^{+2} & 2 & & 1 \\
 & & & & 2 & \\
2\,H^+ + 2 \times 1e^- \longrightarrow 2\overset{0}{H} & \longrightarrow & H_2 & 2 & & 1
\end{array}$$

The complex compound disodium tetrahydroxyzincate is obtained. The same compound is also obtained by reacting zinc dihydroxide with sodium base.

8. Rate of Chemical Processes

Law of Action of Masses. Effect of Concentration of Reactants on Rate. Mechanism of Chemical Reactions. Dependence of Rate on Temperature. Activation Energy. Speed-Determining Stage.

The rate of chemical processes and the factors that affect the rate are studied by chemical kinetics.

Different chemical processes occur at different rates. Thus, hydrogen and chlorine react instantaneously under strong illumination. Ionic reactions proceed rapidly. Processes occurring in a heterogeneous medium are slower. There are also very slow processes that take hours, days, or even years.

The rate of chemical processes is measured by the change in concentration of the reactants or reaction products per unit of time.

For example:

$$2\,CO + O_2 \longrightarrow 2CO_2$$

$$V_1 = -\frac{\Delta c\,(CO)}{\Delta t};\quad V_2 = -\frac{\Delta c\,(O_2)}{\Delta t};\quad V_3 = +\frac{\Delta c(CO_2)}{\Delta t}$$

The Δc denotes the change in concentration of the substance and the **Δt** the time interval over which the change in concentration has taken place. The rates thus defined are not equal. Therefore, although it is possible to define the rate by the change in any concentration, it must be clearly stated on the basis of which concentration the measurement is.

The concentration of the starting substances decreases in the course of the process, therefore a minus sign is written in the expression for the rate. The concentration of the reaction products increases and therefore a plus sign is written. The rate of chemical processes is always a positive quantity. In the most general case, the rate is given by the expression:

$$v = \pm\frac{\Delta c}{\Delta t}$$

The rate of chemical reactions changes, so the defined rates are average rates over a time interval Δt. At the beginning of the process, the rate is greatest and as the interaction proceeds, it decreases as the concentration of the starting substances decreases. This relationship is expressed by the following graph:

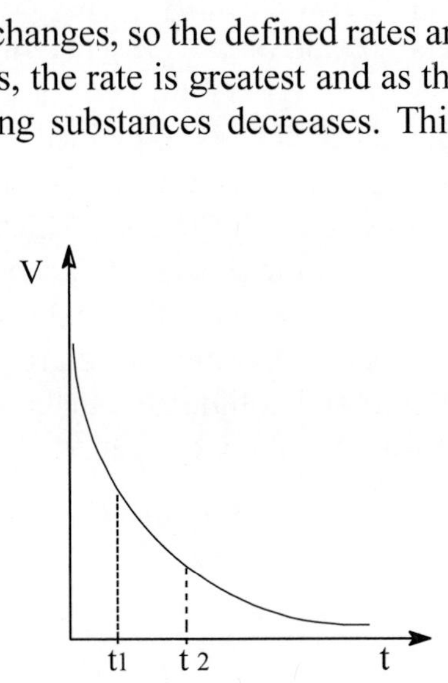

Experimental results can be observed if a piece of zinc is placed in a solution of sulfuric acid. At first, a violent evolution of hydrogen begins and after a while the evolution of hydrogen bubbles gradually slows down. At the end of the reaction, this almost stops due to the gradual depletion of starting substances.

$$Zn + H_2SO_4 \longrightarrow ZnSO_4 + H_2\uparrow$$

$$\begin{array}{lll|c|c|c} Zn^{0} - 2\bar{e} & \longrightarrow & Zn^{+2} & 2 & & 1 \\ & & & & 2 & \\ 2H^{+} + 2 \times 1\bar{e} \longrightarrow 2\overset{0}{H} & \longrightarrow & H_2 & 2 & & 1 \end{array}$$

Law of the action of the masses. Effect of concentration of reactants on velocity

As the concentration of reactants increases, the rate increases. The dependence of velocity on the concentration of reactants is given by the law of action of masses discovered by Guldberg and Waage in 1864:

The rate of chemical reactions is proportional to the product of the concentrations of the reactants raised to a power equal to the number of moles with which these substances participate in the reaction.

For example, when substances A and B react to produce C:

$$A + B \longrightarrow C$$

$$v = k.c(A).c(B)$$

For a reaction of a general kind:

$$n_1 A_1 + n_2 A_2 + n_3 A_3 + \ldots\ldots \longrightarrow$$

$$V = k.\ c(A_1)^{n_1} .\ c(A_2)^{n_2} .\ C(A_3)^{n_3} \ldots\ldots\ldots\ldots\ldots \text{(kinetic equation)}$$

The proportionality factor **k** is called the **rate constant**.

At C = 1 mol/l (unilamellar concentrations) k = v.

The rate constant therefore represents the rate of reaction at unimolecular concentrations of the reactants. The rate constant is a characteristic quantity of any chemical process. It depends on the nature of the reactants, on the temperature and on the presence of catalysts, but does not depend on the concentration of the reactants.

The interaction between inorganic substances is explained by the Arrhenius collision theory. According to it, for two particles to react, they must meet and collide. They must also be suitably oriented and the impact must be sufficiently effective. This collision and its accompanying interactions are called the **elementary act** of the process. In such a case, the rate of reaction will be greater the greater the number of impacts between the molecules per unit time.

The greater the number of particles per unit volume (the greater the concentration), the greater the probability that the particles will meet and collide, i.e., that an interaction will take place.

The dependence of the reaction rate V on the concentration C is given by the following graph:

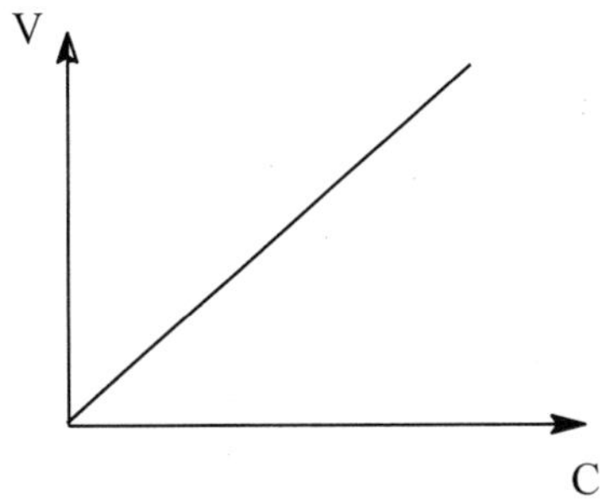

The rate increases in direct proportion to the increase in concentration.

Mechanism of chemical reactions. Rate-determining step

When a reaction proceeds in several stages, its rate is determined by the rate of the slowest stage, which requires the highest activation energy E_a. This stage is called the **rate-determining stage.**

The law of mass action applies to the rate-determining stage. For example, the decay of dinitrogen pentoxide (N_2O_5) to dinitrogen tetroxide (N_2O_4) is given by the following sum equation:

$$2\,N_2O_5 \longrightarrow 2\,N_2O_4 + O_2$$

In fact, the reaction proceeds through the following stages:

$$N_2O_5 \xrightarrow{\text{slow}} N_2O_3 + O_2$$

$$\left.\begin{aligned} N_2O_3 &\longrightarrow NO + NO_2 \\ NO + N_2O_5 &\longrightarrow 3\,NO_2 \\ 2\,NO_2 &\longrightarrow N_2O_4 \end{aligned}\right\} \text{quick}$$

Of all the successive stages through which a chemical reaction proceeds, there is always one which is the slowest and is called the rate-determining one, to which the law of the action of masses applies. In this case it is:

$$v = k.c(N_2O_5)$$

The sum equation only gives the material balance of the reaction - what starting substances will yield certain amounts of reaction products. It shows only the initial and final states of the system, but does not show the path from the starting substances to the products. The law of mass action cannot be applied formally. It is necessary to know which is the rate-determining stage.

The course of a reaction is expressed by its mechanism. By the mechanism of a chemical reaction is meant that the set of successive acts affecting the breaking of old and the formation of new chemical bonds leads to the conversion of the starting substances into the products of the reaction.

In order to reveal the mechanism of a reaction, all the factors that influence the process, such as reaction kinetics, heat effect, etc., must be identified. The molecularity of a reaction is determined by the number of molecules involved in the rate-determining step. For example, the decomposition of dinitrogen pentoxide (N_2O_5) to dinitrogen tetroxide (N_2O_4) is a **monomolecular** reaction.

Bimolecular reactions - two molecules are involved in the rate-determining step:

$$AgNO_3 + NaCl \longrightarrow AgCl\downarrow + NaNO_3$$

$$R\text{—}Cl + NaOH \longrightarrow ROH + NaCl$$

Trimolecular reactions are very rare, as three molecules are unlikely to meet and react simultaneously.

The order of a reaction is determined by the sum of the powers of the concentrations of the reactants in the experimentally found kinetic equation.

For **first order** reactions, the equation will have the form: $v = k.c$;

For **second-order** reactions: $v = k.c_1 .c_2$ or $v = k.c^2$;

For **third-order** reactions: $v = k.c_1 .c_2 .c_3$ or $v = k.c^3$.

Molecularity and order do not always coincide. When they do not coincide, it is an indication that the reaction is proceeding by a more complex mechanism. For example, the hydrolysis of esters is a bimolecular reaction, but it is a first-order reaction.

$$CH_3COOC_2H_5 + H_2O \rightleftharpoons CH_3COOH + C_2H_5OH$$

The change in the amount of water does not affect the velocity and is not involved in the experimentally found kinetic equation

$$v = k.c(CH_3COOC_2H_5)$$

For a reaction of the most general kind

$$n_1 A_1 + n_2 A_2 + n_3 A_3 + \ldots\ldots\ldots\ldots$$

the experimentally found kinetic equation is

$$v = k.c^{m_1}(A_1).\ c^{m_2}(A_2)\ .\ c^{m_3}(A_3) \ldots\ldots\ldots$$

where m_1, m_2, m_3 are the experimentally found number of moles of the substances involved in the reaction and they do not always coincide with n_1, n_2, etc.

Depending on their mechanism, reactions can be:

- **chain-radical** - accompanied by homolytic breakage of covalent bonds;
- **reactions proceeding by an ionic mechanism**. They involve the breaking of ionic bonds (ionic reactions) or the heterolytic breaking of covalent bonds to form ions.

Dependence of velocity on temperature. Activating energy

For two particles to react, they must be sufficiently energy rich, then the impact between them will be effective enough. The bonds between the atoms in the parent compounds will be able to be broken and the atoms will regroup into new molecules.

At any temperature, there is a definite distribution of molecules by kinetic energy (E_K).

A small number of molecules have very low or very high E_K. Molecules with E_K close to some average energy (E_1) predominate. The distribution of gas molecules by E_K can be represented graphically:

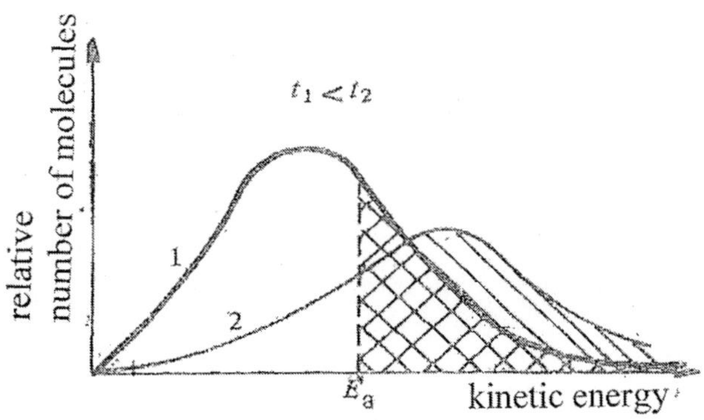

The shaded area gives the relative number of molecules that have $E_K = E_1$ or greater. As the temperature increases, the number of molecules with $E > E_1$ increases. The number of molecules that are sufficiently energy rich grows.

The rate of chemical reactions is directly proportional only to the number of collisions between energy-rich molecules. These molecules are called **active**, and the shocks between them are called **active shocks**.

The minimum energy that molecules must possess for an interaction to take place between them is a specific quantity for each chemical process and is called the activation energy (E_a).

It takes energy to break chemical bonds. Activation energy is needed to loosen the bonds between atoms so that, at the next moment, these bonds can be broken and the atoms regrouped from parent compounds into reaction products.

The energy diagram of the process gives us an idea of the energy variation during the process as well as the E_a and the thermal effect Q. For example

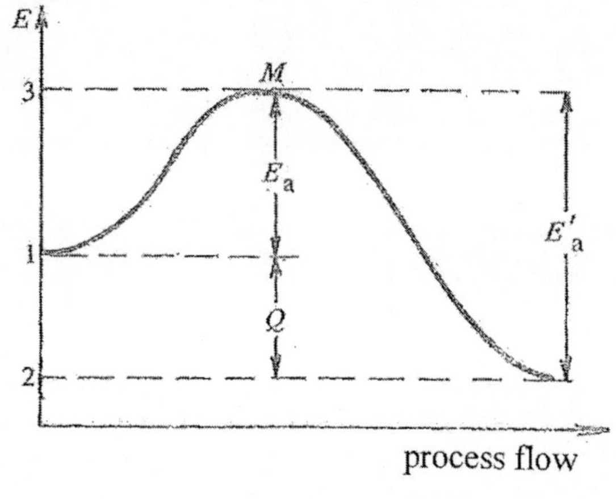

The production of iodohydrogen (hydrogen iodide) is an exothermic reaction because it is accompanied by heat evolution (there is a positive heat effect + Q). The reverse process, the decomposition of HI, is endothermic (- Q).

$$2\,HI \longrightarrow H_2 + I_2 - Q$$

The energy variation in this case is given by the same energy diagram but considered in the opposite direction. The activation energy is E_{a1}.

$$E_{a1} = E_a + Q$$

The heat effect in endothermic reactions is part of the activation energy.

The dependence of velocity on temperature is determined by the velocity constant **k**, which is larger, and hence the velocity is larger, as E_a is smaller and the temperature is higher.

The most common form of energy input in chemical reactions is heat.

The activation of molecules can also occur by irradiation with light (hv), as in many radical chain reactions; by radioactive rays, by electrical discharge, etc.

The rate of chemical reactions also depends on: the homogeneity of the system, particle size distribution, diffusion, agitation, type of solvent, etc.

The study of the kinetics of chemical reactions and the discovery of all the factors that influence the rate of chemical reactions and chemical equilibrium allows the optimum conditions for chemical processes to be selected.

The kinetics and mechanism of reactions are of great importance in biochemical processes.

9. CATALYSIS

HOMOGENEOUS AND HETEROGENEOUS CATALYSIS. BIOCATALYSTS. GENERAL FEATURES OF CATALYTIC PROCESSES. TYPES OF CATALYSTS - POSITIVE AND NEGATIVE. HOMOGENEOUS AND HETEROGENEOUS CATALYSIS. AUTOCATALYSIS. NATURE OF CATALYTIC ACTION. IMPORTANCE OF CATALYSTS FOR INDUSTRY.

The phenomenon of catalysis was first described by the Swedish chemist Berzelius (1835). He observed a marked increase in the rate of formation of water from hydrogen and oxygen at ordinary temperature in the presence of spongy platinum.

$$2\,H_2 + O_2 \xrightarrow{Pt} 2\,H_2O$$

$$\begin{array}{lllll|c|c|c} H_2 & \longrightarrow & 2H^0 - 2 \times 1\bar{e} & \longrightarrow & 2H^+ & 2 & & 2 \\ & & & & & & 4 & \\ O_2 & \longrightarrow & 2O^0 + 2 \times 2\bar{e} & \longrightarrow & 2O^{-2} & 4 & & 1 \end{array}$$

In the absence of platinum, the reaction proceeds immeasurably slowly at room temperature.

The decomposition of bertolite salt (potassium chlorate) ($KClO_3$) with the evolution of oxygen takes place at high temperature, after fusion. Under these conditions the reaction rate is small. If a small amount of pyrolusite (manganese dioxide) (MnO_2) is added to $KClO_3$, a violent evolution of oxygen starts, i.e., the reaction rate increases many times. After the decomposition of $KClO_3$, it is found that the manganese dioxide has remained unchanged. Manganese dioxide is a catalyst.

$$4\,KClO_3 \xrightarrow{MnO_2} 4\,KCl + 6O_2$$

$$\begin{array}{lllll|c|c|c} & & Cl^{+5} + 6\bar{e} & \longrightarrow & Cl^- & 6 & & 2 \\ & & & & & & 12 & \\ 2O^{-2} - 2 \times 2\bar{e} & \longrightarrow & 2O^0 & \longrightarrow & O_2 & 4 & & 3 \end{array}$$

Catalysts are substances that change the rate of a chemical process, remaining quantitatively and qualitatively unchanged at the end of the process. The phenomenon is called catalysis. Catalysts are not written in the overall equation of the reaction, but are denoted above the direction sign of the process connecting the starting substances and the products of the reaction.

Catalysts not only increase the rate of chemical reactions, there are also those that reduce it. Catalysts are positive when they speed up the process and negative (inhibitors) when they slow it down.

Biocatalysts

In plant and animal organisms, a large number of biochemical processes occur at relatively low temperatures - usually below 37° C. This is only possible under the action of biocatalysts, so-called enzymes. Enzymes are protein substances. Most often they are made up of a protein component, the apoenzyme, and a non-protein component, the cofactor. Although the catalytic action of enzymes is attributed to the cofactor, it is unable to exhibit activity on its own. The participation of the apoenzyme is also necessary. This means that the apoenzyme plays the role of an activator.

Cofactors are most often metal ions, phosphate residues, etc. They often also contain some vitamins. Enzymes are remarkable for their extreme specificity of action and their sensitivity to catalytic poisons.

General features of catalytic processes

The following general features were found:

1. Catalysts change the rate of these reactions, which **are possible** without a catalyst.

2. Catalysts change the **rate of the forward and reverse reaction equally**. They do not affect the chemical equilibrium and the value of the equilibrium constant. The presence of a catalyst in the reaction system does not affect the thermal effect of the reaction.

By altering the rate of processes in a reversible reaction, the catalyst alters the time to reach chemical equilibrium in the system. For example, if the catalyst increases the rate of the forward and reverse reactions, the time to reach equilibrium decreases.

3. Catalysts have a **specific action. Universal catalysts do not exist**. A given chemical reaction (or type of reaction) is catalyzed only by a particular catalyst (or group of catalysts). When several parallel reactions are possible in a given reaction mixture, producing different products, the use of a catalyst changes the rate of only one of the possible reactions, i.e., the catalyst has a specific action. For example, methane and water or methanol can be produced from water gas (CO and H_2).

$$CO + H_2 \xrightarrow{Ni} CH_4 + H_2O$$

$$CO + H_2 \xrightarrow{ZnO + Cr_2O_3} CH_3OH$$

The two processes run in parallel. With nickel catalyst, only the reaction to form methane and water is accelerated. Using a mixed zinc oxide and dichromium trioxide catalyst increases the rate of the reaction to form methanol.

4. **The catalyst is recovered** in the course of the chemical reaction. Small amounts of the catalyst change the rate of interaction of large amounts of the reactants.

Types of catalysts

According to the reaction rate variation, **catalysis and catalysts are positive and negative**.

Positive catalysts

In positive catalysis, the catalyst increases the rate of reactions and decreases the time to reach equilibrium in the system in reversible reactions. For example,

$$2\ H_2O_2 \xrightarrow{MnO_2} 2\ H_2O + O_2$$

in the reaction manganese dioxide is a positive catalyst. It increases the decomposition rate of hydrogen peroxide.

Negative catalysts

When the reaction rate decreases, the catalysts are negative. They are called inhibitors.

$$2\ H_2O_2 \xrightarrow{\text{Barbituric acid}} 2\ H_2O + O_2$$

The rate of reaction decreases in the presence of small amounts of phosphoric or barbituric acids. The addition of small amounts of substances which do not exhibit catalytic action can significantly alter the activity of the catalyst.

When these substances cause an increase in its activity, they are called activators or **promoters**. For example, the iron catalyst for ammonia synthesis contains dipotassium oxide and aluminium trioxide as promoters.

When substances lower the activity of the catalyst, they are called **catalytic poisons (inhibitors)**. In the synthesis of ammonia, the presence of traces of oxygen, sulfur or arsenic compounds in the nitrogen-hydrogen mixture poisons the catalyst.

Types of catalysis

Depending on the type of system in which the catalytic process takes place, catalysis is homogeneous and heterogeneous.

Homogeneous catalysis

If a dilute solution of ferric trichloride is added to a hydrogen peroxide solution, oxygen evolution is observed. Catalytic decomposition of the peroxide takes place under the action of Fe^{+3}. The reaction mixture and the catalyst are in a homogeneous system. The process in which the reaction mixture and catalyst form a homogeneous system is called homogeneous-catalytic (**homogeneous catalysis**). In homogeneous catalysis, the reactants and the catalyst are in the same phase. The rate of homogeneous-catalytic processes is proportional to the concentration of the catalyst.

Heterogeneous catalysis

The process in which the reaction mixture and catalyst form a heterogeneous system is called heterogeneous-catalytic (**heterogeneous catalysis**). In order to provide as large a surface area as possible in heterogeneous-catalytic processes, the catalyst is often used in the form of networks or applied in a thin layer on inert supports with a highly developed surface. The most commonly used catalysts in heterogeneous catalysis are metals - platinum, palladium, rhenium, nickel, silver, copper, etc., or metal oxides - vanadium pentoxide, ferric trioxide, nickel oxide, etc. The rate of heterogeneous-catalytic processes is proportional to the surface area of the catalyst.

Autocatalysis

Another type of catalysis is **autocatalysis**. In some reactions, the catalyst is one of the products of the reaction. Such processes are called autocatalytic and the catalyst is called the autocatalyst. For example, the reaction of copper with nitric acid is an autocatalytic process:

$$Cu + 4\,HNO_3 \xrightarrow{NO_2} Cu(NO_3)_2 + 2\,H_2O + 2\,NO_2$$

The autocatalyst is nitrogen dioxide.

A characteristic feature of autocatalytic reactions is the passage of the reaction rate through a maximum. At the beginning of the process, the rate of the autocatalytic reaction is small. Over time it increases as the catalyst concentration increases and reaches a maximum at an optimum ratio between the concentration of autocatalyst and reactants. After this point, the rate decreases as the concentration of reactants decreases.

Nature of catalytic action

There is no single theory to explain the mechanism of catalytic action. It has been shown that the catalyst changes the value of the rate constant of the process and thus affects the rate of reaction.

The interaction of substances A and B (with or without catalyst) gives the reaction product C:

$$A + B \longrightarrow C$$

According to the law of the action of masses, the speed of the process is:

$$v = k.c(A).c(B)$$

The speed can be increased by introducing a catalyst. This is due to the increase in the rate constant **k**, since the catalyst does not affect the concentration of the substances. The effect of the catalyst on the value of the rate constant **k is to** change the activation energy of the reaction.

According to the so-called intermediate theory, the catalyst is involved in intermediate reactions with the starting substances. Intermediate compounds are produced which decompose at the end of the process, the catalyst is released and regenerated and can participate again in further interaction. The reaction between the vesicles A and B and the catalyst K proceeds as follows:

$$A + K \longrightarrow AK$$

$$AK + B \longrightarrow C + K$$

This is a two-stage process and its speed is determined by the speed of the slower stage. When a positive catalyst is used, this slowest stage is significantly faster than the non-catalyzed process. An idea of the energy changes of the system during the process is given by the curve:

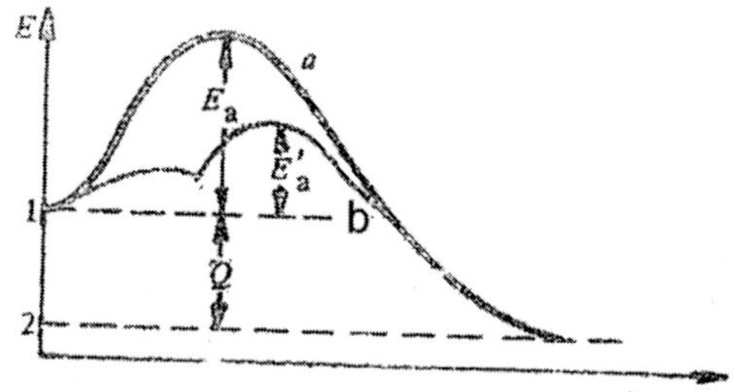

The energy of the system is plotted on the ordinate axis and the process is plotted on the abscissa. The transition from state **1** to state **2** (curve **a**) corresponds to the uncatalysed reaction. The height of the barrier expresses the activation energy of the process, E_a. In the presence of a catalyst, the reaction passes through intermediates (curve **b**) and the barrier splits into two lower barriers. These correspond to the two intermediate processes. The height of the higher of the two barriers expresses the activation energy of the catalyzed process E_a'. From the energy curve it can be estimated that the catalyst changes the activation energy but not the thermal effect of the process. The catalyst reduces the activation energy not only of the straight reaction (the transition from state **1** to state **2**) but also of the reverse reaction (from state **2** to state **1**). The theory of intermediates is generally accepted. In a number of cases, intermediates between the reactants and the catalyst have been isolated and demonstrated. For example, the oxidation of sulfur dioxide to sulfur trioxide, which is catalyzed by nitrogen oxides, can be expressed as follows:

$$NO + \frac{1}{2}O_2 \longrightarrow NO_2$$

$$SO_2 + NO_2 \longrightarrow SO_3 + NO$$

which illustrates the general scheme of catalyst action from the point of view of the theory of intermediate compounds. Here the catalyst is nitric oxide and the intermediate (isolated) compound is nitrogen dioxide.

In a number of cases, most often in heterogeneous catalysis, the intermediates do not have the character of definite chemical compounds but are surface adsorption complexes. Adsorption plays a major role in the activation of the reactants in these cases. The state of the adsorbed molecules is different from that of the molecules in the gas phase. Under the action of the adsorbent catalyst, the bonds between the building blocks in the adsorbed gas molecules are significantly weakened. This leads to a reduction in the activation energy of the reaction, since it is precisely the energy required to reduce the strength of the chemical bonds in the molecules. By way of adsorption, the catalyst reduces the strength of the bonds and prepares the molecules for further interaction. The energy curve of this process is as follows:

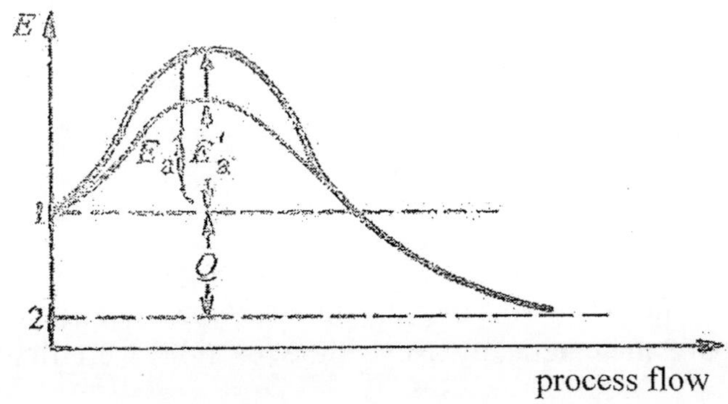

The energy barrier is not split into two lower ones, but is deformed altogether - it is lowered. The activating energy decreases.

A catalyst accelerates chemical reactions by lowering their activation energy, either by altering the reaction pathway or by activating the molecules of the interacting substances and making them more reactive.

The action of negative catalysts is usually attributed to the poisoning of an active positive catalyst.

Importance of catalysts for industry

Catalysts are a powerful means of influencing the rate of chemical processes. Most technological processes in the chemical industry are carried out with their participation. The production of ammonia, sulfuric and nitric acids, various nitrogenous fertilizers, etc., are major industries involving the use of catalysts.

Under the action of catalysts, carbon monoxide combines with hydrogen to produce methanol. The oxidation of methanol to methanal (formaldehyde) is successfully carried out over silver, copper and other catalysts. Formaldehyde is a major feedstock in organic synthesis with wide applications in the preparation of some plastics.

Synthetic motor fuels are produced from the same starting materials, carbon monoxide and hydrogen, but using different catalysts.

Catalysis is widely applied in the petroleum refining industry, in the hydrogenation of fats, in the production of high-molecular compounds, in the **automotive** industry in order to protect the environment from harmful gases.

10. Chemical Equilibrium

Reversible and irreversible processes. Features of chemical equilibrium. Equilibrium constant for homogeneous and heterogeneous systems. Le Chatelier-Brown principle. Equilibrium constant. Factors affecting chemical equilibrium (concentration, temperature and pressure).

Reversible and irreversible processes Irreversible processes

A number of chemical processes proceed in such a way that the reactants are completely converted into reaction products. In such a case the **reaction is said to proceed to completion**,i.e., it cannot **proceed in the opposite direction**. For example, on heating, the bertolite salt (potassium chlorate - $KClO_3$) is converted to potassium chloride and oxygen is evolved.

$$2\,KClO_3 \longrightarrow 2\,KCl + 3\,O_2$$

$$\begin{array}{lll|c|c|c} & Cl^{+5} + 6\bar{e} \longrightarrow & Cl^{-} & 6 & & 2 \\ & & & & 12 & \\ 2\,O^{-2} - 2\times 2\bar{e} \longrightarrow & 2\,O^{0} \longrightarrow & O_2 & 4 & & 3 \end{array}$$

The reverse reaction - the production of potassium chlorate from potassium chloride and oxygen is impossible. Historically, the practically irreversible reactions were first discovered and studied.

Such reactions are the **precipitation, neutralization and gas evolution reactions**:

$$AgNO_3 + NaCl \longrightarrow AgCl\downarrow + NaNO_3$$

$$\underbrace{Ag^{+} + NO_3^{-}} + \underbrace{Na^{+} + Cl^{-}} \longrightarrow AgCl\downarrow + \underbrace{Na^{+} + NO_3^{-}}$$

$$Ag^{+} + Cl^{-} \longrightarrow AgCl\downarrow$$

$$NaOH + HNO_3 \longrightarrow NaNO_3 + H_2O$$

$$\underbrace{Na^{+} + OH^{-}} + \underbrace{H^{+} + NO_3^{-}} \longrightarrow \underbrace{Na^{+} + NO_3^{-}} + H_2O$$

$$H^{+} + OH^{-} \longrightarrow H_2O$$

$$Na_2CO_3 + 2\,HCl \longrightarrow 2\,NaCl + H_2CO_3$$

$$H_2CO_3 \rightleftharpoons H_2O + CO_2\uparrow$$

$$\underbrace{2\,Na^+ + CO_3^{-2}} + \underbrace{2\,H^+ + 2\,Cl^-} \longrightarrow \underbrace{2\,Na^+ + 2\,Cl^-} + \begin{array}{c} H_2CO_3 \\ \updownarrows \\ H_2O + CO_2\uparrow \end{array}$$

$$2\,H^+ + CO_3^{-2} \longrightarrow \begin{array}{c} H_2CO_3 \\ \updownarrows \\ H_2O + CO_2\uparrow \end{array}$$

Reversible processes

Many processes **are also going in the opposite direction**. Chemical processes which, depending on the conditions, can run in both **directions are reversible**.

$$H_2 + I_2 \rightleftharpoons 2\,HI$$

$$\begin{array}{llll|c|c|c} H_2 & \longrightarrow 2\,H^0 - 2 \times 1e^- & \longrightarrow & 2\,H^+ & 2 & \multirow{2}{*}{2} & 1 \\ I_2 & \longrightarrow 2\,I^0 + 2 \times 1e^- & \longrightarrow & 2\,I^- & 2 & & 1 \end{array}$$

$$N_2 + 3\,H_2 \rightleftharpoons 2\,NH_3$$

$$\begin{array}{llll|c|c|c} H_2 & \longrightarrow 2\,H^0 - 2 \times 1e^- & \longrightarrow & 2\,H^+ & 2 & \multirow{2}{*}{6} & 3 \\ N_2 & \longrightarrow 2\,N^0 + 2 \times 3e^- & \longrightarrow & 2\,N^{-3} & 6 & & 1 \end{array}$$

$$CH_3COOH + C_2H_5OH \rightleftharpoons CH_3COOC_2H_5 + H_2O$$

The sign $\rightleftharpoons$ is used to indicate reversible reactions. There are no absolutely irreversible reactions.

Under the right conditions, a number of processes known as irreversible become reversible. For example

$$2\,NO + O_2 \longrightarrow 2\,NO_2$$

$$\begin{array}{llll|c|c|c} & N^{+2} - 2\,e^- & \longrightarrow & N^{+4} & 2 & \multirow{2}{*}{4} & 2 \\ O_2 & \longrightarrow 2\,O^0 + 2 \times 2\,e^- & \longrightarrow & 2\,O^{-2} & 4 & & 1 \end{array}$$

takes place completely at room temperature, and at t° > 670 ° C the opposite process takes place. In a certain temperature range, a chemical equilibrium is established where the system is a homogeneous mixture of nitric oxide, nitrogen dioxide and oxygen.

$$2\,NO + O_2 \rightleftharpoons 2\,NO_2$$

Features of chemical equilibrium

In reversible reactions, chemical equilibrium can be established.

Chemical equilibrium is a state in which the concentrations of all the substances involved in a reaction under given conditions remain constant for a long time.

The chemical equilibrium is dynamic. The concentrations of all substances in the reaction mixture under given conditions are kept constant, but this does not mean that no process is taking place. On the contrary, both the forward and reverse reactions take place continuously. The concentrations are kept constant because these two opposite processes proceed at the same rates.

Chemical equilibrium depends on the conditions under which the system is found, but **not on the path** by which it is reached. For example, in an equilibrium system

$$H_2 + I_2 \rightleftharpoons 2\,HI$$

at a given temperature, the ratio between the concentrations of iodine, hydrogen and iodohydrogen (hydrogen iodide) is constant, i.e., the equilibrium constant is the same.

The chemical equilibrium is mobile. When conditions in the system change, the equilibrium is disturbed, a process in one of the two directions takes place, and a new equilibrium is reached, a new ratio of concentrations characteristic of the new conditions. The principle of mobile equilibrium, known as the Le Chatelier-Brown principle, is used to qualitatively assess the influence that external conditions have on chemical equilibrium. It states:

If an external influence is exerted on an equilibrium system, changes occur in the system and the equilibrium shifts in such a direction that the effect of the external influence is reduced. Or processes take place in the system which partially compensate for the impact.

Equilibrium constant

The equilibrium constant serves as a quantitative characteristic of the equilibrium.

Consider the interaction between hydrogen and iodine

$$H_2 + I_2 \rightleftharpoons 2\,HI$$

The rate of this interaction is greater at the beginning of the process, as this is when the concentration of reactants is greatest. According to the law of mass action

$$v_1 = k_1.c(H_2).c(I_2)$$
,

where k_1 is the rate constant of the straight reaction and **c** is the concentration of the reactants. Over time, the rate v_1 decreases due to decreasing hydrogen and iodine concentrations.

As the process progresses, the concentration of iodine-hydrogen increases and hence the rate v_2 of the reverse reaction.

$$v_2 = k_2.c^2(HI)$$

At some point the two speeds equalise

$$v_1 = v_2$$

This is the point at which chemical equilibrium is established. The rates in the equation are replaced by their equal magnitudes

$$k_1.c(H_2).c(I_2) = k_2.c^2(HI),$$

and after conversion it is obtained

$$K = \frac{k_1}{k_2} = \frac{c^2(HI)}{c(H_2).c(I_2)}$$

The ratio between the two rate constants is also a constant at a given temperature and is called the **equilibrium constant**.

The equilibrium constant is a fundamental characteristic of chemical equilibrium. It depends only on the temperature and nature of the substances involved in the equilibrium, but does not depend on their concentrations, on the pressure, nor on the presence of catalysts. This is because the equilibrium constant is derived from the ratio of rate constants, which also do not depend on concentrations but only on the temperature and nature of the substances.

The variation of the straight and back reaction rates with time is graphically represented as follows:

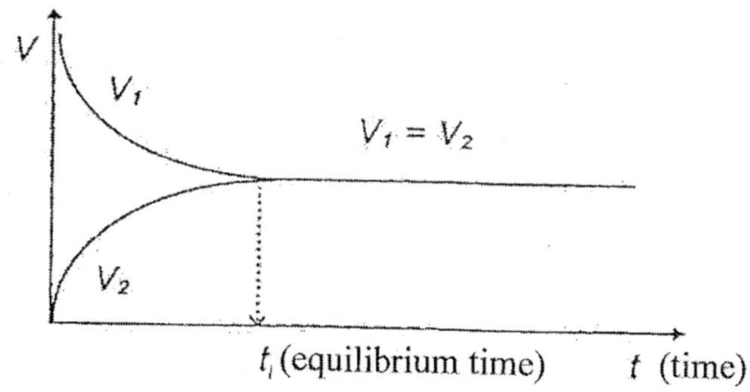

V_1 decreases, V_2 increases and at $V_1 = V_2$ chemical equilibrium occurs. The chemical equation of the process is used to write the expression for the equilibrium constant.

In **homogeneous systems,** the equilibrium constant is expressed by the concentrations of all substances involved in the equilibrium process. The example of the equilibrium system for H_2, I_2 and HI discussed above refers to homogeneous systems.

When equilibrium is established in a **heterogeneous system**, for example, between gases and solids, the concentration of the solids has no specific physical meaning and is not involved in the expression for the equilibrium constant. For the process:

$$C + CO_2 \rightleftharpoons 2\ CO$$

the equilibrium constant is given by the expression

$$K = \frac{c^2(CO)}{c(CO_2)}$$

Factors that affect the chemical equilibrium

Effect of concentration on equilibrium. This relationship can be experimentally established by examining the interaction between ferric trichloride and potassium thiocyanide (rhodanide):

$$FeCl_3 + 3\ KCNS \rightleftharpoons Fe(CNS)_3 + 3\ KCl$$

The equilibrium constant of the process is

$$K = \frac{c[Fe(CNS)_3] \cdot c^3(KCl)}{c(FeCl_3) \cdot c^3(KCNS)}$$

The equilibrium constant maintains its constant value at a given temperature regardless of changes in individual concentrations. If the concentration of ferric chloride increases, this will cause the denominator in the equilibrium constant expression to increase, hence the value of this constant to decrease. Since this is not acceptable, a process will predominantly take place in which the numerator will also increase. This means that the straight reaction will take place. The same will happen if the concentration of potassium thiocyanide is increased. If the concentration of potassium chloride is increased, the opposite process will take place. Therefore, under the given conditions, as the concentration of either reactant increases, the equilibrium shifts in the direction of the straight reaction. The equilibrium shifts in the same direction when some of the reaction products are removed. As the concentration of the reaction products increases, the equilibrium shifts in the direction of the reverse reaction. When the concentration of potassium rhodanide is increased, the reaction mixture turns an intense red (this is the color of ferric trithiocyanide), and when the concentration of potassium chloride is increased, the reaction mixture loses the intense red color.

Effect of pressure on equilibrium. This influence is observed in processes involving gases, and then when a change in the volume of the system occurs in the course of the process. Most often, homogeneous gas reactions are at the same time processes in which there is a change in the number of moles of reaction products with respect to the number of moles of starting substances. As the pressure on the reaction gas mixture increases, a decrease in its volume occurs. This leads to an increase in the concentration of the gas molecules. This effect must be reduced. This means that a process will take place in which the concentration of the gas molecules will decrease. Such is the process associated with a reduction in the number of molecules of substances in the system under consideration. This means that **an increase in pressure above an equilibrium gas system causes a shift of equilibrium in such a direction that a decrease in the number of moles in the system takes place.** Consider the process

$$N_2 + 3\,H_2 \rightleftharpoons 2\,NH_3$$

The mixture of N_2, H_2 and NH_3 is in equilibrium at a certain temperature. Without changing the temperature, the pressure increases twice. From the Boyle-Marriott law (p.V = const), it follows that the volume of the system will decrease by a factor of two. The number of gas molecules have not changed, but they are now in twice the volume. The molar concentration has doubled. This change occurs with the concentration of both the reactants and the products of the reaction. In doing so, the ratio between the rates of the forward and reverse reactions will change – equilibrium is broken. Until the pressure increases, the equilibrium concentrations are $c(H_2)$, $c(N_2)$ and $c(NH_3)$, and the rates of the forward and reverse reactions are given by the equations:

$$v_1 = k_1 \cdot c(N_2) \cdot c^3(H_2)\ ; \qquad v_2 = k_2 \cdot c^2(NH_3)$$

After increasing the pressure twice, the gas concentrations will be

$$2c(H_2),\ 2c(N_2) \text{ and } 2c(NH_3).$$

In this case, the new velocities v_1' and v_2' will be determined from the equations:

$$v_1' = k_1.2c(N_2)\cdot 2^3 c^3(H_2) \qquad v_2' = k_2.2^2.c^2(NH_3)$$

At equilibrium $v_1' = v_2'$ and after substitution we obtain

$$v_1' = k_1.2c(N_2)\cdot 2^3 c^3(H_2) = 16\,k_1.c(N_2).c^3(H_2) = 16\,v_1$$

$$v_2' = k_2.2^2 c^2(NH_3) = 4k_2\,c^2(NH_3) = 4\,v_2$$

Or for the equilibrium constant, we get

$$K^1 = \frac{2^2 c^2(NH_3)}{2c(N_2)\cdot 2^3 c^3(H_2)} = \frac{4c^2(NH_3)}{16\,c(N_2).c^3(H_2)}$$

Thus, as a result of an increase in pressure by a factor of two, the rate of the straight reaction increases by a factor of 16, and that of the reverse - by a factor of 4. The equilibrium in the system is disturbed: the denominator of the fraction has increased. A process must take place preferentially where the numerator increases (or the denominator decreases) in order for the constant to retain its value. The equilibrium will shift to the right to produce new amounts of ammonia.

A similar calculation for equilibrium processes that proceed without volume change shows that the pressure change has no effect. For example, in the interaction of iodine with hydrogen

$$H_2 + I_2 \rightleftharpoons 2\,HI$$

the equilibrium constant of the process is:

$$K = \frac{c^2(HI)}{c(H_2)\cdot c(I_2)}$$

If the pressure on the equilibrium system is doubled, the equilibrium concentrations will also double. Then:

$$v_1 = k_1.2c(H_2)\cdot 2\,c(I_2) = 4k_1.c(H_2).c(I_2)$$

$$v_2 = k_2.2c^2(HI) = 4\,k_2.c^2(HI)$$

For the equilibrium constant:

$$K = \frac{2^2 c^2(HI)}{2\,c(H_2).2\,c(I_2)} = \frac{4\,c^2(HI)}{4\,c(H_2).c(I_2)}$$

i.e., the speed of the straight and back reactions will increase equally and the equilibrium will not be disturbed.

Therefore, equilibrium processes that proceed without volume change are not affected by pressure changes.

Effect of temperature on equilibrium. This influence is observed in processes accompanied by a heat effect. The temperature increase is usually achieved by introducing a certain amount of heat into the system.

The direction in which the equilibrium of the system will shift as its temperature changes can be predicted based on the Le Chatelier-Brown principle. According to this principle, an increase in the temperature of the equilibrium system should shift the equilibrium in the direction of the reaction that will lower the temperature of the system, i.e., in the direction of the endothermic reaction that proceeds with heat absorption. The opposite is the case when the equilibrium system temperature is lowered. Then the reaction will take place which will raise the temperature of the system. Therefore, when the temperature is lowered, the equilibrium will shift in the direction of the exothermic reaction because it proceeds with heat release.

$$H_2 + I_2 \rightleftharpoons 2\,HI - Q$$

When hydrogen and iodine combine to form iodohydrogen (hydrogen iodide), heat is absorbed, i.e., the straight reaction is endothermic, and the decomposition of iodohydrogen to hydrogen and iodine occurs with the release of heat, i.e., the reverse reaction is exothermic.

The introduction of heat into this equilibrium system will facilitate the endothermic reaction (i.e., the production of iodohydrogen) because it, by absorbing heat, counteracts the external change.

When the temperature of the system is lowered, heat is removed from the system. This will facilitate the exothermic reaction to take place because by releasing heat this reaction counteracts the external influence (change).

Sample problems and solutions

Task 1.

When carbon interacts with carbon dioxide in a closed vessel, a reversible process takes place which is endothermic in the straight direction. Under certain conditions, chemical equilibrium is established. In which direction will the equilibrium be shifted: (a) by increasing the temperature; (b) by increasing the pressure; (c) by doubling the amount of carbon; (d) by introducing a catalyst; (e) by introducing sodium chloride; (f) by introducing disodium oxide?

Exemplary solution

Subject to

$$C + CO_2 \rightleftharpoons 2\,CO - Q$$

When equilibrium is established in a **heterogeneous system**, the concentration of solids has no specific physical meaning and does not participate in the expression for the equilibrium constant. For the above process the equilibrium constant is given by the expression:

$$K = \frac{c^2\,(CO)}{c\,(CO_2)}$$

a) Based on the Le Chatelier-Brown principle, raising the temperature of the equilibrium system should shift the equilibrium in the direction of the reaction that will lower the temperature of the system, i.e., in the direction of the endothermic reaction occurring with heat absorption. So as the temperature rises, the equilibrium shifts in the direction of the straight reaction, i.e., towards the production of carbon monoxide.

b) Increasing the pressure above an equilibrium gas system causes a shift of the equilibrium in such a direction that a decrease in the number of moles in the system occurs. In this case, the equilibrium shifts in the direction of the opposite reaction, towards a decrease in the number of moles in the system (formation of carbon dioxide).

c) A twofold increase in the amount of carbon does not affect the chemical equilibrium. In a **heterogeneous system,** the concentration of solids has no concrete physical meaning and does not participate in the expression for the equilibrium constant.

(d) The introduction of a catalyst into the system does not alter the equilibrium state. The catalyst affects both the forward and reverse reactions equally.

(e) When sodium chloride is introduced, the state of chemical equilibrium remains unchanged. The compound does not interact with the components of the equilibrium system.

(f) Disodium oxide affects the chemical equilibrium as an interaction takes place between an acid and a basic oxide in Eq:

$$Na_2O + CO_2 \longrightarrow Na_2CO_3$$

The binding of carbon dioxide in disodium carbonate lowers the concentration of acid oxide in the equilibrium system, therefore the equilibrium shifts to the left.

11. SOLUTIONS

CLASSIFICATION OF DISPERSED SYSTEMS. NATURE OF DISSOLUTION. TYPES OF SOLUTIONS - SATURATED, UNSATURATED AND SUPERSATURATED. SOLUBILITY OF SUBSTANCES. DEPENDENCE OF SOLUBILITY OF SUBSTANCES ON NATURE OF SOLVENT AND SOLUTE, TEMPERATURE AND PRESSURE. CONCENTRATION OF SOLUTIONS. PROPERTIES OF SOLUTIONS - DIFFUSION, OSMOSIS AND OSMOTIC PRESSURE, VAPOR PRESSURE, BOILING AND FREEZING OF SOLUTIONS.

Systems of two or more substances in which the particles of one substance are uniformly dispersed (distributed) among the particles of the other are called **dispersed systems**. The substance which is dispersed is denoted as the **dispersed phase** and the substance in which the phase is distributed as the **dispersed medium**. This separation is largely arbitrary, since each of the substances can be considered as both a phase and a medium. It is customary to consider as the dispersed medium that substance which has the same aggregate state with the dispersed system. If the dispersion medium and the dispersed phase are in the same aggregate state, the component which is in the greater amount is defined as the dispersion medium.

The dispersion systems are homogeneous and heterogeneous. If the dispersion system is a single-phase system, then it is a **homogeneous** dispersion system. If two or more phases are present, even though the substances may be in the same aggregate state, the system is called **heterogeneous**.

The dispersed phase and the dispersion medium can be in all three aggregate states, which determines the wide variety of dispersion systems.

Classification of dispersed systems

According to the aggregate state of the dispersed phase and the dispersion medium, dispersion systems are of nine types:

Dispersion medium	Dispersed phase	Dispersion system – examples
Gas	Gas	Air, gas mixtures
Gas	Liquid	Aerosols, fog
Gas	Solid	Smoke
Liquid	Gas	Foam, CO_2 in carbonated beverages
Liquid	Liquid	Solutions of miscible liquids; emulsions - mixtures of mutually insoluble liquids
Liquid	Solid	Suspensions - liquid with dispersed insoluble solid
Solid	Gas	Pumice stone, foam board
Solid	Liquid	Crystal hydrates: $CaCl_2 .2H_2O$, $CuSO_4 .5H_2O$; $FeSO_4 .7H_2O$; $Na_2 SO_4 .10H_2O$.
Solid	Solid	Coloured glasses, minerals, alloys

The main feature that determines the properties of dispersed systems is the particle size of the dispersed substance. The classification of dispersed systems according to this feature is as follows:

Dispersion systems	Particle size	Characteristic features	
-	-	Uniformity	Stability
Coarsely dispersible	Above 100 nm	Heterogeneous	Unstable
Colloidally dispersible	1 - 100 nm	Microheterogeneous	More stable
Molecularly (ionically) dispersed	> 1 nm	Homogeneous	Stable

True solutions are homogeneous dispersed systems of two or more substances, the amounts of which may vary within certain limits.

The component of a solution that is in the larger amount and its aggregate state matches that of the solution is called the **solvent**, and the remaining components are **solutes**. The most commonly used solvent is **water**, but other substances are also used as solvents, for example, hydrocarbons - benzene, toluene, alkanes, etc., alcohols - methanol, ethanol, propanol, etc., ketones - acetone, etc., esters - ethyl acetate, etc.

Essence of dissolution

Dissolution is a complex physicochemical process. To dissolve a substance, it is necessary to overcome both the forces of attraction acting between its building blocks and the forces acting between the solvent molecules.

When a piece of a substance is placed in a liquid in which it can dissolve, it breaks down to ions or molecules of the low molecular compound. The decomposition of the solute and the distribution of its particles throughout the volume is accompanied by a negative heat effect ($- Q_1$). Simultaneously with the decomposition, an interaction takes place between the solvent particles and the solute. Perishable compounds of indeterminate quantitative composition called **solvates** are formed. The process is called **solvation**. When the solvent is water, the compounds are called **hydrates** and the process is called **hydration**. This process (solvation or hydration) occurs with a **positive heat effect** ($+ Q_2$).

The total heat effect of the dissolution of a substance is the algebraic sum of the heat effects of the individual processes

$$Q = (- Q_1) + (+ Q_2)$$

or

$$Q = Q_2 - Q_1$$

The thermal effect of dissolution can be both positive and negative, i.e., the dissolution process of a substance can be **exothermic or endothermic.**

The thermal effect of dissolving some solids such as KOH, NaOH, $Ca(OH)_2$, $AlCl_3$, etc. is **positive**. This indicates that **less heat is absorbed** in the decomposition of their crystal lattices than is released **in the hydration of their particles.**

The thermal dissolution effect can also be **negative**. For example, the dissolution of KNO_3, $CuSO_4 .5H_2O$, NH_4Cl, etc. is accompanied by a negative heat effect. This indicates that the crystal lattices of these compounds are sufficiently strong and their decomposition is associated with the **absorption of a large amount of heat, which cannot be compensated by the solvation energy.**

In some cases, the dissolution process proceeds **without apparent energy change** and the thermal effect is zero. An example is the dissolution of sodium chloride in water.

The reason for the different thermal effects of dissolution lies in the nature of the solvent, the solute and their interaction.

The solvates (hydrates) formed during dissolution are usually of variable composition. For example, hydrates of the following composition have been isolated from a sulfuric acid solution: H_2SO_4, H_2O, $H_2SO_4 .2H_2O$, $H_2SO_4 .4H_2O$, etc. These groupings are **perishable** and their composition depends of the concentration and temperature of the solution.

In some cases, **permanent hydrates** are also formed, which persist even after the solute has crystallized, e.g., $CuSO_4 .5H_2O$, $FeSO_4 .7H_2O$; $Na_2SO_4 .10H_2O$, etc. Such crystalline substances are called **crystal hydrates**.

In order to dissolve a substance, it is necessary that the bonds between its particles correspond in type and strength to the bonds between the solvent particles. It has been found experimentally that **polar and ionic compounds** dissolve well in **polar solvents** and non-polar compounds dissolve well in **non-polar solvents**, i.e., **like dissolves like.**

Water (a polar compound) is a very good solvent of **acids, bases and salts**, which are mostly ionic compounds, and benzene (a non-polar compound) is a good solvent of a large number of non-polar organic compounds. Taking this fact into account it becomes possible to select the most suitable solvent for a given substance.

The dissolution of any solid takes place as a result of the destruction of its crystal lattice. The particles which have passed into solution, striking the surface of the substance not yet dissolved, may again be retained on its surface, i.e., a process of crystallization takes place. This means that in any solution of a solid, two processes take place: **dissolution and crystallization**. When the rates of the two processes equalise, dynamic equilibrium is established in the solution:

$$\text{undissolved substance} \rightleftharpoons \text{substance in solution}$$

Per unit time, as many particles pass from the undissolved substance into the solution as are separated from the solution and retained on its surface, i.e., a dynamic equilibrium is established between the two processes.

Types of solutions

A solution which at a given temperature has a constant and strictly defined concentration as a result of the dynamic equilibrium established between the undissolved and dissolved matter is called **saturated**.

An unsaturated solution is one in which the amount of solute can be varied within certain limits for a given temperature.

If the solution contains a greater amount of solute than is contained in the corresponding saturated solution at a given temperature, it is called **supersaturated.** Such solutions are very unstable. They contain no undissolved matter, but under external influence, e.g., small crystals, dust, agitation, etc., a large proportion of the solute is released into it and the super saturated solution becomes saturated.

Solubility of substances. Dependence of the solubility of substances on the nature of the solvent and the solute, temperature and pressure

The property of substances to form solutions in suitable solvents is called **solubility.** The solubility of a substance in a given solvent is measured by the concentration of its saturated solution at a given temperature. It is most commonly expressed in grams of solute in 100 g of solvent at 20 ^{o} C.

According to their solubility, substances are divided into very soluble, moderately soluble and slightly soluble in a given solvent. For example, KNO_3, NaCl, KI, NaOH, etc. are very soluble in water; $Ca(OH)_2$ is moderately soluble, and AgI, AgBr, $BaSO_4$, FeS, etc. are slightly soluble in water. There are no absolutely insoluble substances. Obviously, the solubility limits (slightly, medium and very insoluble) are not very strict and one can speak of relative limits. Nevertheless, the solubility of any substance is a strictly quantitative characteristic and a strictly individual property. Dissolution of substances is an **equilibrium process**. The solubility of solids depends on temperature and can vary over a wide range.

The different dependence of solubility on temperature is determined by the thermal effect of dissolution. If the heat effect is positive, dissolution is an exothermic process, and if the heat effect is negative, dissolution is an endothermic process. Since the Le Chatelier-Brown principle can be applied to any equilibrium system, as **the temperature increases the equilibrium shifts in the direction of the endothermic process.** If the production of a saturated solution is an endothermic process, the solubility of a substance increases with increasing temperature. Such is the case in the dissolution of KNO_3, $Pb(NO_3)_2$, NH_4Cl, etc. in water.

Conversely, if the production of a saturated solution is an exothermic process, the solubility of the substance decreases with increasing temperature. Examples of this case are alkaline bases

(NaOH, KOH), alkaline earth bases - $Ca(OH)_2$, some chlorides - $AlCl_3$, etc.

If dissolution is not associated with heat release or absorption, the temperature dependence of solubility is negligible. This is the case for sodium chloride (NaCl).

The solubility of liquid substances in liquid solvents varies. Some liquids have unlimited solubility in their respective solvents, e.g., ethanol, glycerol, sulfuric acid, etc. in water. For some liquids this solubility is limited. For example, diethyl ether, ethyl acetate, benzene, aniline, etc. When equal amounts of one of the above liquids and water are mixed, two layers are obtained. The upper layer is, for example, a saturated solution of water in aniline and the lower layer is a saturated solution of aniline in water.

The solubility of gases in liquids varies over a very wide range and depends on **temperature and pressure**. The dissolution of gases in water is usually an exothermic process and decreases with increasing temperature. At constant temperature, the solubility of a gas in a given liquid is directly proportional to the pressure of the gas above the liquid, provided that **no interaction takes place** between the gas and the liquid,

$$c = k.p, \ T = const$$

where **c** is the concentration of the gas in the saturated solution, p - its partial pressure in the system, k - a constant that characterizes the nature of the components forming the solution. If p = 1 atm, then k = c.

Concentration of solutions

The quantitative composition of the solution is expressed by its concentration. The concentration is denoted by **c**. It gives the ratio between the amount of solute and the amount of solvent.

The concentration of solutions is expressed in several ways:

The percentage concentration (**c**) is expressed in grams of solute in 100 g of solution.

It is calculated by the formula:

$$c = \frac{m_1}{m} \cdot 100$$

where m_1 is the mass of the solute, m - the mass of the solution, expressed in grams.

Molar concentration (c_M) - is expressed by the number of moles of solute (n) in 1 L (or 1000 cm^3) of solution.

To calculate the molar concentration, the formulas:

$$c_m = \frac{n}{V}$$

if the volume is measured in litres and if the volume is measured in cm^3.

$$c_m = \frac{n}{V} \cdot 1000$$

The number of moles (n) is determined by dividing the mass of the solute (m_1) by its molecular mass (M), i.e.

$$n = \frac{m_1}{M}$$

After substituting this expression in the above formulas for with M, we get the following:

$$c_M = \frac{m_1}{M.V}$$

and

$$c_m = \frac{m_1}{M \bullet V} \bullet 1000$$

Molal concentration - expressed as the number of moles of solute in 1000 g of solvent. For example, a one molar solution of sucrose contains 1 mol of the substance dissolved in 1000 g of water. This way of expressing the concentration of solutions is used in elucidating the effect of the solute on the physical properties of the solvent.

Properties of solutions - diffusion, osmosis and osmotic pressure, steam pressure, boiling and freezing of solutions

Each solution has specific properties that are determined by the nature of the solute and the nature of the solvent used. For example, the properties of an aqueous solution of sulfuric acid are determined by the presence of hydrated H^+ and SO^{-2}, etc., in it.

Along with specific properties, solutions have some general properties that depend on the concentration of the solutes. Such properties are vapor pressure, osmotic pressure, boiling and freezing temperatures, diffusion. These are best studied in highly dilute solutions of low- molecular-weight nonelectrolytes, where the solutes are in the form of molecules and are therefore called molecular (or simple) solutions. In these, owing to the great dilution, the specific features due to the properties of the solute are lost, and only those properties are retained which depend on the number of its particles.

The process of equalizing the concentration of a component at all points of a solution without external interference is called **diffusion**. It is caused by the thermal movement of particles in all directions. Initially, the particles move from areas of greater concentration to areas of lesser concentration. Their movement causes the concentrations to equalise. The particles then continue to move chaotically throughout the volume of the solution maintaining their uniform distribution. Diffusion is a spontaneous process. The rate of diffusion is proportional to the temperature and to the difference in concentrations of the two solutions.

If a semi-permeable barrier is placed between two solutions of the same substance of different concentrations, through which only the solvent molecules pass, the concentrations on either side of the semi-permeable membrane are equalized. This is achieved by the solvent molecules moving from the solution of lower concentration to the solution of higher concentration.

The process of unidirectional passage of solvent molecules through a semipermeable barrier between two solutions of different concentrations of the same substance is called **osmosis**. A special instrument called an osmometer is used to measure the osmotic pressure of solutions.

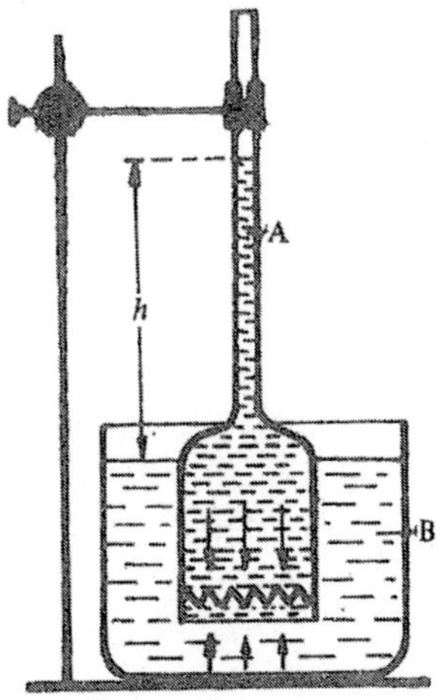

The test solution is placed in an expansion tube which is closed with a semi-permeable septum. The tube is immersed in a beaker of pure solvent so that the levels of the solution in the beaker and of the pure solvent are at the same height. The concentration of water molecules in the beaker is greater than in the osmometer. From the beaker, water passes through the semipermeable membrane into the osmometer and raises the liquid level in the tube. At a certain height (h) of the solution column in the tubing, the rates of water passage from the beaker into the osmometer and vice versa, from the osmometer into the beaker, equilibrate and a dynamic equilibrium is established between the solvent molecules on either side of the semipermeable barrier. The liquid level in the tube ceases to rise.

The pressure of a column of liquid that stops osmosis is called osmotic pressure.

The first studies of osmotic pressure were made by Pfeffer, who found that:

- At constant temperature, the osmotic pressure is proportional to the concentration of the solution;
- At constant solution concentration, the osmotic pressure is proportional to the absolute temperature.

In 1886, Vant Hoff expressed with a general equation the osmotic pressure laws established by Pfeffer

$$\pi = c_M.RT,$$

where c_M is the molar concentration of the solution, R is the gas constant. This equation is analogous to the ideal gas equation of state $pV = nRT$, where n is the number of moles of solute. From this equation the pressure can be determined, i.e.:

$$p = \frac{n}{V}.RT$$

because

$$\frac{n}{V} = c_M$$

then

$$p = c_M.RT$$

i.e., the expression of the osmotic pressure is analogous to the expression of the ideal gas pressure. Hence Want Hoff concludes that the **osmotic pressure** of a solution is equal to the pressure which the solute would exert on the walls of the vessel if it were in a gaseous state and occupied at the same temperature a volume equal to the volume of the solution.

It follows from the Want Hoff equation that solutions of the same molar concentration have the same osmotic pressure at the same temperature.

Vital processes such as food absorption, metabolism, etc. are associated with osmosis.

Any liquid is in equilibrium with its vapor when the number of molecules evaporating per unit time from the surface equals the number of molecules that are retained on its surface. The vapors that are in dynamic equilibrium with a liquid are its saturated vapors, and the pressure they exert over it is **vapor pressure**. As the temperature in a liquid increases, the number of molecules with sufficient kinetic energy to overcome intermolecular interactions and separate from the liquid increases. As a consequence, the vapor pressure above it gradually rises and **when it equals the atmospheric pressure, the liquid begins to boil.**

When a non-volatile substance is dissolved in a liquid, the pressure of the saturated solvent vapor above the solution is always **lower** than the pressure above the pure solvent. This is because part of the surface of the solution is occupied by solvated particles of the non-volatile solute, and the number of evaporating solvent molecules per unit area of solution per unit time is less than the number of evaporating molecules from the same area of pure solvent. The equilibrium state for the solution is therefore established at a lower vapor pressure than for the pure solvent. The decrease in vapor pressure is greater the more concentrated the solution.

All pure solvents have strictly defined and constant boiling and freezing temperatures. The presence of a solute in a liquid affects its vapor pressure, and this changes its **boiling and freezing points**. The lowering of the freezing temperature can be represented by the following diagram:

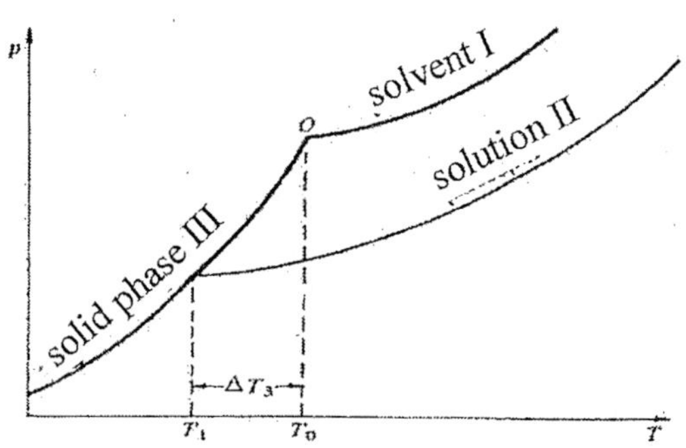

I - vapor pressure curve of the pure solvent; II - solution vapor pressure curve; III - sublimation curve of ice.

Where the vapor pressure curve of the pure solvent intersects the sublimation curve of the ice, there is the v.t. of the pure solvent T_o; where the vapor pressure curve of the solution crosses the sublimation curve of the ice is the freezing point of the solution (T_1).

$$T_o > T_1$$

For the solution, the equilibration of its vapor pressure with the external pressure occurs at a higher temperature (point T_1) than for the pure solvent (point T_o). For solutions of higher concentration ($C_2 > C_1$) the boiling point is even higher.

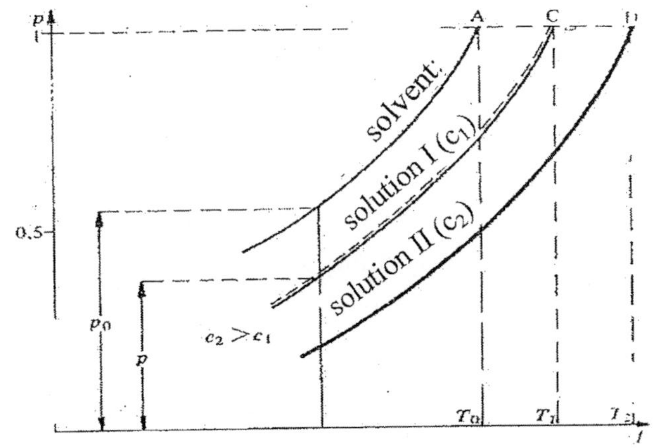

When the vapor pressure above the liquid equals atmospheric pressure, the liquid boils.

The lowering of the freezing point and the raising of the boiling point of a solution compared with the corresponding temperatures of the pure solvent are proportional to the concentration of the solution.

All of the above laws for solutions only occur when the low molecular weight solute is a nonelectrolyte and the solutions are sufficiently dilute.

Sample problems and solutions

Task 1.

Express with chemical equations and explain the transitions:

$$K \longrightarrow KCl \longrightarrow KOH \longrightarrow K_2CO_3 \longrightarrow KNO_3 \longrightarrow O_2 \longrightarrow$$

$$\longrightarrow ZnO \longrightarrow ZnSO_4 \longrightarrow Zn(OH)_2 \longrightarrow Na_2[Zn(OH)_4]$$

The osmotic pressure will be the same or different in solutions of K_2CO_3 and KNO_3 with a concentration of 0.01 mol/L and $\alpha \approx 1$. Determine the oxidant and reductant in the decomposition of KNO_3. Specify the conditions for the transition

$$ZnSO_4 \longrightarrow Zn(OH)_2$$

Exemplary solution

$$2\,K + Cl_2 = 2\,KCl$$

$$K^0 - e^- \longrightarrow K^+ \quad | \; 1 \; | \; 2 \; | \; 2$$

$$Cl_2 \longrightarrow 2Cl^0 + 2 \times 1\,e^- \longrightarrow 2\,Cl^- \quad | \; 2 \; | \; \; | \; 1$$

As a typical metal, potassium interacts very intensively with non-metals. The process is redox. The process of giving up electrons is called oxidation and the process of taking up electrons is called reduction. The substance that gives off electrons is oxidized and is a reductant, and the substance that accepts electrons is reduced and is an oxidizer.

Alkaline bases are produced by electrolysis of an aqueous solution of their salt, usually a chloride. Electrolysis is an oxidation-reduction process that takes place in a solution or melt of an electrolyte while passing a constant electric current. Reduction takes place at the cathode and oxidation at the anode.

$$KCl \longrightarrow K^+ + Cl^-$$

$$H_2O \rightleftharpoons H^+ + OH^-$$

Potassium chloride is a strong electrolyte and in aqueous solution completely dissociates into ions. Water is a weak electrolyte and dissociates reversibly. In an aqueous solution of potassium chloride there are two types of cations and two types of anions. When a direct electric current is passed through the solution, potassium and hydrogen cations are directed to the cathode and chloride and hydroxide anions are directed to the anode. At the cathode, the better oxidizer, i.e., the ion that is further to the right in the Order of Relative Activity of the Metals, is reduced, and at the anode, the better reducer, i.e., the anion that is further to the left in the order of relative activity of the anions, is oxidized:

$$(C^-) \quad 2\,H^+ + 2\,x\,1\,e^- \longrightarrow 2\,H^o \longrightarrow H_2$$

$$(A^+) \quad 2\,Cl^- - 2\,x\,1\,e^- \longrightarrow 2\,Cl^o \longrightarrow Cl_2$$

The potassium hydroxide ions remain in the electrolysis bath.

$$2\,KOH + H_2CO_3 \rightleftharpoons K_2CO_3 + 2\,H_2O$$

$$\underbrace{2\,K^+ + 2\bar{O}H} + \underbrace{2H^+ + CO_3^{-2}} \rightleftharpoons \underbrace{2\,K^+ + CO_3^{-2}} + 2\,H_2O$$

$$H^+ + \bar{O}H \longrightarrow H_2O$$

The process is ion exchange, neutralization.

$$K_2CO_3 + 2\,HNO_3 \longrightarrow 2\,KNO_3 + H_2CO_3$$

$$H_2CO_3 \rightleftharpoons H_2O + CO_2$$

$$\underbrace{2\,K^+ + CO_3^{-2}} + \underbrace{2\,H^+ + 2\,NO_3^-} \longrightarrow \underbrace{2\,K^+ + 2\,NO_3^-} + H_2CO_3$$

$$2H^+ + CO_3^{-2} \longrightarrow H_2CO_3 \rightleftharpoons H_2O + CO_2$$

Nitrates are thermally unstable and decompose on heating. The process is known as the intramolecular redox process:

$$2\,KNO_3 \xrightarrow{t^o} 2\,KNO_2 + O_2$$

$$\begin{array}{rcl|c|c|c} N^{+5} + 2\,e^- & \longrightarrow & N^{+3} & 2 & & 2 \\ & & & & 4 & \\ 2O^{-2} - 2\,x2\,e^- \longrightarrow 2\,O^o & \longrightarrow & O_2 & 4 & & 1 \end{array}$$

When heated, zinc oxidizes.

$$2\,Zn + O_2 \longrightarrow 2\,ZnO$$

$$\begin{array}{lll|c|c|c} & Zn^{0} - 2e^{-} & \longrightarrow Zn^{+2} & 2 & & 2 \\ & & & & 4 & \\ O_2 \longrightarrow & 2\,O^{0} + 2\times 2\bar{e} & \longrightarrow 2O^{-2} & 4 & & 1 \end{array}$$

$$ZnO + H_2SO_4 \longrightarrow ZnSO_4 + H_2O$$

It is an exchange, a neutralization process.

$$ZnSO_4 + 2\,NaOH \longrightarrow Zn(OH)_2\downarrow + Na_2SO_4$$

$$\underbrace{Zn^{+2} + SO_4^{-2}} + \underbrace{2Na^{+} + 2\bar{O}H} \longrightarrow Zn(OH)_2\downarrow + \underbrace{2Na^{+} + SO_4^{-2}}$$

$$Zn^{+2} + 2\bar{O}H \longrightarrow Zn(OH)_2\downarrow$$

Due to the amphoteric nature of zinc dihydroxide, the base should not be added in excess as the dihydroxide will react with the base and the precipitate will dissolve.

$$Zn(OH)_2 + 2\,NaOH \longrightarrow Na_2[Zn(OH)_4]$$

disodium tetrahydroxyzincate

The reaction of amphoteric hydroxides with alkaline bases in aqueous media produces complex compounds.

The osmotic pressure of electrolytes is calculated by the formula $\pi = icRT$, where i is the isotonic coefficient, which indicates how many times the number of particles in the solution of a given electrolyte has increased compared to those originally imported. When dissociation is complete ($\alpha = 1$), the numerical value of i coincides with the number of ions to which the electrolyte dissociates. In the present case

$$i_{KNO_3} \approx 2; \quad i_{K_2CO_3} \approx 3$$

At the same molar concentration (c) and the same temperature (T), the isotonic coefficient is the determinant of the osmotic pressure. In this case

$$\pi_{KNO_3} = 2cRT; \ \pi_{K_2CO_3} = 3cRT.$$

Apparently $\pi_{K_2CO_3} > \pi_{KNO_3}$

Task 2.

Suggest a way of proving aqueous solutions of disodium carbonate, zinc dibromide, dipotassium sulfate and calcium dichloride using an alcohol lamp and suitable reactions for their cationic and anionic moieties. Represent the hydrolysis of the starting salts.

Exemplary solution

Proof of aqueous solutions:

$$Na_2CO_3 + 2\,HCl \longrightarrow 2\,NaCl + H_2CO_3 \rightleftharpoons H_2O + CO_2\uparrow$$

$$\underbrace{2\,Na^+ + CO_3^{-2}} + \underbrace{2\,H^+ + 2\,Cl^-} \longrightarrow \underbrace{2\,Na^+ + 2\,Cl^-} + H_2CO_3 \rightleftharpoons H_2O + CO_2\uparrow$$

$$2\,H^+ + CO_3^{-2} \rightleftharpoons H_2CO_3 \rightleftharpoons H_2O + CO_2\uparrow$$

The carbonate anion is proved by the evolution of carbon dioxide on the addition of hydrochloric acid to the sample of disodium carbonate under study, known in chemistry as the "noisy sample".

The aqueous solution of disodium carbonate colors the flame of the spirit lamp yellow due to the presence of sodium cations.

$$ZnBr_2 + 2\,AgNO_3 \longrightarrow Zn(NO_3)_2 + AgBr\downarrow$$

pale yellow precipitate

$$\underbrace{Zn^{+2} + 2\,Br^-} + \underbrace{2\,Ag^+ + 2\,NO_3^-} \longrightarrow \underbrace{Zn^{+2} + 2\,NO_3^-} + AgBr\downarrow$$

$$Ag^+ + Br^- \longrightarrow AgBr\downarrow$$

$$ZnBr_2 + 2\,NaOH \longrightarrow 2NaBr + Zn(OH)_2\downarrow$$

colorless fluffy residue

$$\underbrace{Zn^{+2} + 2\,Br^-} + \underbrace{2\,Na^+ + 2\,OH^-} \longrightarrow \underbrace{2\,Na^+ + 2\,Br^-} + Zn(OH)_2\downarrow$$

$$Zn^{+2} + 2\,OH^- \longrightarrow Zn(OH)_2\downarrow$$

The bromide ions from the zinc dibromide form a pale yellow precipitate of silver bromide, and the zinc ions from the sodium base salt form a pithy precipitate of zinc dihydroxide.

$$K_2SO_4 + BaCl_2 \longrightarrow 2\ KCl + BaSO_4\downarrow$$

$$\underbrace{2K^+ + SO_4^{-2}}{} + \underbrace{Ba^{+2} + 2Cl^-}{} \longrightarrow \underbrace{2K^+ + 2Cl^-}{} + BaSO_4\downarrow$$

$$Ba^{+2} + SO_4^{-2} \longrightarrow BaSO_4\downarrow$$

Soluble sulfates and sulfuric acid are proved with barium ions. A white precipitate of barium sulfate is separated. The potassium ions color the flame of the spirit lamp violet.

$$CaCl_2 + 2\ AgNO_3 \longrightarrow Ca(NO_3)_2 + 2\ AgCl\downarrow$$

$$\underbrace{Ca^{+2} + 2Cl^-}{} + \underbrace{2\ Ag^+ + 2NO_3^-}{} \longrightarrow \underbrace{Ca^{+2} + 2NO_3^-}{} + 2\ AgCl\downarrow$$

$$Ag^+ + Cl^- \longrightarrow AgCl\downarrow$$

Halogenated anions form precipitates with silver ions. The chloride anion forms a white precipitate with the silver cations, which rapidly changes color under the action of light.

Calcium ions color the flame of a spirit lamp tile red. Chemical reactions could also be used, e.g., with phosphate anions a yellow precipitate of tricalcium diphosphate is given off.

$$3\ CaCl_2 + 2\ Na_3PO_4 \longrightarrow Ca_3(PO_4)_2\downarrow$$

$$\underbrace{3Ca^{+2} + 6Cl^-}{} \quad \underbrace{6Na^+ + 2PO_4^{-3}}{} \longrightarrow \underbrace{6Na^+ + 6Cl^-}{} + Ca_3(PO_4)_2\downarrow$$

$$3Ca^{+2} + 2PO_4^{-3} \longrightarrow Ca_3(PO_4)_2\downarrow$$

Dipotassium sulfate and calcium dichloride do not hydrolyze because they are salts of strong bases and strong acids.

Disodium carbonate is the salt of the strong sodium base and the weak and perishable carbonic acid. The salt hydrolyzes.

Hydrolysis is the reversible interaction of salt ions with water ions, changing the concentration of hydrogen and hydroxide ions.

$$\underbrace{2Na^+ + CO_3^{-2}}{} + \underbrace{2H^+ + 2\ OH^-}{\uparrow\downarrow\ 2H_2O} \rightleftharpoons \underbrace{2Na^+ + 2OH^-}{} + H_2CO_3$$

Carbonate ions from the salt bind with hydrogen ions from the water in the weak electrolyte carbonic acid. In excess, the hydroxide ions remain in solution, which determine the basic character of the salt solution. The color of the litmus changes to blue. pH > 7

Zinc dibromide is a salt of the weak zinc dihydroxide and the strong hydrobromic acid.

The salt hydrolyzes.

$$\underbrace{Zn^{+2} + 2\,Br^{-}} + \underbrace{2\,H^{+} + 2\,OH^{-}} \rightleftharpoons \underbrace{2\,H^{+} + 2\,Br^{-}} + Zn(OH)_2$$

$$\updownarrow$$

$$2\,H_2O$$

The zinc ions from the salt bind to the hydroxide ions from the water in the weak electrolyte zinc dihydroxide. Hydrogen ions remain in excess in the solution and determine the acidic character of the solution. The litmus is colored red, pH < 7.

Task 3.

When carbon reacts with carbon dioxide in a closed vessel, a reversible process takes place which is endothermic in the straight direction. Under certain conditions, chemical equilibrium is established. In which direction will the equilibrium be shifted: (a) by increasing the temperature; (b) by increasing the pressure; (c) by doubling the amount of carbon; (d) by introducing a catalyst; (e) by introducing sodium chloride; (f) by introducing disodium oxide?

Required elements in solving the problem

- Representing the interaction with a chemical equation:

$$C + CO_2 \rightleftharpoons 2\,CO - Q$$

- Definition of chemical equilibrium and expression of equilibrium process constant;
- Definition of the Le Chatelier-Brown principle of mobile equilibrium;
- Explanations related to shifts in chemical equilibrium:
  a) The balance shifts to the right;
  b) The balance shifts to the left;
  c) Equilibrium is not affected;
  d) Equilibrium is not affected;
  e) The equilibrium is not affected;
  f) The equilibrium shifts to the left (chemical equation for the interaction).

$$Na_2O + CO_2 \longrightarrow Na_2CO_3$$

12. Theory of Electrolytic Dissociation

Electrolytes and non- electrolytes. Deviations in the properties of solutions of electrolytes from the properties of nonelectrolytes. Mechanism of dissociation of substances with ionic and polar covalent bond. Dissociation of substances with ionic bond. Dissociation of substances with polar bond. Types of ions and their properties. Dissociation as a reversible and equilibrium process. Degree of electrolytic dissociation. Dissociation constant. Strong electrolytes. Debye and Hueckel theory. Acids, bases and salts from the point of view of the theory of electrolytic dissociation. Electrolytes and non-electrolytes

In the early 19th century, Michael Faraday discovered that aqueous solutions of certain substances (acids, bases, salts) conduct an electric current. He called these substances **electrolytes**. Other substances that do not conduct an electric current (some organic compounds - sugar, alcohols, etc.) Faraday called **non-electrolytes**.

According to Faraday, when an electric current is passed through a solution of an electrolyte, the latter decomposes into ions, which are attracted by the corresponding electrodes and there separated. Later it was found that it is **not the electric current** that causes the ions to appear in the electrolyte but, on the contrary, the presence of free ions determines its electrical conductivity. When an electric current is passed through the electrolyte solution, the anions (negatively charged ions) are directed to the anode and the cations (positively charged ions) to the cathode. As a result of the directed movement of the ions, an electric current flows.

Deviations in the properties of electrolyte solutions from the properties of non-electrolytes

Substances that do not conduct an electric current in aqueous solution or in a melt are **nonelectrolytes**. The properties of dilute solutions of nonelectrolytes obey the laws of Pfeffer, Vant Hoff, and Raoul. In general, the properties of dilute solutions of nonelectrolytes change in **proportion to the molar concentration of the solute.**

Substances which in aqueous solution or melt conduct electric current are called **electrolytes**. Their solutions have a **higher osmotic pressure, a lower vapor pressure, and an elevated boiling point, a lower freezing point,** than follows according to the laws of solution properties of nonelectrolytes. This indicates that a greater number of particles are produced when the electrolytes dissolve or melt.

To apply the law of osmotic pressure to electrolyte solutions, Want Hoff introduced a correction factor called the **isotonic coefficient i.**

$$\pi = i.C.RT$$

The factor i indicates how many times the number of particles in an electrolyte solution increases compared to the number of particles in a non-electrolyte solution of the same molar concentration. For example, for a solution of sodium base (NaOH), hydrochloric acid (HCl), acetic acid (CH_3 COOH) i ~ 2, and for a solution of calcium dichloride ($CaCl_2$), sulfuric acid (H_2SO_4) i ~ 3, etc.

Mechanism of dissociation of substances with ionic and polar covalent bond

In order to explain the deviations in the properties of electrolytes from the laws of dilute solutions of nonelectrolytes, the Swedish chemist Svante Arenius established the theory of electrolytic dissociation (1887). He assumed that electrolytes in aqueous solution or in a melt dissociate, dissociate on dissolving or melting themselves. Electrolytic dissociation is the spontaneous process of electrolytes in solution or in a melt dissociating to differently charged ions. This results in an increase in the number of particles in solution, hence the increase in osmotic pressure and other observed relationships. The coefficient **i** takes on a value greater than 1. For example, in the dissociation of hydrogen chloride in water the number of particles is doubled.

$$HCl \longrightarrow H^+ + Cl^-$$

Dissociation of ionic bond substances

Substances with an ionic bond have an ionic crystal lattice. When such compounds dissolve under the action of dipole molecules of water, a gradual disintegration of the crystal lattice begins as a result of the impact of the dipole molecules of water on the ions from the surface of the crystal and due to the occurrence of attraction between the dipoles and the ions from the crystal. This interaction leads to the overcoming of the forces of attraction between the ions in the crystal lattice. The ions detached from the surface of the crystal lattice in solution remain surrounded by the dipole water molecules. This forms complex groupings called hydrates. The process by which polar water molecules surround solute particles and form complex groupings with them, **hydrates**, is called **hydration**. Hydrates do not have a well-defined composition and therefore hydration is not expressed by chemical equations.

For substances with an ionic crystal lattice, the dissolution process is also an electrolytic dissociation process.

$$NaCl \longrightarrow Na^+ + Cl^-$$

Aqueous solutions of ionocrystalline substances contain only ions that are hydrated.

Dissociation of such substances takes place completely and is an irreversible process.

Substances with an ionic crystal lattice in the molten state also decompose to ions under the action of high temperature.

Dissociation of substances with a polar bond

Substances with polar bonds are made up of polar molecules. When dissolved in water, the polar molecules of a substance are hydrated by the dipolar molecules of water. Forces of attraction (dipole-dipole interaction) occur between the suitably oriented dipoles. As a result, further polarization of the polar molecule of the substance occurs. The chemical bond gradually becomes labile and breaks. Ions are formed which are hydrated.

If the solvent is not water but is a polar organic solvent, a solvation process takes place. Solvation is an exothermic process. The heat released facilitates the breaking of polar bonds or the breakdown of the crystal lattice in ionic compounds.

A process in which substances in solution or in a melt decompose to ions under the action of polar solvent molecules or at high temperature is called **electrolytic dissociation**. The greater the polarity of the solvent molecules, the greater their dissociating action.

Compounds with polar covalent bonds and weakly polar or nonpolar molecules do not dissociate. Their solutions contain solvated molecules.

Substances whose solutions or melts do not contain ions are called nonelectrolytes.

If there are bonds of different polarity in an electrolyte molecule, electrolytic dissociation affects the most polar bond first. For example, the dissociation of acetic acid proceeds according to the equation:

$$CH_3COOH \rightleftharpoons CH_3COO^- + H^+$$

Obviously, the bond between the hydrogen and oxygen in the hydroxyl group as the most highly polar bond in the molecule breaks down.

Types of ions and their properties

When an electric current is passed through a solution or melt of electrolytes, the positive ions are directed to the negative pole (cathode), which is why they are called **cations**. The negative ions are directed to the anode and are called **anions**. The total electrical load of the cations and anions in solution is equal, so their charges balance each other.

Ions can be **simple** or **complex**.

Simple ions are derived from a single atom, e.g., Na^+, K^+, Ca^{+2}, Cl^-, I^-, etc. **Complex ions** are formed by the electrolytic dissociation of more complex compounds. They are derived from several atoms. Complex ions are e.g., NH_4^+, NO_3^-, SO_4^{-2}, etc.

Many of the ions are colorless, as are the compounds from which they are derived. Some ions have a characteristic color which they also impart to the compounds in which they are involved. For example, MnO_4^- (permanganate ions) are violet, and so is the color of potassium permanganate $KMnO_4$.

Many ions are colored only when hydrated. For example, copper sulfate crystallizes with five molecules of water and has a blue color. When heated, it is dehydrated and copper sulfate is colorless.

$$\underset{\text{blue}}{CuSO_4.5H_2O} \xrightarrow{t^o} \underset{\text{colorless}}{CuSO_4 + 5H_2O}$$

Due to the presence of an electrical charge, ions have properties different from those of the simple substances of their constituent elements. For example, Cl^- has neither the color nor the odor and other properties of chlorine gas, which is poisonous, whereas chloride ions are not poisonous.

Solutions and compounds that contain a particular type of ion have properties specific to those ions. For example, hydrogen ions (protons) give an acid taste, chloride ions a salty taste, magnesium ions a bitter taste, etc.

Dissociation as a reversible and equilibrium process. Degree of electrolytic dissociation. Dissociation constant

Dissociation is a reversible process. Simultaneously with the dissociation of molecules to ions, the reverse process of binding of ions to whole molecules takes place. This process is called molarization.

$$CH_3COOH \underset{\text{molarization}}{\overset{\text{dissociation}}{\rightleftharpoons}} CH_3COO^- + H^+$$

A quantitative measure of how much dissociation of a compound has taken place is the **degree of electrolytic dissociation** (α). It is given by the ratio of dissociated (n) to the total number of dissolved molecules (n_o).

$$\alpha = \frac{n}{n_o}$$

α is expressed as a fraction or percentage.

$$\alpha = \frac{n}{n_o} . 100$$

Depending on the degree of electrolytic dissociation, electrolytes are:

Weak: $\alpha < 3$ % – H_2S, HCN, H_2O;
Medium: 3 % < α < 30 % – H_3PO_4, HCOOH;
Strong: $\alpha > 30$ % – HCl, HNO_3, NaOH, NaCl etc.

The degree of electrolytic dissociation depends on the nature of the electrolytes. **Solutions of electrolytes** with an ionic crystal structure **contain only ions**. For them $\alpha \sim 100\%$. They are strong electrolytes and have large electrical conductivity.

In **compounds with polar covalent bonds,** α varies depending on the polarity of the bonds. For less polar compounds α is small. A small fraction of the molecules decompose to ions. The solution is dominated by undissociated molecules.

$$H_2S \rightleftharpoons 2\,H^+ + S^{-2} \qquad \alpha = 4\,\%$$

Therefore, **strong electrolytes** are those whose solutions contain only ions or a large amount of ions and a small number of polar molecules. In the dissociation of strong electrolytes, the dissociation equilibrium is practically shifted to the right toward dissociation. **Weak electrolytes** are those whose solutions contain a small amount of ions and a large amount of polar molecules. In these the dissociation is an equilibrium process.

Dissociation equilibrium is characterized by an equilibrium constant, which is called the **dissociation constant K.** For example, the dissociation constant for the dissociation of acetic acid is:

$$K = \frac{c(CH_3COO^-) \cdot c(H^+)}{c(CH_3COOH)}$$

The dissociation constant is a **constant quantity** for a given electrolyte at a given temperature and does not depend on the concentration of the solution. It depends on the nature of the electrolyte and solvent.

The dissociation equilibrium may shift in one direction or the other in accordance with the Le Chatelier-Brown principle of mobile equilibrium.

As **the temperature increases,** the equilibrium shifts in the direction of **dissociation of the weak electrolyte**. Dissociation is an endothermic process.

When a **weak electrolyte solution is diluted**, the rate of molarization decreases because the distance between ions in solution increases significantly and the probability of them meeting and interacting is reduced. The equilibrium shifts in the direction of dissociation. When the solution is diluted, the electrolyte dissociation rate α increases, and the dissociation constant is not affected by dilution.

Strong electrolytes. Theory of Debye and Hueckel

The aqueous solutions of strong electrolytes, and particularly those of an ionic crystal structure, contain a large amount of ions which are closely spaced because of their high concentration. Electrostatic forces of attraction occur between oppositely charged ions. In solution, an ionic atmosphere of oppositely charged ions forms around each ion. This leads to a decrease in the mobility of the ions. They move more slowly to their respective electrodes and show a lower electrical conductivity than corresponds to their complete dissociation. This also explains the smaller value of α determined by the electrical conductivity. For solutions of strong electrolytes, an apparent degree of electrolyte dissociation is introduced which is lower than the actual one. When such solutions are diluted, the ionic atmosphere around the ions decreases and the apparent degree of electrolytic dissociation increases.

To quantify the properties of solutions of strong electrolytes, Lewis (1923) introduced the quantity of ion activity. It represents the conditional concentration of ions manifested in action. The activity is denoted by **a**. The relationship between it and the concentration of ions in solution is given by the expression:

$$a = f.c$$

f - activity coefficient

The activity coefficient is a quantity that summatively reflects all the force field interactions existing in the electrolyte solution. f - depends on concentration, temperature, etc.

At very high dilution of the solution the attractive forces between the ions weaken and then $a = c, f = 1$.

Acids, bases and salts from the point of view of the theory of electrolytic dissociation

One of the great achievements of the theory of electrolytic dissociation is the theory of acids, bases, and salts established on its basis by Arrhenius.

Acids are all compounds which, in aqueous solution, dissociate into hydrogen cations and acid anions and yield no other kind of cation.

$$HCl \longrightarrow H^+ + Cl^-$$

$$H_2SO_4 \longrightarrow 2H^+ + SO_4^{-2}$$

In fact, it is not free H^+, but hydroxonium cations H_3O^+ contained in the solution as a result of the interaction:

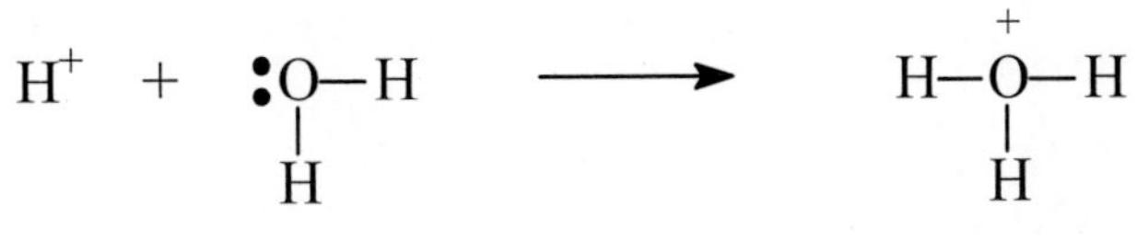

or

$$H^+ + H_2O: \longrightarrow H_3\overset{+}{O}$$

The presence of H^+ or H_3O^+ in acid solutions determines their general properties: acid taste, change the color of blue litmus to red, interact with active metals, with basic oxides, hydroxides, etc.

The number of H^+, which are formed in the dissociation of an acid molecule, determine its **basicity**. Depending on this, acids can be:

single-base: HCl, HBr, HNO_3, HCOOH, CH_3COOH;
Bivariate: **H_2SO_4, H_2CO_3, $H_2C_2O_4$;**
three main ones: H_3PO_4
multifactorial.

The dissociation of dibasic and polybasic acids takes place in steps. An equilibrium is established for each degree, which is characterized by a separate dissociation constant:

$$H_2SO_4 \longrightarrow H^+ + HSO_4^-$$

(hydrogensulfate anion)

$$HSO_4^- \longrightarrow H^+ + SO_4^{-2}$$

(sulfate anion)

In the first step, the dissociation constant is the largest

$$K_1 = \frac{c(H^+).c(HSO_4^-)}{c(H_2SO_4)}$$

$$K_1 > K_2$$

$$K_2 = \frac{c(H^+).c(SO_4^{-2})}{c(HSO_4^-)}$$

Bases - Compounds which on dissociation give hydroxide anions $\overline{O}H$ and metal or NH_4^+ cations.

$$KOH \longrightarrow K^+ + \overline{O}H$$

$$Ba(OH)_2 \longrightarrow Ba^{+2} + 2\overline{O}H$$

The general properties of bases are determined by the presence of hydroxide ions in solution. According to the **number of hydroxide groups formed by** the dissociation of one mole of base, bases are:

univalent: NaOH, KOH
bivalent: $Ca(OH)_2$, $Mg(OH)_2$, $Ba(OH)_2$
trivalent and polyvalent

Dissociation of multivalent bases also proceeds in steps, with the first step being the largest.

$$Ca(OH)_2 \longrightarrow \overset{+}{Ca}(OH) + \overline{O}H$$

$$\overset{+}{Ca}(OH) \longrightarrow Ca^{+2} + \overline{O}H$$

Weak electrolytes that can dissociate as both acids and bases, i.e., give the characteristic ions for acids H^+ and for bases HO^-, are called **ampholytes**. An ampholyte is, for example, water.

$$H_2O \rightleftharpoons \overset{+}{H} + \overset{-}{O}H \; ;$$

$$K = \frac{c(H^+).c(\overline{O}H)}{c(H_2O)}$$

The number of dissociated molecules is negligible, therefore it can be assumed $c(H_2O)$ is practically constant. In such a case, one can multiply K by $c(H_2O)$ and denote the new constant by $K(H_2O)$. It is called the ionic product of water and is equal to the product of the equilibrium concentrations of hydrogen and hydroxide ions.

$$K(H_2O) = K.c(H_2O) = c(H^+).c(\overline{O}H)$$

The ion product of water has a well-defined value for a given temperature. For example, for 22 °C

$$K(H_2O) = 10^{-14} \text{ mol}^2/L^2$$

Since one hydrogen cation (H^+) and one hydroxide anion (HO^-) are produced from one molecule of water by dissociation, their concentration is equal to

$$c(H^+) = c(\overline{O}H) = 10^{-7} \text{ mol/L},$$

which is valid for pure water. If this equilibrium is disturbed, the medium becomes acidic or basic:

- $c(H^+) > 10^{-7}$ mol/L **acidic character**
- $c(H^+) < 10^{-7}$ mol/L **basic character**

In many cases, the nature of a solution or medium needs to be determined. For convenience, instead of working with very small numbers, a **hydrogen pH index** has been introduced which is equal to:

$$pH = -\lg c(H^+)$$

This is how the nature of the solution changes depending on the pH:

1	2	3	4	5	6	7	8	9	10	11	12	13	14
acid						neutral	basic						

Salts - they are strong electrolytes and their dissociation proceeds in one step to produce **metal (or ammonium) cations and acid** anions

$$NaNO_3 \longrightarrow Na^+ + NO_3^-$$

These salts are **normal.**

Salts which, on dissociation, give off, besides metal (or ammonium) cations, hydrogen cations and acid anions are called **hydrogen salts**.

$$Na_2HPO_4 \longrightarrow 2\,Na^+ + HPO_4^{-2}$$

$$HPO_4^{-2} \rightleftharpoons H^+ + PO_4^{-3}$$

If the dissociation of salts also produces hydroxide anions (HO–) the salts are called **basic**. The hydrogen salts and basic salts are dissociated into grades with the first grade being referred to as strong and the second grade as weak electrolytes.

$$Mg(OH)Cl \longrightarrow \overset{+}{Mg}(OH) + Cl^-$$

$$\overset{+}{Mg}(OH) \rightleftharpoons Mg^{+2} + {}^-OH$$

When the dissociation of a salt produces **two types of cations other than H$^+$, the** salt is called **a double**. These salts dissociate to one degree and are strong electrolytes.

$$KAl(SO_4)_2 \longrightarrow K^+ + Al^{+3} + 2\,SO_4^{-2}$$

Salts which on dissociation give **complex cations or complex anions** are called **complex salts.**

$$K_4[Fe(CN)_6] \rightleftharpoons 4\,K^+ + [Fe(CN)_6]^{-4}$$

The resulting complex anion dissociates to a negligible extent, therefore Fe^{+2} and CN^- ions cannot be detected in solution.

$$[Fe(CN)_6]^{-4} \rightleftharpoons Fe^{+2} + 6\,CN^-$$

The balance is shifted strongly to the left.

How a compound dissociates - as an acid, base or salt - depends on its structure and the type of bonds in the molecule. For example, calcium dihydroxide and sulfuric acid each have two hydroxyl groups and the following structure:

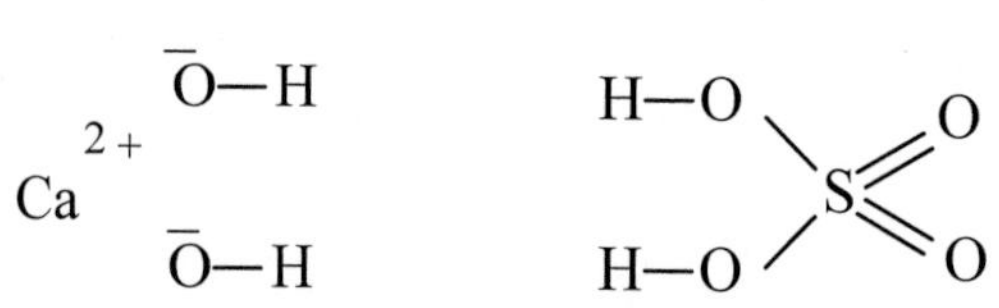

In calcium dihydroxide, the calcium-oxygen bonds are ionic and it is these bonds that are broken during dissociation. Calcium dihydroxide dissociates as a base.

In the molecules of sulfuric acid, there are only covalent bonds, and from them the most polar H-O are broken during dissociation, so the compound dissociates as an acid.

The theory of electrolytic dissociation has provided a scientific explanation for a number of concepts related to electrolytes. It has become clear that reactions between electrolytes are ionic reactions that proceed rapidly. They are equilibrium reactions, but can proceed to completion in the following cases:

When a **hard soluble substance is formed**:

$$AgNO_3 + KCl \longrightarrow AgCl\downarrow + KNO_3$$

$$\underbrace{Ag^+ + NO_3^-} + \underbrace{K^+ + Cl^-} \longrightarrow AgCl\downarrow + \underbrace{K^+ + NO_3^-}$$

or

$$Ag^+ + Cl^- \longrightarrow AgCl\downarrow$$

When a **weak electrolyte is formed:**

$$HCl + NaOH \longrightarrow NaCl + H_2O$$

$$\underbrace{H^+ + Cl^-} + \underbrace{Na^+ + OH^-} \longrightarrow \underbrace{Na^+ + Cl^-} + H_2O$$

or

$$H^+ + OH^- \longrightarrow H_2O$$

The interaction releases **a gas which takes off**

$$Na_2S + 2HCl \longrightarrow 2NaCl + H_2S\uparrow$$

$$\underbrace{2Na^+ + S^{-2}} + \underbrace{2H^+ + 2Cl^-} \longrightarrow \underbrace{2Na^+ + 2Cl^-} + H_2S\uparrow$$

$$2H^+ + S^{-2} \longrightarrow H_2S\uparrow$$

The later generalized theory of acids and bases (by Brønstedt and Lowry) does not contradict Arrhenius' theory, but incorporates it. According to this generalized conception:

Acids are any compounds or ions that can give off hydrogen ions (H^+). Bases are all compounds or ions that can join hydrogen ions (H^+):

$$\underset{\text{(acid)}}{NH_4^+} \longrightarrow \underset{\text{(base)}}{:NH_3} + H^+$$

$$\underset{\text{(acid)}}{CH_3COOH} \longrightarrow \underset{\text{(base)}}{CH_3COO^-} + H^+$$

Sample problems and solutions

Task 1.

You have aqueous solutions of the substances $MgCl_2$, $BaCl_2$, and $CaCl_2$, each containing 5 g of salt in 1 liter of solution. Arrange the solutions by:

a) lowering their osmotic pressure at constant temperature;
b) lowering their boiling point;
c) lowering their freezing temperature. Justify your answers.

Exemplary solution

A process in which substances in solution or in a melt decompose to ions under the action of polar solvent molecules or at high temperature is called **electrolytic dissociation**. The greater the polarity of the solvent molecules, the greater their dissociating action.

Aqueous solutions of the salts of elements of group IIA of the Periodic Table are ionic compounds having the same anionic moiety - chloride anion and different divalent cations - magnesium, calcium and barium. All three salts are completely dissociated in water.

$$MgCl_2 \longrightarrow \overset{+2}{Mg} + 2\overset{-}{Cl}$$

$$BaCl_2 \longrightarrow \overset{+2}{Ba} + 2\overset{-}{Cl}$$

$$CaCl_2 \longrightarrow Ca^{+2} + 2\overset{-}{Cl}$$

As a result of dissociation, all three salts release three particles – one cation and two anions.

In order to answer the condition of the problem it is necessary to represent the molar concentrations of the solutions. The three solutions each contain 5 g of salt in 1 L of solution. The molar mass of magnesium chloride is 95.3 g and the concentration of the solution is 0.052 mol/L. The molar mass of calcium dichloride is 111 g and the concentration of the solution is 0.045 mol/L. The molar mass of barium dichloride is 208 g and the concentration of the solution is 0.024 mol/L.

Therefore, by lowering the osmotic pressure at constant temperature, the solutions are ordered:

$$MgCl_2 > CaCl_2 > BaCl_2$$

In the same sequence, the boiling point of the solutions changes - the boiling point of barium dichloride is the lowest:

$$MgCl_2 > CaCl_2 > BaCl_2$$

The higher the concentration of the salt solution, the lower the freezing point.

$$BaCl_2 > CaCl_2 > MgCl_2$$

13. HYDROLYSIS PROCESSES

WATER AS A WEAK ELECTROLYTE. REVERSIBLE AND IRREVERSIBLE HYDROLYSIS PROCESSES. TYPES OF HYDROLYTIC PROCESSES OF SALTS ACCORDING TO THE STRENGTH OF THE ACID AND THE BASE FROM WHICH THEY ARE DERIVED. HYDROLYSIS OF SALTS DERIVED FROM WEAK ACID AND STRONG BASE. HYDROLYSIS OF SALTS DERIVED FROM STRONG ACIDS AND WEAK BASES. HYDROLYSIS OF SALTS DERIVED FROM WEAK ACIDS AND WEAK BASES. HYDROLYSIS OF SALTS DERIVED FROM STRONG ACID AND STRONG BASE. HYDROLYSIS OF HYDROGEN SALTS. HYDROLYSIS OF SOAPS. HYDROLYSIS OF ESTERS. HYDROLYSIS CONSTANT AND DEGREE OF HYDROLYSIS. IRREVERSIBLE HYDROLYSIS PROCESSES. HYDROLYSIS OF PROTEINS. HYDROLYSIS OF LACTAMS. IMPORTANCE OF HYDROLYSIS.

Hydrolysis is a special case of a more general phenomenon – solvolysis. Solvolysis is the interaction of molecules or ions of a solute with solvent molecules (from solvent, Latin).

Hydrolysis processes are for example:

$$CH_3COONa + H_2O \rightleftharpoons CH_3COOH + NaOH$$

$$KCN + H_2O \rightleftharpoons HCN + KOH$$

Substances that can be hydrolysed belong to different groups of compounds: salts, esters and fats, nitriles, lactams, carbohydrates, proteins, etc.

Water as a weak electrolyte

Although weak, pure water conducts electricity. The electrical resistivity is large, which means that water is a weak electrolyte. The electrical conductivity of pure water is due to hydrogen cations (H^+) and hydroxide anions (-OH) produced by the dissociation of water molecules.

$$H_2O \rightleftharpoons H^+ + \bar{O}H$$

or

$$2\,H_2O \rightleftharpoons H_3O^+ + {}^-OH$$

hydroxonium ion

Electrolytic dissociation of water is a reversible and equilibrium process. The dissociation constant is as follows:

$$K = \frac{[H^+].[\bar{O}H]}{[H_2O]}$$

Due to the negligible dissociation of water, it $[H_2O]$ can be assumed to be a constant quantity,

hence

$$K.[H_2O] = [H^+].[\bar{O}H]$$

or the new constant:

$$K(H_2O) = [H^+].[\bar{O}H]$$

$K(H_2O)$ is called the ionic product of water. It shows that in both pure water and aqueous solutions of acids, bases, and salts, the product of the concentration of hydrogen and hydroxide ions for a given temperature is a constant quantity. Determining the concentration of one ion gives information about the other ion.

When:

$[H^+] = [OH]$ - the medium is neutral;
$[H^+] > [OH^-]$ - the environment is acidic;
$[H^+] < [OH^-]$ - the medium is alkaline.

The concentration of hydrogen ions is usually determined $[H^+]$. In order to avoid small numerical values such as 10^{-1}, 10^{-2}, 10^{-3}, etc. g.ion/L. a hydrogen pH is introduced, which is:

$$pH = - \lg [H^+]$$

The pH is a numerical order from 0 to 14.

At pH 0 to 7 - acid reaction,

pH = 7 - neutral reaction,

pH 7 to 14 - alkaline reaction.

Reversible and irreversible hydrolysis processes

Hydrolysis processes are divided into reversible and irreversible.

The process of interaction of ions of salts with ions of water, in which the pH of the medium changes, is called hydrolysis. This process is reversible and equilibrium. Typical reversible hydrolysis processes are those of salts and esters.

The explanation according to the classical theory of Arrhenius is as follows: water is a weak electrolyte, and although it dissociates to a minor extent according to the equations above, the concentration of H^+ and OH^- is the same, which determines the neutral character of water. When a normal salt is dissolved, one of its ions may combine with one of the ions of water in a weak electrolyte or gas. New water molecules dissociate in order to restore equilibrium (Le Chatelier-Brown principle). As a result, ions opposite to those bound by the water accumulate in the solution and it becomes acidic or alkaline.

Types of hydrolysis processes of salts according to strength of the acid and base from which they are derived

Several cases of hydrolysis of salts are distinguished:

Hydrolysis of salts derived from weak acid and strong base

$$\underbrace{CH_3CO\bar{O} + \overset{+}{Na}} + \underbrace{\overset{+}{H} + \bar{O}H}_{\substack{\upharpoonleft\downharpoonright \\ H_2O}} \rightleftharpoons CH_3COOH + \underbrace{\overset{+}{Na} + \bar{O}H}$$

Sodium acetate is a salt obtained by the neutralization of weak acetic acid (CH_3COOH) and strong sodium base (NaOH). When the acetate anion reacts with the hydrogen cation, whole molecules of acetic acid are produced. The hydrogen ion concentration decreases. This results in a shift in the equilibrium of the dissociation of water to ions. The concentration of hydroxide ions increases and the medium becomes alkaline, i.e.,

$$[OH^-] > [H^+]$$

Hydrolysis of salts derived from strong acids and weak bases

$$\underbrace{\overset{+}{NH_4} + \overset{-}{Cl}} + \underbrace{\overset{+}{H} + \overset{-}{OH}}_{\underset{H_2O}{\upharpoonleft\downharpoonright}} \rightleftharpoons \underbrace{\overset{+}{H} + \overset{-}{Cl}} + \underset{\underset{NH_3\,+\,H_2O}{\upharpoonleft\downharpoonright}}{NH_4OH}$$

In this case, ammonium cations and hydroxide anions have interacted to produce the weak and perishable electrolyte ammonium hydroxide. It decomposes with the evolution of ammonia and water. The concentration of H^+ increases and that of OH^- decreases because the ionic equilibrium of water is disturbed.

The same process takes place with salts:

$$\underbrace{2\overset{+}{NH_4} + \overset{-2}{SO_4}} + \underbrace{2\overset{+}{H} + 2\overset{-}{OH}}_{\underset{2\,H_2O}{\upharpoonleft\downharpoonright}} \rightleftharpoons \underbrace{2\overset{+}{H} + \overset{-2}{SO_4}} + \underset{\underset{2\,NH_3\,+\,2\,H_2O}{\upharpoonleft\downharpoonright}}{2\,NH_4OH}$$

$$\underbrace{Al^{+3} + 3\,\overset{-}{Cl}} + \underbrace{3\overset{+}{H} + 3\overset{-}{OH}}_{\underset{3\,H_2O}{\upharpoonleft\downharpoonright}} \rightleftharpoons Al(OH)_3 + \underbrace{3\overset{+}{H} + 3Cl^-}$$

Aqueous solutions of salts derived from strong acids and weak bases are acidic in nature: as

$$[H^+] > [OH^-] \text{ and } pH < 7$$

Hydrolysis of salts derived from weak acids and weak bases

$$\underbrace{\overset{+}{NH_4} + CH_3CO\overset{-}{O}} + \underbrace{\overset{+}{H} + \overset{-}{OH}}_{\underset{H_2O}{\upharpoonleft\downharpoonright}} \rightleftharpoons CH_3COOH + \underset{\underset{NH_3\,+\,H_2O}{\upharpoonleft\downharpoonright}}{NH_4OH}$$

All ions are involved in the interaction, i.e., both types of water ions are involved. The reaction of the medium depends on the relative strength of the acid and base produced. In this case, the reaction of the medium is neutral, since the dissociation constants of acetic acid and ammonium hydroxide are almost the same.

Hydrolysis also occurs with diammonium carbonate:

$$\underbrace{2\overset{+}{NH_4} + \overset{-2}{CO_3}} + \underbrace{2\overset{+}{H} + 2\overset{-}{OH}}_{\substack{\updownarrow\\ 2H_2O}} \rightleftharpoons \underset{\substack{\updownarrow\\ H_2O\ +\ CO_2}}{H_2CO_3} + \underset{\substack{\updownarrow\\ 2NH_3\ +\ 2H_2O}}{2NH_4OH}$$

The aqueous solution of diammonium carbonate has a basic character, that is, the concentration of hydroxide anions is greater than the concentration of hydrogen cations. This indicates that carbonic acid is a weaker electrolyte than ammonium hydroxide.

Aqueous solutions of salts derived from weak acids and weak bases have a slightly acidic or slightly basic reaction depending on whether the acid or base produced in the hydrolysis is the stronger electrolyte.

Hydrolysis of salts derived from strong acid and strong base

The salt ions produced in electrolytic dissociation do not bind to the water ions in a weak electrolyte molecule. The ratio between the concentration of hydrogen cations and hydroxide anions in solution does not change, remaining as in water, namely:

$$[H^+] = [OH^- = 10^{-7} \quad \text{and } pH = 7$$

Salts derived from strong acids and strong bases are practically not hydrolyzed. Aqueous solutions of these salts (NaCl, KCl, Na_2SO_4, etc.) are neutral.

The hydrolysis of normal salts produced by the interaction of polybasic acids with strong bases and polyvalent bases with strong acids proceeds stepwise. Salts such as trisodium phosphate (Na_3PO_4, Na_2CO_3, $Bi(NO_3)_3$, etc. hydrolyze stepwise. At moderate dilutions and ordinary temperature, hydrolysis is mainly limited to the first step yielding hydrogen cations, respectively hydrogen salts and hydroxo cations, respectively hydroxo salts.

$$\underbrace{2\overset{+}{Na} + \overset{-2}{CO_3}} + \underbrace{\overset{+}{H} + \overset{-}{OH}}_{\substack{\updownarrow\\ H_2O}} \rightleftharpoons \underbrace{\overset{+}{Na} + \overset{-}{HCO_3}} + \underbrace{\overset{+}{Na} + \overset{-}{OH}}$$

$$\overset{-2}{CO_3} + \underbrace{\overset{+}{H} + \overset{-}{OH}} \rightleftharpoons \overset{-}{HCO_3} + \overset{-}{OH}$$

$$\underbrace{\overset{+}{NH_4} + \overset{-}{HCO_3}} + \underbrace{\overset{+}{H} + \overset{-}{OH}}_{\substack{\updownarrow\\ H_2O}} \rightleftharpoons \underbrace{\overset{+}{Na} + \overset{-}{OH}} + \underset{\substack{\updownarrow\\ H_2O\ +\ CO_2}}{H_2CO_3}$$

Hydrolysis of hydrogen salts

In these salts, in addition to the strength of the starting acid and the starting base, the fact that their anions undergo electrolytic dissociation and release (liberate) hydrogen cations must be taken into account. This leads to a lowering of the pH of the medium and it is possible for a salt solution obtained from a strong base and a weaker acid to have an acid reaction, while the markedly alkaline character of certain normal salts weakens when they are converted into hydrogen salts, as is the case, for example, with sodium dihydrogen phosphate, sodium hydrogen carbonate, etc.

The nature of the solution of these salts is determined for a given temperature by the following equilibrium processes:

$$NaH_2PO_4 + H_2O \rightleftharpoons H_3PO_4 + NaOH \quad (1)$$

$$H_2PO_4^- \rightleftharpoons H^+ + HPO_4^{-2} \quad (2)$$

According to (1) the medium must be alkaline. However, the process proceeds to a very small extent (h = 0.0001 %) and the concentration of sodium base in solution is many times lower than that of the dihydrogen phosphate anion. This also determines the lower concentration of hydroxide anions despite the fact that sodium hydroxide is a much stronger electrolyte than the dihydrogen phosphate anion, i.e., the medium remains acidic.

In the aqueous solution of sodium hydrogen carbonate, equilibria are present,

$$NaHCO_3 + H_2O \rightleftharpoons NaOH + \underset{\substack{\updownarrow \\ H_2O + CO_2}}{H_2CO_3} \quad (1)$$

$$HCO_3^- \rightleftharpoons H^+ + CO_3^{-2} \quad (2)$$

the reaction of the solution is alkaline.

Hydrolysis of soaps

Soaps are alkaline salts of the higher fatty acids

$$\underbrace{C_{15}H_{31}COO^- + Na^+}_{\text{sodium palmitate}} + \underset{\substack{\updownarrow \\ H_2O}}{\underbrace{H^+ + OH^-}} \rightleftharpoons \underset{\substack{\text{palmitic acid} \\ \text{(hexadecanoic acid)}}}{C_{15}H_{31}COOH} + \underbrace{Na^+ + OH^-}$$

sodium palmitate *palmitic acid (hexadecanoic acid)*

Due to the alkaline nature of the solution, these soaps are not suitable for washing wool and silk textiles.

Hydrolysis of esters

$$R{-}COOR_1 + H_2O \xrightleftharpoons{H^+} RCOOH + R_1OH$$

$$C_3H_5(OCOC_{17}H_{35})_3 + 3\ H_2O \xrightleftharpoons{H^+} C_3H_5(OH)_3 + 3C_{17}H_{35}COOH$$

stearic acid
(octadecanoic acid)

An important chemical property of fats, as of all esters, is their ability to hydrolyse. The process takes place under heating (220 ° C) and elevated pressure in the presence of a catalyst.

$$C_3H_5(OCOR)_3 + 3\ H_2O \rightleftharpoons C_3H_5(OH)_3 + 3\ RCOOH$$

Fat hydrolysis is a reversible process.

In the body of higher animals and man, fats are hydrolysed in the small intestine by the catalytic action of the enzyme lipase.

Hydrolysis constant and degree of hydrolysis

Hydrolysis is quantitatively characterized by the hydrolysis constant – K_h. The latter is defined as the ratio of the number of hydrolyzed molecules to the total number of dissolved molecules, and it is expressed as a percentage. The weaker the electrolyte (or electrolytes) formed during hydrolysis, the greater the degree of hydrolysis. For the hydrolysis process

$$CH_3COONa + H_2O \xrightleftharpoons{H^+} CH_3COOH + NaOH$$

$$\underbrace{CH_3CO\overset{-}{O} + \overset{+}{Na}} + \underbrace{\overset{+}{H} + \overset{-}{OH}}_{\substack{\uparrow\downarrow \\ H_2O}} \rightleftharpoons CH_3COOH + \underbrace{\overset{+}{Na} + \overset{-}{OH}}$$

The hydrolysis constant has the form

$$Kh = \frac{[NaOH].\ [CH_3COOH]}{[CH_3COONa].[H_2O]}$$

At **high temperature and high dilutions,** the hydrolysis rate increases, while at concentrated solutions and low temperature, it decreases.

According to the protolytic theory of Brønstedt and Lowry, the hydrolysis of salts is an acid-base interaction, which is reduced to the transfer of protons from water molecules to salt ions or vice versa (water is an ampholyte). For example:

$$\underset{\text{base I}}{CH_3COO^-} + \underset{\text{acid II}}{H_2O} \xrightleftharpoons{H^+} \underset{\text{acid I}}{CH_3COOH} + \underset{\text{base II}}{{}^-OH}$$

In the hydrolysis of ammonium salts (NH_4Cl):

$$\underset{\text{acid I}}{NH_4^+} + \underset{\text{base II}}{H_2O} \xrightleftharpoons{H^+} \underset{\text{base I}}{NH_3} + \underset{\text{acid II}}{H_3O^+}$$

Irreversible hydrolysis processes

They are typical of a large number of organic compounds. They boil down to the breaking of bonds in molecules and the formation of two or more compounds involving parts of the water molecule. They proceed at low rates and are usually catalyzed by H^+ (acids), by ^-OH (bases), and by enzymes. For example, hydrolysis of carbohydrates:

$$\underset{\textit{starch}}{(C_6H_{10}O_5)n} + n\,H_2O \xrightarrow{(H^+)} \underset{\textit{glucose}}{n\,C_6H_{12}O_6}$$

$$\underset{\textit{sucrose}}{C_{12}H_{22}O_{11}} + H_2O \xrightarrow{(H^+)} \underset{\textit{glucose}}{C_6H_{12}O_6} + \underset{\textit{fructose}}{C_6H_{12}O_6}$$

Hydrolysis of proteins

$$H_2N{-}\underset{R_1}{CH}{-}CO{-}NH{-}\underset{R_2}{CH}{-}CO{-}NH{-}\underset{R_3}{CH}{-}CO{-} \xrightarrow[(H^+)]{H_2O}$$

$$H_2N{-}\underset{R_1}{CH}{-}COOH + H_2N{-}\underset{R_2}{CH}{-}COOH + H_2N{-}\underset{R_3}{CH}{-}COOH + \cdots\cdots$$

Hydrolysis of proteins can take place under the action of proteolytic enzymes, for example, protease:

$$H_2N{-}\underset{R_1}{CH}{-}CO{-}HN{-}\underset{R_2}{CH}{-}COOH \xrightarrow[\textit{protease}]{H_2O} H_2N{-}\underset{R_1}{CH}{-}COOH + H_2N{-}\underset{R_2}{CH}{-}COOH$$

This enzyme causes the peptide bond to break. In living organisms, enzymatic hydrolysis proceeds stepwise:

protein ⟶ polypeptides ⟶ α-amino acids

Hydrolysis of lactams

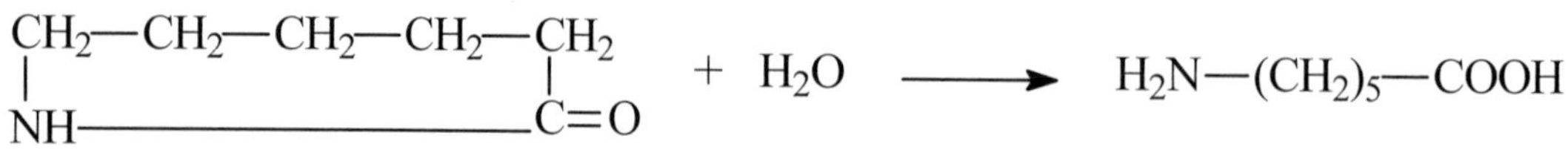

ε-caprolactam *ε-aminocaproic acid*

Importance of hydrolysis

Many industries are based on hydrolysis processes – production of glucose, glycerine, higher fatty acids, antibiotics, vitamins, etc.

Along with its purely chemical application, hydrolysis is of great importance for many processes occurring in living organisms.

For example, sodium hydrogen carbonate and disodium hydrogen phosphate, entering into the composition of blood, play a role in maintaining a constant blood pH, which is 7.4.

Sample problems and solutions

Task 1.

Aqueous solutions of which of the substances - sulfuric acid, calcium dichloride, ammonium acetate, potassium base, sodium sulfide, barium sulfate and ammonium nitrate, are hydrolyzed. Express the hydrolysis processes by chemical equations and determine the nature of the solution (pH). Explain the factors that influence the hydrolysis process.

Exemplary solution

Hydrolysis is the process of reversible interaction of ions of salts with ions of water in which the ratio between hydrogen and hydroxide ions changes. Salts of strong acids and weak bases, strong bases and weak acids, weak acids and weak bases hydrolyze. Salts produced by neutralization of strong acid and strong base, as well as salts insoluble in water, do not hydrolyze.

The sulfuric acid and potassium base mentioned in the problem condition do not hydrolyze. They dissociate as strong electrolytes. Calcium dichloride does not hydrolyse because it is a salt of a strong acid and a strong base – their aqueous solutions are neutral in nature. Barium sulfate does not hydrolyze because it is a salt of a strong base and a strong acid, and on the other hand is insoluble in water. Barium ions are used as a reagent for proving sulfuric acid and soluble sulfates.

Ammonium acetate is a salt of the weak ammonium base and weak acetic acid, and in aqueous solution hydrolyzes.

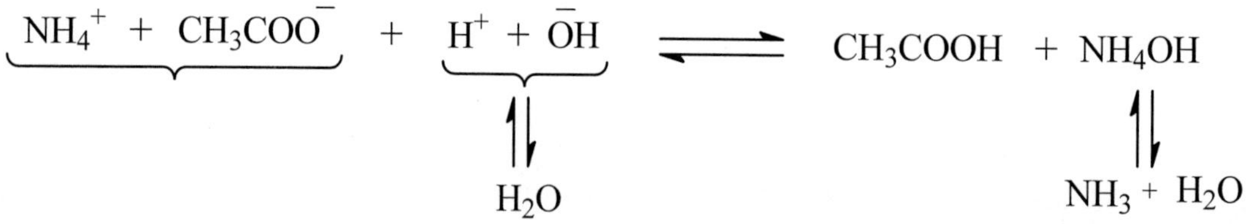

The acetate ions from the salt bind with the hydrogen ions from the water in the weak electrolyte acetic acid. The ammonium ions from the salt and the hydroxide ions from the water form the weak and perishable electrolyte ammonium hydroxide, which decomposes into ammonia and water. The nature of the medium (acidic, neutral or basic) depends on the relative strength of the two weak electrolytes.

Sodium sulfide is the salt of the strong sodium base and the weak hydrogen sulfide. In aqueous solution it hydrolyzes.

$$\underbrace{2Na^+ + \overset{-2}{S}} + \underbrace{2H^+ + 2\overset{-}{O}H}_{\substack{\uparrow\downarrow \\ 2H_2O}} \rightleftharpoons H_2S\uparrow + \underbrace{2Na^+ + 2\overset{-}{O}H}$$

The sulfide ions from the salt bind to the hydrogen ions from the water in the weak electrolyte hydrogen sulfide. In excess, the hydroxide ions remain in solution. These determine the basic character of the medium.

Ammonium nitrate is a salt of the weak electrolyte ammonium hydroxide and strong nitric acid, and in aqueous solution hydrolyzes.

$$\underbrace{NH_4^+ + \overset{-}{NO_3}} + \underbrace{H^+ + \overset{-}{O}H}_{\substack{\uparrow\downarrow \\ H_2O}} \rightleftharpoons \underset{\substack{\uparrow\downarrow \\ NH_3 + H_2O}}{NH_4OH} + \underbrace{H^+ + \overset{-}{NO_3}}$$

The ammonium ions from the salt bind with the hydroxide ions from the water in the weak and perishable electrolyte ammonium hydroxide. Hydrogen ions remain in excess in the solution and this determines the acidic character of the solution.

A quantitative characteristic of hydrolysis is the hydrolysis constant - K_h. The latter is defined as the ratio of the number of hydrolyzed molecules to the total number of dissolved molecules and is expressed as a percentage. The weaker the electrolyte (or electrolytes) formed during hydrolysis, the greater the degree of hydrolysis. For the hydrolysis process

$$\underbrace{NH_4^+ + CH_3COO^-} + \underbrace{H^+ + \overset{-}{O}H}_{\substack{\uparrow\downarrow \\ H_2O}} \rightleftharpoons CH_3COOH + \underset{\substack{\uparrow\downarrow \\ NH_3 + H_2O}}{NH_4OH}$$

the constant has the form

$$Kh = \frac{[NH_4OH].\,[CH_3COOH]}{[CH_3COONH_4].[H_2O]}$$

At **high temperature and high dilutions,** the hydrolysis rate increases, while at concentrated solutions and low temperature, it decreases.

14. COLLOID-DISPERSED SYSTE MS

PREPARATION OF COLLOIDAL SOLUTIONS. STRUCTURE OF COLLOIDAL PARTICLES. PROPERTIES OF COLLOIDAL SOLUTIONS - KINETIC, OPTICAL AND ELECTRICAL. STABILITY OF COLLOIDAL SOLUTIONS - COAGULATION AND SEDIMENTATION. SIGNIFICANCE AND USE OF COLLOIDAL SOLUTIONS

Colloidal solutions are such dispersed systems in which the particles of the dispersed substance have sizes from 1 to 100 nm. The particles of the dispersed phase are separated by a boundary surface from the dispersion medium, but cannot be observed immediately because they are small. They are microheterogeneous systems that have more specific properties and this necessitates the separation of a branch of chemical science for their study called **colloid chemistry**. The founder of colloid chemistry was Graham, who found that aqueous solutions of substances such as sodium chloride, magnesium sulfate, sugar, etc., readily pass through plant or animal semipermeable barriers, while others, such as Tutkal, egg yolk, gelatine, do not. Graham calls the former substances crystalloids because they crystallize readily, and the latter colloids (a combination of the Greek words "kola" = glue and "dos" = species, i.e., a substance that forms a glue-like solution). Colloids are also called **colloidal solutions** or **ash**. If the dispersed phase is water, they are called **hydrosols**.

Colloid-disperse systems are microheterogeneous systems. It has been found that the same substance, depending on conditions, manifests itself as a crystalloid or as a colloid. For example, in aqueous solution sodium chloride is a crystalloid and in benzene it is a colloid.

Depending on the aggregate state of the dispersion medium, colloidal solutions are aerosols, lyosols and solid ash.

Depending on the structure of the dispersed phase particles, colloids are divided into association (phase particles are groupings of atoms, ions or molecules) and molecular (phase particles are individual macromolecules).

Molecular colloids are prepared analogously to true solutions by dissolving the corresponding high molecular weight substance in a suitable solvent. Thus, each molecule of the solute is a colloidal particle because its size ranges from 1 to 100 nm. Molecular colloidal solutions of gelatine in water, of natural rubber in benzene, of egg white in water, etc., are thus obtained.

Association colloids are produced by two main methods: dispersion and condensation:

	Dispersion methods Condensation methods	
Coarsely dispersed systems (particles larger than 100 nm)	↘ ↙ Colloid-dispersed systems (particles from 1 to 100 nm)	Molecular (ionic) dispersed systems or true solutions (particles smaller than 1 nm)

The dispersion method consists of crushing the starting substance to particles of colloidal size. The comminution is achieved in several ways: by direct mechanical grinding of the substances in colloid mills, by using ultrasound, by electrical pulverization (when obtaining ashes of metals), etc.

Crushing does not change the chemical composition of the starting substance.

The condensation method for the preparation of colloidal solutions is limited to the reduction of the particles (atoms, molecules or ions) of the solute to colloidal size. This nucleation is achieved in several ways:

1. When lowering the temperature of gas mixtures – so the fog is formed;

2. Precipitation to colloidal particle sizes can also be carried out in a solution of a low molecular weight compound by reducing the solute solubility in the solvent of interest. For example, if water is added to a solution of sulfur in ethyl alcohol, the solubility of the sulfur in the ethyl alcohol and water mixture is reduced, giving a sulfur ash.

Chemical reactions resulting in the formation of a small solute also produce colloidal solutions.

3. Moderate oxidation of hydrogen sulfide produces sulfur ash:

$$2H_2S + O_2 \longrightarrow 2H_2O + 2S\downarrow$$

$$\begin{array}{llll} S^{-2} - 2\bar{e} \longrightarrow S^{0} & 2 & \multirow{2}{*}{4} & 2 \\ O_2 \longrightarrow 2O^{0} + 2\times 2\bar{e} \longrightarrow 2O^{-2} & 4 & & 1 \end{array}$$

In the "silver mirror" reaction, the aldehyde group is oxidized by the weak oxidant silver ammonia hydroxide, and the silver ion is reduced to colloidal silver:

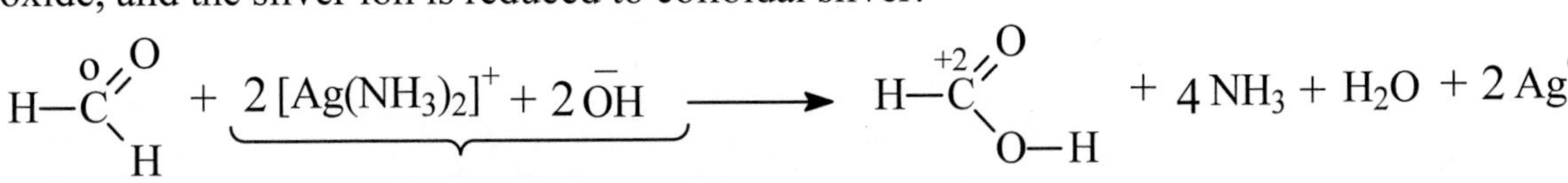

$$\begin{array}{llll} C^{0} - 2\bar{e} \longrightarrow C^{+2} & 2 & \multirow{2}{*}{2} & 1 \\ Ag^{+} + \bar{e} \longrightarrow Ag^{0} & 1 & & 2 \end{array}$$

4. Hydrolysis of salts can also produce colloidal solutions when the products of the interaction are insoluble hydroxides.

$$\underbrace{Fe^{+3} + 3\bar{Cl}} + \underbrace{3H^{+} + 3\bar{OH}}_{\substack{\uparrow\downarrow \\ 3H_2O}} \longrightarrow \underbrace{3\overset{+}{H} + 3\bar{Cl}} + Fe(OH)_3$$

5. Similarly, ion exchange reactions can produce ashes of insoluble sulfides, halides, carbonates, etc., for example:

$$AgNO_3 + KI \longrightarrow KNO_3 + AgI\downarrow$$

$$\underbrace{Ag^{+} + NO_3^{-}} + \underbrace{K^{+} + I^{-}} \longrightarrow \underbrace{K^{+} + NO_3^{-}} + AgI\downarrow$$

$$Ag^{+} + I^{-} \longrightarrow AgI\downarrow$$

When zol is obtained by the above equation, the ions of potassium nitrate remain in solution.

In chemical reactions in which few solutes are formed, it is necessary to dilute the starting solutions to obtain colloidal solutions.

Colloidal solutions are purified of ions remaining in the medium using semipermeable barriers through which ions and molecules of low molecular weight compounds pass.

The ash to be purified shall be placed in a vessel, the bottom of which shall be a semi- permeable barrier. This vessel is placed in another, larger vessel with distilled water. The ions and molecules of low molecular weight compounds pass through the barrier and into the distilled water, which is continuously replaced until the impurities are removed.

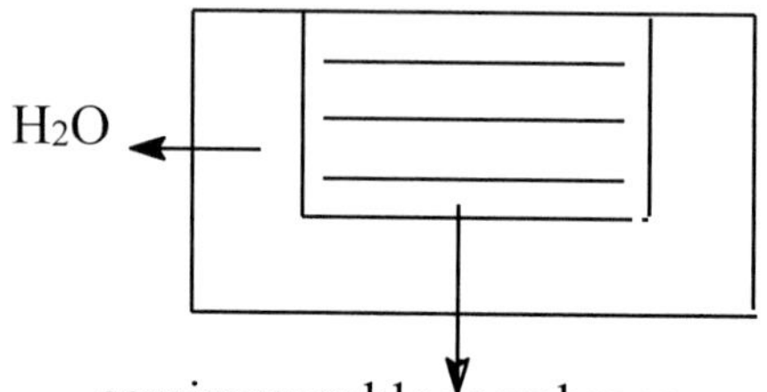

This method of purifying the ash is called dialysis, and the instruments used are dialyzers. Dialysis is a slow process.

Modern dialyzers combine conventional dialysis with electrolysis and are therefore called electrodialyzers.

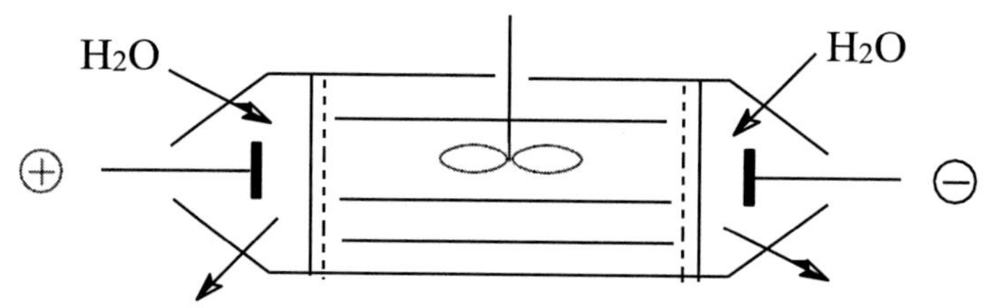

Conventional electrodialyzers are three-chamber. The electrodes – cathode and anode – are immersed in the water of the two end chambers. In the middle chamber, the colloidal solution is placed. The middle chamber is separated from the end chambers by semipermeable membranes. Under the action of the electric field created between the electrodes, ions from the middle chamber pass into the end chambers, depending on their charge. With the aid of the mechanical stirrer, the process is considerably accelerated.

Structure of colloidal particles

Research has shown that colloidal particles derived from low molecular weight compounds (associative colloids) have complex structures. For example, the composition and structure of a silver iodide ash colloidal particle produced by the interaction of silver nitrate and potassium iodide (in excess) can be represented as follows:

$$\underbrace{Ag^{+} + NO_3^{-}} + \underbrace{K^{+} + I^{-}} \longrightarrow \underbrace{K^{+} + NO_3^{-}} + AgI$$

A collection of insoluble silver iodide particles forms the nucleus of each colloidal particle in this ash. Iodide ions are adsorbed on the surface of the nucleus. The adsorbed ions in turn attract ions of opposite charge, called counterions (in this case K^{+}), from solution. The adsorbed ions and counter ions form the adsorption layer of the colloidal particles. The remaining counterions in the liquid, mobile relative to the adsorption layer, form the diffusion layer. The nucleus together with the adsorption layer form a particle which is called **a granule** and has a definite electric charge. In this case it is negative. The granule and the counterions of the diffusion layer form the so-called **micelle**.

The micelle is an electroneutral particle in the colloidal solution.

Structure of colloidal particles

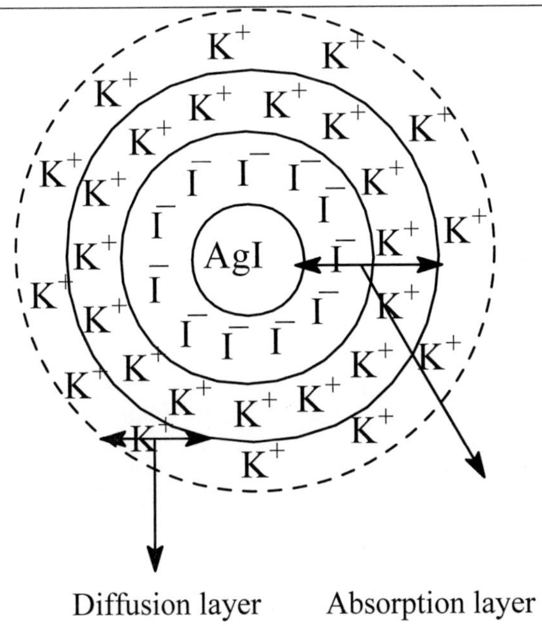

(A particle with a negative charge obtained in excess of potassium iodide)

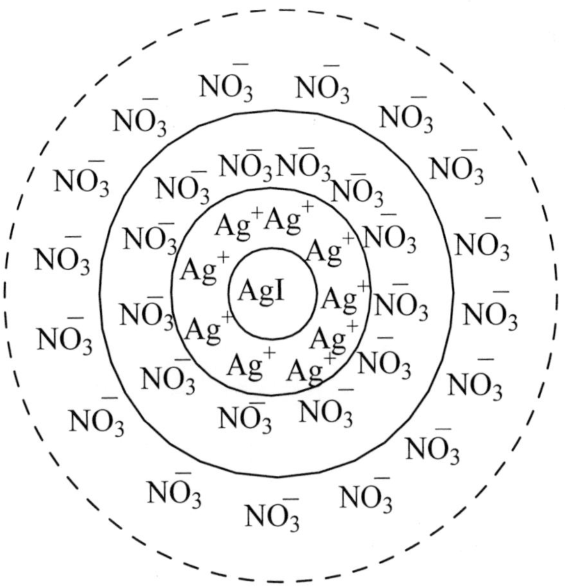

(A particle with a positive charge, obtained in excess of silver nitrate)

The composition of colloidal particles can be represented by qualitative formulas, which in this example have the form:

$$\underbrace{\underbrace{\underbrace{\{m[AgI]}_{\text{core}}\ \underbrace{n\bar{I}\ (n-x)K^{+}\}^{x-}}_{\text{adsorption layer}}}_{\text{granule}} \cdot \underbrace{xK^{+}}_{\text{diffusion layer}}}_{\text{micelle}}$$

If a silver iodide pellet is prepared in excess of silver nitrate, it adsorbs mainly silver ions, which determine the positive charge of the pellet.

$$\underbrace{\underbrace{\left\{\underbrace{m[AgI]}_{\text{core}}\ \underbrace{nAg^{+}(n-x)NO_3^{-}}_{\text{adsorption layer}}\right\}^{x+}}_{\text{granule}} \bullet\ \underbrace{xNO_3^{-}}_{\text{diffusion layer}}}_{\text{micelle}}$$

For example, the colloidal particle of barium sulfate precipitate can be obtained in two ways:

(a) in excess of barium dichloride:

$$BaCl_2 + Na_2SO_4 \longrightarrow BaSO_4\downarrow + 2\,NaCl$$

$$\underbrace{Ba^{+2} + 2Cl^{-}} + \underbrace{2Na^{+} + SO_4^{-2}} \longrightarrow BaSO_4\downarrow + \underbrace{2Na^{+} + 2\,Cl^{-}}$$

$$\{m[BaSO_4]\bullet nBa^{+2}\bullet 2\,(n-x)\,Cl^{-}\}^{2\,x^{+}}\bullet 2\,xCl^{-}$$

(b) in excess of disodium sulphate:

$$Na_2SO_4 + BaCl_2 \longrightarrow BaSO_4\downarrow + 2\,NaCl$$

$$\underbrace{2Na^{+} + SO_4^{-2}} + \underbrace{Ba^{+2} + 2Cl^{-}} \longrightarrow BaSO_4\downarrow + \underbrace{2Na^{+} + 2\,Cl^{-}}$$

$$\{m[BaSO_4]\bullet nSO_4^{-2}\bullet 2\,(n-x)Na^{+}\}^{2\,x^{-}}\bullet 2\,xNa^{+}$$

Colloidal particles relate differently to solvent molecules. If the particles of a given sol **interact with solvent molecules** to form a solvate (hydrated in solvent water) shell, the colloids are called **lyophilic (hydrophilic).** To this type of colloids belong the sol derived from gelatin, starch, proteinaceous substances, silicic acid (H_2SiO_3), etc.

If the particles of a sol do **not interact with the solvent molecules**, the colloids are called **lyophobic (hydrophobic).** Such are the ashes derived from metals, sulphides and other salts insoluble in water.

Properties of colloidal solutions

The properties of colloidal solutions are determined by the complex structure and the specific sizes of the colloidal particles. They can be grouped into three: kinetic, optical and electrical.

The kinetic properties of colloidal solutions differ from those of true solutions. The particles of colloidal solutions are in a continuous motion called **Brownian motion**. It results from the chaotic shocks that the particles experience from the moving molecules of the medium. This causes each colloidal particle to travel a very complex path.

Colloidal particles have much larger particle sizes than particles in ordinary solutions, so they move much more slowly than they do. This accounts for the significantly slower diffusion in colloidal solutions.

At the same concentration, the osmotic pressure of colloidal solutions is lower than the osmotic pressure of ordinary solutions. This is due to the smaller number of particles in colloidal solutions.

The decrease in solvent vapor pressure above the colloid solution, the increase in boiling point, and the decrease in freezing point of colloid solutions are **insignificant** compared to the changes that occur in ordinary solutions, again due to the smaller number of particles in colloid solutions.

Colloidal solutions exhibit some **specific optical properties**. These are due to the larger particle sizes of the colloids compared to the particle sizes in ordinary solutions. The sizes of colloidal particles are commensurate with the wavelengths of light in the visible part of the spectrum.

If a light beam from the visible part of the spectrum passes through a solution, it can be absorbed, reflected or scattered depending on the size of the particles in the solution.

The most characteristic **optical property of** colloidal solutions is the scattering of light in all directions. This property was first studied by the English physicist **D. Tyndall**. He found that if a colloidal solution is illuminated with a bright beam of light and its path is observed laterally, the beam is seen as a glowing cone in the solution. This phenomenon is called the **Tyndall effect** and is due to the diffraction of light between colloidal particles.

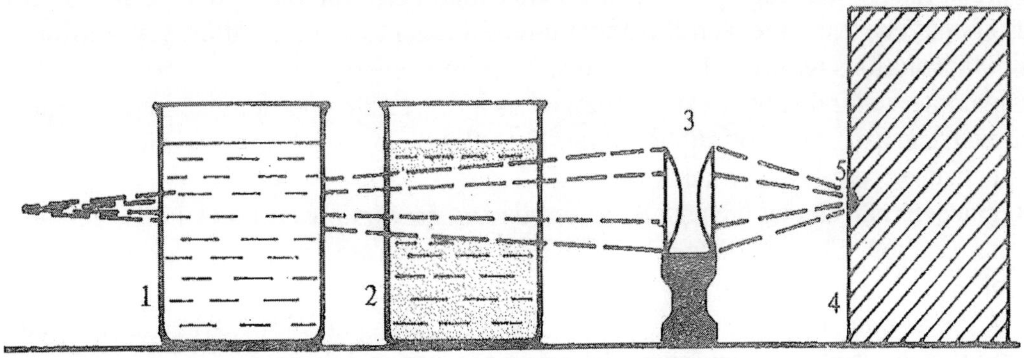

No Tyndall effect is observed when light is passed through pure water or a true solution.

Some **electrical properties** are also inherent in the ashes. Colloidal particles have a significant boundary surface and can adsorb ions from the medium in which they are located. The **ions adsorbed** on the nucleus are **mainly those** which form insoluble compounds with ions of opposite charge from the nucleus. If a silver iodide pellet is prepared in excess of silver nitrate, silver ions will be adsorbed which will determine the positive charge of the pellet. If the preparation of the same ash is carried out in excess of potassium iodide, iodide ions will adsorb and the granule will have a negative charge. This means that the charge of the granule depends on whether the reaction took place in excess of one or the other starting substance.

Since colloidal particles are carriers of certain electric charges, under the action of a constant electric current, they acquire a directed motion and move towards the opposite charged pole of the field. This motion is called **electrophoresis**. When a constant electric current is passed through a colloidal solution, the granule and diffusion layer move in opposite directions. Therefore, the charge of the colloidal particle is understood as the charge of the granule.

Electrophoresis finds application in medicine (clinical chemistry, immunohematology, etc.).

Stability of colloidal solutions

A major factor that determines the stability of colloidal solutions is the charge of the colloidal particles. In a sol, all the granules have the same charge and continuously repel each other as they meet as a result of Brownian motion. On the other hand, the solvation of counterions in the diffusion layer is also a factor in preventing the colloidal particles from clumping together and thus assists in increasing the stability of the sol. The greater the degree of solvation of the counterions in the diffusion layer, the bulkier the diffusion layer and the denser the solvate envelope around the monovalent loaded particles becomes, resulting in a more stable colloidal solution.

The factors that affect the stability of association colloids are various: temperature, ash concentration, various mechanical effects, introduction of electrolytes, other ashes, etc.

The process of nucleation of the colloidal particles is called **coagulation**, and sedimentation under the action of their gravity - **sedimentation**.

Coagulation takes place under the action of various factors. When the ash is heated by the high temperature, the desorption of adsorbed ions is accelerated, and with this the particles lose their charge, the solvate shell is destroyed, due to which the colloidal particles are more easily joined and nucleated.

The addition of a sol of opposite charge also accelerates coagulation as the particles from the two solutions neutralize each other's charges.

The addition of electrolytes causes coagulation of the colloidal solution. Concentrated electrolyte solutions are used for coagulation of hydrophilic colloids. The electrolyte takes away the hydrate shell of the colloidal particles, after which the ash coagulates. This process is called **desalination**. Desalination is the process of separating the dispersed phase when a large amount of salt is added to the colloid. The desalting action of salts consists in their own solvation at the expense of the desalvation of macromolecules. Desalination has important practical significance in the manufacture of soaps.

The precipitates produced by coagulation are called **gels**. Some gels may re-dissolve to ash. Such colloids are called reversible, and the transition of the gel to a sol is called **peptization**.

Colloidal solutions whose gels cannot dissolve are irreversible.

For example, natural rubber can be viewed as a reversible colloid. The latex extracted from the rubber tree is the ash of natural rubber. After coagulation, a gel is formed which can be re-dissolved (e.g., in benzene, petrol) to form ash.

Proteinaceous substances are usually referred to as irreversible colloids. For example, if a solution of a proteinaceous substance coagulates at high temperature, re-dissolution of the gel formed is impossible.

Under the action of the Earth's gravity, the colloidal particles settle. This process is called sedimentation and is more pronounced the larger the size and density of the colloidal particles.

Meaning and use

Colloidal solutions are widespread in nature and are used in many industries. Colloidal solutions are milk, blood, protoplasm of living cells, lymph, plant juices, etc., which as ash provide life to organisms. In the form of gels, muscle fibers, tendons, connective tissue, etc., they perform certain functions in living organisms.

The minerals opal, agate, ruby, etc. are solid colloidal solutions. Clay and other sedimentary minerals are also colloidal solutions.

Colloid solutions are used in a number of production processes of the food, textile, fur, rubber and other branches of industry.

Sample problems and solutions

Task 1:

Suggest a binary (two-element) alkali salt and a salt of an element of group IVA, the interaction of whose aqueous solutions can produce a colloidal solution. Give reasons for your choice. What is this method of making colloids called? Explain with a drawing the structure and electrical charge of the colloidal particles depending on the conditions for their preparation.

Exemplary solution

A necessary condition for obtaining lyophobic colloidal solutions (as is the case here) is that the dispersed phase be practically insoluble in the dispersion medium. This means that the reaction sought must give a slightly soluble salt. The condition of the problem implies variation in the choice of the starting alkali salt. It can be a halide or a sulfide. The other salt is soluble lead dinitrate - $Pb(NO_3)_2$. Then one of the possible reactions would be:

$$Na_2S + Pb(NO_3)_2 \longrightarrow PbS + NaNO_3$$

The full ionic equation of the reaction is:

$$\underbrace{2\,Na^{+} + S^{-2}} + \underbrace{Pb^{+2} + 2\,NO_3^{-}} \longrightarrow PbS + \underbrace{2\,Na^{+} + 2\,NO_3^{-}}$$

Abbreviated ionic equation:

$$Pb^{+2} + S^{-2} \longrightarrow PbS$$

The reaction used refers to condensation methods, since larger particles (microcrystals), which have the size of colloidal particles, are produced from ions by crystallization (condensation).

The particles of the lyophobic colloidal solutions are composed of a core, an adsorption and a diffusion layer. The core is a microcrystal of the slightly soluble substance (in this case PbS). It adsorbs selectively on its surface one of the constituent ions (lead or sulfide). The particular type of these ions is determined by the conditions of preparation. The ion initially adsorbed is the one which in the precipitation process turns out to be in excess. Counterions are attracted around these initially adsorbed ions and the resulting double electric layer is called adsorptive.

The core and the adsorption layer form the so-called granule. Around the granule are further counterions which are not tightly bound to it and this last ionic shell constitutes the diffusion layer. The granule and this layer are called the mycelium.

The ions initially adsorbed on the nucleus determine the electric charge of the particles and are called potential-determining ions.

For the reaction under consideration, the structure of the colloidal particles can schematically be presented as follows:

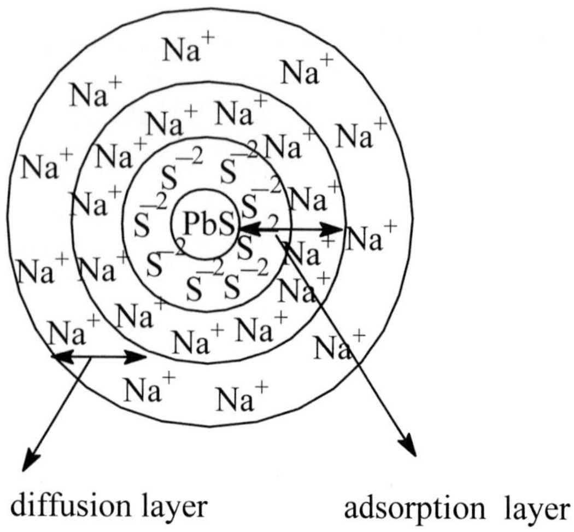

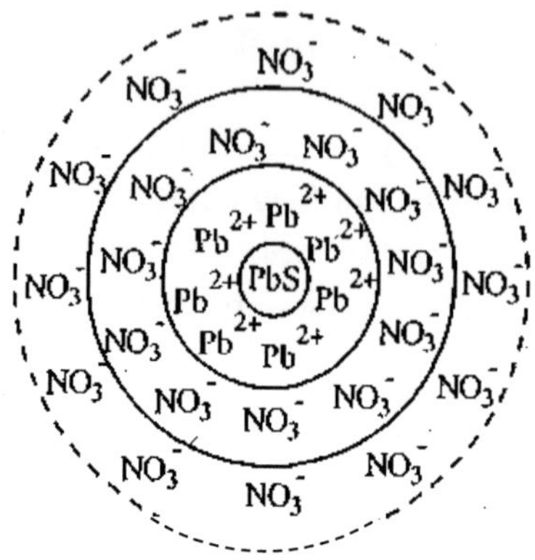

Particle with negative charge, obtained in excess of Na_2S

Particle with positive charge, obtained in excess of $Pb(NO_3)_2$

The composition of colloidal particles can be represented by qualitative formulas, which for this case have the form:

$$\left\{ m[PbS]nS^{-2}\ 2\ (n-x)Na^{+} \right\}^{2x^{-}} \cdot 2xNa^{+}$$

$$\left\{ m[PbS]nPb^{+2}\ 2\ (n-x)NO_3^{-} \right\} \cdot 2xNO_3^{-}$$

15. SORPTION

BASIC CONCEPTS. NATURE OF ADSORPTION FORCES. PHYSICAL AND CHEMICAL ADSORPTION. FEATURES OF ADSORPTION. ADSORPTION ISOTHERM. APPLICATION OF ADSORPTION.

A number of chemical and chemical-technological processes take place in heterogeneous systems. At the interface between the different phases there is often an accumulation of particles contained in one phase. Sometimes such particles also penetrate and persist in the volume of the other phase.

The uptake of one substance by another in a different phase is a physicochemical process called **sorption**. It consists in a change in the concentration of a substance at a boundary surface with respect to its concentration in one of two phases, i.e., retention, absorption of one substance by another substance in another phase.

Sorption is a typical phenomenon in all heterogeneous systems.

When the sorbed substance is retained at the interface between the two phases in the system, the process is called **adsorption**, and when absorption occurs throughout the volume of one of the phases, **absorption**.

The substance that is retained on its surface or absorbed is called the **sorbent (adsorbent, absorbent)**, and that which is retained or absorbed by the sorbent is called the **sorbate (adsorbate, absorbate).**

Often the two processes - adsorption and absorption - occur together. Substances are first adsorbed and then penetrate into the interior (absorbed). Absorption therefore always starts from the surface as adsorption.

Nature of adsorption forces

Adsorption is due to the different state of the particles on the surface of the adsorbent and those located in its volume.

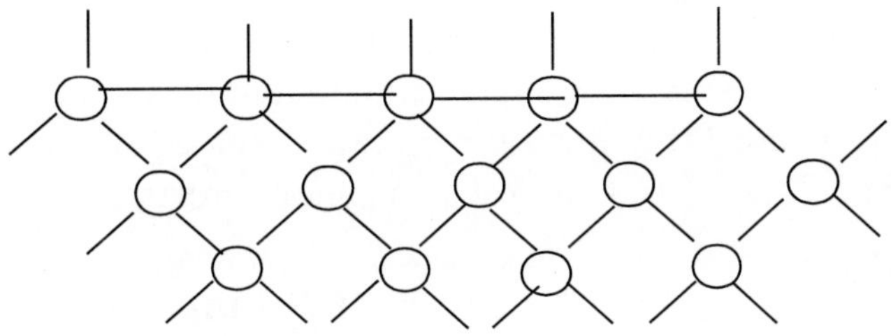

A force field exists around each building block. The force fields of the particles inside the crystal saturate each other. These fields, which correspond to the particles on the surface of the body, attract and hold any particles that come within the reach of the force field formed on the surface. These adsorption forces are distinct from chemical bonding forces.

Adsorption forces do not have the property of saturation. A large number of adsorbent molecules can be retained (adsorbed) on the surface. Several molecular layers can even be adsorbed.

If the forces of attraction between the adsorbent and adsorbate are weak and no chemical interaction takes place, the phenomenon is called physical adsorption. This is, for example, the adsorption of ammonia by activated carbon.

In chemical adsorption (chemisorption), an electron exchange takes place between the adsorbent and the adsorbate, the bond formed is strong and in some cases even a chemical compound is formed.

The adsorption of water by diphosphorus pentoxide or by calcium oxide during drying of gases can be considered as chemical adsorption

$$P_2O_5 + 3H_2O \longrightarrow 2\,H_3PO_4$$

$$CaO + H_2O \longrightarrow Ca(OH)_2$$

as well as the adsorption of hydrogen chloride, nitrogen dioxide, sulfur trioxide from water to produce the corresponding acids:

$$2\,NO_2 + H_2O \longrightarrow HNO_3 + HNO_2$$

and

$$4\,NO_2 + 2H_2O + O_2 \longrightarrow 4\,HNO_3\;;$$

$$SO_3 + H_2O \longrightarrow H_2SO_4$$

Physical and chemical adsorption

There is no sharp boundary between physical and chemical adsorption.

Adsorption of electrolytes by solid adsorbents is the basis of such important processes as uptake of nutrients from the soil, transfer of mineral substances from the soil into the cells of plant organisms, etc. Adsorption of electrolytes is of great importance in many phenomena that occur in colloid-dispersed systems.

Of the ions present in a solution, mainly those ions which are involved in the composition of the adsorbent and can complete its crystal lattice will be adsorbed on the surface of the adsorbent. For example, if silver iodide crystals are obtained by Eq,

$$\underbrace{Ag^{+} + NO_3^{-}} + \underbrace{K^{+} + I^{-}} \longrightarrow AgI + \underbrace{K^{+} + NO_3^{-}}$$

but in excess of silver nitrate, then of the ions present in the medium (K^+, NO_3^-, Ag^+) on the surface of the adsorbent will adsorb mainly Ag^+. This type of adsorption is selective in nature and is called selective or specific adsorption.

The essence of ion-exchange adsorption is that the adsorbent absorbs a certain type of ion from solution and at the same time releases an equivalent amount of another type of ion with the same charge into solution. The adsorbents in this case are called ionites. Ionites that adsorb and exchange cations are called cationites, and those that exchange anions are anionites.

Features of adsorption

Adsorption is an equilibrium process. Along with it, the opposite process takes place - desorption, in which the adsorbed particles leave the adsorbent surface. Desorption is due to the thermal movement of the particles. If an adsorbed particle acquires a kinetic energy greater than the energy with which it is retained by the adsorbent, it is detached from its surface. Adsorption is therefore a reversible process.

$$\text{Adsorption} \rightleftharpoons \text{Desorption}$$

Adsorption is almost always an exothermic process. In accordance with the Le Chatelier-Brown principle, as temperature increases, the desorption process is enhanced, and as temperature decreases, adsorption is enhanced.

If a system is at a constant temperature, an adsorption equilibrium is established in which the adsorption rate equals the desorption rate. Per unit time, as many particles are adsorbed as are desorbed. The adsorption equilibrium is ***dynamic and mobile***.

The amount of substance adsorbed per unit amount of adsorbent depends on the nature and boundary surface of the adsorbent and the conditions under which the process takes place – temperature and pressure.

The relationship between the amount adsorbed and the pressure (concentration) of the adsorbing gas at constant temperature is given by the **adsorption isotherm**.

Application of adsorption

Adsorption is used for the purification of natural products, gases and liquids, in the preparation of pure substances, in the separation of mixtures, in catalytic processes, in gas protection, etc.

An important method of separation of substances based on adsorption is chromatography. It is a physicochemical method for separating substances from mixtures by exploiting differences in their adsorption properties. Chromatography can be column, paper, thin layer, liquid, gas etc. The method was first proposed by the Russian botanist M. Tsvet. The principle of the method is as follows:

A glass column is filled with adsorbent. A small portion of the mixture to be separated is introduced into the upper end of the column. It is adsorbed there. The column is then washed with a slow stream of suitable solvent (eluent). The less adsorbed substances are desorbed and moved down the column with the solvent. As pure adsorbent is encountered here, they are adsorbed again. Thus, as the column is washed, there is continuous desorption, movement down the length of the column, adsorption, desorption again, new adsorption, and so on. The various components of the mixture move downward, and more slowly, the more firmly the adsorbent holds them. A portion containing the least adsorbed substance will come out of the bottom of the column first, then a second portion of a more strongly adsorbed substance, and so on. The substances are separated according to the strength of their interaction with the adsorbent.

Chromatography is used to separate complex mixtures of substances, e.g., mixtures of vitamins, amino acids, enzymes, drugs, etc.

Ion exchange chromatography is widely used in engineering. Sodium cationite, which is involved in the exchange process, is used to soften hard natural water:

$$2[\text{Cationite}]Na^{+} + Ca^{+2} \rightleftharpoons [\text{Cationite}]_2\,Ca^{+2} + 2\,Na^{+}$$

Ionites are used to purify sugar juice in sugar production, to purify wine from metal ions that cause it to cloud.

Molecular sieves are also widely used. They serve to separate mixtures of substances into fractions according to the size of their molecules. Molecular sieves are adsorbents with a particularly highly developed surface, as they contain numerous pores, slits, labyrinths. The capabilities of a molecular sieve are determined by the sizes of these pores. When a mixture is passed through a molecular sieve, only those molecules that are smaller in diameter and can enter the capillaries of the sieve adsorbent are adsorbed. The other molecules pass through the sieve without being retained. A sieve with slightly wider pores is then used. These will adsorb and separate the next molecules in size and so on. The individual fractions can be obtained after desorption.

Molecular sieves are natural zeolites.

16. Hydrogen Compounds

Types. Ionic (Salt) Hydrogen Compounds (Hydrides). Covalent Hydrogen Compounds. Metallic Hydrogen Compounds. Preparation of Hydrogen Compounds.

Hydrogen compounds are chemical compounds of the elements with hydrogen. Depending on the element bonded to the hydrogen, and more specifically the electronic structure of its atoms, hydrogen can be in the form of a proton (H^+) and a hydride anion (H^-).

A large number of elements form polymeric hydrogen compounds and hydrogen compounds of variable composition (solid solutions).

Types

Hydrogen compounds are classified on the basis of the nature of the chemical bond connecting the hydrogen to the element. They are **ionic (salt-like), covalent** and **metal-like**.

Ionic hydrogen compounds (hydrides) correspond to the s-elements of the Periodic System, **excluding the elements beryllium and magnesium**. Their general formula for:

alkaline elements: $E^+ H^-$

alkaline earth elements: $E^{+2}\, 2H^-$

Their composition includes metal cations and hydride anions. They are called **salts** because they are similar in structure to the alkali and alkaline earth halides (NaCl, CaF_2). They have an ionic crystalline lattice and, under ordinary conditions, are solid colorless substances. They have a high melting point and can conduct an electric current, proving their ionic structure. When heated to temperatures above 400° C in the absence of air, hydrides decompose to metal and hydrogen, except for **lithium hydride (LiH)**, which melts at 650° C without decomposition.

Alkaline hydrides:

$$2\, E^+ H^- \xrightarrow{t^o} 2\, E^o + H_2^o \uparrow$$

Alkaline earth hydrides:

$$E^{+2} 2H^- \xrightarrow{t^o} E^o + H_2^o \uparrow$$

These are redox processes in which the alkali or alkaline earth ion **is reduced to free metal** and the hydride ion **is oxidized to elemental hydrogen**. As the sequence number of the element in the IA and IIA groups increases, the thermal stability of the corresponding hydrides decreases. **Hydrides are very strong reducers due to the instability of the hydride** ion:

$$2\, H^- - 2 \times 1\, e^- \longrightarrow 2\, H^o \longrightarrow H_2 \uparrow$$

The hydride ion acts as a reducing agent and is oxidized. The interaction with water proceeds vigorously with hydrogen evolution and base formation:

$$E^{+}H^{-} + \underbrace{H^{+} + \bar{O}H} \longrightarrow \underbrace{E^{+} + \bar{O}H} + \overset{0}{H_2}\uparrow$$

$$\left.\begin{array}{l} H^{-} - e^{-} \longrightarrow H^{0} \\ H^{+} + e^{-} \longrightarrow H^{0} \end{array}\right\} \longrightarrow H_2\uparrow$$

$$E^{+2}2H^{-} + \underbrace{2H^{+} + 2\bar{O}H} \longrightarrow E^{+2}(\bar{O}H)_2 + 2H_2\uparrow$$

Hydrogen compounds of the alkali and alkaline earth elements react violently with acids:

$$2Na^{+}H^{-} + \underbrace{2H^{+} + 2Cl^{-}} \longrightarrow \underbrace{2Na^{+} + 2Cl^{-}} + H_2\uparrow$$

Hydrides are strong reducers and can reduce metals by decomposing their oxides:

$$2Na^{+}H^{-} + MoO \longrightarrow Mo + Na_2O + H_2\uparrow$$

$$\begin{array}{ll|c|c|c} Mo^{+2} + 2e^{-} \longrightarrow & \overset{0}{Mo} & 2 & & 1 \\ & & & 2 & \\ 2H^{-} - 2\times1e^{-} \longrightarrow 2H^{0} \longrightarrow & H_2\uparrow & 2 & & 1 \end{array}$$

Heated in the presence of oxygen, hydrides burn to the corresponding oxide and water:

$$4Na^{+}H^{-} + CO_2 \longrightarrow 2Na_2O + C + 2H_2\uparrow$$

$$\begin{array}{ll|c|c|c} 2H^{-} - 2\times1\bar{e} \longrightarrow 2H^{0} \longrightarrow & H_2\uparrow & 2 & & 2 \\ & & & 4 & \\ C^{+4} + 4e^{-} \longrightarrow & C^{0} & 4 & & 1 \end{array}$$

The reducing power of hydrides is so great that they can reduce even carbon dioxide to elemental carbon:

$$4Na^{+}H^{-} + CO_2 \longrightarrow 2Na_2O + C + 2H_2\uparrow$$

$$\begin{array}{ll|c|c|c} 2H^{-} - 2\bar{e} \longrightarrow 2H^{0} \longrightarrow & H_2\uparrow & 2 & & 2 \\ & & & 4 & \\ C^{+4} + 4e^{-} \longrightarrow & C^{0} & 4 & & 1 \end{array}$$

In the case of group IIA elements, the activity of their simple substances towards hydrogen decreases with increasing sequence number.

The covalent hydrogen compounds correspond to the p-elements of groups IVA to VIIA of the Periodic Table. Their general formulae are: IVA - $E^{-4}\,4H^{+}$; VA -$E^{-3}\,3H^{+}$;

VIA – $2H^{+}\,E^{-2}$; VIIA - $H^{+}\,E^{-}$.

They are characterized by **molecular structure**. The degree of polarization of the E-H bond depends on the electronegativity of the hydrogen bonded element. Under ordinary conditions, **covalent hydrogen compounds are gaseous or liquid**. They are volatile substances and some of them show sharp variations in boiling point. H_2O, HF, and NH_3 have higher boiling points than their counterparts. This is due to the association of molecules in the liquid state due to the formation of intermolecular hydrogen bonds. As the sequence number of the p-elements increases, the thermal stability of the corresponding covalent hydrogen compounds decreases. Covalent hydrogen compounds exhibit reducing properties towards halogen elements and oxygen.

$$2\,NH_3 + 3\,Cl_2 \longrightarrow N_2 + 6\,HCl$$

$$\begin{array}{llll|c|c|c} 2N^{-3} - 2\times 3\bar{e} \longrightarrow 2N^{0} \longrightarrow & N_2 & & & 6 & & 1 \\ & & & & & 6 & \\ Cl_2 \longrightarrow 2Cl^{0} + 2\times 1\bar{e} \longrightarrow & 2Cl^{-} & & & 2 & & 3 \end{array}$$

Ammonia is oxidised in two ways depending on whether a catalyst is used or not.
Without catalyst:

$$4NH_3 + 3O_2 \longrightarrow 2N_2 + 6\,H_2O + Q$$

$$\begin{array}{ll|c|c|c} 2N^{-3} - 2\times 3\bar{e} \longrightarrow 2N^{0} \longrightarrow & N_2 & 6 & & 2 \\ & & & 12 & \\ O_2 \longrightarrow 2O^{0} + 2\times 2e^{-} \longrightarrow & 2O^{-2} & 4 & & 3 \end{array}$$

In the air, hydrogen sulfide burns with a sky-blue flame to sulfur dioxide:

$$2H_2S + 3O_2 \longrightarrow 2\,SO_2 + 2\,H_2O$$

$$\begin{array}{ll|c|c|c} S^{-2} - 6\bar{e} \longrightarrow & S^{+4} & 6 & & 2 \\ & & & 12 & \\ O_2 \longrightarrow 2O^{0} + 2\times 2e^{-} \longrightarrow & 2O^{-2} & 4 & & 3 \end{array}$$

Covalent hydrogen compounds treat water differently. For example, the hydrogen compounds of boron (B – group IIIA) and silicon (Si – group IVA) react with water, decomposing with the evolution of hydrogen to form the corresponding acids:

$$SiH_4 + 3\,H_2O \longrightarrow H_2SiO_3 + 4\,H_2\uparrow$$

$$\begin{array}{lll|c|c|c}
 & \overset{-4}{Si} - 8\,\bar{e} & \longrightarrow \overset{+4}{Si} & 8 & & 1 \\
 & & & & 8 & \\
2H^+ + 2 \times 1\,\bar{e} & \longrightarrow 2\,H^o & \longrightarrow H_2\uparrow & 2 & & 4
\end{array}$$

$$BH_3 + 3H_2O \longrightarrow H_3BO_3 + 3\,H_2\uparrow$$

$$\begin{array}{lll|c|c|c}
 & \overset{-3}{B} - 6\,\bar{e} & \longrightarrow \overset{+3}{B} & 6 & & 1 \\
 & & & & 6 & \\
2H^+ + 2 \times 1\,\bar{e} & \longrightarrow 2\,H^o & \longrightarrow H_2\uparrow & 2 & & 3
\end{array}$$

Hydrogen compounds of the elements arsenic (As) and antimony (Sb) of group VA do not react with water. In the hydrogen compounds of nitrogen and phosphorus, protons join and the compounds act as weak bases (ammonia is a weak base, and phosphine (hydrogen phosphide) reacts only with very strong acids:

$$NH_3 + \underbrace{H^+ + \overset{-}{OH}} \longrightarrow \underbrace{\overset{+}{NH_4} + \overset{-}{OH}}$$

$$PH_3 + \underbrace{\overset{+}{H} + ClO_4^-} \longrightarrow \underbrace{\overset{+}{PH_4} + ClO_4^-}_{\text{phosphonium perchlorate}}$$

Salt is hydrolysed when water is added, releasing phosphine (a fishy smelling gas).

Of the hydrogen compounds of the group VIA elements, oxygen (O_2), sulfur (S), selenium (Se), and tellurium (Te), water is an ampholyte, and the remaining compounds act as weak dibasic, oxygen-free acids, such as hydrogen sulfide.

Hydrogen compounds of the elements of group VIIA – fluorine, chlorine, bromine, iodine – in aqueous solution dissociate as acids. They are gaseous substances with a pungent, irritating odor. Hydrogen fluoride is highly poisonous. The hydrogen fluoride molecule is the most strongly polar but the least polarizable compared to the molecules of other halogenated hydrocarbons. This is the reason hydrochloric acid is moderately strong and weaker than hydrofluoric acid. Aqueous dipoles have difficulty polarizing the hydrogen fluoride molecule and therefore its dissociation occurs to a lesser degree. The molecules of HCl, HBr, HJ are less polar but more easily polarizable under the influence of water molecules and are stronger acids. The strongest of these is hydroiodic acid, but it is also the most perishable because hydroiodic is a good reducing agent.

HF, HCl, HBr, HJ

← Polarity

→ Polarizability

→ Acidity of aqueous solutions

→ Perishability of acids

As the ordinal number of the element increases (increments), its reducing power (by group) also increases.

In the compounds of beryllium (Be) and magnesium (Mg) of the IIA and aluminium (Al), boron (B), gallium (Ga), indium (In) of the IIIA groups, most of which are obtained indirectly, the molecules are connected by hydrogen bridges (hydrogen bonds) – stronger than usual. For this reason, they are polymeric.

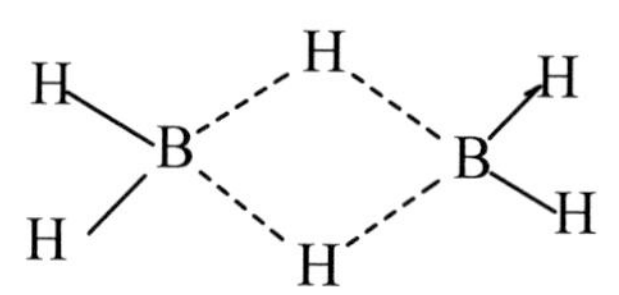

B_2H_6 (diboran)

These hydrogen compounds can react with ionic hydrides to produce complex hydrogen compounds:

$$2\ LiH + B_2H_6 \longrightarrow 2\ \overset{+}{Li}[BH_4]^-$$

The relationship is donor-acceptor.

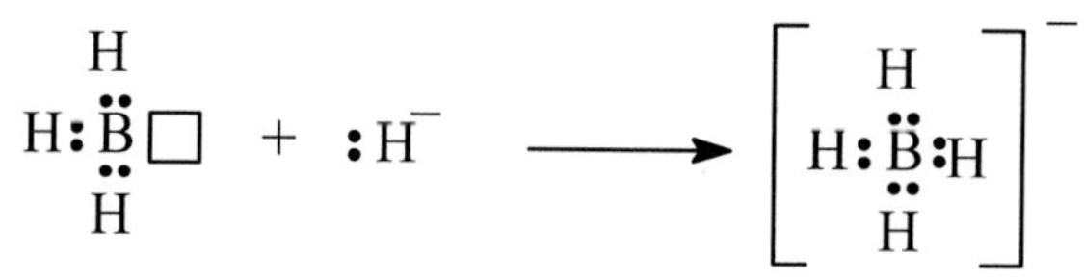

The hydrides of aluminium, gallium, indium of group IIIA are polymers of composition EH_3. They are slightly stable, solid amphoteric substances.

Period by period, with an increasing order number of elements from IA to VIIA groups, a gradual transition from ionic to covalent hydrogen compounds takes place. For the elements of period III, the boundary between the two classes (types) of hydrogen compounds is at magnesium and aluminium, to which correspond compounds with a polymeric structure.

The metallic hydrogen compounds correspond to the d-elements of groups IIIB to VIIIB, as well as the lanthanides and actinides. The simple substances of these elements absorb reversibly a considerable amount of hydrogen with heat evolution, often accompanied by an increase in the volume of the metal. The amount of hydrogen absorbed depends on partial pressure and temperature. Metals (Pd) that dissolve hydrogen **swell** considerably, become brittle and are covered with cracks. The dissolved hydrogen can be completely separated when the metal is heated in a vacuum to 100° C. The resulting hydrogen compounds are called **metalloids** because they are similar to metals in the nature of the chemical bonds between the element and the hydrogen. The hydrogen in the metal crystal lattice is in the form of protons or hydride anions. Metallic hydrogen compounds are powdery with considerable thermal and electrical conductivity and have a metallic lustre. Most metallic hydrogen compounds are of variable stoichiometric composition.

Preparation of hydrogen compounds

Hydrogen compounds are produced in two ways:

- ***In direct interaction of simple substances:***

$$2\,Na + H_2 \xrightarrow[p]{t^o} 2\,NaH$$

$$Ca + H_2 \xrightarrow[p]{t^o} CaH_2$$

- **By indirect route:**

$$Na_2S + H_2SO_4 \longrightarrow H_2S\uparrow + Na_2SO_4$$

Sample problems and solutions

Task 1.

Explain the type of chemical bonds in hydrogen compounds of chemical elements with the outer electron layer configuration: a) $3s^1$; b) $2s^2\,2p^3$; c) $2s^2\,2p^4$; d) $3s^2\,3p^4$ and e) $3s^2\,3p^5$. Express with chemical equations the possible interactions between these compounds.

Exemplary solution

Electronic configuration of the outer electron layer $3s^1$ (**a**) belongs to the element **sodium**. It is a third period, group IA, sequence number 11, alkali metal. The compounds of metals with hydrogen are called hydrides. They are solids, readily dissolve in water, and have markedly high reactivity. In these reactions, hydrogen appears as an oxidizing agent.

$$2\,Na + H_2 \longrightarrow \overset{+}{Na}\overset{-}{H}$$

$$\overset{o}{Na} - 1\bar{e} \longrightarrow \overset{+}{Na} \quad \Big|\, 1 \,\Big|\, 2 \,\Big|\, 2$$

$$H_2 \longrightarrow 2\,\overset{o}{H} + 2x1\,\bar{e} \longrightarrow 2\,\overset{-}{H} \quad \Big|\, 2 \,\Big|\, \;\Big|\, 1$$

Ionic hydrogen compounds (hydrides) correspond to the s-elements of the Periodic Table.

Electronic configuration of the outer electron layer $2s^2\,2p^3$ (**b**) belongs to the element **nitrogen**. It is of the second period, group VA, sequence number 7. In the ammonia molecule, the nitrogen-hydrogen bonds are covalent polar. Ammonia is a gas.

$$N_2 + 3\,H_2 \longrightarrow 2\,NH_3 + Q$$

$$N_2 \longrightarrow 2\,\overset{o}{N} + 2\,x\,3\bar{e} \longrightarrow 2N^{-3} \quad \Big|\, 6 \,\Big|\, 6 \,\Big|\, 1$$

$$H_2 \longrightarrow 2\,\overset{o}{H} - 2\,x\,1e^- \longrightarrow 2\,H^+ \quad \Big|\, 2 \,\Big|\, \;\Big|\, 3$$

The hydrogen compounds of the non-metallic elements have a molecular structure. Reaction with hydrogen is more characteristic of non-metals since hydrogen has an electronic configuration similar to metals.

The electronic configuration of the outer electron layer $2s^2\ 2p^4$ (**c**) belongs to the element **oxygen**. It is located in the second period, group VIA, sequence number 8. The chemical bonds are covalent polar. The aggregate state of the hydrogen compound of oxygen is liquid.

$$O_2 + 2H_2 \longrightarrow 2H_2O$$

$$\begin{array}{lll|c|c|c} O_2 \longrightarrow 2\overset{0}{O} + 2 \times 2e^- & \longrightarrow & 2\overset{-2}{O} & 4 & \multirow{2}{*}{4} & 1 \\ H_2 \longrightarrow 2\overset{0}{H} - 2 \times 1e^- & \longrightarrow & 2H^+ & 2 & & 2 \end{array}$$

The electronic configuration of the outer electron layer $3s^2\ 3p^4$ (**d**) belongs to the element **sulfur**. Sulfur is a third period element, group VIA, sequence number 16. With hydrogen it forms a gaseous compound with the smell of rotten eggs.

$$S + H_2 \longrightarrow H_2S$$

$$\begin{array}{lll|c|c|c} S^0 + 2e^- & \longrightarrow & S^{-2} & 2 & \multirow{2}{*}{2} & 1 \\ H_2 \longrightarrow 2H^0 - 2 \times 1e^- & \longrightarrow & 2H^+ & 2 & & 1 \end{array}$$

Hydrogen sulfide is the most important hydride of sulfur. It is industrially produced in the interaction of sulfur with hydrogen:

$$S + H_2 \longrightarrow H_2S$$

$$\begin{array}{lll|c|c|c} S^0 + 2e^- & \longrightarrow & S^{-2} & 2 & \multirow{2}{*}{2} & 1 \\ H_2 \longrightarrow 2H^0 - 2 \times 1e^- & \longrightarrow & 2H^+ & 2 & & 1 \end{array}$$

Laboratory **hydrogen sulfide** is produced from metal sulfides and mineral acids:

$$FeS + 2HCl \longrightarrow H_2S + FeCl_2$$

$$Na_2S + 2HCl \longrightarrow H_2S + 2NaCl$$

Hydrogen sulfide occurs in volcanic gases and in some thermal waters. It has a pungent odor of rotten eggs and is poisonous.

Hydrogen sulfide has a reducing effect. In an air environment it burns according to the equation:

$$2\,H_2S + 3\,O_2 \longrightarrow 2\,H_2O + 2\,SO_2$$

$$S^{-2} - 6\,e^- \longrightarrow S^{+4} \quad | \; 6 \; | \; 12 \; | \; 2$$

$$O_2 \longrightarrow 2\,O^0 + 2 \times 2\,e^- \longrightarrow 2\,O^{-2} \quad | \; 4 \; | \; \; | \; 3$$

Hydrogen sulfide dissolves well in water. Its aqueous solution is a weak dibasic acid which yields two types of salts, **hydrogen sulfides** and **sulfides**:

$$H_2S + H_2O \rightleftharpoons H_3O^+ + HS^-$$

$$HS^- + H_2O \rightleftharpoons H_3O^+ + S^{-2}$$

The sulfide ion is a very strong reducing agent. Even from atmospheric oxygen, elemental sulfur is released from its solutions:

$$2\,H_2S + O_2 \longrightarrow 2\,S + 2\,H_2O$$

$$S^{-2} - 2\,e^- \longrightarrow S^0 \quad | \; 2 \; | \; 4 \; | \; 2$$

$$O_2 \longrightarrow 2\,O^0 + 2 \times 2\,e^- \longrightarrow 2\,O^{-2} \quad | \; 4 \; | \; \; | \; 1$$

Electronic configuration of the outer electron layer $3s^2\,3p^5$ (**e**) belongs to the element **chlorine**. Chlorine is a third period element, group VIIA, sequence number 17. With hydrogen it forms a gaseous compound with a polar covalent bond.

$$Cl_2 + H_2 \longrightarrow 2\,HCl$$

$$H_2 \longrightarrow 2\,H^0 - 2 \times 1\,\bar{e} \longrightarrow 2\,H^+ \quad | \; 2 \; | \; 2 \; | \; 1$$

$$Cl_2 \longrightarrow 2\,Cl^0 + 2 \times 1\,\bar{e} \longrightarrow 2\,Cl^- \quad | \; 2 \; | \; \; | \; 1$$

Of the five compounds mentioned, sodium hydride does not react with ammonia. With water it reacts violently. A redox process takes place – disproportionation.

$$NaH + H{-}OH \longrightarrow NaOH + H_2\uparrow$$

$$\underbrace{Na^+ + H^-} + \underbrace{H^+ + \bar{O}H}_{\rightleftharpoons\,H_2O} \longrightarrow \underbrace{Na^+ + \bar{O}H} + H_2\uparrow$$

$$H^- - \bar{e} \longrightarrow H^0$$

$$H^+ + \bar{e} \longrightarrow H^0$$

$$H^0 + H^0 \longrightarrow H_2\uparrow$$

Sodium hydride also interacts with the weak, dibasic hydrogen sulfide.
Disproportionation takes place here as well. The interaction takes place in two stages.

$$NaH + H_2S \longrightarrow NaHS + H_2\uparrow$$

(stage I)

$$2\,NaH + H_2S \longrightarrow Na_2S + 2\,H_2\uparrow$$

(stage II)

$$\begin{array}{lll} H^{-} - \bar{e} \longrightarrow \overset{0}{H} & \searrow & \\ & & H_2\uparrow \\ H^{+} + \bar{e} \longrightarrow \overset{0}{H} & \nearrow & \end{array}$$

Sodium hydride also interacts with hydrogen chloride. The process is redox, disproportionation.

$$NaH + HCl \longrightarrow NaCl + H_2\uparrow$$

Ammonia dissolves in water and reacts with it. The result is the weak is a non-volatile ammonium hydroxide by the equation:

$$NH_3 + H_2O \rightleftharpoons NH_4OH$$

A proton from the water molecule forms a covalent polar bond with the uncharged electron pair of the nitrogen atom from the ammonia molecule (by the mechanism of donor-acceptor interaction) and forms an ammonium cation, which with the hydroxide anion from the water molecule forms the perishable ammonium hydroxide.

Ammonia reacts with hydrogen sulfide in a two-step process (hydrogen sulfide acid is dibasic). Two salts are produced, ammonium hydrogen sulfide and diammonium sulfide.

$$NH_3 + H_2S \longrightarrow NH_4HS \quad (I)$$

$$2\,NH_3 + H_2S \longrightarrow (NH_4)_2S \quad (II)$$

Ammonia also reacts with hydrogen chloride to form the salt ammonium chloride, used in paediatric practice to clean purulent deposits on a sore throat.

$$NH_3 + HCl \longrightarrow NH_4Cl$$

The water molecules dissolve the hydrogen sulfide to form hydrogen sulfide acid, and dissolving hydrogen chloride in water produces hydrochloric acid.

Hydrogen sulfide and hydrogen chloride do not interact because both substances are acidic.

17. Oxide.

Composition and structure of oxides. Classification of oxides - oxides with ionic, oxides with covalent bond. Basic, acid, amphoteric and neutral oxides. Chemical properties of oxides. Application of oxides in practice.

Oxides are one of the most abundant chemical compounds in nature and the most widely used in practice. Their theoretical importance is also great. Knowledge of oxides allows chemical elements to be more fully characterised and hydroxides and oxygen-containing acids (oxoacids) to be more accurately defined.

Composition and structure of oxides

Oxides are compounds of chemical elements with oxygen. Unlike peroxides (also oxygen compounds of elements), oxides do not contain chemical bonds between oxygen atoms. All elements form oxides except **helium, neon, and argon**.

In the composition of oxides, oxygen exhibits **a constant value** and a **negative sign of oxidation** (-2). There is a genetic relationship between the chemical elements and their oxides and hydroxides – the chemical elements form directly or indirectly one or more oxides to which correspond hydroxides:

Chemical element → Oxide of element → Hydroxide of element

Oxides are composed of elements with different atomic structures and properties. This affects the **nature of the chemical bond** between oxygen and the other elements in the oxides. Depending on the type of this bond, **oxides with an ionic bond** and **oxides with a covalent bond** are distinguished.

The chemical nature of oxides is determined by the **nature of the chemical elements** in their composition. At the same time, the chemical nature of oxides is an important characteristic of the elements themselves.

The properties of oxides change periodically as the atomic mass (atomic numbers) of the elements increase. By period and group, the properties of oxides change in a regular way: by period, the basic properties of oxides weaken, go through amphoteric, and at the end of periods, acidic properties strengthen; in the main groups, acidic properties weaken and basic properties of oxides strengthen.

Under ordinary conditions, oxides occur in all three aggregate states. Oxides with ionic chemical bonding are crystalline solids (CaO, MgO, Na_2O). Those with a covalent chemical bond are solids (SiO_2, Al_2O_3, As_2O_3, P_2O_5), liquids (H_2O, Mn_2O_7), or gases (SO_3, CO_2, NO). Oxides with an ionic crystal structure are characterized by a high melting point and low volatility.

The change in **the chemical character of the oxides** is most clearly observed in the elements of the third period of the Mendeleyev table. Na2O, MgO, A12O3, SiO2, P2O5, SO3, C12O7. The gradual transition from the typically basic oxide of sodium to the typically acidic oxides of sulfur and chlorine is evident. This also corresponds to an increase in the oxidation state of the elements forming oxides in their higher valence.

Correspondence is also observed when considering the **properties of oxides of the same element in different degrees of oxidation**, e.g., in the order MnO, Mn_2O_3, MnO_2, MnO_3, $Mn_2O_7$the gradual transition from the basic oxide of divalent manganese to the acidic oxide of sevenvalent manganese. Similar changes in properties are observed for oxides of variable valence metals. In the case indicated with the oxides of manganese, MnO and Mn_2O_3 are basic oxides, MnO_2 is amphoteric, and MnO_3 and Mn_2O_7 are acid oxides. The valence of an element in its highest oxide usually corresponds to the number of the group in the Periodic Table in which it is found.

Exceptions are the elements of group IB, e.g., CuO, Cu_2O, and also some elements of group VIIIB - e.g., Fe_2O_3, FeO.

Classification of oxides

Depending on the type of chemical bond, oxides are

Oxides with ionic bonding - oxides of the s-elements, of the p-elements with metallic properties in their lower valence, and of the d- and f-elements in their lower valence (**Na_2O, Cs_2O, PbO, MgO, CaO, MnO**, etc.) have ionic chemical bonding.

Oxides with covalent bonding - Covalent bonding is present in oxides of p-elements with nonmetallic properties and of d- and f-elements in their intermediate and higher valence (**Cl_2O_7, MnO_2, Mn_2O_7, SO_2, SO_3, CO_2, SiO_2, P_2O_5, H_2O**, etc.).

The classification of oxides is based on the properties of the hydroxides they immediately form or correspond to.

Oxides that form, or to which they correspond, basic hydroxides are called **basic oxides (CaO, MgO, CrO, MnO, Li_2O, Na_2O**). The **oxides of the s-elements, of the d- and f- elements in their lowest valence, of the p-elements of metallic character also in their lowest valence, are basic**. The basic oxides have an ionic chemical bond.

Oxides that form or to which correspond acid hydroxides are called **acid oxides (CO_2, SO_2, SO_3, SiO_2**, etc.). The **oxides of the p-elements of non-metallic character and of the d- and f-elements in their higher valence** are acidic. Acid oxides have a covalent chemical bond.

Oxides to which correspond **amphoteric** hydroxides are called amphoteric oxides (**Al_2O_3, MnO_2, ZnO**). **Amphoteric are the oxides of the d- and f-elements in their intermediate valence and of some p-elements with a weakly pronounced metallic and non-metallic character.** The bonding in amphoteric oxides is covalent.

Oxides to which hydroxides do not correspond are **neutral oxides**. Neutral oxides (e.g., **NO, N_2O, CO**, etc.) form certain p-elements in certain valence states. The bond in their molecules is covalent.

Oxides of many chemical elements are involved in redox processes in which the oxygen-bound element changes its valence.

Chemical properties of oxides

The basic oxides of the s-elements react with water to produce basic hydroxides:

$$Na_2O + H_2O \longrightarrow 2NaOH$$

The reason for these interactions is the instability of the oxygen ion (O^{-2}) in aqueous solution:

$$\overset{-2}{O} + H_2O \longrightarrow 2\overset{-}{O}H$$

All basic oxides react with acids and acid oxides to form their respective salts:

$$Na_2O + H_2SO_4 \longrightarrow Na_2SO_4 + H_2O$$

$$CaO + H_2SO_4 \longrightarrow CaSO_4 + H_2O$$

$$CaO + CO_2 \longrightarrow CaCO_3$$

$$MgO + CO_2 \longrightarrow MgCO_3$$

Some **acid oxides** with water form acid hydroxides:

$$SO_2 + 3\,H_2O \longrightarrow 2\,H_3\overset{+}{O} + \overset{-2}{SO_3}$$

All acid oxides react with bases and basic oxides to produce salts:

$$CO_2 + 2\,NaOH \longrightarrow Na_2CO_3 + H_2O$$

$$SiO_2 + CaO \xrightarrow{1200^oC} \underset{\text{calcium metasilicate}}{CaSiO_3}$$

Amphoteric oxides react with both acids, bases, acid oxides and basic oxides. These interactions produce salts.

$$ZnO + \underbrace{2\,H^+ + 2\,Cl^-} \longrightarrow \underbrace{Zn^{+2} + 2\,Cl^-} + H_2O$$

$$ZnO + \underbrace{2\,Na^+ + 2\,OH^-} + H_2O \longrightarrow \underbrace{2\,Na^+ + [Zn(OH)_4]^{-2}}$$

Neutral oxides do not interact with water, acids or bases. Such are CO, NO, N_2O, etc.

The participation of oxides in the above chemical reactions does not lead to a change in the valence of the oxygen-bound element.

However, the oxides of many chemical elements are involved in redox processes in which the oxygen-bound element changes its valence.

$$Fe_2O_3 + 3H_2 \longrightarrow 2\,Fe + 3\,H_2O$$

$$\begin{array}{l|c|c|c}
Fe^{+3} + 3\,\bar{e} \longrightarrow Fe^o & 3 & & 2 \\
 & & 6 & \\
H_2 \longrightarrow 2\,H^o - 2 \times 1\,\bar{e} \longrightarrow 2\,H^+ & 2 & & 3
\end{array}$$

$$3\,Fe_3O_4 + 8\,Al \longrightarrow 4\,Al_2O_3 + 9\,Fe$$

$$\begin{array}{l|c|c|c}
Al^o - 3\,\bar{e} \longrightarrow Al^{+3} & 3 & & 2 \\
Fe^{+2} + 2\,\bar{e} \longrightarrow Fe^o & 2 & 6 & 3 \\
Fe^{+3} + 3\,\bar{e} \longrightarrow Fe^o & 3 & & 2
\end{array}$$

$$3\,V_2O_5 + 10\,Al \longrightarrow 5Al_2O_3 + 6\,V$$

$$\begin{array}{lcl|c|c|c}
Al^{0} - 3\bar{e} & \longrightarrow & Al^{+3} & 3 & \multirow{2}{*}{15} & 5 \\
V^{+5} + 5\bar{e} & \longrightarrow & V^{0} & 5 & & 3
\end{array}$$

$$CO_2 + C \longrightarrow 2\,CO$$

$$\begin{array}{lcl|c|c|c}
C^{0} - 2\bar{e} & \longrightarrow & C^{+2} & 2 & \multirow{2}{*}{2} & 1 \\
C^{+4} + 2\bar{e} & \longrightarrow & C^{+2} & 2 & & 1
\end{array}$$

$$ZnO + C \longrightarrow Zn + CO$$

$$\begin{array}{lcl|c|c|c}
Zn^{+2} + 2\bar{e} & \longrightarrow & Zn^{0} & 2 & \multirow{2}{*}{2} & 1 \\
C^{0} - 2\bar{e} & \longrightarrow & C^{+2} & 2 & & 1
\end{array}$$

$$PbO + C \longrightarrow Pb + CO$$

$$\begin{array}{lcl|c|c|c}
Pb^{+2} + 2\bar{e} & \longrightarrow & Pb^{0} & 2 & \multirow{2}{*}{2} & 1 \\
C^{0} - 2\bar{e} & \longrightarrow & C^{+2} & 2 & & 1
\end{array}$$

$$2\,NO + O_2 \longrightarrow 2\,NO_2$$

$$\begin{array}{lcl|c|c|c}
N^{+2} - 2\bar{e} & \longrightarrow & N^{+4} & 2 & \multirow{2}{*}{4} & 2 \\
O_2 \longrightarrow 2O^{0} + 2 \times 2\bar{e} & \longrightarrow & 2O^{-2} & 4 & & 1
\end{array}$$

Water can be viewed as a hydride of oxygen and as an oxide of hydrogen. It is the most important compound of both chemical elements. Its importance to life on Earth is enormous.

Water is a transparent liquid without color or odor. Due to its polarity and significant dielectric constant, water is a good dissociating solvent. It interacts with dissolved substances, molecules and ions by hydrating them. The presence of free electron pairs determines its donor properties and its propensity to participate in addition and complexation processes. In the form of crystallization, water remains in the composition of many salts when they crystallize from aqueous solutions –

$CuSO_4 . 5H_2O$, $FeSO_4 . 7H_2O$, etc.

The water molecule is resistant to heating.

Water interacts with many chemical elements. It reacts with the most active metals at room temperature and with others when heated. Its capabilities in this respect can be judged by the position of the metal in the electroaffinity order (the Order of Relative Activity of the Metals), taking into account the influence of the hydrogen ion concentration.

Water reacts with soluble oxides to give acids and bases. It participates in almost all redox or neutralization-type interactions by being present either between the reactants or between the reaction products. Water is the natural medium in which most chemical reactions take place and often plays a catalytic role.

Its electrolytic dissociation characterizes it as an amphoteric electrolyte.

Application of oxides in practice

The properties of oxides determine the diverse areas of their practical application:

- in construction - CaO, SiO_2 etc.;
- in the chemical industry as raw materials and intermediates in the production of hydroxides, acids, salts, polymers, etc., in the composition of catalysts in inorganic and organic synthesis (V_2O_5, MnO_2, Cr_2O_3, etc.), in the production of mineral paints (ZnO - zincweisse, Fe_3O_4 - ochre, Pb_3O_4 - minium, Cr_2O_3 - chrome green, etc.);
- in metallurgy as basic raw materials in the production of metals;
- obtained in the form of single crystals, many oxides are used as materials in the strategic directions of scientific and technological progress;
- SiO_2 – in resonators for communication and hydroacoustic devices, ultrasonic frequency generators, electromechanical transducers and filters, modulators in laser technology, etc.;
- TeO_2 – in optics and optoelectronics due to its unique acousto-optic properties;
- Oxides of s- and p-elements (among them SiO_2 and GeO_2 are particularly promising) are used to make optical fibers as substitutes for metallic wires for long-distance information transmission;
- Oxides of elements of IIA, IIIA and IVA groups are used in laser technology, and oxides of some f-elements – as phosphors;
- Based on some oxides - (e.g., WO_3, TiO_2), ceramic materials are made for the needs of electronics, etc.

18. HYDROXIDES AND OXOACIDS

DEFINITION. CLASSIFICATION. STRUCTURE AND PROPERTIES. APPLICATION IN PRACTICE. OXOACIDS. GENERAL PROPERTIES OF OXOACIDS. METHODS OF PREPARATION OF HYDROXIDES.

Hydroxides are a class of inorganic compounds of great importance to science and practice. On the one hand, knowledge of them contributes to a fuller knowledge and more accurate characterization of the elements, oxides, oxygen-containing (oxoacids) acids and salts. On the other hand, without hydroxides, one can hardly imagine the development of the chemical and pharmaceutical industries and the realization of a full-fledged environmentally friendly chemicalization of the other branches of the economy.

Classification of hydroxides:

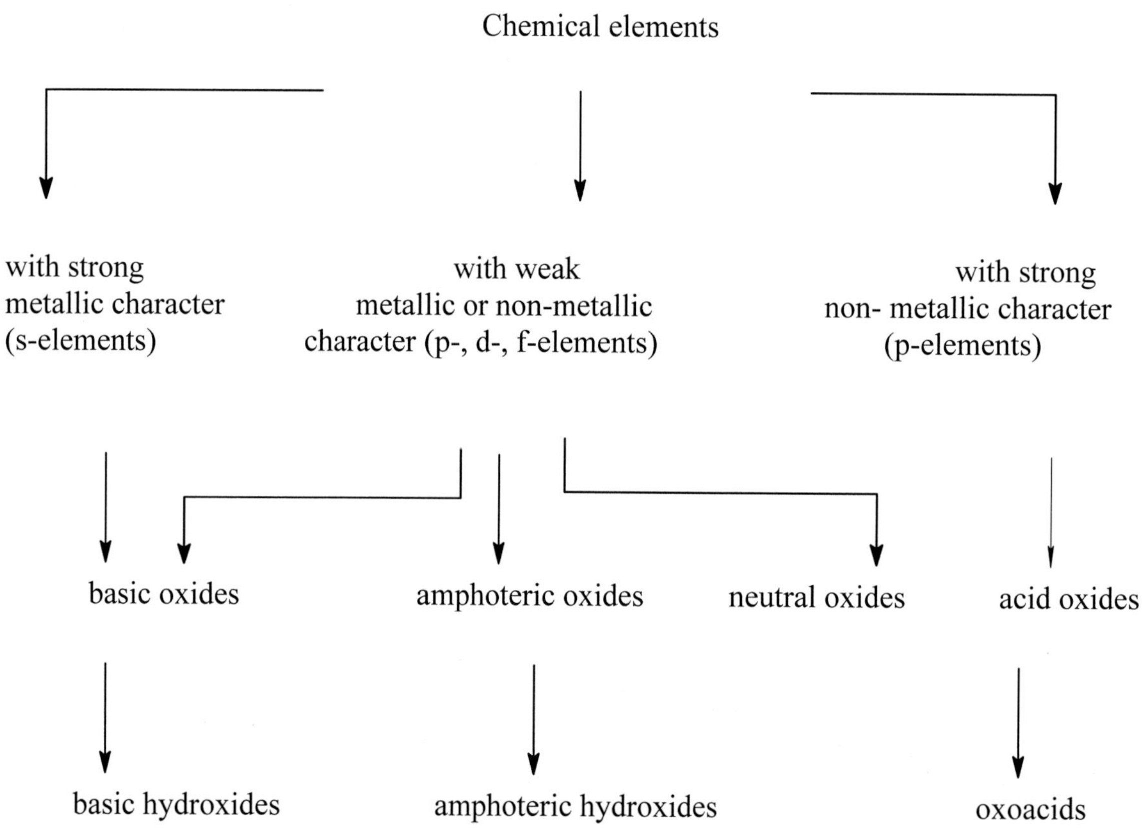

If the classification is based on the relationship between the oxides and the compounds that form with water or that correspond to them, the scheme has the following form:

Oxides Basic → basic hydroxides
Oxides Amphoteri → amphoteric hydroxides
Oxides Acidic → oxoacids

The basic hydroxides correspond to the basic oxides. They are hydroxides of elements of metallic character (s-, p-, d- and f-elements in their lowest valence) - NaOH, KOH, $Ca(OH)_2$, $Mn(OH)_2$, $Cr(OH)_2$, etc.

Amphoteric hydroxides correspond to amphoteric oxides. They are hydroxides of elements with metallic character (d- and f-elements) in their intermediate valence and of some p-elements with weakly pronounced metallic or non-metallic character - $Pb(OH)_2$, $Al(OH)_3$, $Mn(OH)_4$, $Zn(OH)_2$, etc.

In the composition of each hydroxide a certain element is involved, which is directly bonded to one or several hydroxide groups - $E(OH)_n$. The hydroxides of the s-elements and of some p-, d- and f-elements in their lower oxidation states are basic. In the composition of amphoteric hydroxides, p- and d-elements are most often involved in their intermediate oxidation states and f-elements in their intermediate and higher oxidation states.

Structure and properties

The chemical nature of the elements in the composition of hydroxides determines the nature of the chemical bonds in them. The nature, as well as the degree of polarity and polarizability, of the E-O and O-H chemical bonds are important for the structure and properties of hydroxides. The bond in the hydroxide anion is a covalent polar bond, insignificantly polarized due to the presence of a negative charge at the oxygen atom. Depending on the nature of the element E, the E - OH bond is either strongly ionic or covalently polar with varying degrees of ionicity or covalency. When E is an s-element, the E - OH bond is strongly ionic. This is due to the large difference in electronegativity of the s-element and oxygen. When E is a p-, d-, or f-element in a lower oxidation state, the E - OH bond is ionic with some degree of polarity or polar with a significant degree of ionicity. When these elements are in their intermediate degrees of oxidation, the E - OH bond is covalent polar.

Basic hydroxides are primarily ionic compounds, while amphoteric hydroxides are compounds with a covalent polar bond.

Under ordinary conditions, basic and amphoteric hydroxides are solids, colorless (NaOH, KOH, $Al(OH)_3$, $Mg(OH)_2$) or characteristically colored - $Cu(OH)_2$ - blue, $Ni(OH)_2$ - green, $Fe(OH)_2$ - pale green, $Fe(OH)_3$ - brown, $Cr(OH)_2$ - yellow, etc. Most of the hydroxides are insoluble in water. The exceptions are the hydroxides of the alkali and alkaline earth elements – the alkali hydroxides are very soluble, and the alkaline earth hydroxides are moderately soluble in water.

The hydroxides of the alkali and of the alkaline earth elements (except $Be(OH)_2$) are strong electrolytes. They are called bases. In aqueous solution and in melt they dissociate completely into ions:

$$KOH \longrightarrow \overset{+}{K} + \overset{-}{O}H$$

$$Ca(OH)_2 \longrightarrow Ca^{+2} + 2\,\overset{-}{O}H$$

The number of hydroxide ions released in the dissociation of a mole of hydroxide determines its valence. Calcium hydroxide is divalent and potassium hydroxide is monovalent.

Amphoteric hydroxides in aqueous solution dissociate as weak acids or weak bases depending on the conditions:

$$H^+ + E\overset{-}{O} \rightleftharpoons E\ OH \rightleftharpoons E^+ + \overset{-}{O}H$$

$$2\,H^+ + Zn\overset{-2}{O_2} \rightleftharpoons Zn(OH)_2 \rightleftharpoons Zn^{+2} + 2\,\overset{-}{O}H$$

The most characteristic general chemical properties of the basic hydroxides are interaction with acid oxide, with acid and with salt. For example:

$$Ca(OH)_2 + CO_2 \longrightarrow CaCO_3 + H_2O$$

$$2KOH + H_2SO_4 \longrightarrow K_2SO_4 + 2H_2O$$

$$2\,NaOH + CuCl_2 \longrightarrow Cu(OH)_2 + 2NaCl$$

Amphoteric hydroxides react with both acids and bases to produce salts and water. They participate in the above interactions only with strong acids and with strong bases:

$$Zn(OH)_2 + \underbrace{2\,H^+ + 2\,Cl^-} + 2\,H_2O \longrightarrow [Zn(H_2O)_4]^{+2} + 2\,Cl^-$$

tetraaqua zinc dichloride

$$Zn(OH)_2 + \underbrace{2\,Na^+ + 2\,OH^-} \longrightarrow 2\,Na^+ + [Zn(OH)_4]^{-2}$$

disodium tetrahydroxyzincate

$$Cr(OH)_3 + \underbrace{3\,Na^+ + 3\,OH^-} \longrightarrow 3\,Na^+ + [Cr(OH)_6]^{-3}$$

trisodium hexahydroxychromate

$$Cr(OH)_3 + \underbrace{3\,H^+ + 3\,Cl^-} + 3\,H_2O \longrightarrow [Cr(H_2O)_6]^{+3} + 3\,Cl^-$$

hexaquachrome trichloride

These salts belong to the so-called complex salts, which are characterized by the fact that they release complex ions upon dissociation.

At high temperature, the hydroxides of the p-, d- and f-elements decompose - the corresponding oxides are produced and water is released.

$$2\,Al(OH)_3 \xrightarrow{t^o} Al_2O_3 + 3\,H_2O$$

$$Cu(OH)_2 \xrightarrow{t^o} CuO + H_2O$$

In the chemical properties presented, the element bonded to the hydroxide group in the composition of the hydroxides does not change its degree of oxidation. Hydroxides may also participate in redox processes associated with a change in the degree of oxidation of the element concerned:

$$4\,Fe(OH)_2 + 2H_2O + O_2 \longrightarrow 4\,Fe(OH)_3$$

$$\begin{array}{lll|c|c|c} & Fe^{+2} - e^- & \longrightarrow Fe^{+3} & 1 & & 4 \\ & & & & 4 & \\ O_2 \longrightarrow & 2O^0 + 2 \times 2e^- & \longrightarrow 2O^{-2} & 4 & & 1 \end{array}$$

$$4\,Cr(OH)_2 + O_2 + 2\,H_2O \longrightarrow 4\,Cr(OH)_3$$

$$\begin{array}{lll|c|c|c} & Cr^{+2} - e^- & \longrightarrow Cr^{+3} & 1 & & 4 \\ & & & & 4 & \\ O_2 \longrightarrow & 2O^0 + 2 \times 2e^- & \longrightarrow 2O^{-2} & 4 & & 1 \end{array}$$

The product of the most characteristic chemical reactions involving basic and amphoteric hydroxides are salts.

The most accessible, cheapest and most widely used in practice are sodium and calcium bases. Sodium base is one of the most popular reagents in chemical laboratories. It is used in enormous quantities in the chemical industry in the manufacturing of artificial chemical fibres and dyes, of soap and detergents, of paper and drugs, and in petroleum refining. Calcium base is used in construction under the name 'slaked lime', in agriculture for the preparation of bordol solution for the control of plant pests, in the sugar industry, in the leather industry, etc.

Oxoacids

Oxoacids are chemical compounds that can be derived from acid oxides and water, or that correspond to acid oxides - H_2SO_4, HNO_3, $HClO_3$ (chlorine), $HMnO_4$ (permanganate), etc. They are compounds of elements of non-metallic character (p-elements) and of elements of metallic character (d- and f-elements) in their higher valence - HNO_2, H_3PO_2, HNO_3, H_3PO_3, H_3PO_4, H_2CrO_4, $HClO_4$, $HMnO_4$, etc.

In contrast to oxygen-free acids (HCl, HBr, H_2S, etc.) oxoacids are oxygen-containing (oxo- from oxygenium - oxygen). They contain in their molecule one or several hydroxyl groups bonded to a central atom, which may also be bonded to other oxygen atoms:

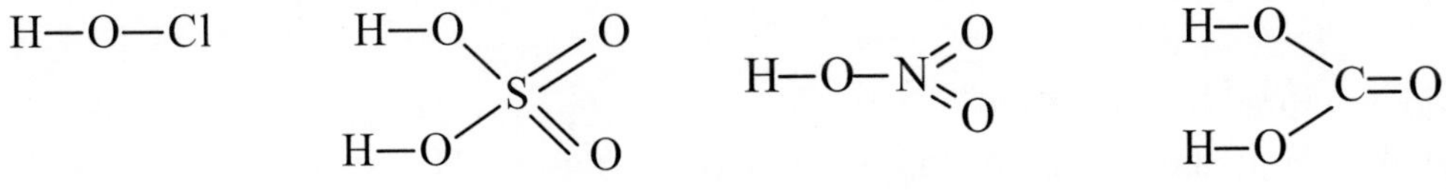

Oxoacids are compounds of the p-elements with a pronounced non-metallic character in almost all oxidation states permissible for them and of the d-elements in their higher oxidation states. The f-elements do not form oxoacids. They form only hydroxides.

Chemical elements	Oxoacids				
	Degrees of oxidation of the elements				
	+1	+3	+5	+6	+7
p-elements					
N	–	HNO_2 nitrous acid	HNO_3 nitric acid	–	–
P	H_3PO_2 hypophosphorous acid	H_3PO_3 phosphoristic acid	H_3PO_4 phosphoric acid	–	–
Cl	$HClO$ hypochlorous acid	$HClO_2$ chlorous acid	$HClO_3$ chloric acid	–	$HClO_4$ perchloric dcid
d-elements					
Cr	–	–	–	H_2CrO_4 chromic acid	–
Mn	–	–	–	H_2MnO_4 manganic acid	$HMnO_4$ permanganic acid
Re	–	–	–	H_2ReO_4 rhenium acid	$HReO_4$ perrhenic acid

Oxoacids are compounds with a molecular structure with polar covalent bonds between atoms in the molecules. The strength of their acid action depends on the degree of polarity of the covalent bonds in their molecules, which in turn is determined mainly by the electronegativity of the central atom E.

$$E-O-H$$

As a rule, the more electronegative is the element E, the more strongly polar is the O - H bond in the hydroxyl group; the more readily in aqueous solution H^+ is separated from it; the more pronounced are the acid properties of the oxoacid.

In addition to the electronegativity of element E, the strength of oxoacids depends on the ability of other atoms bonded to element E to attract electrons to themselves. For example, the strength of chloric acids increases as the degree of oxidation of chlorine increases, i.e., as the number of oxygen atoms bonded to chlorine increases.

Acids of chlorine

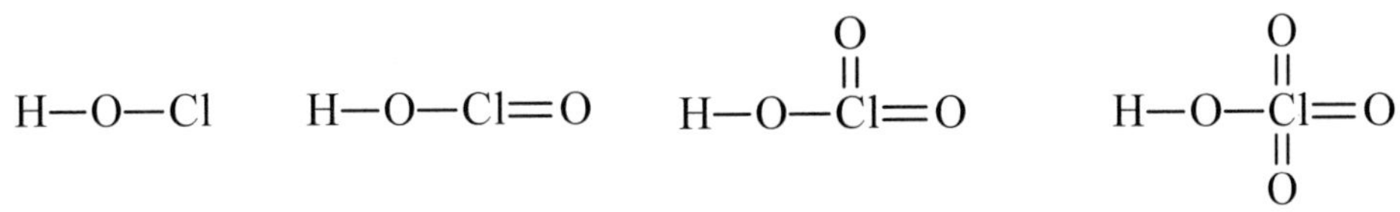

Degree of oxidation of the chlorine atom	+ 1	+ 3	+ 5	+ 7

Strength of the acids $\longrightarrow$

Towards perchloric acid, the ability of the chlorine atom to attract electrons from the Cl - O bond increases, resulting in the polarity of the H - O bond in the hydroxyl group increasing. This is explained by the greater number of oxygen atoms attracting electrons to themselves from their bonds with the chlorine atom, resulting in a significant decrease in the electron density around it.

The electronegativity of the p-elements increases with increasing periodic sequence number, leading to an enhancement of the acidity of the oxoacids. In groups with increasing atomic radius (element number) the acidic properties weaken, go through amphoteric and finally the basic properties of hydroxides are manifested.

Under ordinary conditions, acids are in a liquid or solid aggregate state. There are no gaseous oxoacids. Liquid oxoacids are those of the p-elements with a distinct non-metallic character – sulphur, chlorine, nitrogen. Some p-element oxoacids are known only in aqueous solution – e.g., H_2CO_3, H_2SO_3, etc. The oxoacids of the p-elements are colorless, while those of the d-elements are mostly colored in characteristic colors.

In aqueous solution, oxoacids dissociate into hydrogen cations and acid anions:

$$HNO_3 \longrightarrow H^+ + NO_3^-$$

$$H_2SO_4 \longrightarrow 2\,H^+ + SO_4^{-2}$$

The number of hydrogen ions released in the dissociation of a mole of acid determines its basicity. Polybasic acids are dissociated in steps:

$$H_2SO_4 \longrightarrow H^+ + HSO_4^-$$

$$HSO_4^- \longrightarrow H^+ + SO_4^{-2}$$

$$H_3PO_4 \rightleftharpoons H^+ + H_2PO_4^-$$

$$H_2PO_4^- \rightleftharpoons H^+ + HPO_4^{-2}$$

$$HPO_4^{-2} \rightleftharpoons H^+ + PO_4^{-3}$$

Sulfuric and nitric acids are strong oxoacids. Carbonic acid (H_2CO_3), nitrous acid (HNO_2), boric acid (H_3BO_3) are weak oxoacids.

In aqueous solution, all oxoacids have common properties: metal salt and hydrogen

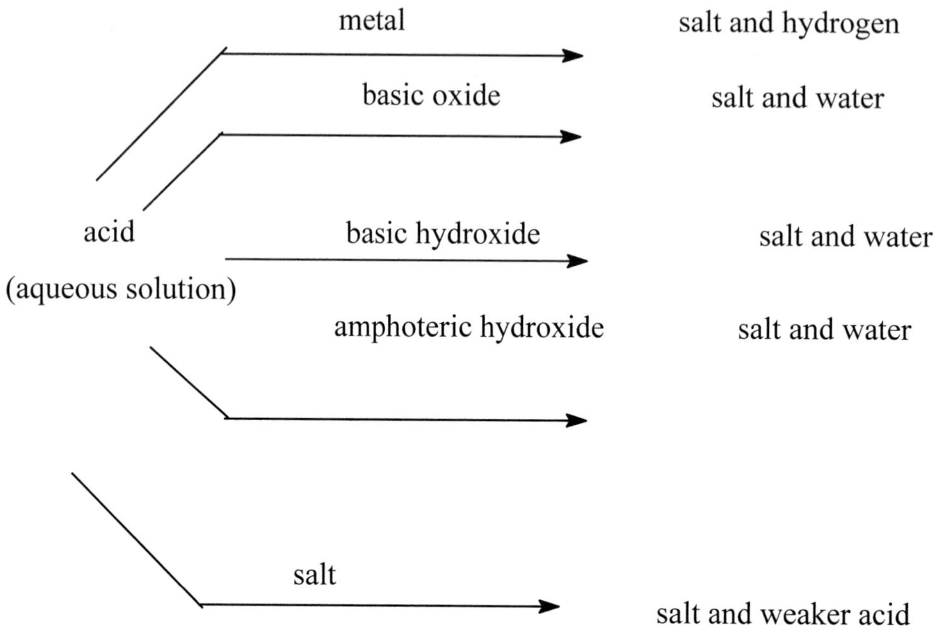

$$2\,HCl + Fe \longrightarrow FeCl_2 + H_2$$

$$Fe^{0} - 2\bar{e} \longrightarrow Fe^{+2} \quad \left|\ 2\ \right|\ 2\ \left|\ 1\right.$$

$$2\,H^{+} + 2 \times 1\,\bar{e} \longrightarrow 2\,H^{0} \longrightarrow H_2 \quad \left|\ 2\ \right|\quad \left|\ 1\right.$$

$$H_2SO_4 + CaO \longrightarrow CaSO_4 + H_2O$$

$$H_2SO_4 + ZnO \longrightarrow ZnSO_4 + H_2O$$

$$H_2SO_4 + Zn(OH)_2 \longrightarrow ZnSO_4 + H_2O$$

$$H_2SO_4 + 2\,NaOH \longrightarrow Na_2SO_4 + H_2O$$

$$H_2SO_4 + CaCO_3 \longrightarrow CaSO_4 + H_2CO_3$$

$$H_2CO_3 \rightleftharpoons H_2O + CO_2\uparrow$$

Only metals that are ahead of hydrogen in the Order of Relative Activity can interact with oxoacids in solution.

Oxoacids can only react with salts of acids weaker than them.

The general properties of oxoacids in aqueous solution are due to the common hydrogen ions they contain.

Oxoacids in their pure state exhibit some specific properties:

Involved in redox processes associated with a change in the degree of oxidation of the central element E:

$$2H_2\overset{+6}{S}O_4 + \overset{0}{Cu} \longrightarrow \overset{+2}{Cu}SO_4 + 2H_2O + \overset{+4}{S}O_2\uparrow$$

$$Cu + H_2SO_4 \longrightarrow CuO + SO_2 + H_2O$$

$$\begin{array}{l|c|c|c} Cu^{0} - 2e^{-} \longrightarrow Cu^{+2} & 2 & 2 & 1 \\ S^{+6} + 2e^{-} \longrightarrow S^{+4} & 2 & & 1 \end{array}$$

$$3Ag + 4HNO_3 \longrightarrow 3AgNO_3 + NO + 2H_2O$$

$$\begin{array}{l|c|c|c} Ag^{0} - e^{-} \longrightarrow Ag^{+} & 1 & 3 & 3 \\ N^{+5} + 3e^{-} \longrightarrow N^{+2} & 3 & & 1 \end{array}$$

Another option for the interaction of metals located after hydrogen in the Order of Relative Activity of Metals with concentrated oxidizing oxoacids.

$$2HNO_3 + Ag \longrightarrow AgNO_3 + NO_2 + H_2O$$

$$\begin{array}{l|c|c|c} Ag^{0} - e^{-} \longrightarrow Ag^{+} & 1 & 1 & 1 \\ N^{+5} + e^{-} \longrightarrow N^{+4} & 1 & & 1 \end{array}$$

They decompose under ordinary conditions or when heated:

$$H_2CO_3 \longrightarrow H_2O + CO_2\uparrow$$

$$H_2SO_3 \longrightarrow H_2O + SO_2\uparrow$$

Oxoacids are primarily of interest to the chemical industry as feedstocks in the production of other substances and as raw materials for the purification of a variety of products.

Methods for the preparation of hydroxides

Interaction of basic oxides with water

Only hydroxides of the s-elements, whose oxides are ionic compounds, are obtained by this method. The oxygen ions of the oxides are unstable in aqueous media – with water they form hydroxide ions:

$$K_2O + H_2O \longrightarrow 2KOH$$

$$CaO + H_2O \longrightarrow Ca(OH)_2$$

$$O^{-2} + H_2O \longrightarrow 2\overline{O}H$$

For example, calcium base is obtained by this method:

$$CaO + H_2O \longrightarrow Ca(OH)_2$$

Interaction of salts with bases

This method is most often used to obtain insoluble hydroxides – from soluble salts of the corresponding element and a solution of the base:

$$AlCl_3 + 3NaOH \longrightarrow Al(OH)_3\downarrow + 3NaCl$$

$$\underbrace{Al^{+3} + 3Cl^{-}} + \underbrace{3Na^{+} + 3\overline{O}H} \longrightarrow Al(OH)_3\downarrow + \underbrace{3Na^{+} + 3Cl^{-}}$$

$$Al^{+3} + 3\overline{O}H \longrightarrow Al(OH)_3\downarrow$$

$$CuCl_2 + 2NaOH \longrightarrow Cu(OH)_2\downarrow + 2NaCl$$

$$\underbrace{Cu^{+2} + 2Cl^{-}} + \underbrace{2Na^{+} + 2\overline{O}H} \longrightarrow Cu(OH)_2\downarrow + \underbrace{2Na^{+} + 2Cl^{-}}$$

$$Cu^{+2} + 2\overline{O}H \longrightarrow Cu(OH)_2\downarrow$$

$$CaCl_2 + 2NaOH \longrightarrow Ca(OH)_2 + 2NaCl$$

$$\underbrace{Ca^{+2} + 2Cl^{-}} + \underbrace{2Na^{+} + 2\overline{O}H} \longrightarrow Ca(OH)_2 + \underbrace{2Na^{+} + 2Cl^{-}}$$

$$\underbrace{Ca^{+2} + 2\overline{O}H} \longrightarrow Ca(OH)_2$$

$$FeCl_3 + 3NaOH \longrightarrow Fe(OH)_3\downarrow + 3NaCl$$

$$\underbrace{Fe^{+3} + 3Cl^{-}} + \underbrace{3Na^{+} + 3\overline{O}H} \longrightarrow Fe(OH)_3\downarrow + \underbrace{3Na^{+} + 3Cl^{-}}$$

$$\underbrace{Fe^{+3} + 3\,\overline{O}H} \longrightarrow Fe(OH)_3\downarrow$$

Hydrolysis of salts

$$FeCl_3 + 3\,H_2O \longrightarrow Fe(OH)_3 + 3\,HCl$$

$$\underbrace{Fe^{+3} + 3\,\overline{C}l} + \underbrace{3\,H^+ + 3\,\overline{O}H}_{\substack{\uparrow\downarrow \\ 3\,H_2O}} \rightleftharpoons Fe(OH)_3 + \underbrace{3\,H^+ + 3\,Cl^-}$$

$$CuCl_2 + 2\,H_2O \longrightarrow Cu(OH)_2 + 2\,HCl$$

$$\underbrace{Cu^{+2} + 2\,\overline{C}l} + \underbrace{2\,H^+ + 2\,\overline{O}H}_{\substack{\uparrow\downarrow \\ 2\,H_2O}} \rightleftharpoons Cu(OH)_2 + \underbrace{2\,H^+ + 2\,Cl^-}$$

Electrolysis of aqueous solutions of salts

The chloralkali electrolysis method produces sodium base from a concentrated aqueous solution of sodium chloride.

Ions in solution (before electrolysis)

$$Na^+,\ H^+,\ Cl^-,\ \overline{O}H$$

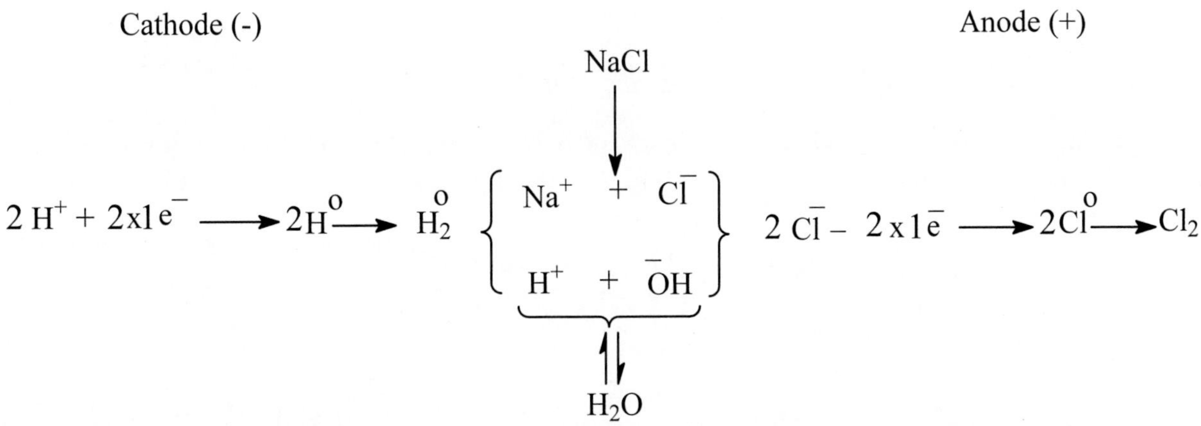

In the cathode space remain Na+ In the anode space remain $\overline{O}H$

$$Na^+ + \overline{O}H \rightleftharpoons NaOH$$

There are four types of ions contained in aqueous sodium chloride solution.

$$NaCl \longrightarrow Na^+ + Cl^-$$

$$H_2O \rightleftharpoons H^+ + \bar{O}H$$

When a direct electric current is passed through the sodium chloride solution, the ions acquire directional motion. The cations are directed towards the cathode and the anions towards the anode. At the anode the chloride anions give up electrons and are electroneutralised. Chlorine gas is released at the anode.

$$(A^+) \quad 2\,Cl^- - 2 \times 1\,\bar{e} \longrightarrow 2\,Cl^0 \longrightarrow Cl_2\uparrow$$

At the cathode, hydrogen cations are electroneutralised and hydrogen gas is released.

$$(C^-) \quad 2\,H^+ + 2 \times 1\,\bar{e} \longrightarrow 2\,H^0 \longrightarrow H_2\uparrow$$

The sodium base ions remain in the electrolysis bath. The process is known as chloralkali electrolysis, an industrial method for producing hydrogen, chlorine and sodium base.

The processes involved in the preparation of caesium hydroxide (CsOH) by electrolysis of an aqueous solution of caesium chloride (CsCl) are analogous.

Sample problems and solutions

Task 1.

On heating, solid **A** reacts with colorless gas **B** to produce gas **C** with a pungent irritating odor, which on further interaction with **B**, is converted to substance **D**. This substance reacts with water to form compound **D**, the aqueous solution of which changes the blue litmus to red. With some metals of group IIA of the periodic system, **D** forms slightly soluble compounds. Identify substances **A**, **B**, **C**, **D** and **E**, describe and explain the processes.

Exemplary solution

The solid **A** is sulfur (**S)** and the colorless gas **B** is oxygen (O_2). The interaction of the two substances produces sulfur dioxide **C** with a pungent irritating odor.

$$S + O_2 \longrightarrow SO_2$$

$$S^0 - 4\,e^- \longrightarrow S^{+4} \quad \Big| \; 4 \; \Big| \; \; \Big| \; 1$$

$$\qquad \qquad \qquad \qquad \qquad \qquad \quad 4$$

$$O_2 \longrightarrow 2\,O^0 + 2 \times 2\,\bar{e} \longrightarrow 2\,O^{-2} \quad \Big| \; 4 \; \Big| \; \; \Big| \; 1$$

The process is redox. The sulfur atom gives up four electrons and is a reductant, and the oxygen atoms take up two electrons and are oxidants.

Sulfur dioxide reacts with additional oxygen to form sulfur trioxide (**D**).

$$2\,SO_2 + O_2 \longrightarrow 2\,SO_3$$

$$\begin{array}{lll|l|l|l} S^{+4} - 2e^- & \longrightarrow & S^{+6} & 2 & & 2 \\ & & & & 4 & \\ O_2 \longrightarrow 2O^0 + 2 \times 2e^- & \longrightarrow & 2O^{-2} & 4 & & 1 \end{array}$$

The reaction of sulfur trioxide (**D**) with water produces sulfuric acid (**E**).

$$SO_3 + H_2O \longrightarrow H_2SO_4$$

In aqueous solution, sulfuric acid dissociates into hydrogen cations (H^+) and acid (sulfate) anions.

$$H_2SO_4 \longrightarrow 2H^+ + SO_4^{-2}$$

The aqueous sulfuric acid solution changes the color of the blue litmus to red.

Sulfuric acid forms slightly soluble compounds with some metals of group IIA of the Periodic Table. On this property is based the application of barium ions for the demonstration of sulfuric acid and its soluble salts. On mixing, a precipitate of barium sulfate is separated.

$$Ba + H_2SO_4 \longrightarrow BaSO_4\downarrow + H_2\uparrow$$

Task 2.

Barium dihydroxide reacts with an aqueous solution of substance **A** to produce a white precipitate **B** and to give off gas **C** with a pungent specific odor. The aqueous solution of **A** changes the color of the litmus to red and that of **B** to blue. From **C** obtain gas **D**, which is a mixed anhydride. Neutralise the solution of **D** with a solution of **C**. What is the nature of the aqueous solutions of the resulting salts?

Exemplary solution:

Substance **A** is $(NH_4)_2SO_4$ a salt of a strong acid and a weak base. In aqueous solution, diammonium sulfate hydrolyzes.

$$\underbrace{2NH_4^+ + SO_4^{-2}} + \underbrace{2H^+ + 2\overline{O}H}_{\underset{2\,H_2O}{\upharpoonleft\downharpoonright}} \rightleftharpoons \underbrace{2H^+ + SO_4^{-2}} + \underset{\underset{2\,H_2O + 2\,NH_3\uparrow}{\upharpoonleft\downharpoonright}}{2\,NH_4OH}$$

Hydrolysis is a reversible interaction of salt ions and water ions, in which the ratio between hydrogen and hydroxide ions changes, i.e., the pH of the medium. The hydroxide ions from the water bind with the ammonium ions from the salt to form the weak electrolite ammonium hydroxide. In excess, hydrogen ions remain in solution, which colors the blue litmus red.

$$\underbrace{Ba^{+2} + 2\,\bar{O}H}\; + \;\underbrace{2NH_4^{+} + SO_4^{-2}} \longrightarrow BaSO_4\downarrow \; + \; 2\,NH_4OH$$

$$2\,NH_4OH \rightleftharpoons 2\,H_2O + 2\,NH_3\uparrow$$

Barium sulfate does not hydrolyze, also because it is insoluble in water. The white precipitate is $BaSO_4$ (**B**) and the gas with a pungent specific odor is ammonia (**C**). The aqueous solution of ammonia stains the litmus blue.

With a catalyst (the reaction has great practical application) ammonia oxidizes to NO:

$$4\,NH_3 + 5O_2 \xrightarrow{Pt} 4\,NO + 6\,H_2O + Q$$

$$\begin{array}{lll|c|c|c}
N^{-3} - 5\,\bar{e} & \longrightarrow & N^{+2} & 5 & \multirow{2}{*}{20} & 4 \\
O_2 \longrightarrow 2O^{0} + 2\,x\,2\,e^{-} & \longrightarrow & 2O^{-2} & 4 & & 5
\end{array}$$

Oxidation of nitric oxide to nitrogen dioxide:

$$2\,NO + O_2 \rightleftharpoons 2NO_2 \;(\mathbf{D})$$

$$\begin{array}{lll|c|c|c}
N^{+2} - 2e^{-} & \longrightarrow & N^{+4} & 2 & \multirow{2}{*}{4} & 2 \\
O_2 \longrightarrow 2O^{0} + 2\,x\,2\,e^{-} & \longrightarrow & 2O^{-2} & 4 & & 1
\end{array}$$

Nitrogen dioxide NO_2 is a brown-colored, highly poisonous gas. It is produced by the oxidation of NO by oxygen in the air. It interacts vigorously with water to form a mixture of nitric and nitrous acids. The process is disproportionation.

$$2NO_2 + H_2O \longrightarrow HNO_2 + HNO_3$$

$$\begin{array}{lll|c|c|c}
N^{+4} + e^{-} & \longrightarrow & N^{+3} & 1 & \multirow{2}{*}{1} & 1 \\
N^{+4} - e^{-} & \longrightarrow & N^{+5} & 1 & & 1
\end{array}$$

It is therefore a mixed anhydride. In the presence of oxygen, it reacts with water to form nitric acid:

$$4\,NO_2 + O_2 + 2\,H_2O \longrightarrow 4\,HNO_3$$

$$\begin{array}{lll|c|c|c}
N^{+4} - e^{-} & \longrightarrow & N^{+5} & 1 & \multirow{2}{*}{4} & 4 \\
O_2 \longrightarrow 2O^{0} + 2\,x\,2\,e^{-} & \longrightarrow & 2O^{-2} & 4 & & 1
\end{array}$$

The reaction is a more special case of an oxidation-reduction process, since atoms of the same element (the nitrogen atom of the NO molecule) are oxidants and reducers.

$$NH_4OH + HNO_2 \rightleftharpoons NH_4NO_2 + H_2O$$

ammonium nitrite

$$\underbrace{NH_4^+ + OH^-} + \underbrace{H^+ + NO_2^-} \rightleftharpoons \underbrace{NH_4^+ + NO_2^-} + H_2O$$

$$H^+ + OH^- \longrightarrow H_2O$$

and

$$NH_4OH + HNO_3 \rightleftharpoons NH_4NO_3 + H_2O$$

$$\underbrace{NH_4^+ + OH^-} + \underbrace{H^+ + NO_3^-} \rightleftharpoons \underbrace{NH_4^+ + NO_3^-} + H_2O$$

$$H^+ + OH^- \longrightarrow H_2O$$

ammonium nitrate

$$\underbrace{NH_4^+ + NO_2^-} + \underbrace{H^+ + OH^-}_{\substack{\upharpoonleft\downharpoonright \\ H_2O}} \rightleftharpoons NH_4OH + HNO_2$$

$$\underbrace{NH_4^+ + NO_3^-} + \underbrace{H^+ + OH^-}_{\substack{\upharpoonleft\downharpoonright \\ H_2O}} \rightleftharpoons NH_4OH + \underbrace{H^+ + NO_3^-}$$

$$pH < 7$$

19. Theory of the Structure of Chemical Compounds

Structural theory. Fundamentals of structural theory. Interaction of atoms in molecules. Structural and rational formulae. Trivial nomenclature. Modern nomenclature (IUPAC). Development of structural theory. Types of isomerism: chain, positional, tautomerism. Stereochemical theory. Spatial structure. Electronic structure of organic compounds. Hybridization of atomic orbitals.

The rapid development of organic chemistry, which began in the middle of the 19th century, was related both to the demands of practice and to the development of other fields of chemistry and other sciences. The real, rapid development of organic chemistry began in the 1870s, when the Russian chemist A.M. Butlerov established the theory of the structure of chemical compounds. This theory, called the structural theory, not only generalized and explained the experimental material gathered up to that time, but opened up new possibilities for the development of organic chemistry.

The division of compounds into organic and inorganic was made in the early 19th century (1808). It was believed that organic compounds could only be isolated from organisms in which they were formed under the action of the so-called 'vital force' – vis vitalis. After the synthesis of a large number of organic compounds in the laboratory, the vitalistic theory was abandoned, but the name organic compounds and organic chemistry have been retained to this day.

Organic chemistry studies the compounds of carbon. The exceptions are carbon monoxide, carbon dioxide, carbonic acid and its salts. No sharp distinction can be drawn between organic and inorganic chemistry, because all substances, whatever their composition and structure, are united by common natural laws.

A particular property of organic compounds is that many of them have covalent bonds and do not dissociate, or the degree of dissociation in aqueous solution is insignificant. At relatively low temperatures, organic compounds decompose and undergo significant changes.

After the introduction and clarification of the concept of **a molecule**, the question of its **"anatomy"** arises, i.e., the way in which atoms are bonded and arranged in the molecule.

In the middle of the nineteenth century, it was not clear whether the molecules of chemical compounds were collections of atoms that moved chaotically without obeying any regularity, or whether the atoms bonded into molecules were arranged in a strictly defined way in space, i.e., the molecules had a well-defined chemical structure.

The solution to this fundamental problem was linked to the answer to the following questions:

1. If a molecule is a system of atoms bonded in a particular way, i.e., if it has a particular structure, what is the main, fundamental feature that defines it?
2. On what do the chemical and physical properties of substances depend?
3. Is it possible to establish the structure of the molecules and if so, by what means, in what way? Can this structure be expressed by a chemical formula?

The basic concept in the theory of the structure of chemical compounds is the concept of **chemical structure**. This concept was introduced into chemistry by Butlerov. Applied to a molecule, it expresses its chemical structure.

Structural theory is a general theory for all chemical compounds regardless of their composition and origin.

The prerequisites for the creation of structural theory are as follows:

1. The discovery of a significant number of organic and inorganic compounds with elucidated composition.
2. The establishment of the doctrine of valence and in particular the establishment of the valence of the carbon atom. In the late 1850s, the German chemist Kekule established that the carbon atom is tetravalent.
3. The discovery of the ability of carbon atoms to form chains, i.e., to link together.
4. The discovery of the phenomenon of isomerism in hydrocarbons.

The ability of carbon atoms to form chains was discovered by the Scottish scientist Cooper in the late 1850s, and isomerism in hydrocarbons by Butler.

Depending on the sequence in which the carbon atoms are linked, the chains are of two types: open and ring (cyclic). In turn, open carbon chains are straight and branched:

Open carbon chains — rights

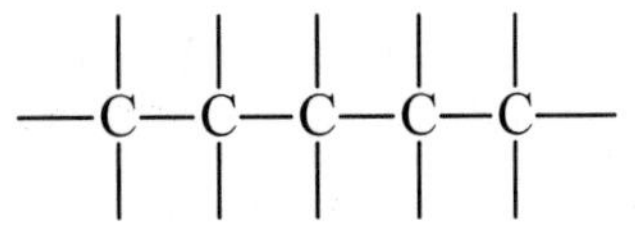

brancing

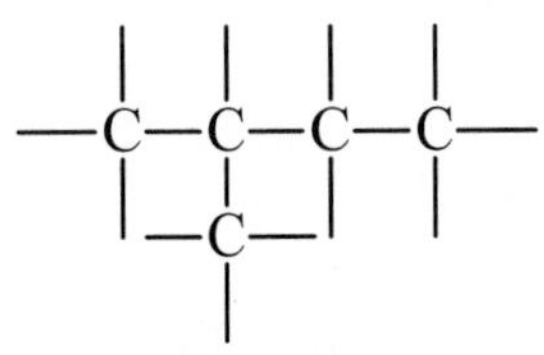

The chain that contains the larger number of carbon atoms is called the main chain and the chain connected to it is called the side chain.

Ring (cyclic) chains are: three-, four-, five-, six- and polyatomic:

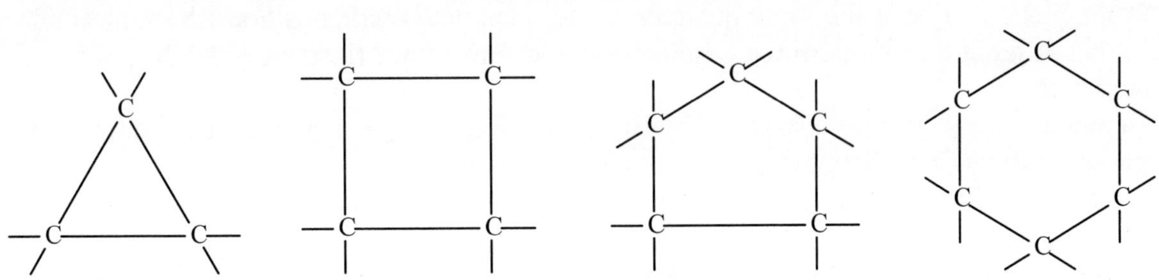

ets

When the chains are made up of carbon atoms only, the compounds are called **carbo- chains**. If they contain another atom other than carbon, they are **heterovalent:** open or closed:

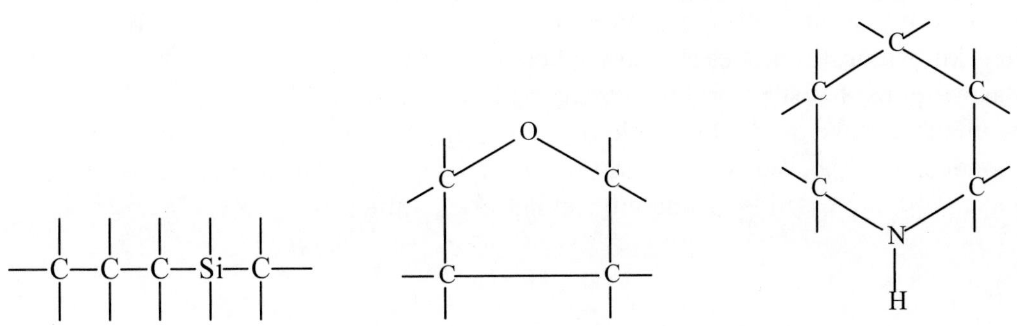

According to the place they occupy in the chain, the carbon atoms connected by simple bonds are: **primary** - if they are immediately connected with only one carbon atom; **secondary** - if they are connected with two carbon atoms; **tertiary and quaternary** - if they are connected with three, respectively with four carbon atoms:

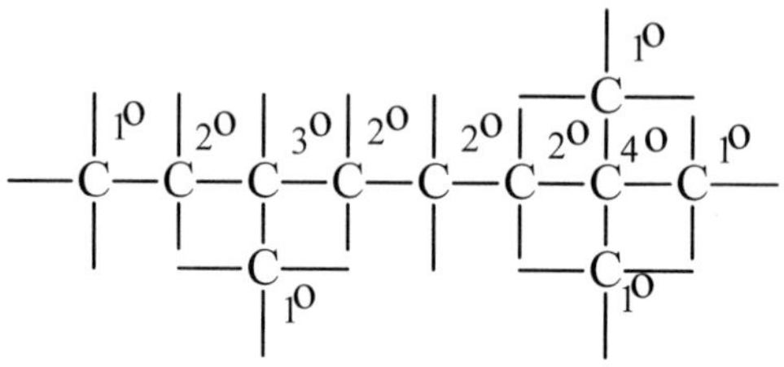

1° - primary carbon atom; 2° - secondary carbon atom;

3° - tertiary carbon atom; 4° - quaternary carbon atom.

Isomerism is a phenomenon that could not be explained by the existing theories and notions in chemistry until Butlerov's time. Its discovery in hydrocarbons contributed most to the establishment of structural theory.

At the end of the 17th century, the law of the constant composition of chemical compounds was discovered. According to this law, every chemical compound has a precise qualitative and quantitative composition (number of atoms in the molecule), which is expressed by the chemical formula. Since its discovery, the law of constant composition has been interpreted to mean that a single chemical compound corresponds to a given composition, a given formula. This statement turned out not to be true. In the early 19th century, substances were discovered that had the same composition but different properties. At the suggestion of Berzelius (1830), the concept of **isomerism** was introduced into chemistry. It is defined as follows:

Compounds that have the same qualitative and quantitative composition (the same molecular mass) but differ in their properties are called isomeric compounds (isomers), and the phenomenon - isomerism.

Isomerism in alkanes was discovered by Butlerov. He found that there were two compounds with empirical formula C_4H_{10}, i.e., two isomers of butane. Later it was shown that there are three isomers in pentane and four in hexane. As the number of carbon atoms increases, the number of isomers increases rapidly.

Fundamentals of structural theory

Butlerov laid out the basic ideas of structural theory in 1861.

According to structural theory, the molecular structure is knowable. It can be determined by studying the physical and chemical properties of substances. Conversely, if the structure of a molecule is known, its properties can be predicted.

The basic premises of structural theory can be formulated as follows:

The atoms that make up the molecule of a compound are not arranged in a disordered manner, but are bonded in a specific way according to their valence. The structure of the molecule of a compound can be expressed by a single formula, for example:

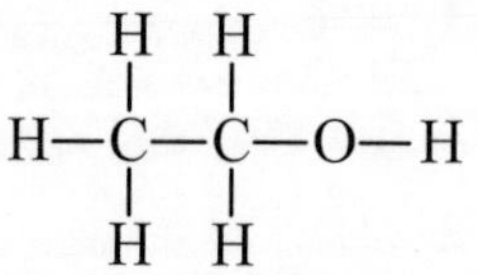

The properties of a substance are determined by the structure of the molecule, not just the type and number of atoms in it.

According to the theories that precede structural theory, chemical properties are determined by the qualitative and quantitative composition of the molecule. The composition of a molecule plays an important role in determining the properties of a molecule, but it is not always sufficient. For example, isomeric compounds have the same qualitative and quantitative composition but different, often radically opposite properties.

The main, most essential feature that determines the structure of molecules, and therefore their properties, is the bonding sequence of the atoms.

Starting from this basic premise one can easily explain the isomerism in butane. The difference in the properties of the two isomers is explained by the different bonding sequence of the carbon atoms in the butane molecule. One of them has a straight carbon chain:

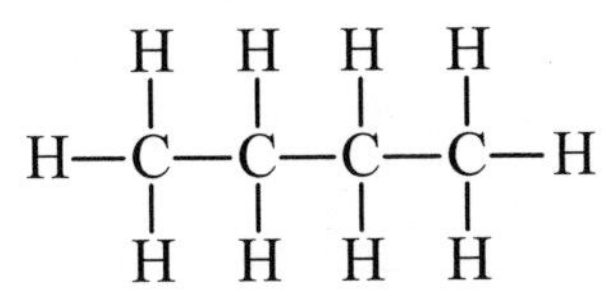

It is called normal butane (n-butane). The second has a branched chain in which the sequence of bonding of carbon atoms is different:

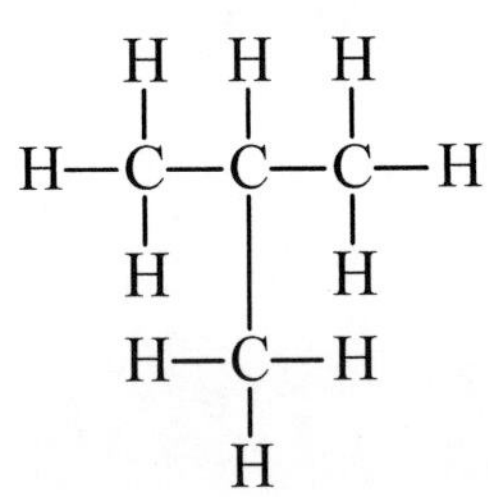

It is called isobutane (i-butane).

The three isomers of pentane C_5H_{12} differ in the structure of their carbon chains, i.e., the bonding sequence of their atoms is different:

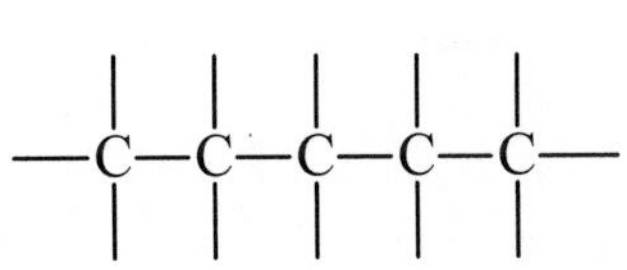

n- pentane

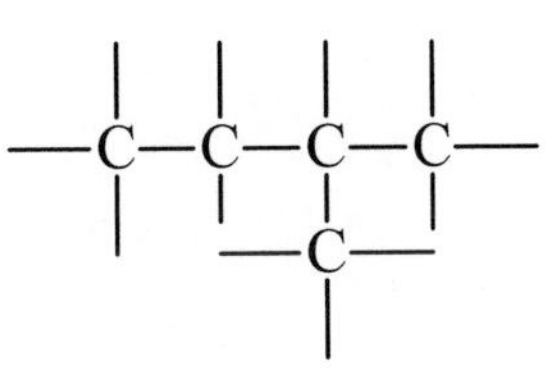

i-pentane

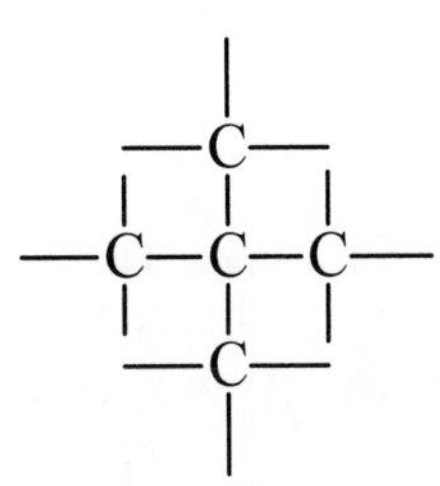

neopentane

The isomerism, which is determined by the different structure of the carbon chain, is called **chain isomerism.**

The chemical affinities of the atoms and atomic groups in a molecule are determined not only by the immediately bonded atoms and atomic groups but also by the more distant atoms and atomic groups.

Atoms in molecules influence each other.

This position of structural theory is known as "mutual influence of atoms and atomic groups". For example, in the chloroethane molecule

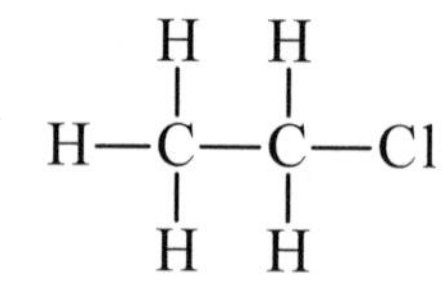

the hydrogen atoms that are located at the carbon atom bonded to the chlorine atom have different properties from the hydrogen atoms bonded to the other carbon atom. This is due to the influence of the chlorine atom.

The influence of different substituents in the aromatic core in electrophilic substitution reactions is the reason for the different orientation of the newly introduced substituent in the aromatic core:

HO—, H_2N—, RHN—, R_2N—, etc. are o⁻ and p⁻orientants;

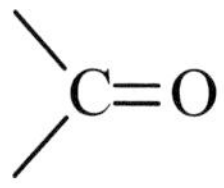, —CHO, —COOH, —COR, —COOR, $—CONH_2$, $—NO_2$

etc. are m-orientants.

Chemical formulae in which the sequence of bonding of atoms in a molecule and their valency are marked by valence traits are called **structural formulae**:

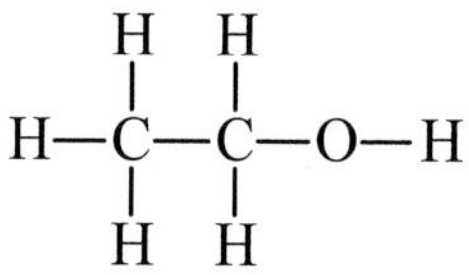

ethyl alcohol

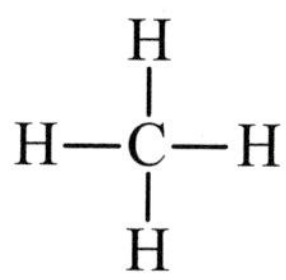

methane

This way of expressing thc structural formulae does not at all accurately reflect the structure and properties of the molecules.

Very often, for convenience, the complete structural formula of a compound's molecule is not written. So-called **rational formulas** are used. In these, not all valence features between atoms are written. For example, the structural and rational formulas of butane are written as follows:

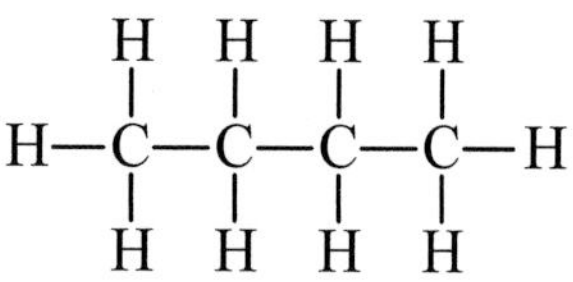

$CH_3—CH_2—CH_2—CH_3$ $CH_3CH_2CH_2CH_3$

structural formula rational formulas

In organic chemistry, the so-called **trivial nomenclature** for naming organic compounds has arisen from the beginning. The trivial names have no direct relation to the composition and structure of the compounds. They are usually related either to some property or to the origin of the compound. These are, for example, the names of formic acid (found in ants, Formica rufa) and of acetic acid (found in vinegar). Many trivial names of organic compounds are still used in the chemical literature.

The foundations of the **modern nomenclature** of organic compounds were laid at the Geneva Congress of Chemists in 1892 and the nomenclature was called the Geneva nomenclature. It was later refined at a congress of the International Union of Pure and Applied Chemistry, or **IUPAC** for short. It defines the names of the simplest organic compounds, the hydrocarbons. These are taken as the basis for the names of more complex organic compounds. Through a system of rules, it is possible to give any organic compound a name that expresses unambiguously the relationship between its composition and structure.

Development of structural theory

Structural theory covers and explains a large part of the properties of organic compounds. Like any scientific theory, it is evolving and contributing ever deeper insights into the structure of molecules. It can explain all kinds of structural isomerism:

chain - different sequence of carbon atoms in the chain;

positional - isomers differ in the location of a substituent or complex bond:

$$CH_3-CH=CH-CH_3 \qquad CH_2=CH-CH_2-CH_3$$

$$CH_3-CH_2-CH_2-OH$$

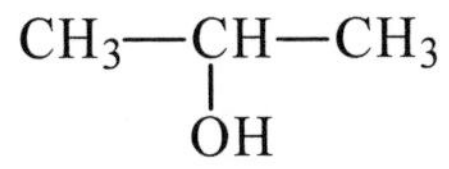

tautomerism - isomers differ in the location of a hydrogen atom relative to a complex bond:

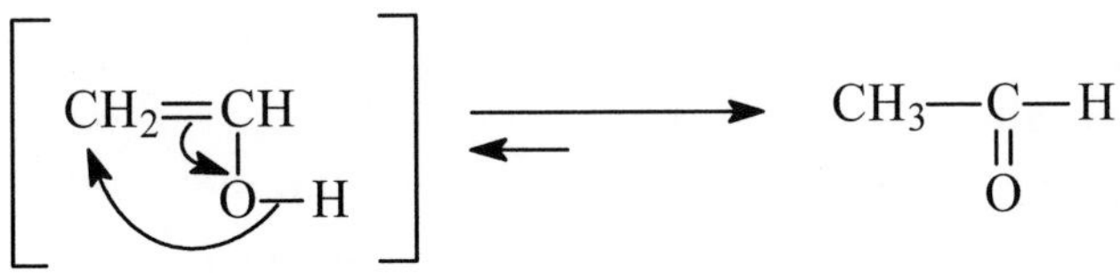

vinyl alcohol (ethenol) acetaldehyde (ethanal)

The structural formulas of substances express the arrangement of atoms in a plane (two-dimensional space). In reality, molecules are not flat but have a three-dimensional structure. The idea of the spatial structure of molecules originated before and immediately with the formation of structural theory (the French physicist Jean-Baptiste Biot and the microbiologist Louis Pasteur). The **stereochemical theory of** Le Bel and Want Hoff was established in 1873, according to which:

The four valence bonds of the carbon atom are oriented in space so that their ends coincide with the vertices of a **regular tetrahedron** in the center of which the carbon atom is located.

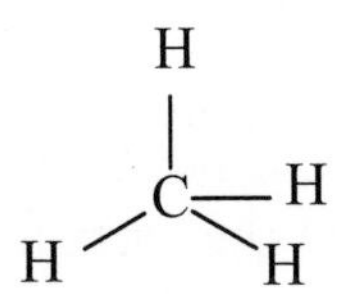

When two carbon atoms are bonded with a simple bond, the tetrahedra have a common tip.

Stereochemical theory explained the spatial isomerism that could not be explained by structural theory.

An idea of the **spatial structure** of molecules is obtained if the chemical signs of the elements in the structural formulae are replaced by spheres of different size and color. In this way, an enlarged model of the real molecule is obtained.

More accurate models are those in which the relative sizes of the atoms forming the molecule and the distances between them are in the same proportion as in the actual molecule.

Although reasonably accurate, the models reflect the actual structure of the molecule only to a certain extent.

Electronic structure of organic compounds

After the discovery of the electronic structure of the atom, electronic concepts were introduced into organic chemistry. They broadened and deepened the knowledge of the structure of organic compounds.

In constructing the electronic shell of the carbon atom, Pauli's principle and Hund's rule are followed.

Carbon has an ordinal number of six (6) in the Periodic Table, i.e., its electron count is equal to 6. When the carbon atom is in the ground state, the distribution of electrons by layers and sublayers is as follows:

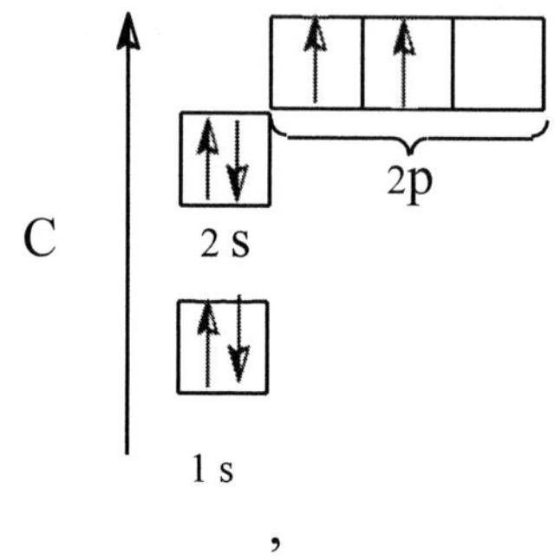

i.e., the electronic configuration is $1s^2\ 2s^2\ 2p^2$.

In this state, it has two unpaired electrons. The carbon atom should therefore be divalent. With few exceptions, the carbon atom is tetravalent. This valence is explained as follows: the energies of the 2s- and 2p-subunits are close in energy. Therefore, in the course of the chemical reaction, the electron pair from the 2s-subunit is broken and one of the electrons passes into the 2p-subunit. Along with this, spin reversal occurs. In the new valence state, the carbon atom has four electrons with unpaired spins, i.e., it is tetravalent.

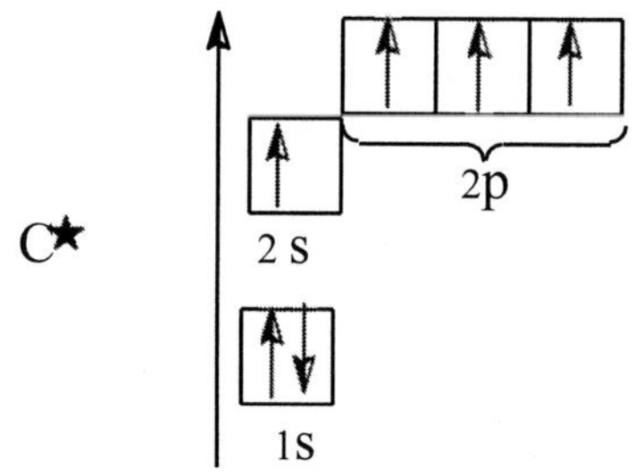

From the structure of the electron shell of the carbon atom in the excited state, it should form two types of bonds with different energies due to the different energy state in which its valence electrons are located - 2s- and 2p-. Furthermore, three of the four valence bonds should make angles of 90° with each other since the 2p-orbitals are mutually perpendicular. In the methane molecule and its homologues, the four bonds are equivalent and their axes make angles equal to 109° 28'. This contradiction is explained by the hybridization of the atomic orbitals.

In the excited carbon atom, three types of **hybridization of** atomic orbitals take place, from which three valence states of the carbon atom follow:

"Mixing" the 2s-orbital with the three 2p-orbitals yields four equal-energy hybrid orbitals with the shape of an irregular volume eight. They are directed with their bulky parts towards the vertices of a regular tetrahedron and their axes make an angle 109° 28'. This type of "mixing" is called **sp^3-hybridization** or **tetragonal.** It is characteristic of saturated compounds (alkanes, cycloalkanes, and for ammonium ions of the atoms of the elements of Group VA of the Periodic Table, for example NH_4^+).

"Mixing" the 2s-orbital with two 2p-orbitals yields three identical in shape and energy **sp^2 -hybrid orbitals**. The hybrid orbitals are arranged in one **plane at an angle of 120°**. The unhybridized $2p_z$ -orbital is positioned perpendicular to the plane of the sp^2-hybrid orbitals. This type of "mixing" of atomic orbitals is called **sp^2 -hybridization or trigonal**. It is characteristic of alkenes, alkadienes, aromatic compounds, and the elements of group IIIA of the Periodic Table.

"Mixing" the 2s orbital with one 2p-orbital produces **two sp-hybrid orbitals located at an angle of 180°.** The two non-hybridized orbitals – $2p_y$ - and $2p_z$ – at each carbon atom are mutually perpendicular. This type of "mixing" of atomic orbitals is called **sp-hybridization or digonal.** This type of carbon atom hybridization explains the linear structure of the alkyne molecule. Digonal hybridization is also characteristic of the elements of group IIA of the Periodic Table.

In almost all organic compounds, the valence electrons of the carbon atom form covalent bonds.

The electronic theory of the covalent chemical bond clarified the nature (essence) of the basic symbol used in writing structural formulas – the valence dash. Each valence bar conventionally expresses an electron pair that connects two atoms. This means that the electronic formula of a chemical compound can be written by replacing the valence bars of the structural formula with pairs of dots representing the electron pairs. For example, the structural and electronic formulas of methane are as follows:

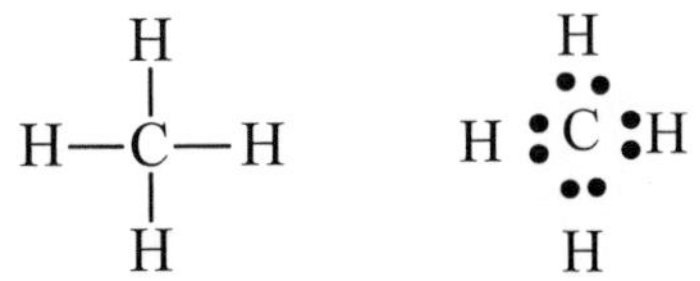

Electrons in electron pairs connecting two atoms have antiparallel spins:

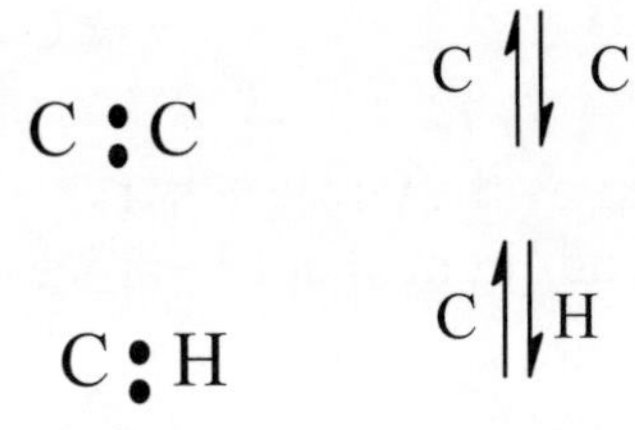

When two identical atoms are bonded, e.g., C - C, H - H, the bond is non-polar. In these cases, the electron pair is located at the geometric center of the chemical bond:

C – C   C : C

When a carbon atom is bonded to an atom of another element, such as oxygen, the covalent bond is polar. This means that the electron pair is pulled towards the oxygen atom, which has a greater electronegativity.

C – O   C :O

As a result of the asymmetric arrangement of the electron pair, the carbon atom acquires a partial positive charge and the oxygen atom a partial negative charge. The polarity of the bond is expressed by an arrow indicating the direction (or atom) in which the electron pair is drawn:

$$\overset{\delta+}{C} \longrightarrow \overset{\delta-}{O}$$

Partial charges are denoted by the Greek letter δ (delta) and the corresponding sign "+" or "-".

If there is one polar bond in a polyatomic molecule, it causes a consistent, decreasing polarization of the adjacent nonpolar covalent bonds. This effect is called the **polarization effect**. For example, in a molecule containing a chlorine atom bonded at the beginning of the carbon chain, the bond

$$\overset{\delta+}{C} \longrightarrow \overset{\delta-}{Cl}$$

is polar and the polarization effect caused by it is expressed as follows:

$$\overset{\delta_3^+}{C} \longrightarrow \overset{\delta_2^+}{C} \longrightarrow \overset{\delta_1^+}{C} \longrightarrow \overset{\delta^-}{Cl}$$

$$\delta_3^+ < \delta_2^+ < \delta_1^+ \text{ , and}$$
$$\delta_{1+} + \delta_{2+} + \delta_{3+} = \delta^+ = -\delta^{--}$$

The chemical bond between two atoms that is made by a single electron pair is called a simple bond. Such are all bonds in methane hydrocarbons.

In many chemical compounds, atoms are bonded by two or more electron pairs. Such bonds are called complex bonds.

They are **double** bonds, for example;

and **triplets:**

—C≡C— —C ⋮⋮⋮C—

Simple covalent bonds are denoted by the Greek letter σ (sigma). Such are the bonds in the hydrogen molecule, as well as the bonds in methane hydrocarbons:

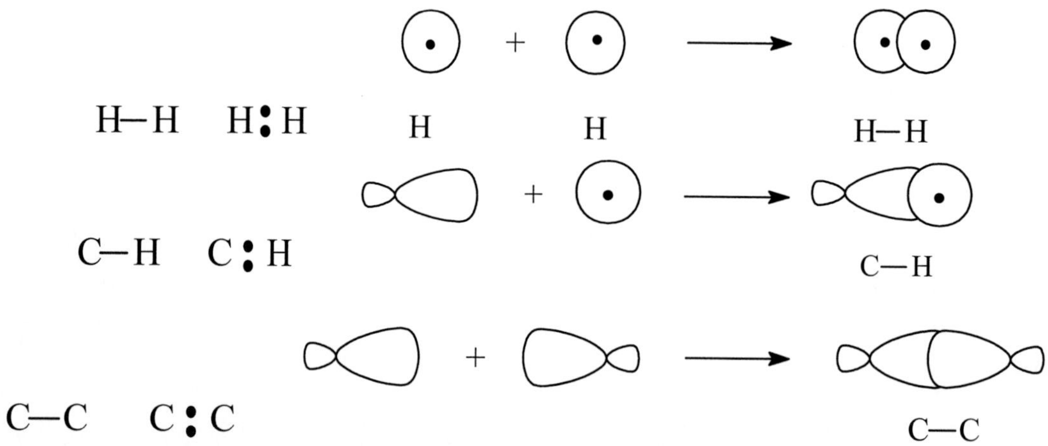

σ-Bonds are characterized by high strength and energy

Bond energy is the energy required to break a bond to produce free radicals or atoms. It is expressed in kJ/mol. When a chemical bond is formed, the same amount of energy is released. Another characteristic feature of σ-bonds is that the electron pair that makes them is localized between the two atoms. This means that the electrons are equally possessed by both atoms.

Free radicals are particles with an odd number of electrons. For example, the methyl radical.

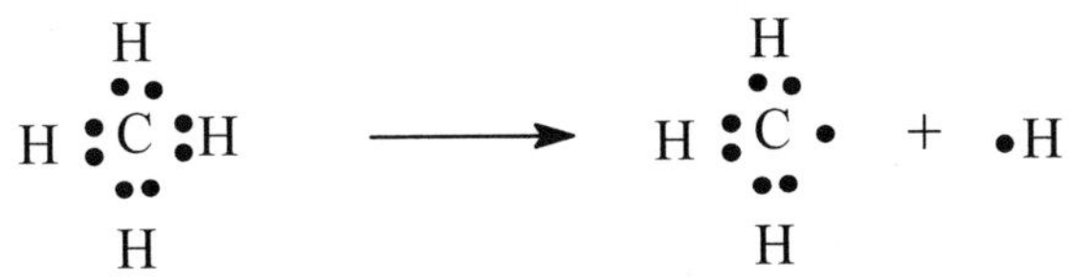

It is univalent because there is one electron with uncompensated spin. For this reason, radicals are very active particles.

With the development of the chemical bond theory and the structure of chemical compounds, the spatial structure was explained. This was also helped by the introduction of the theory of hybridization, which explains very well the tetrahedral structure of compounds with a σ-bond, the location of the π-bond perpendicular to the σ-bond in the

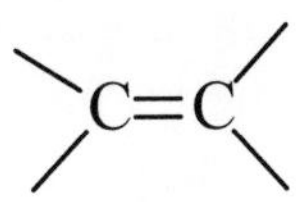

et al.

Hybridization is also possible with other atoms. For example, in NH_4^+, the nitrogen atom is in a sp^3-hybridization (mentioned above).

The modern development of stereochemistry is an extension of the theory of the structure of chemical compounds.

20. SATURATED HYDROCARBONS (ALKANES)

HOMOLOGOUS ORDER OF METHANE HYDROCARBONS. NAMES. ISOMERISM (CHAIN AND OPTICAL). ELECTRONIC STRUCTURE, Σ -BOND. PHYSICAL PROPERTIES OF ALKANES. CHEMICAL PROPERTIES OF ALKANES. METHODS OF PREPARATION OF ALKANES. IMPORTANCE AND USES.

Hydrocarbons are organic compounds made up of carbon and hydrogen, hence their name. The simplest hydrocarbon is methane. Hydrocarbons are known to be similar in their properties to methane, but differ from it in that their molecules are made up of a greater number of carbon and hydrogen atoms. They are called methane hydrocarbons or alkanes. Alkanes are also known as saturated hydrocarbons or paraffins – from the Latin parum affinis – of little affinity.

Homologous order and names

Individual compounds of the alkane group contain different numbers of carbon atoms in their molecule – from one to several dozen. If the alkanes are arranged in ascending order of their molecular masses, a series of compounds is formed that have similar chemical properties, properly varying physical properties, and a common method of preparation. Each successive hydrocarbon in this row has one carbon and two hydrogen atoms more than the previous one, i.e., it differs by a group – CH_2 – called a methylene group.

A series of compounds whose molecules are made up of the same kind of atoms, i.e., have the same qualitative composition, possess similar chemical properties, and any two adjacent members differ by one methylene group, is called a homologous series. The compounds are called homologues and the phenomenon is called homology.

The existence of homologous series facilitates systematics and the study of organic compounds. If the properties of some members of the homologous order are known, the properties of the remaining homologues can be predicted.

Because the number of atoms in the molecules of homologues changes properly, it can be expressed by a general formula. The general formula for alkanes is:

$$C_nH_{2n+2}$$

The homologous order of alkanes is:

CH_4 - methane
C_2H_6 - ethane
C_3H_8 - propane
C_4H_{10} - butane
C_5H_{12} - pentane
C_6H_{14} - hexane

C_7H_{16} - heptane
C_8H_{18} – octane
C_9H_{20} - nonane
$C_{10}H_{22}$ –decane
$C_{11}H_{24}$ undecane
....

$C_{20}N_{42}$ - eicosan
....
C_nH_{2n+2}

The names of the first four homologous members are trivial. The names of the higher homologues are formed according to IUPAC nomenclature by adding the suffix -an to the root of the Greek or Latin numerical name, indicating the number of carbon atoms in the alkane molecule.

According to IUPAC, the names of branched-chain alkanes are formed as follows: the main chain is determined and according to the number of carbon atoms, the corresponding alkane name is assigned. The side chain(s) is regarded as a substituent whose name matches the name of the residue with the same number of carbon atoms. This name is placed as a prefix in front of the main name with a corresponding number. The numbering starts at the end of the main chain where the substituent will be given the lowest number. For example:

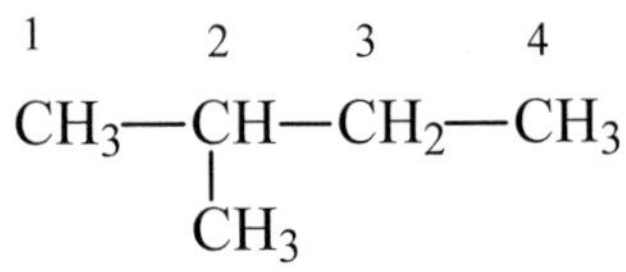

2-methylbutane

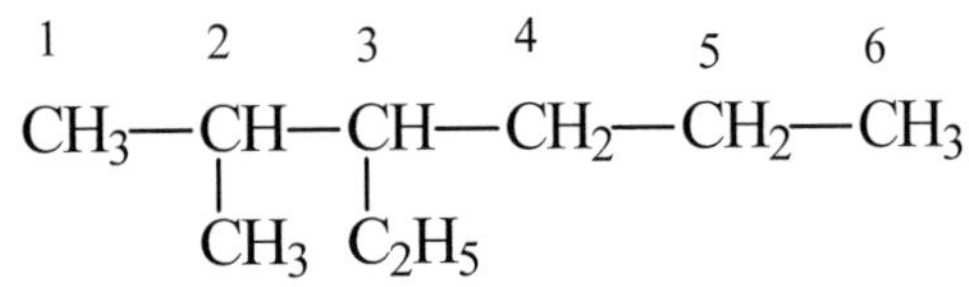

3-ethyl-2-methylhexane

If one hydrogen atom is taken away from the methane hydrocarbon molecules, residues with the general formula C_nH_{2n+1}, called alkyl groups, are obtained, which are denoted by R-. Their names are formed from the name of the given alkane by replacing the ending -an with - yl and the word residue, e.g.:

CH_4 - methane - CH_3 - methyl residue C_2H_6 - ethane - C_2H_5 - ethyl residue

....

C_nH_{2n+2} - alkane - C_nH_{2n+1} - alkyl residue

Isomerism

With the exception of the first three homologues, chain isomerism exists in all other alkanes. It is determined by the way the carbon atoms are linked in the chain. In the molecules of the higher alkanes, the carbon chains may be straight or branched, giving rise to chain isomers. Isomers with straight carbon chains contain only primary and secondary carbon atoms

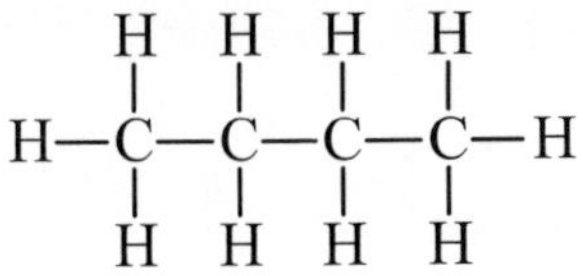

and those with branched carbon chains – and tertiary or quaternary carbon atoms:

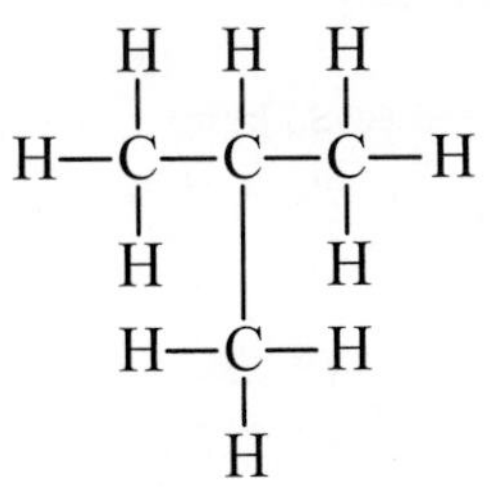

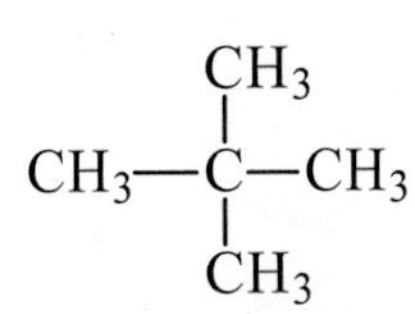

Alkanes that contain a straight carbon chain are called normal, and those that have the grouping in their carbon chain isoalkanes - the prefix iso- is placed in front of the names of the corresponding alkanes.

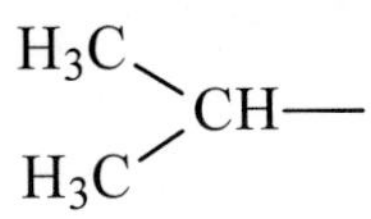

In this grouping, there are two methyl groups as substituents at one carbon atom.
Alkanes containing three methyl groups at one carbon atom are designated as neo- alkanes.

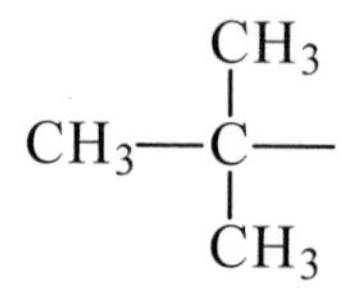

Isomerism in methane hydrocarbons was discovered by Butlerov. He found that there were two compounds with empirical formula C_4H_{10}, i.e., two isomers of butane with different boiling points.

Later it was found that there are three isomers in pentane and four in hexane. As the number of carbon atoms in the chain increases, the number of isomers increases rapidly.

The concept of isomerism is defined as follows:

Compounds that have the same qualitative and quantitative composition (the same molecular mass) but differ in their properties are called isomeric compounds (isomers), and the phenomenon - isomerism.

When an alkane molecule contains an asymmetric carbon atom (with four different substituents), optical isomerism is observed. For example, in 3-methylhexane

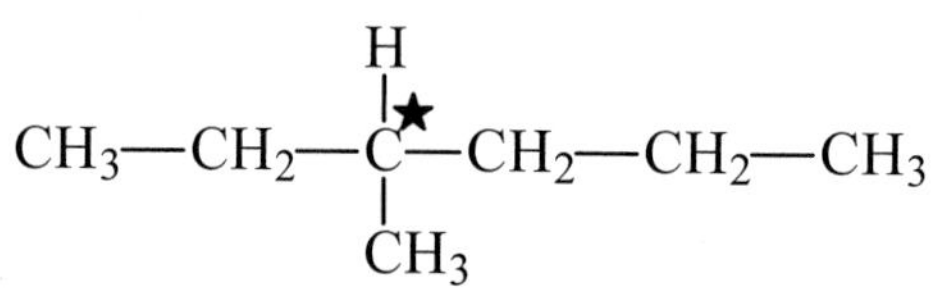

two spatial forms are known, which relate to each other as an object and its mirror image:

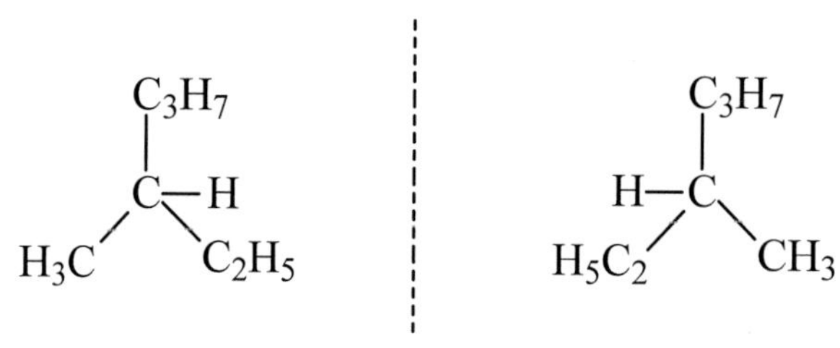

D- L-

Electronic construction

As presented, the carbon atom is tetravalent and in alkanes possesses four sp^3 - hybrid orbitals.

When σ-bonds are formed between carbon atoms, an overlapping of sp^3-hybrid orbitals of two carbon atoms occurs:

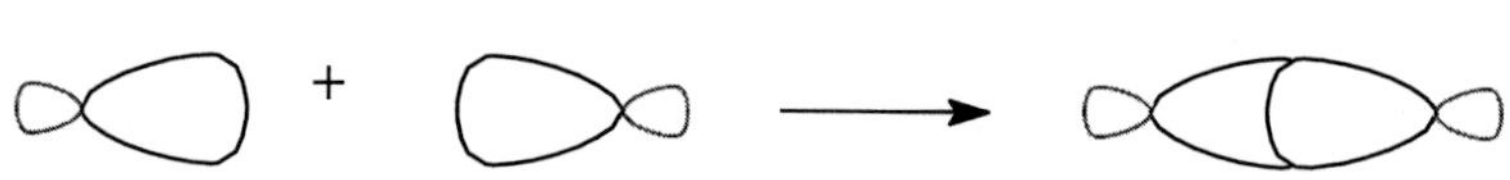

C—C

The formation of σ-bonds between carbon and hydrogen atoms is accomplished by overlapping the 1s-orbital of a hydrogen atom and the sp^3-hybrid orbital of a carbon atom:

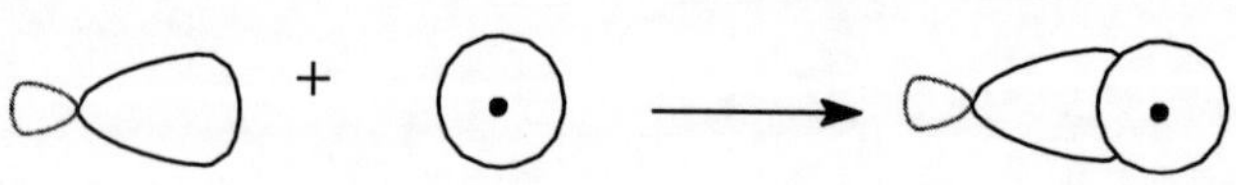

C—H

There are also cycloalkanes. Their constituent carbon atoms are in the sp^3-hybrid state. Cycloalkanes with a three- and four-atom ring have been found to be perishable – the ring opens easily. Stable are cycloalkanes with a six-atom ring.

Physical properties of alkanes

The first four alkanes are gaseous substances at ordinary temperature, the next to the sixteenth are liquids, and the higher ones are solids. Alkanes are insoluble in water. In accordance with the empirical rule "like substances dissolve in like", alkanes dissolve in non-polar solvents. They are themselves good solvents for other non-polar compounds.

The main natural source for the production of alkanes is petroleum.

Chemical properties of alkanes

Under the right conditions, alkanes prove to be relatively reactive compounds.

Under ordinary conditions they are resistant to oxidants. When ignited in the presence of oxygen (air), they burn with a colorless flame, giving off a large amount of heat:

$$CH_4 + 2\,O_2 \longrightarrow CO_2 + 2\,H_2O + Q$$

$$\begin{array}{lll|c|c|c}
 & C^{-4} - 8\,e^- & \longrightarrow C^{+4} & 8 & & 1 \\
 & & & & 8 & \\
O_2 \longrightarrow & 2\,O^0 + 2\times 2\,e^- & \longrightarrow 2\,O^{-2} & 4 & & 2
\end{array}$$

At high temperature (~1000° C) in the absence of oxygen, methane decomposes into carbon and hydrogen:

$$CH_4 \xrightleftharpoons{1000^oC} C + 2\,H_2$$

$$\begin{array}{lll|c|c|c}
 & C^{-4} - 4\,e^- & \longrightarrow C^0 & 4 & & 1 \\
 & & & & 4 & \\
2H^+ + 2\times 1\,e^- & \longrightarrow 2H^0 & \longrightarrow H_2 & 2 & & 2
\end{array}$$

The oxidation of alkanes to oxygen derivatives can be carried out under different conditions: heating, oxidants:

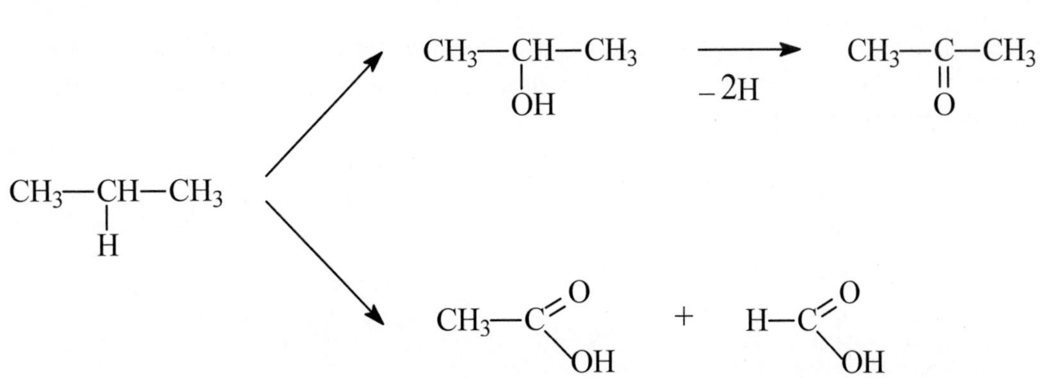

One of the most characteristic reactions for alkanes is their interaction with halogens.

The reaction can be expressed as follows:

$$CH_4 + Cl_2 \longrightarrow CH_3Cl + HCl$$

chloromethane (methyl chloride)

$$CH_3Cl + Cl_2 \longrightarrow CH_2Cl_2 + HCl$$

dichloromethane

$$CH_2Cl_2 + Cl_2 \longrightarrow CHCl_3 + HCl$$

trichloromethane

$$CHCl_3 + Cl_2 \longrightarrow CCl_4 + HCl$$

tetrachloromethane

In these reactions, there is a gradual replacement of hydrogen atoms in the methane molecule by chlorine atoms.

Substitution reactions between methane molecules and halogen molecules are called halogenation (in this case chlorination), and the resulting compounds are called halogen derivatives (respectively chlorine derivatives). The reaction with bromine (bromination) proceeds similarly to chlorination, but more slowly. Fluorine reacts explosively with methane, so fluorine derivatives of methane are obtained by indirect methods.

The reactivity of halogen molecules towards alkanes was found to vary in the following order:

$$F_2 \gg Cl_2 > Br_2 \gg J_2$$

The chlorine molecule has been shown to initially break down into atoms under the action of light (he):

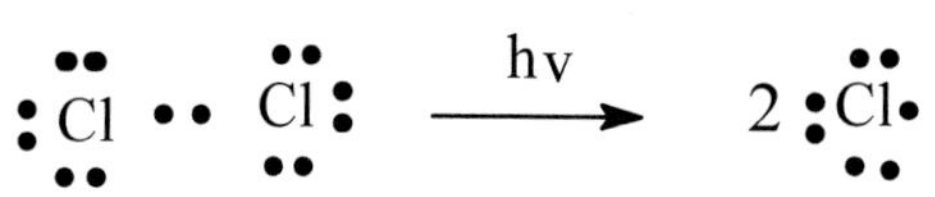

The chlorine atom attacks the methane molecule, tears off one hydrogen atom and forms hydrogen chloride. A methyl radical is produced:

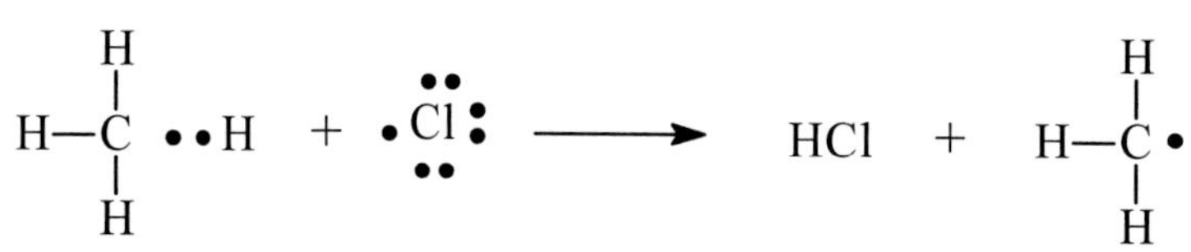

Free radicals are particles that have an odd number of electrons or, in an atom, a single electron with an unpaired spin. Free radicals, like free atoms, are chemically very active particles because they contain one or more unpaired electrons.

The resulting methyl radical reacts with a chlorine molecule:

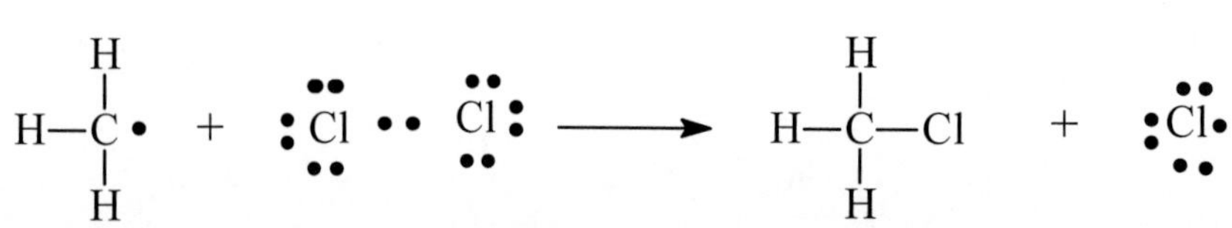

Chloromethane and a new chlorine atom are produced. It reacts with a new methane molecule and the process repeats until the reactants are exhausted (more precisely, until the chlorine is exhausted). The chain-radical reaction can stop at radical depletion, for example:

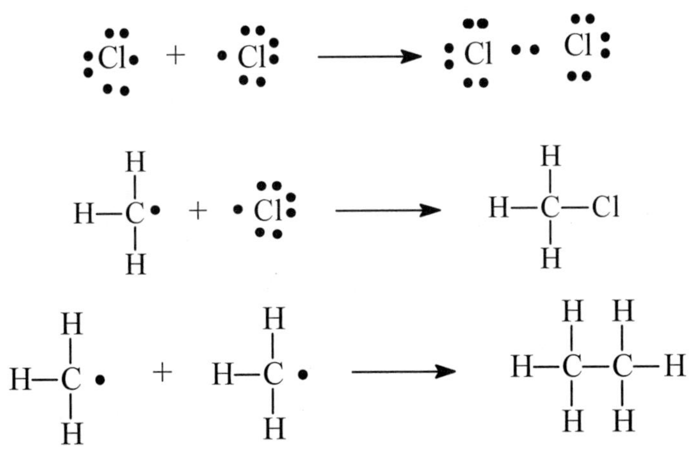

In branched-chain alkanes, a hydrogen atom is most easily substituted at a tertiary carbon atom, more difficult at a secondary, then at a primary, and most difficult in the methane molecule.

The sulfonation of alkanes, the replacement of a hydrogen atom by a sulfonic group, can take place when they interact with sulfur dioxide (SO_2) and oxygen under irradiation with UV light and in the presence of chain reaction initiators:

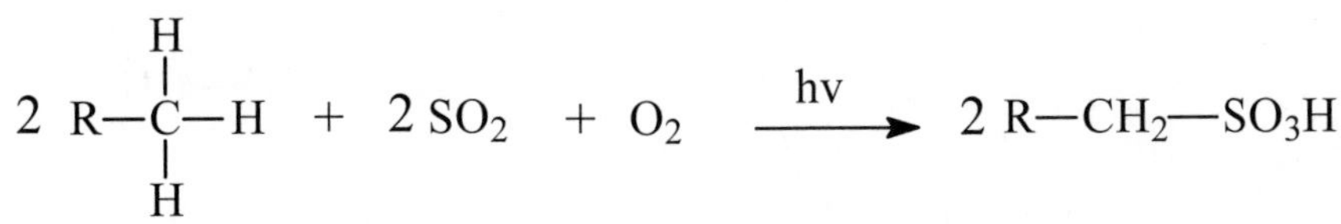

Nitration of alkanes is also a substitution reaction consisting of the replacement of a hydrogen atom by a nitro group (- NO_2):

$$H_3C{-}CH_3 \xrightarrow[-\,H_2O]{HNO_3} H_3C{-}CH_2{-}NO_2$$

nitroethane

In substitution reactions, it is easiest to substitute a hydrogen atom at a tertiary carbon atom, more difficult at a secondary carbon atom, and most difficult at a primary carbon atom. The isomerization reaction is associated with a change in the structure of higher alkanes (carbon chain modification), for example:

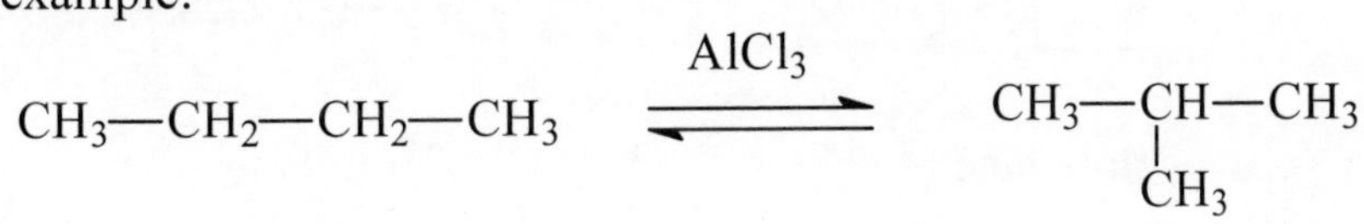

n-butane i-butane (isobutane)

Cracking process (from cracking) is a way of chemical processing of petroleum and petroleum fractions, in which chemical processes result in lower boiling hydrocarbons and increase the amount of gasoline.

Two types of cracking processes are distinguished: thermal cracking, which takes place at high temperatures, and catalytic cracking, which takes place when heated and in the presence of catalysts.

Under thermal cracking conditions (~ 550° C), carbon-hydrogen and carbon-carbon bonds are broken and a mixture of lower saturated and unsaturated hydrocarbons and hydrogen is formed. For example, in the thermal decomposition of the hydrocarbon hexadecane $C_{16}H_{34}$ the following reactions take place:

$$C_{16}H_{34} \longrightarrow \underset{\text{octane}}{C_8H_{18}} + \underset{\text{octene}}{C_8H_{16}}$$

$$C_8H_{18} \longrightarrow \underset{\text{butane}}{C_4H_{10}} + \underset{\text{butene}}{C_4H_8}$$

$$\underset{\text{methane}}{CH_4} + \underset{\text{propene}}{C_3H_6} \longleftarrow \underset{\text{butane}}{C_4H_{10}} \longrightarrow \underset{\text{ethane}}{C_2H_6} + \underset{\text{ethene}}{C_2H_4}$$

$$C_2H_6 \longrightarrow \underset{\text{ethene}}{C_2H_4} + \underset{\text{hydrogen}}{H_2}$$

In catalytic cracking, natural or aluminium silicates are used as catalysts. The process is carried out at lower temperatures. In addition to isomerisation reactions (see above), catalytic cracking also involves cyclisation and dehydrogenation reactions:

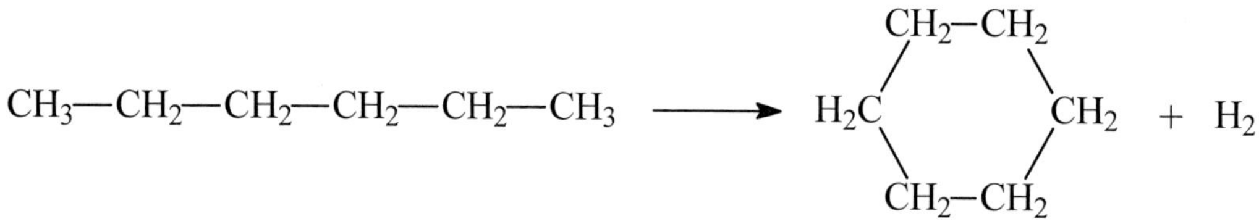

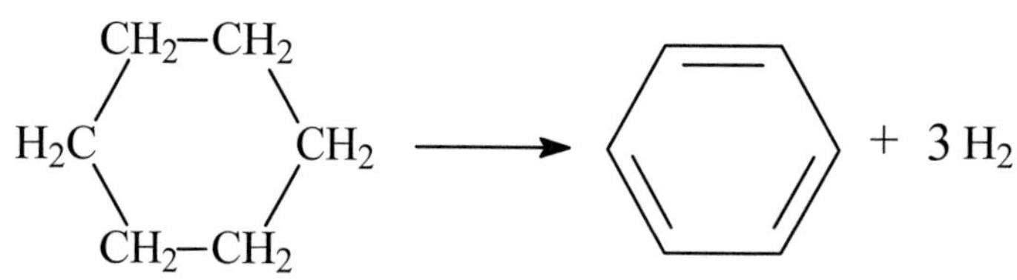

The reactions of dehydrogenation and cyclization often occur together, so thc process is called **dehydrocyclization** and the formation of aromatic hydrocarbons is called **aromatization**.

Small-ring cycloalkanes are characterized by ring-opening reactions – addition reactions:

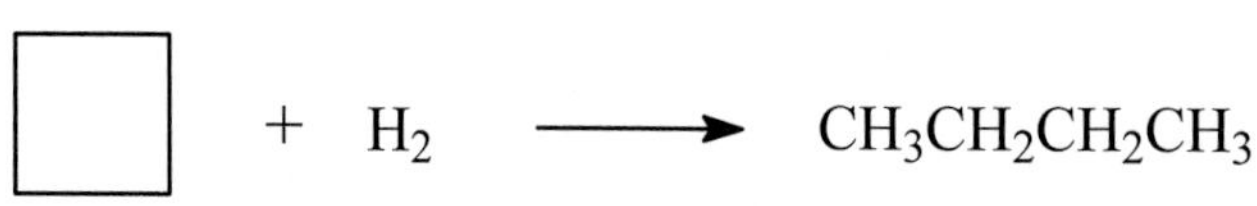

cyclobutane n-butane

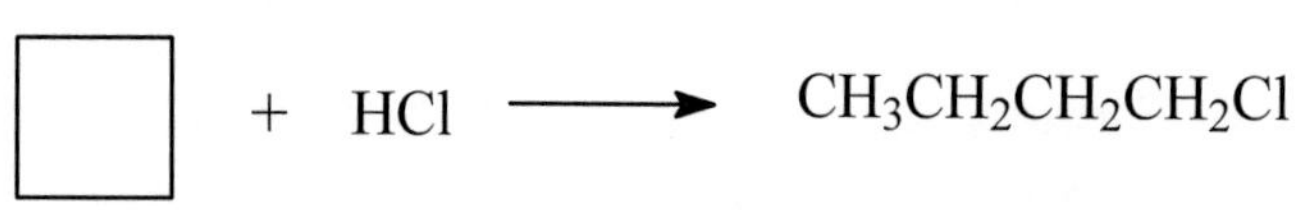

1-chlorobutane (butyl chloride)

Methods for the preparation of alkanes

Alkanes can be produced by various chemical methods:

1. One of the main methods is that of Wurz, which consists in the interaction of halogen derivatives of alkanes with metallic sodium:

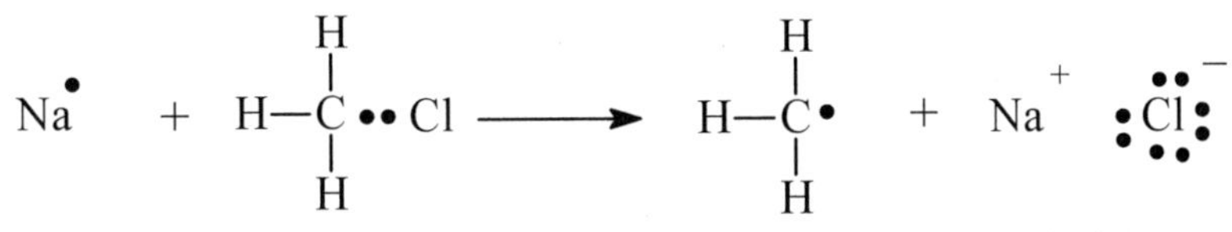

methyl radical

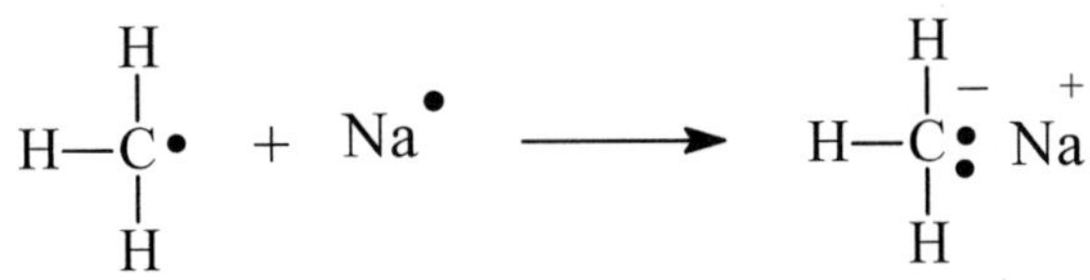

methyl sodium

In sum, the reaction is expressed as follows:

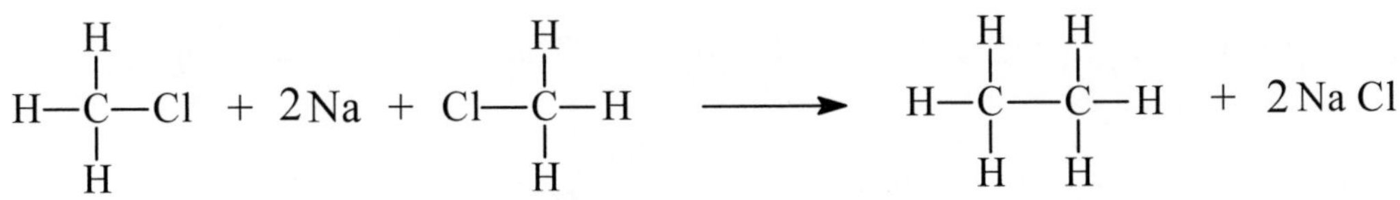

The method makes it possible to obtain a symmetrical alkene with double the number of carbon atoms compared to that in the alkyl residue -R.

Nonsymmetrical alkanes can be produced by the reaction of halogen derivatives of various hydrocarbons with sodium:

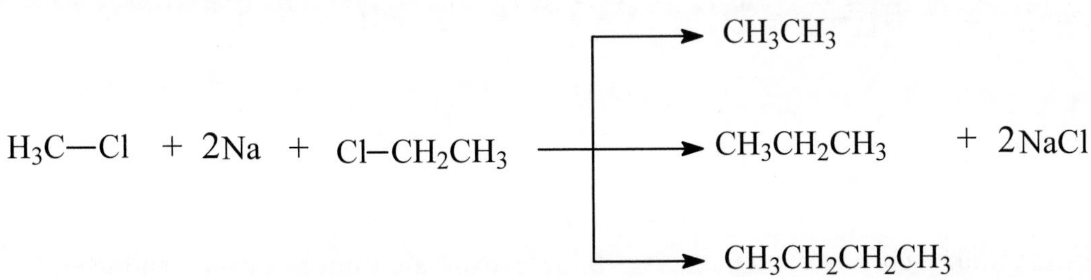

Competitive reactions also take place to form the symmetrical alkanes ethane and butane.

2. Alkanes are also produced by hydrogenation of double and triple bond hydrocarbons in the presence of catalysts (Pt, Ni, etc.):

$$H_2C{=}CH_2 + H_2 \longrightarrow H_3C{-}CH_3$$

$$HC{\equiv}CH + 2H_2 \longrightarrow H_3C{-}CH_3$$

3. Cycloalkanes are produced by dehalogenation of dihalogenated derivatives of alkanes with sodium or zinc:

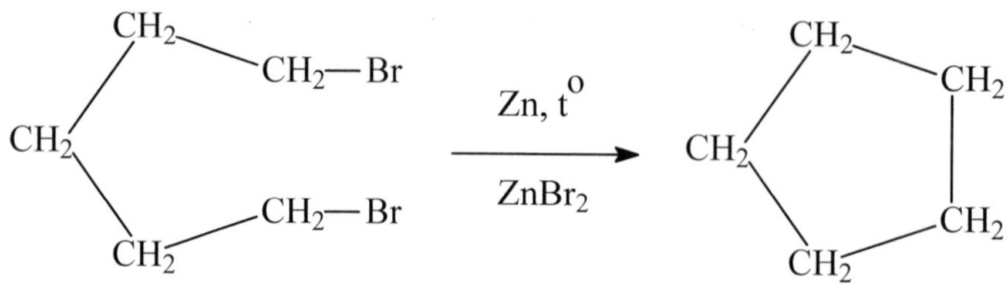

1,5-dibromopentane cyclopentane

4. Methane for industrial use is obtained from water gas under nickel catalyst:

$$CO + 3H_2 \xrightarrow{Ni} CH_4 + H_2O$$

$$C^{+2} + 6e^- \longrightarrow \overset{-4}{C} \quad | \; 6 \; | \; 6 \; | \; 1$$

$$H_2 \longrightarrow 2H^0 - 2 \times 1e^- \longrightarrow 2H^+ \quad | \; 2 \; | \quad | \; 3$$

5. Methane can also be produced by direct synthesis from carbon and hydrogen:

$$C + 2H_2 \xrightarrow[t^o]{Ni} CH_4$$

$$C^0 + 4e^- \longrightarrow \overset{-4}{C} \quad | \; 4 \; | \; 4 \; | \; 1$$

$$H_2 \longrightarrow 2H^0 - 2 \times 1e^- \longrightarrow 2H^+ \quad | \; 2 \; | \quad | \; 2$$

The method is theoretically important because it shows the genetic relationship between inorganic and organic substances.

6. For laboratory purposes, methane is produced from aluminium carbide and water on heating:

$$Al_4C_3 + 12H_2O \xrightarrow{t^o} 3CH_4 + 4Al(OH)_3$$

7. Methane is also formed by dry distillation of wood, coal and peat. It occurs in nature as "swamp gas". It is formed in the bacterial decay of plant materials. In coal mines, the 'mine gas' ('grize gas') emitted also contains methane. It is also a constituent of ground gases.

8. The higher alkanes enter into the composition of the earth's oil.

Sample problems and solutions

Task 1.

From methane, by suitable reactions, obtain ethanol and acetic acid. Their condensation yields a pleasant-smelling substance. If this substance is reacted with ammonia a new product is obtained. How is acetonitrile obtained from it? What is the hybridization of the carbon and nitrogen atoms in the nitrile molecule? Name the products of hydrolysis of acetonitrile.

Exemplary solution

High-temperature methane processing produces acetylene (ethyne) and hydrogen:

$$2CH_4 \xrightarrow{2000^{o}\,C} HC{\equiv}CH \;+ 3\,H_2$$

$$\begin{array}{lcl|c|c|c} \overset{-4}{C} - 3\,e^- & \longrightarrow & \overset{-}{C} & 3 & & 2 \\ & & & & 6 & \\ 2\,H^+ + 2 \times 1\,e^- \longrightarrow 2\,\overset{o}{H} & \longrightarrow & H_2 & 2 & & 3 \end{array}$$

Hydration of the ethyne results in ethanal. Kucherov reaction, catalyzed by divalent mercuric ions in a sulfuric acid medium.

$$HC{\equiv}CH + H_2O \xrightarrow[(H_2SO_4)]{\overset{+2}{Hg}} \left[H_2C{=}CHOH \right] \rightleftarrows CH_3CHO$$

Initially, vinyl alcohol (ethenol), a non-natural product, is produced and rapidly isomerizes to ethanal (acetaldehyde).

Reduction of ethanal to ethanol. The presence of a multiple bond in the structure of the functional aldehyde group allows participation in addition reactions.

$$CH_3{-}C\overset{\nearrow O}{\underset{\searrow H}{}} \;+\; H_2 \xrightarrow{(Ni)} CH_3{-}\underset{H}{\overset{H}{\underset{|}{\overset{|}{C}}}}{-}OH$$

Aldehydes are good organic reducing agents and are oxidized by weak oxidizing agents such as silver ammonia hydroxide (Tolens reagent), Fehling's solution, or freshly precipitated copper dihydroxide.

$$CH_3{-}\overset{+}{C}\overset{\nearrow O}{\underset{\searrow H}{}} + 2\,\overset{+2}{Cu}(OH)_2 \xrightarrow{t^{o}} CH_3{-}\overset{+3}{C}\overset{\nearrow O}{\underset{\searrow OH}{}} + \overset{+}{Cu_2}O\downarrow + 2\,H_2O$$

$$\begin{array}{lcl|c|c|c} C^+ - 2\,e^- & \longrightarrow & C^{+3} & 2 & & 1 \\ & & & & 2 & \\ Cu^{+2} + e^- & \longrightarrow & Cu^+ & 1 & & 2 \end{array}$$

The redox process produces a tile red precipitate of dicopper oxide.

Interaction of ethanol with ethanoic acid and a catalyst of sulfuric acid takes place the esterification process. Esterification is a slow, molecular, reversible process catalyzed by mineral acids.

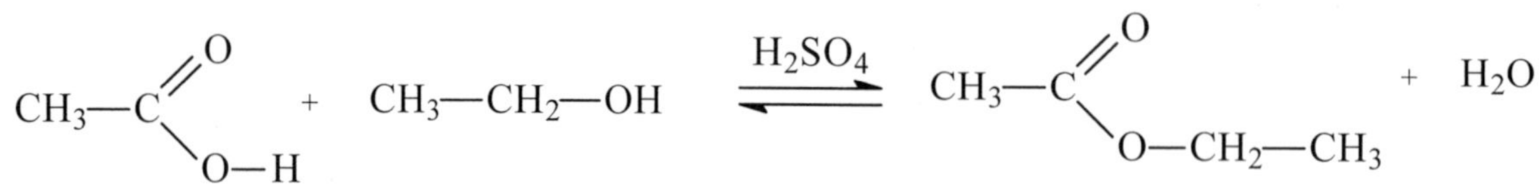

The product is a pleasant-smelling substance and is called ethyl ester of acetic acid (ethyl acetate). When ethyl acetate reacts with ammonia, acetamide is produced and ethanol is released.

$$CH_3COOCH_2CH_3 + NH_3 \longrightarrow \underset{\text{acetamide}}{CH_3CONH_2} + CH_3CH_2OH$$

When acetamide is dehydrated, acetonitrile is produced. The process proceeds in the presence of diphosphoric pentoxide, which binds the reaction water released.

$$CH_3CONH_2 \xrightarrow[-\ H_2O]{P_2O_5,\ t^{o}C} \underset{\text{acetonitrile}}{CH_3CN}$$

The structure of acetonitrile can be represented by the following formula:

$$CH_3{-}C{\equiv}N$$

The presence of a triple bond in the functional group of acetonitrile (nitrile group) indicates that the carbon and nitrogen atoms are in the sp-hybrid state and have a linear structure.
Nitriles hydrolyze in acidic media and in alkaline media. Hydrolysis proceeds stepwise

$$CH_3CN + H{-}OH \underset{}{\overset{H^+ (HCl)}{\rightleftharpoons}} CH_3CONH_2 \quad (I)$$

$$CH_3CONH_2 + H{-}OH \overset{H^+}{\rightleftharpoons} CH_3COOH + NH_4Cl \quad (II)$$

The hydrolysis of nitriles in an alkaline medium is an irreversible process because the alkaline salt of the acid does not interact with ammonia.

$$CH_3CN + 2\ H_2O \xrightarrow{\overline{O}H\ (NaOH)} CH_3COONa + NH_3$$

21. Unsaturated Hydrocarbons (Alkenes). Composition, Isomerism. Physical properties of alkenes. Molecular structure of alkenes, Π-bonding. Homologous order of alkenes. Chemical properties of alkenes. Hydrogenation, halogenation. Addition of polar hydrogen containing molecules. Mechanism of addition reactions - Markovnikov's rule. Polymerization. Oxidation-reduction processes. Preparation. Uses of unsaturated hydrocarbons.

Alkenes are hydrocarbons that contain a carbon-carbon double bond in their molecule.

It determines their chemical properties and is therefore their functional group.

The alkenes form a homologous order.

The name olefins (from the Latin oleum - oil and affinis - close) originates from the oily appearance of the product obtained from the interaction of the simplest representative ethylene (ethene) with chlorine.

In the IUPAC nomenclature, the names of homologues are formed from the names of the corresponding alkane homologues by substituting **-en** for the suffix **-an**. In this way, the common name of ethylene hydrocarbons - alkenes - was formed.

The monovalent alkene residues R

$$—CH{=}CH—$$

are called alkenyl residues, and the ethylene (ethene) residue $H_2C{=}CH–$ is called the ethenyl or vinyl residue.

It is shown that the two relations (σ- and π-) are not equivalent. It takes significantly less energy to break a double bond than to break two σ-bonds. The difference between the two covalent bonds connecting the carbon atoms is due to the difference in the energy state of the two electron pairs. The electron pair that forms the π-bond is energetically richer than the electron pair that forms the σ-bond. Of the two bonds, the π-bond is more unstable and requires less energy to break. The higher energy state of the π-electron pair (π-bond) determines the greater reactivity of double bond molecules.

According to the stereochemical theory, when two carbon atoms are bonded by a double bond the tetrahedra have a common edge.

Isomerism. Types of isomerism - chain, position, geometric (π- diastereomerism)

In ethene and propene, the site of the double bond is uniquely defined. In butene there are two possibilities:

$$\overset{4}{C}H_3—\overset{3}{C}H_2—\overset{2}{C}H{=}\overset{1}{C}H_2 \qquad \overset{4}{C}H_3—\overset{3}{C}H{=}\overset{2}{C}H—\overset{1}{C}H_3$$

1-butene   2-butene

Butene isomers differ in the position of the double bond in the carbon chain. They are called *positional isomers*. To denote positional isomers, the carbon atoms in the chain are numbered starting at the end of the chain to which the double bond is closer. The name of the corresponding hydrocarbon is preceded by a number indicating the number of that carbon atom of the double bond which is closer to the beginning of the carbon chain from which the numbering begins.

In pentene, hexene and other higher alkenes, positional isomers are also possible and their number is greater.

$\overset{6}{C}H_3-\overset{5}{C}H_2-\overset{4}{C}H_2-\overset{3}{C}H_2-\overset{2}{C}H=\overset{1}{C}H_2$

1-hexene

$\overset{6}{C}H_3-\overset{5}{C}H_2-\overset{4}{C}H_2-\overset{3}{C}H=\overset{2}{C}H-\overset{1}{C}H_3$

2-hexene

$\overset{6}{C}H_3-\overset{5}{C}H_2-\overset{4}{C}H=\overset{3}{C}H-\overset{2}{C}H_2-\overset{1}{C}H_3$

3-hexene

In alkenes, as in alkanes, *chain isomerism* also exists. For example, in butene, two chain isomers are possible – n-butene and isobutene:

$CH_3CH_2CH=CH_2$

n-butene (1-butene)

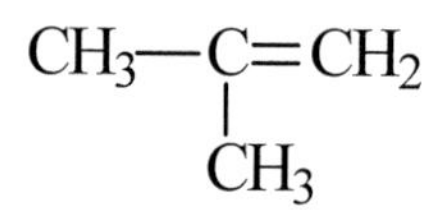

isobutene (2-methylpropene)

In alkenes there is also a *geometric isomerism* (π-diastereomerism), where the π-connection defines a plane. When the different substituents at one carbon atom are respectively the same as those at the other carbon atom, there are two π-diastereomers: a cis- isomer, the two identical substituents are on one side of the π-bond plane; and a trans-isomer, the two identical substituents are on different sides of the π-bond plane:

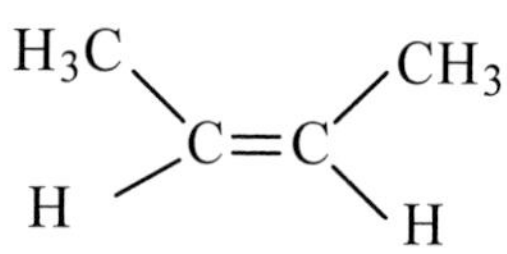

cis-2-butene

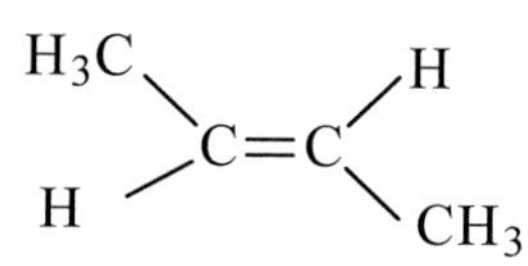

trans-2-butene

Physical properties of alkenes

The first members of the homologous order of alkenes are gases, the next are liquids, and the higher are solids. For straight-chain alkenes there is a fairly regular increase in boiling points. The density of alkenes is greater than that of the corresponding alkanes. The solubility of alkenes is better in polar solvents than that of alkanes because alkenes are more polar compounds (especially cis-isomers).

Construction of alkenes

In the **alkene** molecule, the chemical bond between two of the carbon atoms is made by two electron pairs, i.e., it is double:

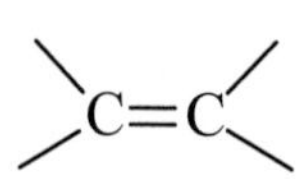

In alkenes, the two carbon atoms between which the carbon-carbon double bond is formed, each have three sp^2-hybrid orbitals and one unhybridized 2p-orbital. The sp^2-orbitals are arranged in one plane and the angle between them is close to 120°. Each of them can participate in the formation of a σ-bond with the orbital of another atom (the s-orbital of a hydrogen atom and the sp^3- or sp^2-orbital of a carbon atom). Such is one bond in the group.

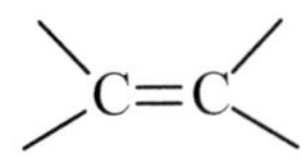

The second bond is a π-bond. It is formed by the two unhybridized 2p- orbitals at the carbon atoms and lies in a plane perpendicular to the plane in which the sp^2 - orbitals, respectively σ-bonds lie.

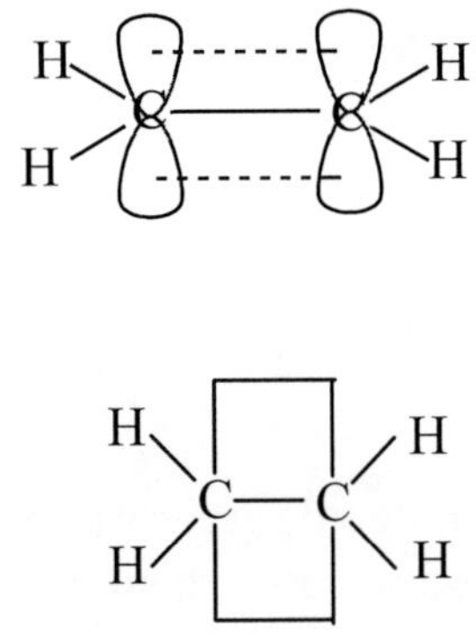

sp^2 -hybrid state (flat building)

Homologous order of alkenes

$H_2C{=}CH_2$	C_2H_4	ethene (ethylene)
$CH_3{-}CH{=}CH_2$	C_3H_6	propene
$CH_3{-}CH_2{-}CH{=}CH_2$	C_4H_8	1– butene
$CH_3{-}CH_2{-}CH_2{-}CH{=}CH_2$	C_5H_{10}	1– pentene
$CH_3{-}CH_2{-}CH_2{-}CH_2{-}CH{=}CH_2$	C_6H_{12}	1– hexene
....		....
CnH_2n		aluene

Chemical properties of alkenes

Addition reactions are most characteristic of alkenes.

Reactions in which molecules of one compound bond to molecules of another compound resulting in new chemical bonds in the single reaction product are called addition reactions.

Alkenes join hydrogen in the presence of a catalyst - Ni, Pt. Alkanes are obtained.

$$CH_2{=}CH_2 + H_2 \xrightarrow{Ni} CH_3{-}CH_3$$

$$\overset{-2}{C} + e^- \longrightarrow \overset{-3}{C} \quad \Big| \; 1 \; \Big| \; 2 \; \Big| \; 2$$

$$H_2 \longrightarrow 2H^o - 2x1\,e^- \longrightarrow 2H^+ \quad \Big| \; 2 \; \Big| \quad \Big| \; 1$$

The reactions of hydrogen addition to the molecules of a compound are called hydrogenation..If ethylene is passed through an aqueous solution of bromine, the solution is discolored. Ethylene reacts under ordinary conditions with bromine. 1,2-dibromoethane is obtained.

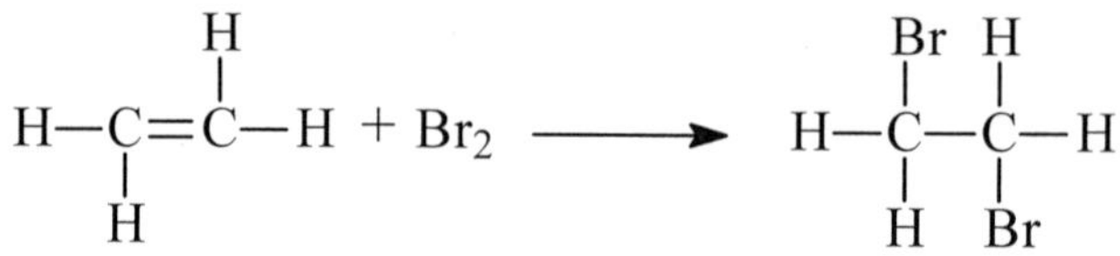

1,2-dibromoethane

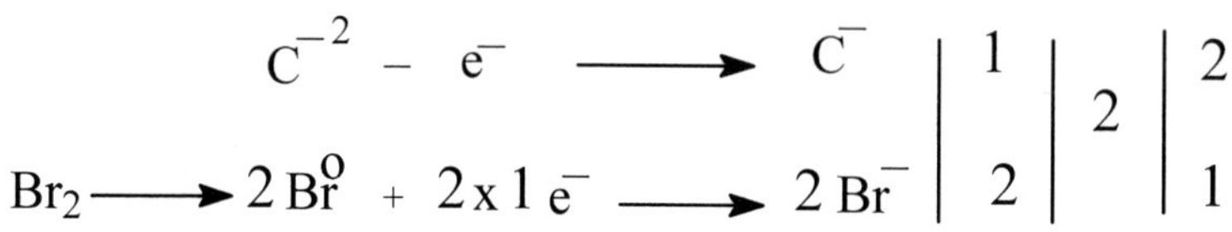

When the bromine joins the ethylene molecule at the same time as the new carbon- bromine bonds are formed, the double bond is broken and becomes a simple bond.

The decolorization of bromine water is a qualitative reaction for the presence of a fold bond in the molecule of the compound under study.

Alkenes join **hydrogen halides**: HCl, HBr, HJ. The addition of HBr to propene theoretically leads to two different isomeric products:

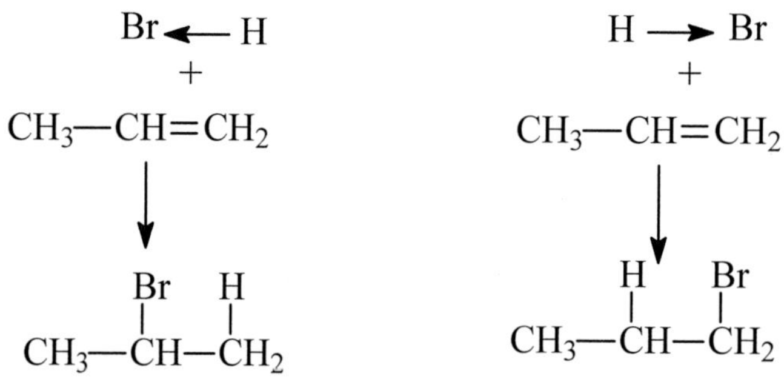

2-bromopropane 1-bromopropane

Vladimir Markovnikov has found that this reaction prouces mainly 2-bromopropane, i.e., the hydrogen cation is attached to the carbon atom at the double bond to which a larger number of hydrogen atoms are bonded, and the bromide anion (i.e., the electronegative atom) - to the carbon atom bonded to a smaller number of hydrogen atoms.

Markovnikov's rule can be explained as follows: the propene molecule is polar. Due to the polarity of the carbon-hydrogen bonds in the methyl group, the π-bond is polarized:

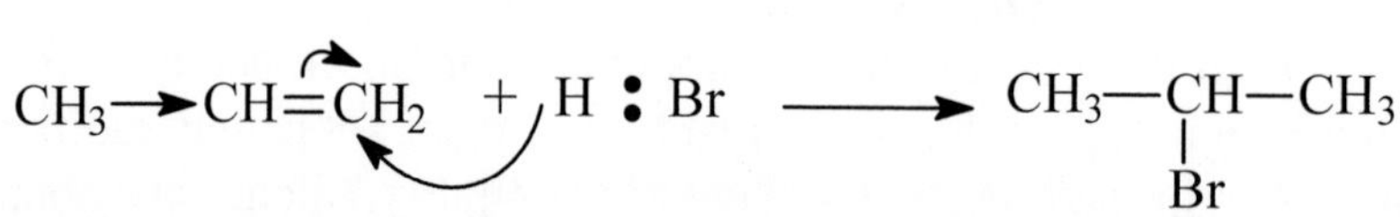

Therefore, the hydrogen cation released from the polar HBr molecule orients and attacks the terminal, partially negatively charged carbon atom in the propene. This results in the formation of 2-bromopropane.

Connecting water

$$CH_3-\underset{\displaystyle CH_3}{\underset{|}{C}}=CH_2 + H-OH \longrightarrow CH_3-\overset{\displaystyle OH}{\overset{|}{\underset{\displaystyle CH_3}{\underset{|}{C}}}}-CH_3$$

tert. butyl alcohol
(2-methyl-2-propanol)

This regularity is known as **Markovnikov's rule**. Markovnikov's rule can be explained as follows: the molecule of 2-methylpropene is polar. Because of this, the π-bond is polarized. Therefore, the hydrogen cation released from the polar H_2O (H-OH) molecule orients and attacks the terminal, partially negatively charged carbon atom in 2-methylpropene. This results in the formation of 2-methyl-2-propanol.

Addition of hydrogen cyanide to alkenes results in alkyl cyanides (nitriles):

$$H_2C=CH_2 \quad + \quad H-C\equiv N \longrightarrow H_3C-CH_2-C\equiv N$$

propanonitrile
(propionitrile, ethyl cyanide)

In unsymmetrical alkenes, the addition of hydrogen cyanide proceeds according to the Markovnikov rule:

$$H_2C=CH-CH_3 \quad + \quad HCN \longrightarrow H_3C-\underset{\displaystyle CN}{\underset{|}{CH}}-CH_3$$

2-methylpropanonitrile

Addition of hydrogen cyanide leads to carbon chain elongation.

Alkenes do not interact with active metals because there are no mobile hydrogen atoms, regardless of the site of the multiple (double) bond.

At high temperature and elevated pressure, or in the presence of catalysts, many ethylene molecules can cross-link each other:

$$n\,CH_2=CH_2 \longrightarrow \left(-CH_2-CH_2-\right)_n$$

Polyethylene is a type of polymer (from Greek "poly" - many and "meros" - part). The process is called polymerization.

Alkenes burn. In the presence of oxygen and high temperature, they oxidize to carbon dioxide and water. Ethylene with air (oxygen) forms explosive mixtures.

$$H_2C=CH_2 \; + 3O_2 \longrightarrow 2CO_2 \; + 2H_2O + Q$$

$$\begin{array}{rl|c|c|c}
 & C^{-2} - 6e^- \longrightarrow C^{+4} & 6 & \multirow{2}{*}{12} & 2 \\
O_2 \longrightarrow & 2O^{0} + 2\times 2e^- \longrightarrow 2O^{-2} & 4 & & 3
\end{array}$$

Unlike saturated hydrocarbons, alkenes oxidize at ordinary temperature. Their oxidation with **hydrogen peroxide** and **potassium permanganate** is typical. Oxidation produces different products depending on the type of alkene and oxidant. Under mild oxidation conditions (H_2O_2 and a 2% solution of $KMnO_4$), the alkenes yield divalent alcohols called glycols.

$$CH_2{=}CH_2 \xrightarrow[(2\%\ KMnO_4)]{H_2O_2} HO{-}CH_2{-}CH_2{-}OH$$

1,2-ethanediol (ethylene glycol)

On vigorous oxidation of alkenes with a solution of potassium permanganate and sulfuric acid, the carbon chain is broken at the double bond to give a mixture of carboxylic acids.

$$CH_3{-}CH_2{-}CH_2{-}CH{=}CH{-}CH_3 \xrightarrow[H_2SO_4]{KMnO_4} CH_3{-}CH_2{-}CH_2{-}COOH + CH_3{-}COOH$$

2- hexene; butanoic acid; ethanoic acid

Catalytic alkylation of benzene yields compounds important for organic synthesis:

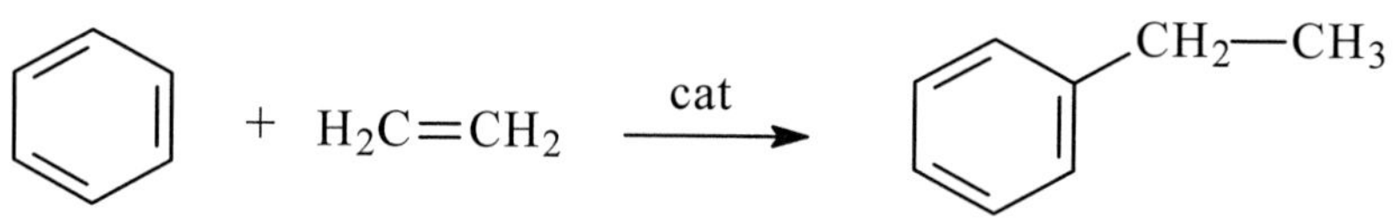

ethylbenzene,
feedstock for styrene and polystyrene;

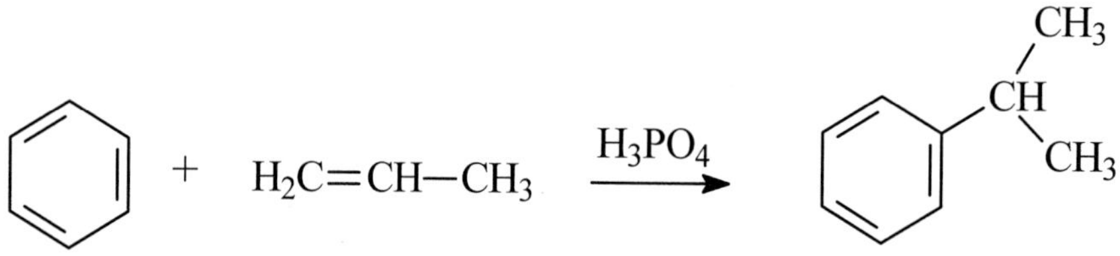

isopropylbenzene (cumene, cumol)

Methods for the preparation of alkenes

- Alkenes are prepared from monohalogenated derivatives of alkanes by acting with an alcoholic solution of an alkaline base (KOH).

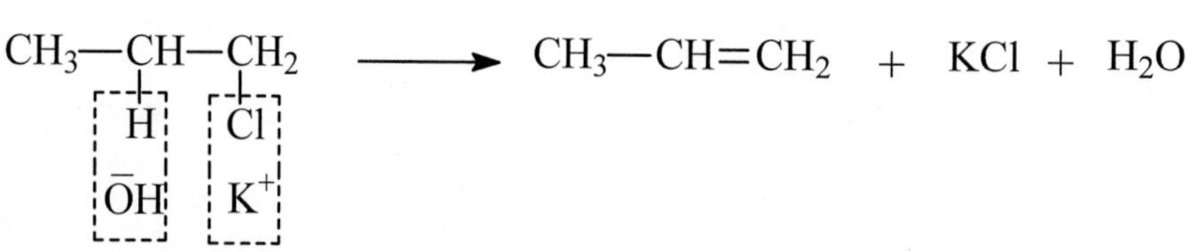

- When 1,2-dihalogenoalkanes are heated with zinc powder, alkenes are produced:

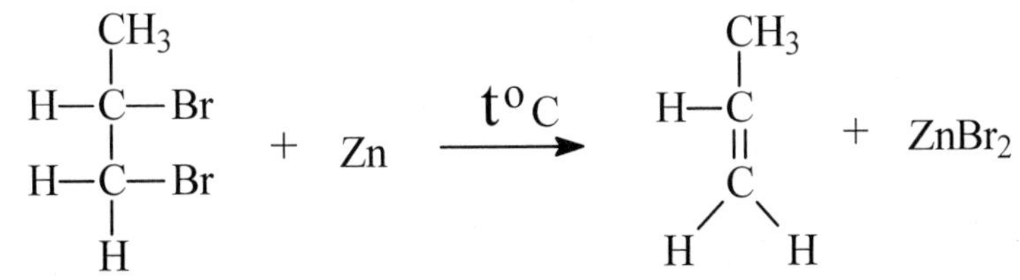

- Alkenes are produced in the refining of petroleum – in the cracking process.

- Under industrial conditions, ethylene is produced by the decomposition of alkanes at high temperature (thermal decomposition). It can be obtained from ethane by dehydrogenation, i.e., by separation of a hydrogen molecule:

$$CH_3{-}CH_3 \xrightarrow[Ni]{500^{o}C} CH_2{=}CH_2 + H_2$$

$$\overset{-3}{C} - e^- \longrightarrow \overset{-2}{C} \quad \Big| \; 1 \; \Big| \; \; 2 \; \Big| \; 2$$

$$2H^+ + 2 \times 1e^- \longrightarrow 2H^o \longrightarrow H_2 \quad \Big| \; 2 \; \Big| \qquad \Big| \; 1$$

- Ethene is produced by the intramolecular dehydration of ethanol when heated to 180° C in the presence of sulfuric acid:

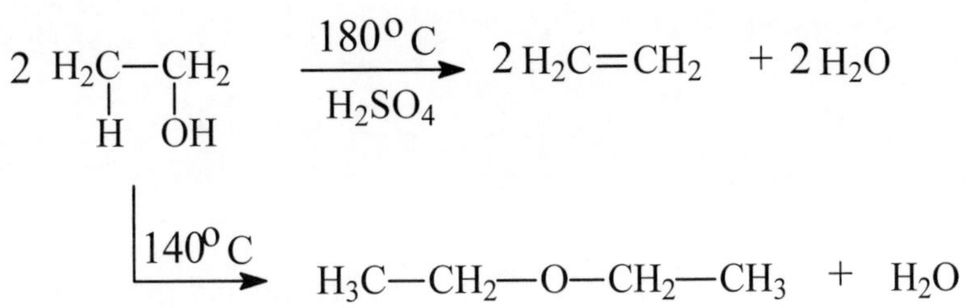

At a temperature of about 140° C, intermolecular dehydration takes place and diethyl ether is produced.

Ethylene, unlike methane, does not occur in nature. It has a higher reactivity and is converted to other compounds under natural conditions.

Sample problems and solutions

Task 1.

Propose a method for distinguishing isomeric hydrocarbons with gross formula C_6H_{12} and structural formulae $CH_3CH_2CH{=}C(CH_3)CH_3$ (**a**) and $CH_3CH_2(CH_3)CH{=}CHCH_3$ (**b**) **(3-methyl-2-pentene)**.

Write the names of the two compounds; Define the phenomenon of isomerism;

What qualitative reactions are used to prove a fold relationship?

Define the rule of addition of polar molecules (water, hydrogen halides) to unsymmetrical alkenes (known in chemistry as Markovnikov's rule) for the compounds **a** and **b** listed and write an example to illustrate.

Exemplary solution

Isomerism is a phenomenon in which two or more substances have the same gross formula but are of different structures, and therefore exhibit different properties.

Isomeric alkenes can be distinguished upon vigorous oxidation. It is carried out using **$KMnO_4$** in **the presence** of H_2SO_4. Different oxidation products are obtained. For example: a)

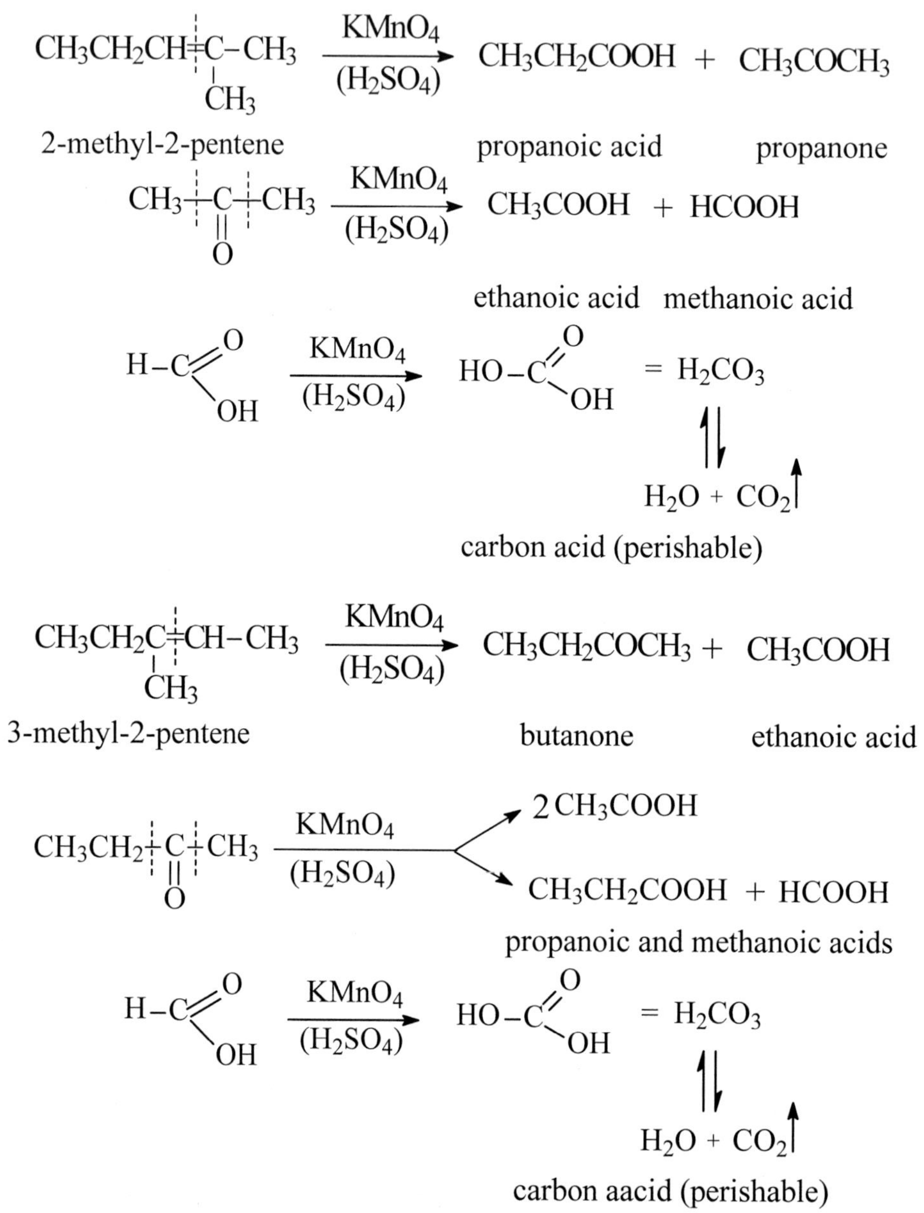

The short-circuit (double or triple) is proved by two reactions: bromine water and a dilute aqueous solution (2 %) of potassium permanganate. The following processes take place:

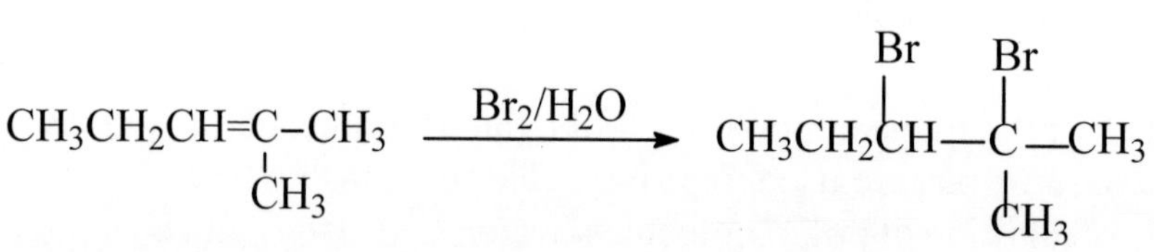

2,3-dibromo-2-methylpentane

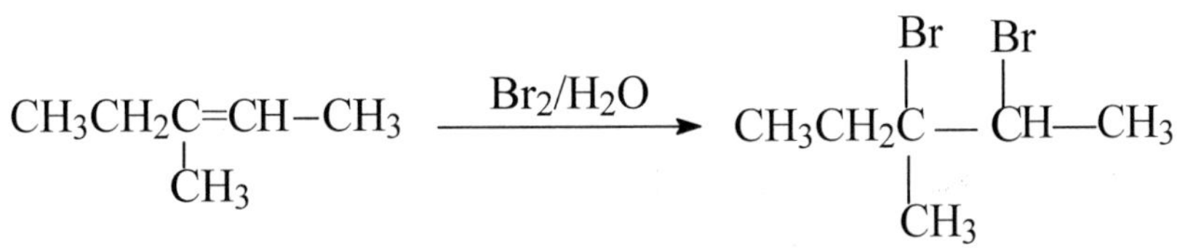

2,3-dibromo-3-methylpentane

The bromine water (pale yellow to brown in color depending on the concentration of the solution) becomes discolored.

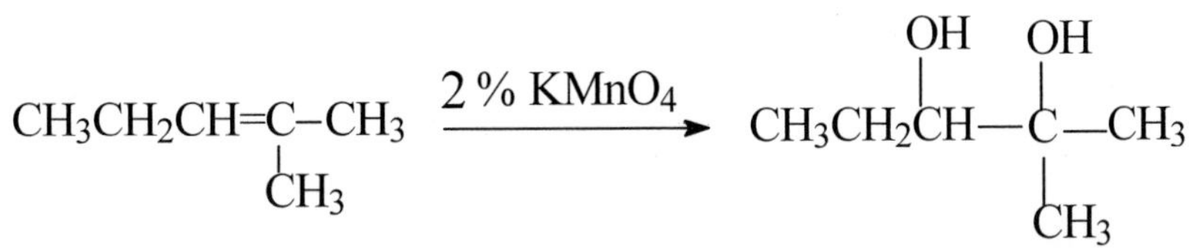

2-methyl-2,3-pentanediol

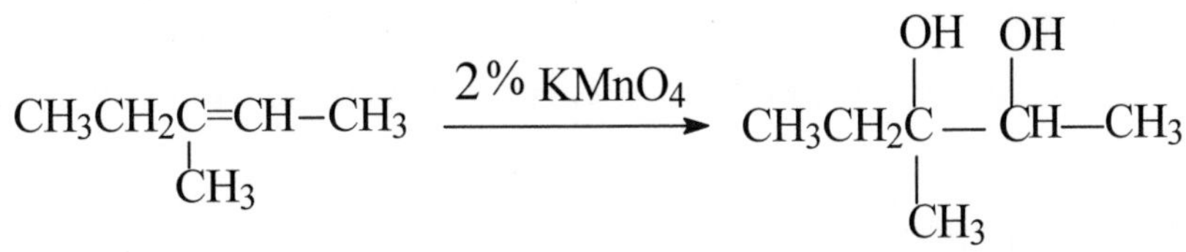

3-methyl-2,3-pentanediol

The pinkish purple color of the dilute aqueous solution of potassium permanganate changes to an earthy brown. As a result of the oxidation-reduction process, manganese dioxide is released and the alkene is converted to divalent alcohol (alcandiol).

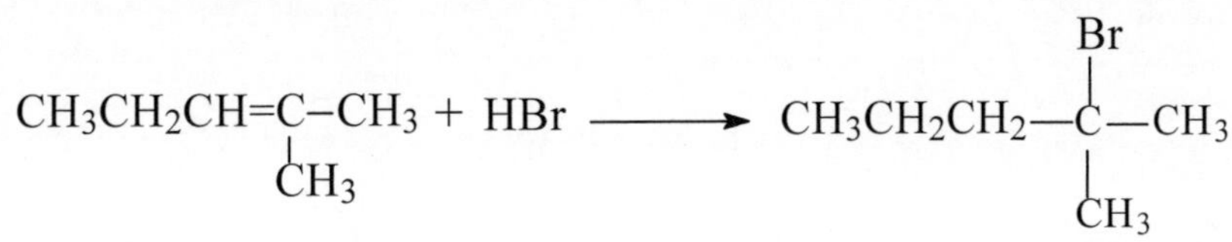

2-bromo-2-methylpentane

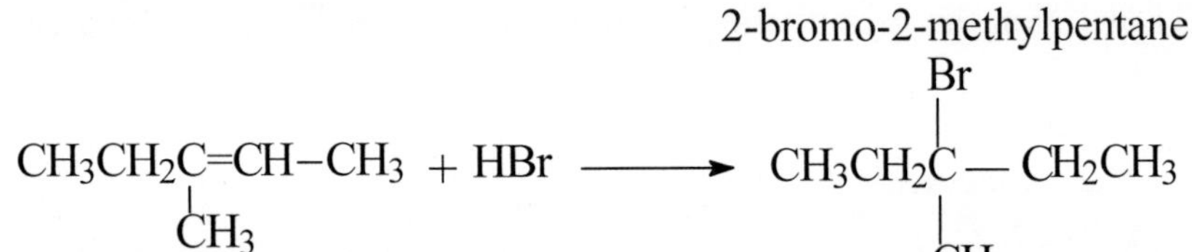

3-bromo-3-methylpentane

Attachment of polar hydrogen-containing molecules to unsymmetrical alkenes proceeds by attaching the electropositive portion of the polar molecule to the carbon atom of the fold bond bonded to the larger number of hydrogen atoms, and the electronegative portion to the carbon atom bonded to the smaller number of hydrogen atoms.

22. UNSATURATED HYDROCARBONS (ALKYNES)

COMPOSITION, NAMES OF ALKYNES. ISOMERISM IN ALKYNES. PHYSICAL PROPERTIES OF ALKYNES. MOLECULAR STRUCTURE OF ALKYNES, Π- BONDS. HOMOLOGOUS ORDER OF ALKYNES. CHEMICAL PROPERTIES OF ALKYNES. HYDROGENATION, HALOGENATION. ADDITION OF POLAR HYDROGEN CONTAINING MOLECULES. MECHANISM OF ADDITION REACTIONS - MARKOVNIKOV'S RULE. POLYMERIZATION. OXIDATION-REDUCTION PROCESSES. PREPARATION. USES OF ALKYNES.

Isomerism in alkynes

The IUPAC names of straight carbon chain **alkynes** are formed from the names of the corresponding alkanes with the same number of carbon atoms by substituting the suffix **-an for -in**.

In alkynes, numbering starts at the end of the chain to which the triple bond is closest, for example:

$$\overset{1}{CH_3}-\overset{2}{C}\equiv\overset{3}{C}-\underset{\displaystyle CH_3}{\overset{4}{C}H}-\overset{5}{CH_2}-\overset{6}{CH_3}$$

4-methyl-2-hexyne

The hydrocarbon residue of ethyne is $HC\equiv C-$ called ethynyl and of its homologues, alkynyl:

$$R-C\equiv C-$$

It has been shown that the π-bond energy in alkynes is smaller than the π-bond energy in alkenes.

In alkynes, positional isomerism is possible according to the location of the triple bond in the carbon skeleton of the molecules:

$$\overset{1}{H}\overset{2}{C}\equiv\overset{3}{C}-\overset{4}{CH_2}-CH_3 \qquad \overset{1}{H_3C}-\overset{2}{C}\equiv\overset{3}{C}-\overset{4}{CH_3}$$

1-butyne — 2-butyne

$$\overset{1}{H}\overset{2}{C}\equiv\overset{3}{C}-\overset{4}{CH_2}-\overset{5}{CH_2}-CH_3 \qquad \overset{1}{H_3C}-\overset{2}{C}\equiv\overset{3}{C}-\overset{4}{CH_2}-\overset{5}{CH_3}$$

1-pentyne — 2-pentyne

Chain isomerism also exists in alkynes due to the differently constructed carbon chain:

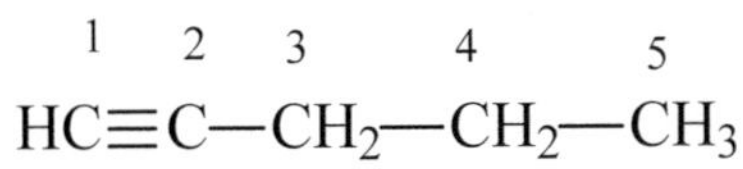

1-pentyne

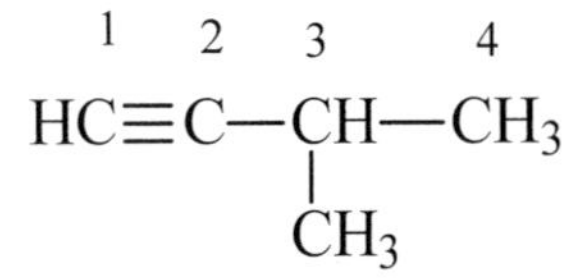

3-methyl-1-butyne

In alkynes, no geometric isomerism (π-diastereomerism) can exist due to the linear arrangement of substituents in sp-hybridized carbon atoms. According to the stereochemical theory, when two carbon atoms are linked by a triple bond, the tetrahedra of the carbon atoms share a common wall.

The functional group of alkynes is the triple carbon-carbon bond.

Physical properties of alkynes

The physical properties of alkynes are similar to those of alkanes and alkenes. The first members are gases, the next are liquids, and the higher members are solids.

Alkynes are slightly soluble in water, but their solubility is better than that of alkenes and alkanes. They are well soluble in less polar organic solvents (ether, benzene, etc.).

Structure of alkynes

—C≡C—

Alkynes are unsaturated hydrocarbons that contain the functional group in their molecules. Compared to alkene and alkane homologues, alkyne homologues with the same number of carbon atoms in the molecule have two and four fewer hydrogen atoms, respectively. There is a genetic relationship between the three types of hydrocarbons which can be expressed by the following scheme:

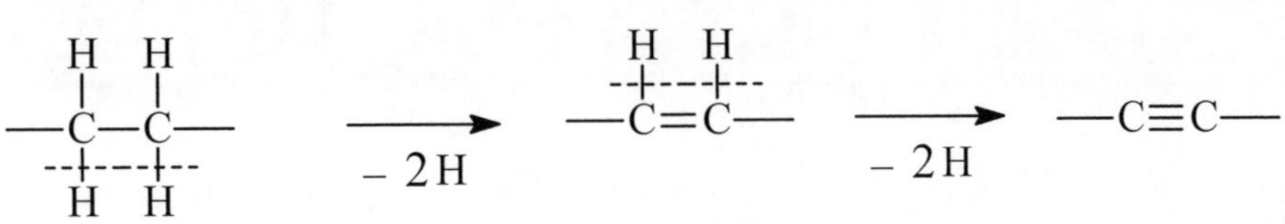

alkane alkene alkyne

The carbon atoms that form a triple bond are in the sp-hybrid state. They each have two hybridized sp-orbitals and two unhybridized 2p-orbitals. With their sp-orbitals, each of the carbon atoms form σ-bonds with another carbon or hydrogen atom. The angle between these orbitals and the σ-bonds formed with them is 180°, i.e., there is a linear arrangement of σ-bonds. The two 2p-orbitals at each of the carbon atoms are arranged mutually perpendicular and through them the atoms form two π-bonds (by lateral overlap) forming two mutually perpendicular planes, where the π-electron cloud consisting of the two π-bonds has cylindrical symmetry. Compared to the double carbon-carbon bond, the triple carbon-carbon bond is shorter and possesses higher energy.

As the s-character of the hybrid carbon atom orbitals ($sp^3 < sp^2 < sp$) increases, the bond lengths decrease due to the decrease in the size of these orbitals.

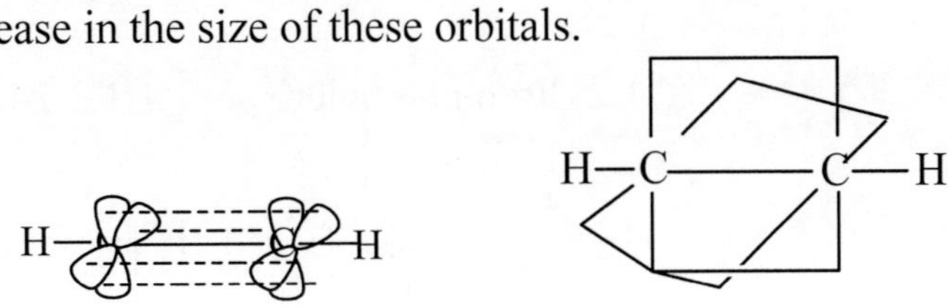

sp-hybrid state (linear structure)

Homologous order of alkynes

$HC{\equiv}CH$	C_2H_2	ethyne (acetylene)
$HC{\equiv}C{-}CH_3$	C_3H_4	propyne
$HC{\equiv}C{-}CH_2{-}CH_3$	C_4H_6	1-butyne
$HC{\equiv}C{-}CH_2{-}CH_2{-}CH_3$	C_5H_8	1-pentyne
$HC{\equiv}C{-}CH_2{-}CH_2{-}CH_2{-}CH_3$	C_6H_{10}	1-hexyne
-------------	----------	
CnH_2n-2	alkyne	

Chemical properties of alkynes

Adjunctive reactions

Reactions can proceed sequentially involving both π-bonds.

The addition of hydrogen to acetylene (hydrogenation) produces ethylene and ethane sequentially. Hydrogenation takes place in the presence of metal catalysts (Ni, Pd, Pt, etc.).

$$HC{\equiv}CH \xrightarrow[(Ni)]{+H_2} H_2C{=}CH_2 \xrightarrow[(Ni)]{+H_2} H_3C{-}CH_3$$

Stage I - hydrogenation to ethene (ethylene):

$$\begin{array}{llll|c|c|c} & C^{-} + e^{-} & \longrightarrow & C^{-2} & 1 & & 2 \\ & & & & & 2 & \\ H_2 \longrightarrow 2H^{o} & - 2\times 1e^{-} & \longrightarrow & 2H^{+} & 2 & & 1 \end{array}$$

Stage II - hydrogenation to ethane:

$$\begin{array}{llll|c|c|c} & C^{-2} + e^{-} & \longrightarrow & C^{-3} & 1 & & 2 \\ & & & & & 2 & \\ H_2 \longrightarrow 2H^{o} & - 2\times 1e^{-} & \longrightarrow & 2H^{+} & 2 & & 1 \end{array}$$

Similarly, bromine is added to acetylene to give successively bromine derivatives of ethylene and ethane:

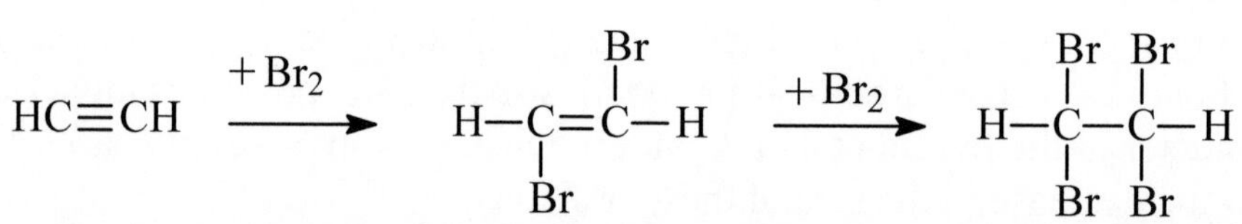

1,2-dibromoethylene 1,1,2,2-tetrabromoethane

$$\begin{array}{llll|c|c|c} & C^{-} - e^{-} & \longrightarrow & C^{o} & 1 & & 2 \\ & & & & & 2 & \\ Br_2 \longrightarrow 2Br^{o} & + 2\times 1e^{-} & \longrightarrow & 2Br^{-} & 2 & & 1 \end{array}$$

Stage I

$$C^{0} - e^{-} \longrightarrow C^{+} \quad \Big| \; 1 \; \Big| \; 2 \; \Big| \; 2$$

$$Br_2 \longrightarrow 2\,Br^{0} + 2 \times 1\,e^{-} \longrightarrow 2\,Br^{-} \quad \Big| \; 2 \; \Big| \; 2 \; \Big| \; 1$$

Stage II

The reaction with bromine is qualitative to prove a fold relationship. Polar molecules, e.g., hydrogen bromide, also join sequentially:

$$HC{\equiv}CH + HBr \longrightarrow H_2C{=}CHBr$$

bromoethylene

$$H_2C{=}CHBr + HBr \longrightarrow H_3C{-}CHBr_2$$

1,1-dibromoethane

The addition of the second hydrogen bromide molecule proceeds according to the Markovnikov's rule.

Hydrogen chloride and hydrogen cyanide are joined in a similar manner: the resulting derivatives are used as monomers in polymerization processes.

Addition of water in the presence of mercuric ions (Hg^{+2}) and sulfuric acid results in the production of the perishable vinyl alcohol, which isomerizes to acetaldehyde (ethanal). The reaction is known as the **Kucherov reaction:**

$$HC{\equiv}CH + H_2O \xrightarrow[(H_2SO_4)]{Hg^{2+}} \left[H_2C{=}CHOH\right] \rightleftarrows CH_3CHO$$

Acetylene homologues join water and the end products are ketones, for example:

$$HC{\equiv}C{-}CH_2{-}CH_3 + H_2O \xrightarrow[(H_2SO_4)]{Hg^{2+}} \left[H_2C{=}C(OH){-}CH_2{-}CH_3\right] \rightleftarrows CH_3{-}C({=}O){-}CH_2{-}CH_3$$

Addition of hydrogen **cyanide** to alkynes produces alkenyl cyanides (nitriles):

$$HC{\equiv}CH + HCN \longrightarrow CH_2{=}CH{-}CN$$

acrylonitrile (vinyl cyanide)

Addition of a second hydrogen cyanide molecule proceeds according to Markovnikov's rule – 2-methyl-1,3-propanedinitrile is obtained:

$$CH_2{=}CH{-}CN \;+\; HCN \longrightarrow CH_3{-}CH(CN)_2$$

Under the influence of the triple bond, the hydrogen atoms in the acetylene molecule (and in molecules of alkynes with a triple bond at the beginning of the chain) become carriers of a partial positive charge. Due to this, acetylene interacts relatively easily with alkali metals (K, Na) as the hydrogen atoms are replaced by metal atoms:

$$HC{\equiv}C{:}H \;+\; {\bullet}Na \longrightarrow [HC{\equiv}C{:}]^{-}\,Na^{+} \;+\; H^{\bullet}$$

The resulting compounds are called acetylenides. They are salts of acetylene with the monovalent metals. Calcium also forms a derivative called a carbide. Carbides are salts of acetylene with the divalent metals. The chemical bond in calcium carbide is ionic:

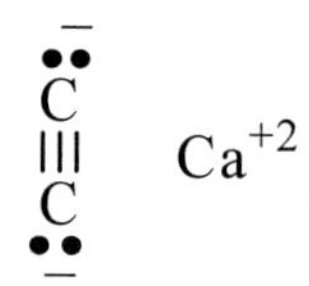

At high temperatures (glowing metal tube) acetylene is converted to benzene. The process is known as cyclic trimerization:

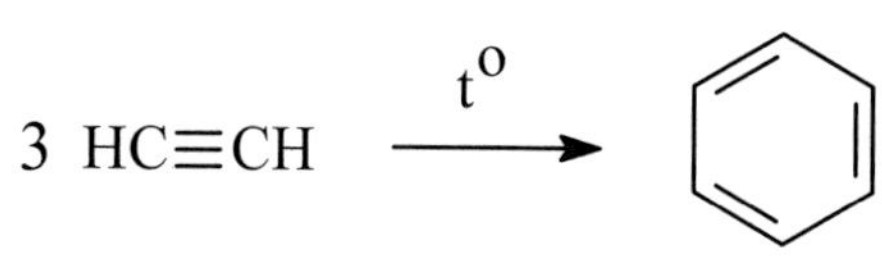

In the presence of oxygen at the appropriate stoichiometric ratio, alkynes burn to carbon dioxide and water. The proccss is highly exothermic (explosive). Mixtures of acetylene with air (oxygen) are explosive.

$$2\,HC{\equiv}CH \;+\; 5O_2 \longrightarrow 4\,CO_2 \;+\; 2\,H_2O \;+\; Q$$

$$\begin{array}{lll|c|c|c} & \overset{-}{C} - 5\,e^{-} \longrightarrow & \overset{+4}{C} & 5 & & 4 \\ & & & & 20 & \\ O_2 \longrightarrow 2\overset{0}{O} + 2\times 2\,e^{-} \longrightarrow & & 2\overset{-2}{O} & 4 & & 5 \end{array}$$

As with alkenes, the complex bond between carbon atoms can also be demonstrated with potassium permanganate solution (reduction to manganese dioxide). The carbon chain is broken at the triple bond and a mixture of carboxylic acids is obtained:

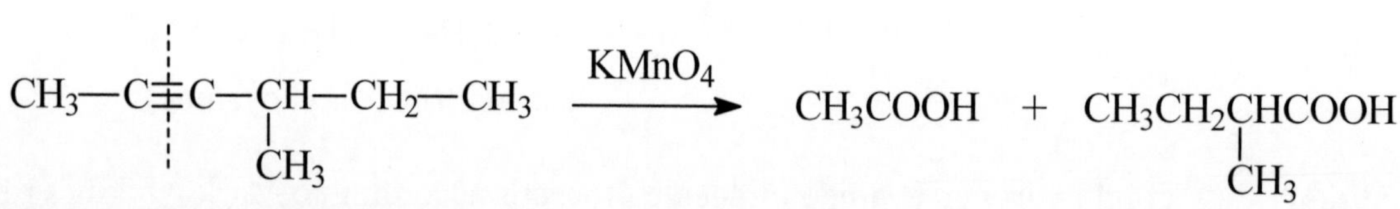

4-methyl-2-hexyne acetic acid 2-methylbutanoic acid

Acetylene reacts with **benzene** to produce styrene, the polymerisation of which results in polystyrene.

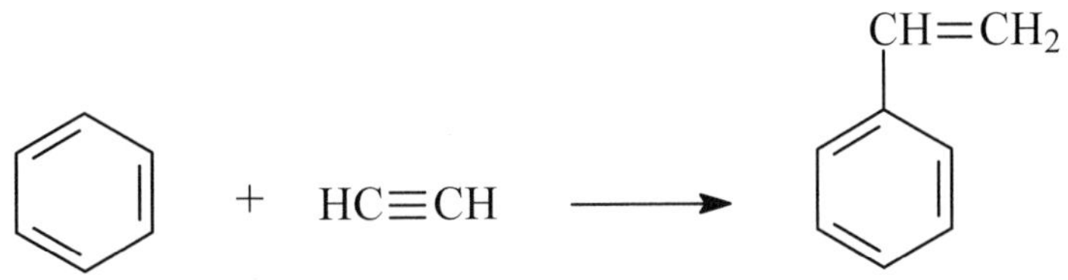

Methods for the preparation of alkynes

A basic method for the preparation of alkynes is by elimination of halogenated hydrocarbons from **dihalogenated derivatives of alkanes**, in the molecules of which the halogen atoms are bonded to two adjacent carbon atoms. The interaction takes place with an alcoholic solution of potassium base:

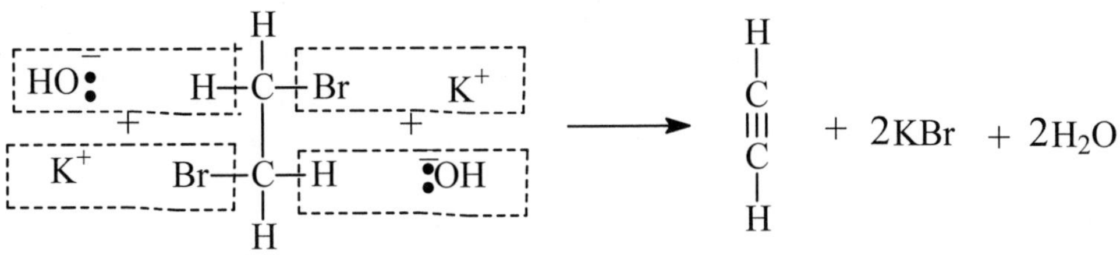

Acetylene can also be obtained by direct synthesis from carbon and hydrogen at high temperature. The synthesis can be carried out by passing hydrogen through an electric arc between graphite electrodes.

Under laboratory and industrial conditions, acetylene is readily produced from calcium carbide and water:

$$CaC_2 + 2H_2O \longrightarrow HC{\equiv}CH + Ca(OH)_2$$

Alkynes are starting substances in the chemical industry. Ethyl alcohol, acetic acid, synthetic rubber and a number of other products are made from acetylene.

Sample problems and solutions

Task 1.

Substance **A** has the gross formula C_5H_8 and reacts with bromine water, potassium permanganate solution and water. Write the structural formula of the hydrocarbon if it is known that a mixture of butanoic acid, carbon dioxide, and water is produced by reaction with a solution of potassium permanganate in the presence of concentrated sulfuric acid.

Write its IUPAC name.

Write the formulas and names of the isomers of **A**. Express the interaction of **A** with bromine water. What product is produced when **A** reacts with water?

Which hydrocarbon isomers interact with active metals?

Exemplary solution

Substance **A** is 1-pentyne. By condition on vigorous oxidation a mixture of butanoic acid and formic acid are obtained. Formic acid contains an aldehyde group in its structure and this makes it vulnerable to even weak oxidants. Under the conditions of the reaction, it is oxidized to the weak and perishable carbonic acid, which decomposes with the evolution of carbon dioxide and water. This is unequivocal evidence that the triple bond is at the beginning of the chain:

$$CH_3CH_2CH_2C\equiv CH \xrightarrow[H_2SO_4]{KMnO_4} CH_3CH_2CH_2COOH + HCOOH$$

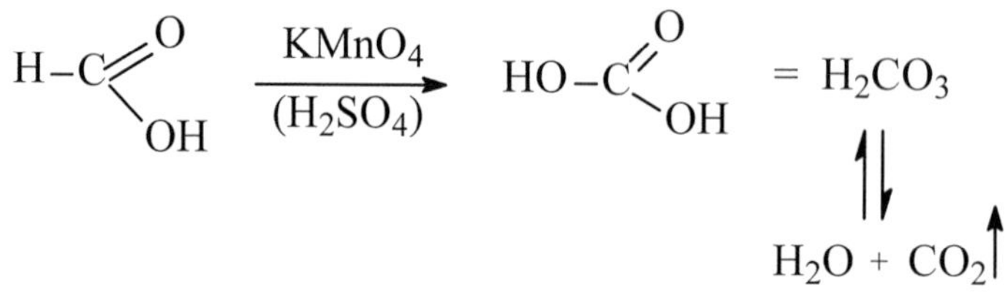

In vigorous oxidation, the carbon chain is broken at the triple bond to give a mixture of carboxylic acids.

The isomers of substance **A** (1-pentyne) are as follows:

2-pentyne 3-methyl-1-butyne

1-Pentyne and 2-pentyne are positional isomers, and 1-pentyne and 3-methyl-1-butyne are chain isomers. The positional isomerism is determined by the location of the functional group. In alkynes, the functional group is the triple bond. The chain isomerism is determined by the sequence of bonding of the carbon atoms in the chain.

Alkynes, as well as alkenes, decolorize bromine water. This color change is used as a qualitative reaction to prove multiple (double or triple) bonding in organic compound molecules. In the case of alkynes, the interaction with halogens proceeds in two steps with the two π-bonds being broken in sequence.

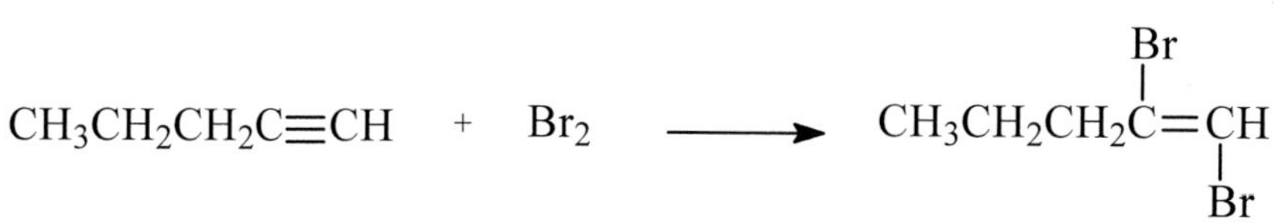

1,2-dibromo-1-pentene

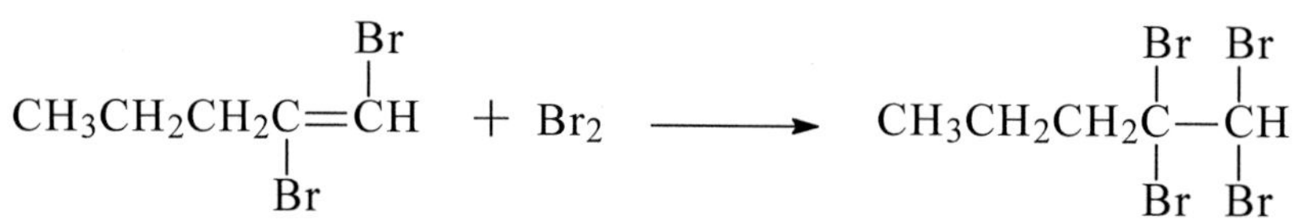

1,1,2,2-tetrabromopentane

Alkynes join water in the presence of mercuric salts in a sulfuric acid medium. The reaction is known as the "Kucherov reaction". The alkyne is unsymmetrical and the addition of the polar water molecule takes place according to the Markovnikov's rule.

1-penten-2-ol 2-pentanone

A neutral 1-penten-2-ol is obtained, which isomerizes to a ketone, 2-pentanone.

Alkynes whose triple bond is at the beginning of the chain interact with active metals.

Sodium (sodium amide) will interact with 1-pentyne and 3-methyl-1-butyne. The resulting products have an ionic structure and are called alkynides (alkynylides).

$$2\,CH_3CH_2CH_2C{\equiv}CH + 2\,Na \longrightarrow 2\,CH_3CH_2CH_2C{\equiv}\bar{C}\!:\!Na^+ \quad H_2$$

$$\begin{array}{lll|c|c|c} & \overset{0}{Na} - e^- & \longrightarrow Na^+ & 1 & & 2 \\ & & & & 2 & \\ 2H^+ + 2 \times 1e^- & \longrightarrow 2\overset{0}{H} & \longrightarrow H_2 & 2 & & 1 \end{array}$$

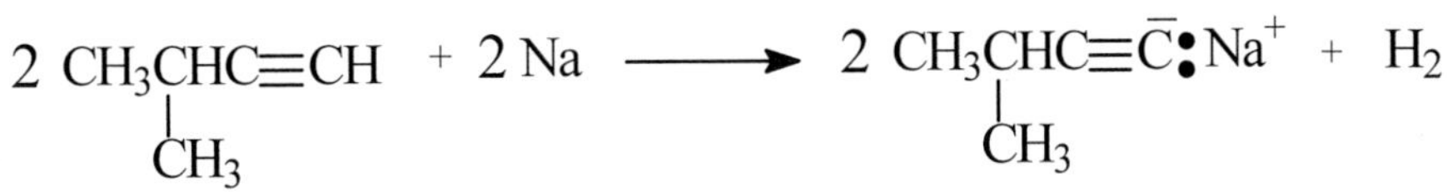

$$\begin{array}{lll|c|c|c} & \overset{0}{Na} - e^- & \longrightarrow Na^+ & 1 & & 2 \\ & & & & 2 & \\ 2H^+ + 2 \times 1e^- & \longrightarrow 2\overset{0}{H} & \longrightarrow H_2 & 2 & & 1 \end{array}$$

$$CH_3C{\equiv}CCH_2CH_3 + Na \not\longrightarrow$$

Task 2.

Write and name the isomers of the fourth members of the homologous series of alkenes and alkynes. Express the chemical properties of one of the isomeric alkynes containing a tertiary carbon atom adjacent to the complex bond in its molecule.

Exemplary solution

Compounds that have the same qualitative and quantitative composition (the same molecular mass) but differ in their properties are called isomers, and the phenomenon - isomerism. At the suggestion of Berzelius in 1830, the concept of isomerism was introduced.

The fourth members of the homologous series of alkenes and alkynes contain five carbon atoms in the chain and the corresponding functional group. For alkenes, a double carbon- carbon bond and for alkynes, a triple carbon-carbon bond.

The isomers of pentene are:

$H_2C{=}CHCH_2CH_2CH_3$

1-pentene

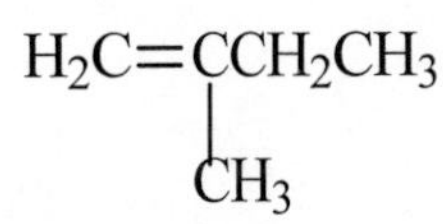

2-methyl-1-butene

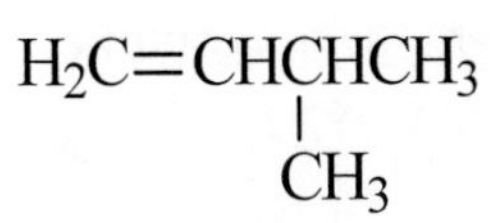

3-methyl-1-butene

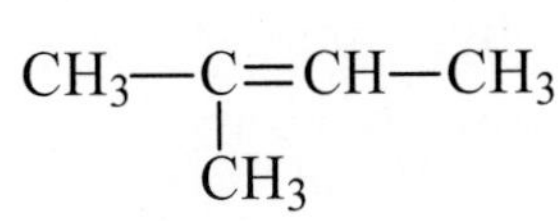

2-methyl-2-butene

$CH_3CH{=}CHCH_2CH_3$

2-pentene

1-Pentene and 2-pentene are positional isomers. The positional isomerism is determined by the location of the functional group in the carbon chain.

2-Methyl-1-butene, 3-methyl-1-butene and 2-methyl-2-butene are chain isomers.

Isomerism, which is determined by the different structure, is the different sequence of bonding of carbon atoms in the chain.

In alkenes, there is also a geometric isometry where the π-connection defines a plane.

cis-2-pentene trans-2-pentene

geometric (cis-trans) isomerism: 2-methyl-2-butene

The isomers of pentyne are:

$$HC{\equiv}CCH_2CH_2CH_3$$

1-pentyne

$$H_3C{-}C{\equiv}CCH_2CH_3$$

2-pentyne

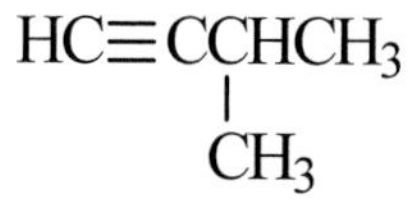

3-methyl-1-butyne

1-Pentyne and 2-pentyne are positional isomers and 1-pentyne and 3-methyl-1-butyne are chain isomers.

The problem statement asks to express the chemical properties of one of the isomeric alkynes containing a tertiary carbon atom adjacent to the complex bond in its molecule. This condition is satisfied by the isomer 3-methyl-1-butyne.

Hydrocarbons with a triple bond in their molecule are involved in addition reactions. Alkynes join hydrogen and the reaction proceeds in two steps, with the two π-bonds being broken in sequence. Hydrogenation proceeds with the help of a catalyst. Hydrogenation is an oxidation-reduction process.

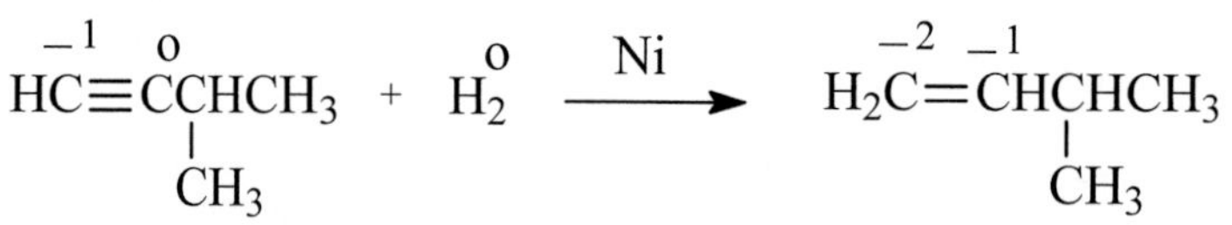

3-methyl-1-butyne 3-methyl-1-butene

$$\overset{-1}{C} + \bar{e} \longrightarrow \overset{-2}{C} \quad 1$$

$$\overset{0}{C} + \bar{e} \longrightarrow \overset{-1}{C} \quad 1 \Big\} 2 \quad 2 \quad 1$$

$$H_2 \longrightarrow 2\overset{0}{H} - 2\times 1\bar{e} \longrightarrow 2H^+ \quad 2 \quad 1$$

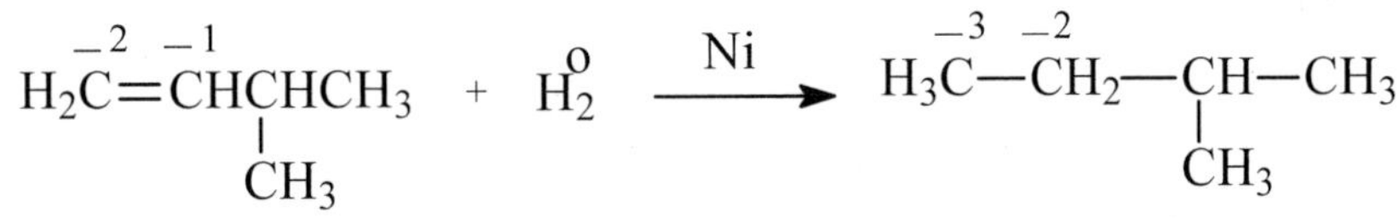

3-methyl-1-butene 2-methylbutane

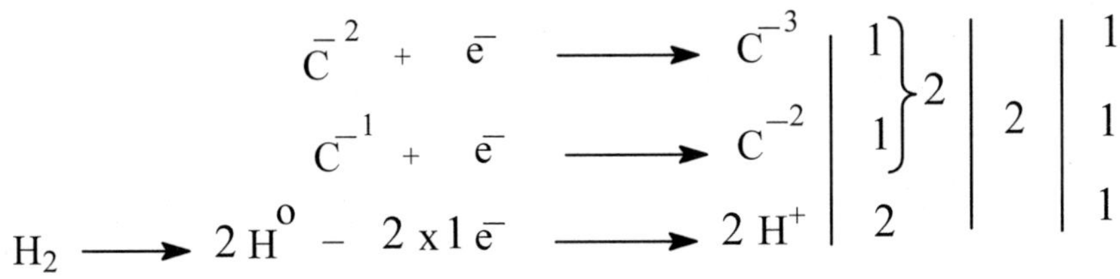

3-Methyl-1-butyne joins halogen elements. On interaction with bromine, bromine water is decolored. The reaction is used to prove the presence of a multiple bond in the hydrocarbon molecule. Halogenation is also an oxidation-reduction process. It proceeds in two steps with successive breaking of the π-bonds in the molecule.

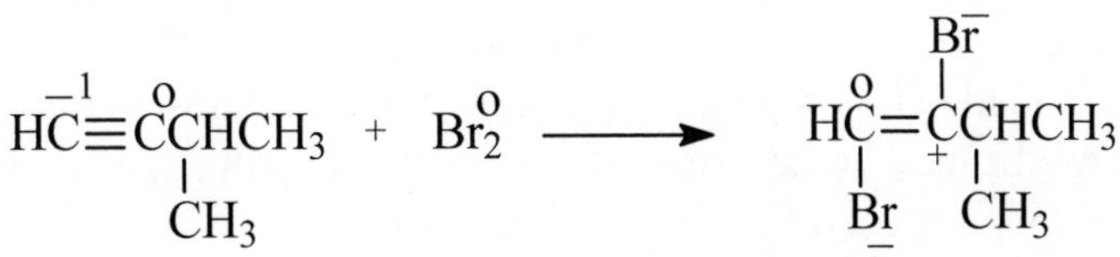

1,2-dibromo-3-methyl-1-butene

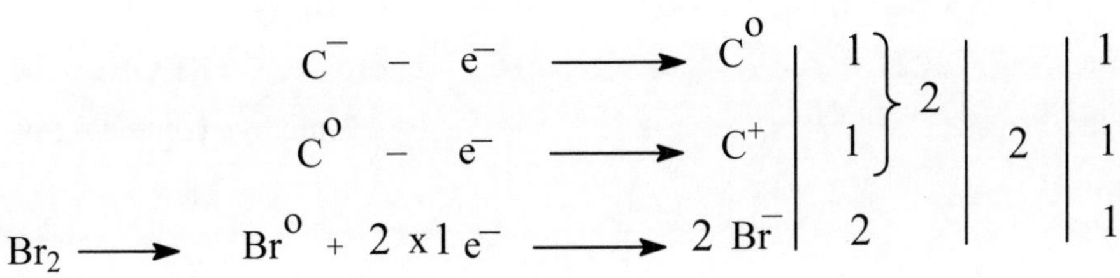

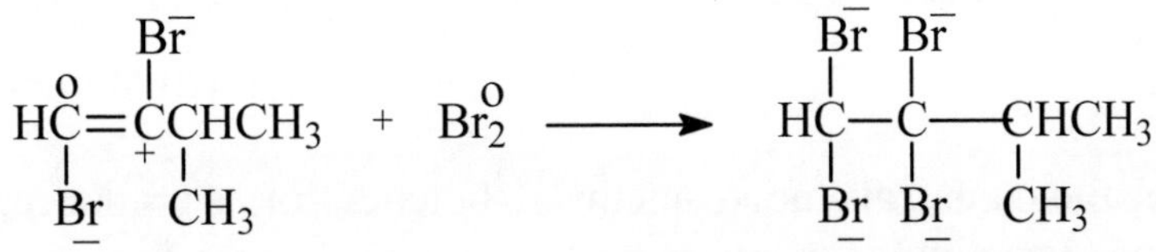

1,1,2,2-tetrabromo-3-methylbutane

Alkynes join hydrocarbon molecules in a two-step process.

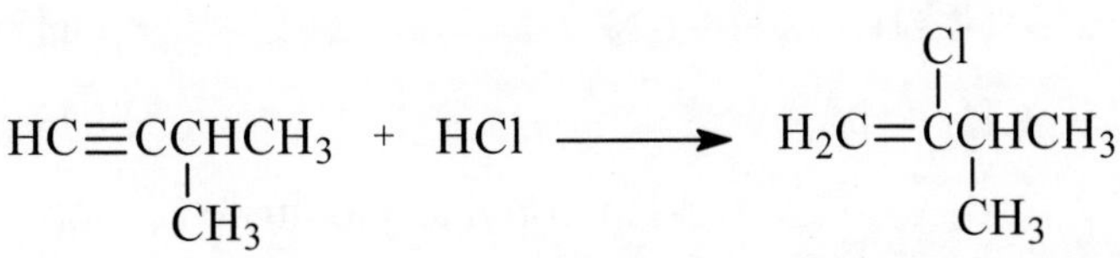

2-chloro-3-methyl-1-butene

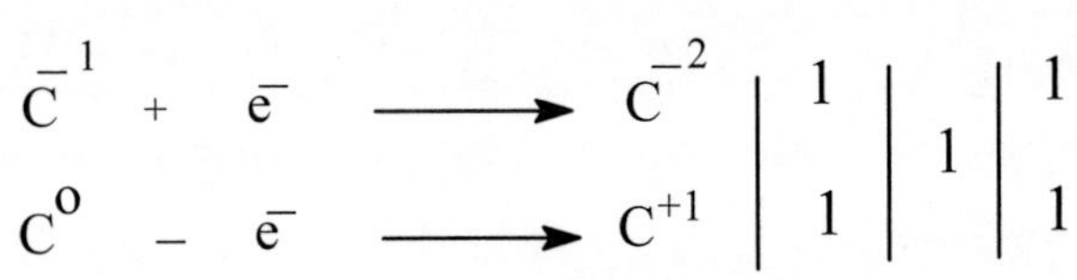

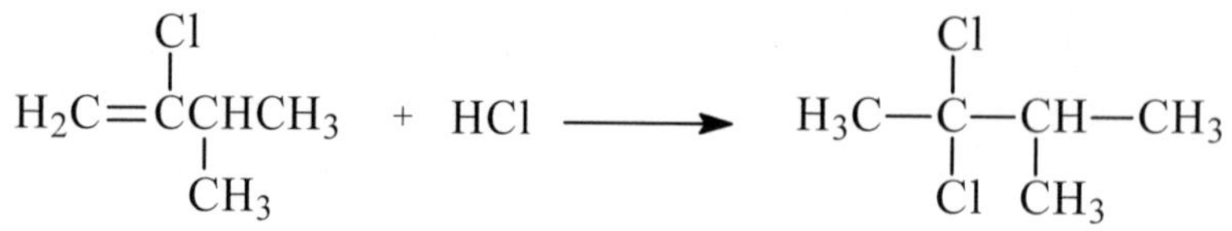

2,2-dichloro-3-methylbutane

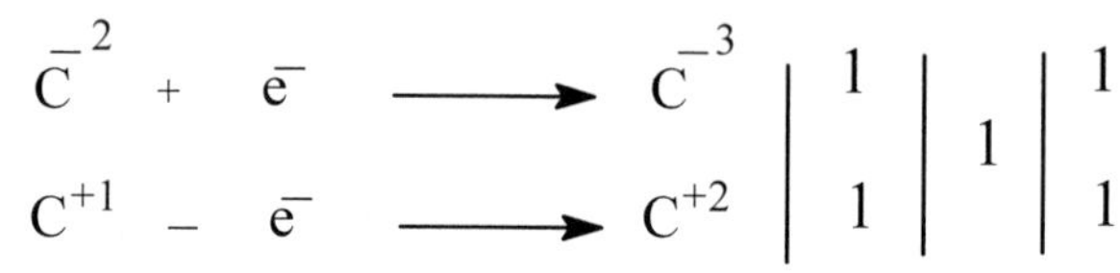

The molecule of 3-methyl-1-butyne is unsymmetrical and the addition of polar hydrogen-containing molecules occurs according to the Markovnikov's rule. According to the rule, the hydrogen cation of the polar molecule is joined to that carbon atom of the multiple (double or triple) bond which is bonded to the larger number of hydrogen atoms, and the electronegative atom (chloride anion) is bonded to the carbon atom which is poorer in hydrogen atoms.

The addition of water to alkynes in the chemical literature is known as the "Kucherov reaction". Hydration of ethyne (acetylene) produces ethanal (acetaldehyde). On hydration of homologues of ethyne, ketones are produced. The hydration of alkynes occurs with a catalyst of divalent mercuric ions in a sulfuric acid medium. The addition of water (polar hydrogen- containing molecule) to an unsymmetrical alkyne proceeds according to the Markovnikov's rule.

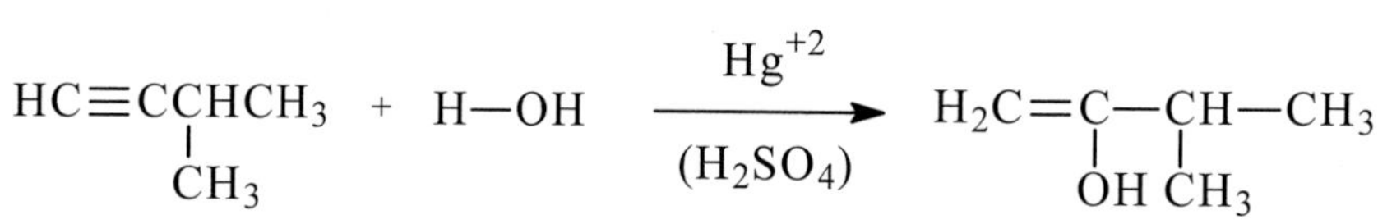

3-methyl-1-butene-2-ol

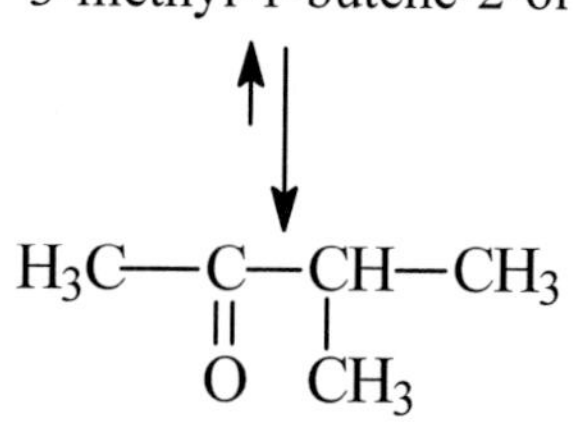

3-methylbutanone

Initially, a non-volatile methyl alkenol (3-methyl-1-butene-2-ol) is produced, which spontaneously isomerizes to a ketone (3-methylbutanone).

In the series of addition reactions, we have to express the interaction of the alkyne with hydrogen cyanide; in this case the reaction proceeds in two steps, and according to Markovnikov's rule:

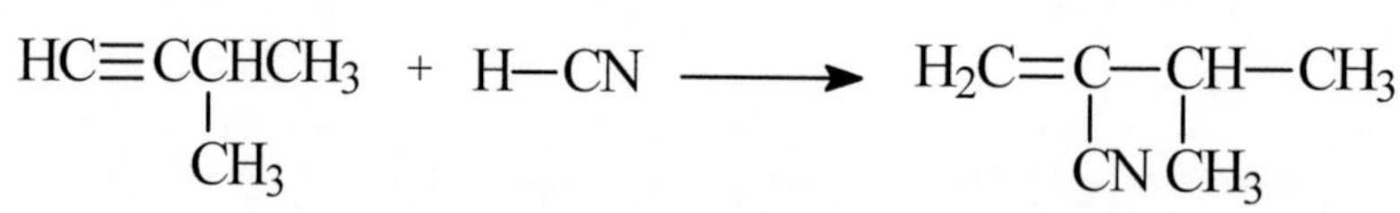

2-isopropylacrylonitrile (2-isopropylvinyl cyanide)

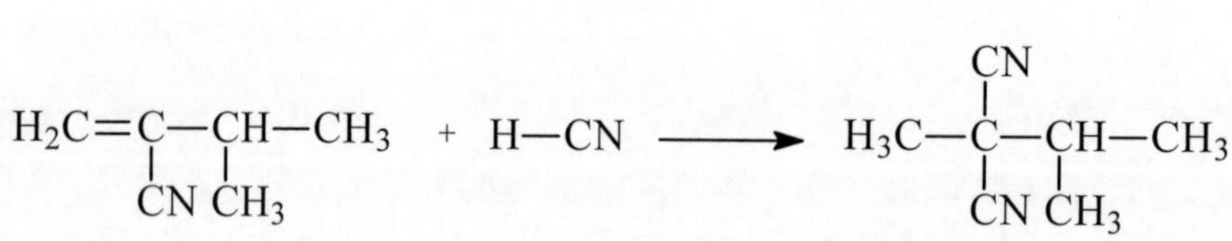

2-isopropyl-2-methyl-1,3-propanedinitrile

In the presence of oxygen at the appropriate stoichiometric ratio, alkynes burn to carbon dioxide and water. The process is highly exothermic (explosive).

As with alkenes, the complex bond between carbon atoms can also be demonstrated with potassium permanganate solution (reduction to manganese dioxide). The carbon chain is broken at the triple bond and a mixture of carboxylic acids is obtained:

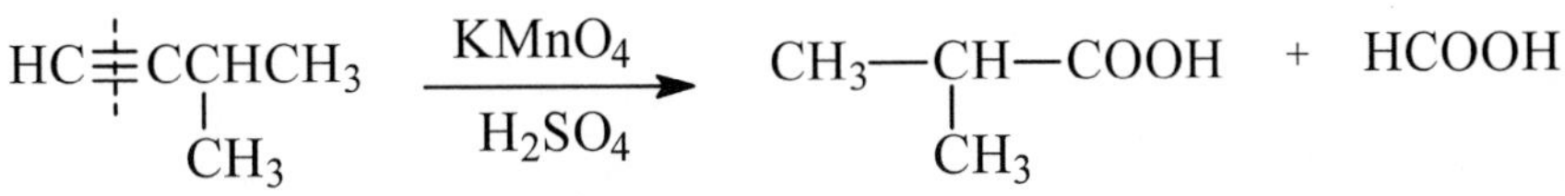

As a result of the redox process, the carbon chain of 3-methyl-1-butyne is broken at the triple bond and a mixture of two acids is obtained: 2-methylpropanoic and methanoic (formic). The formation of formic acid is an indication that the triple bond is at the beginning of the carbon chain.

Formic acid contains in its structure an aldehyde group. Under conditions of vigorous oxidation, it is oxidized to the weak and perishable carbonic acid, which decomposes with the evolution of water and carbon dioxide.

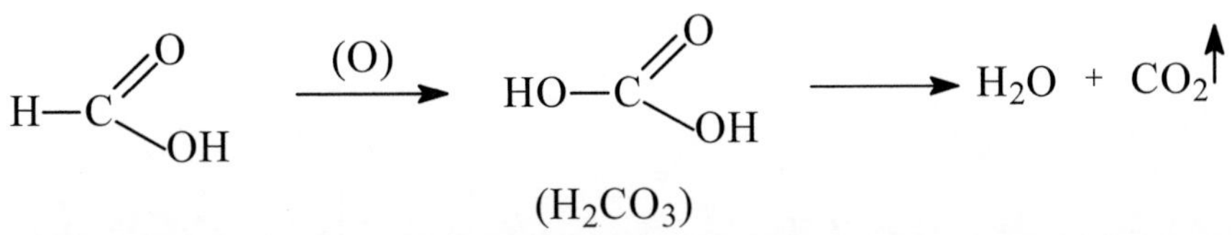

Further evidence of the place of the triple bond in the hydrocarbon chain is the interaction of the alkyne with metallic sodium.

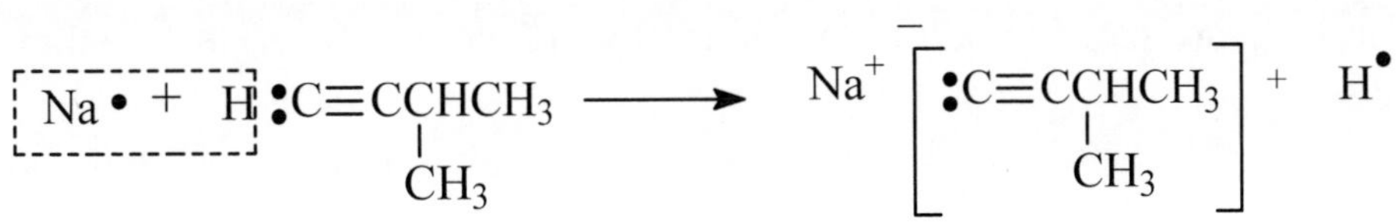

Task 3.

From propine, using suitable inorganic reagents, obtain phenylacetate and 2-methyl-2-hydroxypropionitrile. Express with equations the chemical conversions.

Required elements in solving the problem

preparation of phenol;
preparation of acetic acid anhydride;
preparation of phenylacetate;
acetone production;
preparation of 2-methyl-2-hydroxypropionitrile

23. Aromatic Hydrocarbons (Arenes)

Benzene. Structure and Reactivity. Homologous Order. Methods for the Preparation of Aromatic Hydrocarbons. Comparative Consideration of the Chemical Properties of Arenes, Alkanes and Alkenes as a Function of Their Electronic Structure. Distribution, Preparation and Application of Benzene.

Aromatic hydrocarbons are the simplest representatives of so-called aromatic compounds. This group of substances received its name "aromatic" because of the pleasant smell of the first known representatives. The name has persisted to the present day, although this property is not characteristic of many of these compounds. It is now associated with their structure and chemical properties.

Aromatic hydrocarbons are separated into a separate group because the aromatic ring is a stable cyclic system with a specific nature of bonds and some more special properties. The simplest representative is benzene. Its molecule has the composition C_6H_6. According to classical ideas, the benzene molecule is a hexatomic cycle having three conjugated double bonds. Hence the highly unsaturated nature of the compound. This structure of the benzene molecule is confirmed by its synthesis from acetylene:

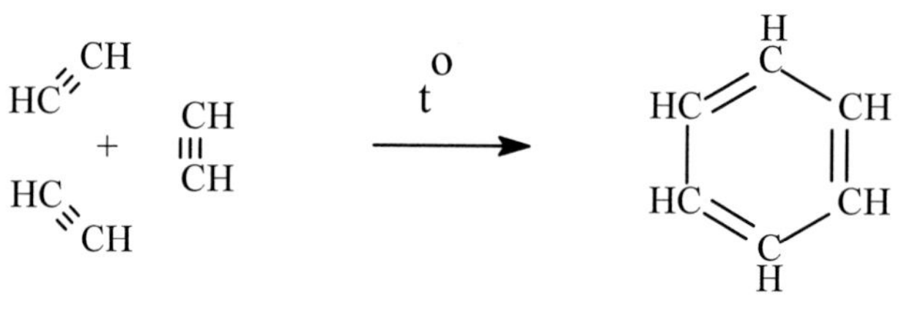

as well as from its hydrogenation (addition of three hydrogen molecules) to cyclohexane using a suitable catalyst:

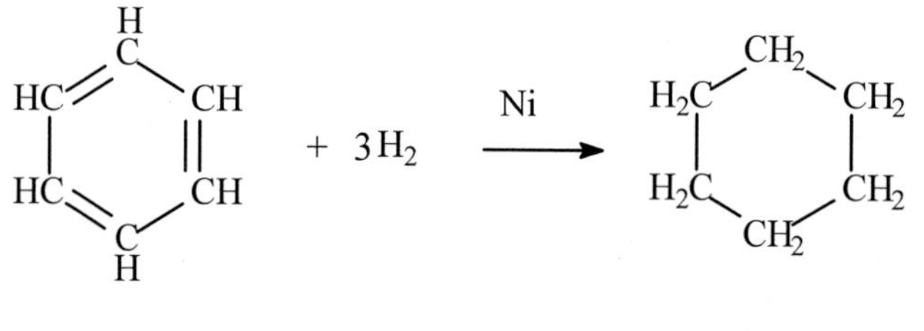

The structural formula of benzene was proposed in 1865 by Kekule. Despite its use today, it does not correctly reflect the properties of benzene. According to the structural formula, there are three conjugated double bonds in the benzene molecule, which means that benzene should be referred to as a highly unsaturated compound. In reality, it is referred to as a saturated hydrocarbon rather than an unsaturated one. Halogen elements break the double bonds and join benzene only under special conditions. Substitution of the hydrogen atoms in the benzene molecule occurs more readily.

If the carbon atoms in the benzene molecule were linked by different chemical bonds (simple and double), substituting hydrogen atoms at adjacent carbon atoms, e.g., with chlorine atoms, should give two isomeric compounds. The isomerism would have been determined by whether the chlorine atoms were located at a simple or at a double bond:

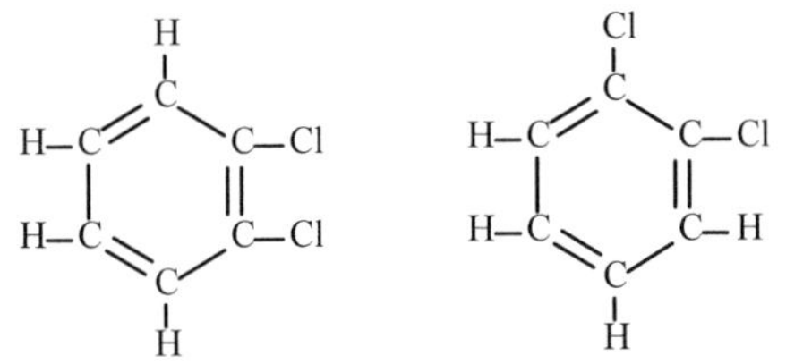

No such isomers have been found, implying that the bonds between adjacent carbon atoms are equivalent.

According to modern ideas, all the carbon atoms in the benzene molecule are in the sp^2-hybrid state. The atoms lie in one plane. The carbon atoms form a regular hexagon and the distance between them is the same and equal to 0.140 nm.

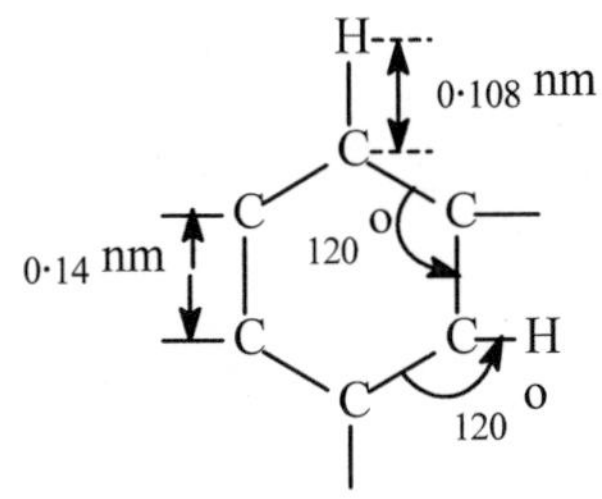

There are two types of bonds in the benzene molecule: σ- and π-. Each carbon atom is bonded to one hydrogen atom by a simple (σ-) bond. It is also bonded to two adjacent carbon atoms, also by one simple (σ-) bond. The σ-bonds in the benzene molecule do not differ in character from the σ-bonds in the homologous series of alkanes, alkenes and alkynes. They are carried by the single electron pair which is localized between the atoms it binds. To form σ- bonds, the carbon atom donates three of its four valence electrons. The remaining (a total of six - one for each carbon atom) valence electrons form a special kind of π-bond.

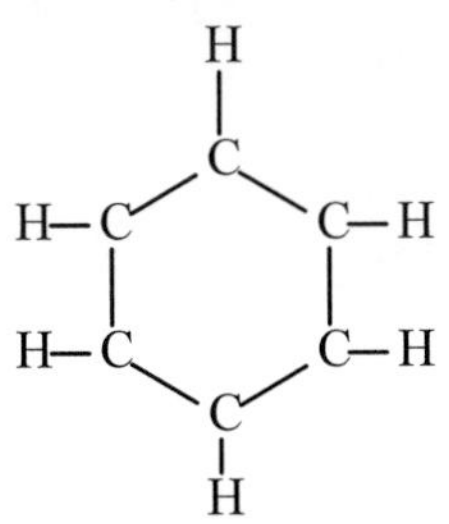

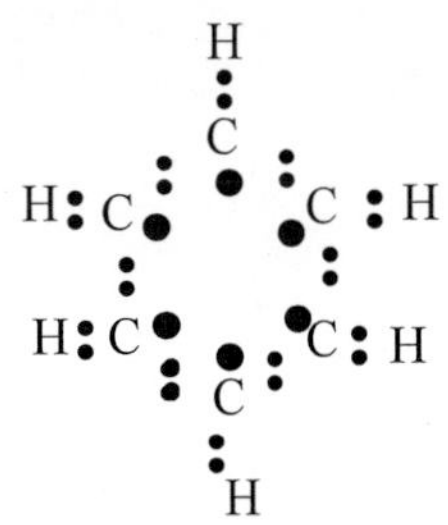

With each carbon atom in the sp^2-hybrid state, one unhybridized 2p-orbital remains, located perpendicular to the σ-skeleton plane of the molecule. The π-bond is not made by electron pairs that are localized between two of the carbon atoms, as is the case with alkenes or alkynes. The π-electrons belong to all the carbon atoms, overlap each other laterally above and below the plane, and bond in a benzene ring.

A chemical bond that is made by a few electrons common to the entire molecule is called delocalized. A delocalized chemical bond is collective (common). In the benzene molecule, it consists of six electrons - a π-set. The π-sextet binds several carbon atoms at the same time (in benzene there are six). The delocalization of the π-electrons determines the properties of benzene. For example, all the bonds in the benzene molecule have the same length because the electrons are evenly distributed over the molecular orbital common to the six carbon atoms. Schematically, this uniform distribution of electrons is expressed as follows:

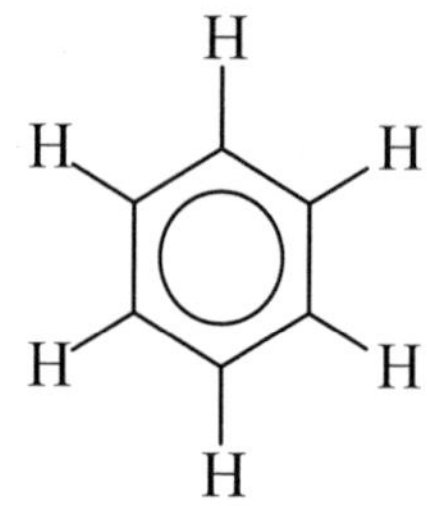

The circle symbolically expresses the relocalization and uniform distribution of electrons.

Hydrocarbons of the **homologous order** of benzene may be regarded as derived from benzene, in the molecule of which one or more hydrogen atoms are substituted by alkyl residues.

C_6H_6 benzene
$C_6H_5CH_3$ methylbenzene (toluene)
$C_6H_5C_2H_5$ ethylbenzene
$C_6H_4(CH_3)_2$ dimethylbenzene (xylene)
....
C_nH_{2n-6} arenes ($n \geq 6$)

According to IUPAC, benzene homologues bear the common name arenes. The names of the individual representatives are formed from the name and number of hydrocarbon residues associated with the nucleus and the word benzene.

When two hydrogen atoms are substituted in the benzene molecule, three **chain isomers** are possible. Depending on the mutual arrangement of the substituents, the isomers are designated ortho- (o-), meta- (m-) and para- (p-). For example, in dimethylbenzene, the following three isomers are possible:

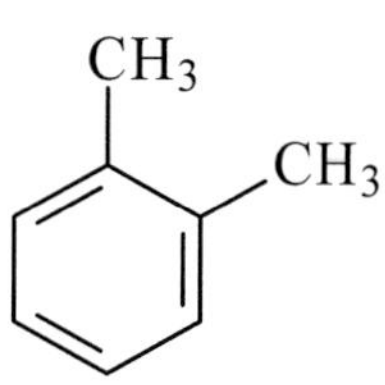

o-dimethylbenzene

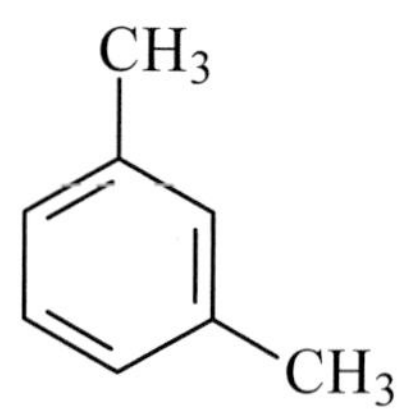

m-dimethylbenzene

p-dimethylbenzene

Aromatic hydrocarbons are known in which the molecule contains two or more benzene nuclei bonded by common carbon atoms. Such hydrocarbons are called aromatic hydrocarbons with fused aromatic nuclei. The number of these nuclei may vary. Of greatest importance are compounds with two and three fused benzene nuclei, which form separate groups of compounds: a naphthalene group, an anthracene group and a phenanthrene group:

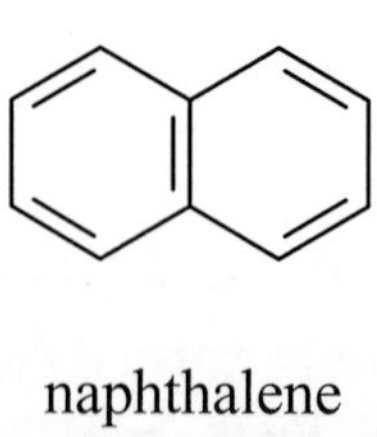

naphthalene

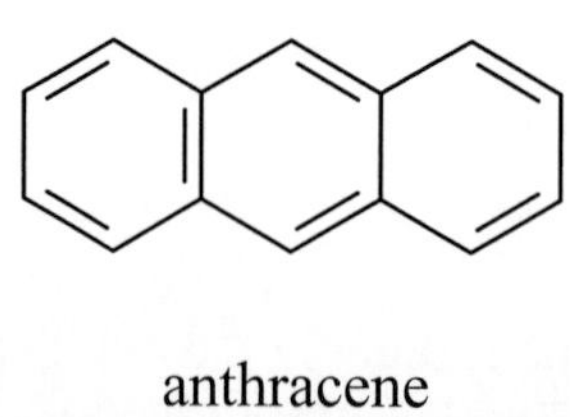

anthracene

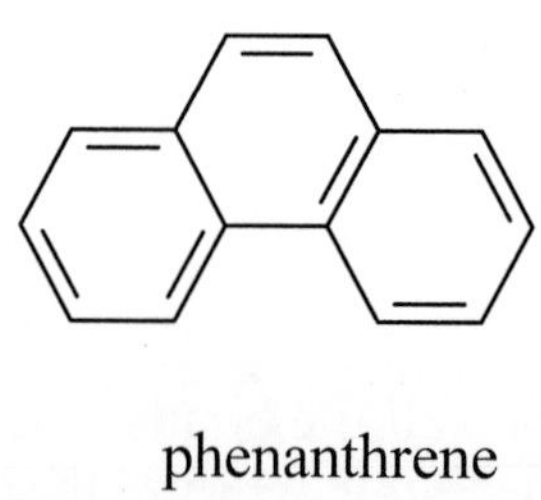

phenanthrene

In compounds with fused nuclei, the aromatic character is less pronounced compared to benzene, i.e., they are more readily involved in addition reactions.

The chemical properties of hydrocarbons are determined by the type of bonds that make up their molecules and the type of carbon chains.

Methods for the preparation of aromatic hydrocarbons

- Arenes are found in certain types of earth oils from which they can be obtained.
- Large quantities of aromatic hydrocarbons are produced from coal tar and catalytic cracking during petroleum processing.
- The cyclic trimerisation of ethyne on heating results in benzene:

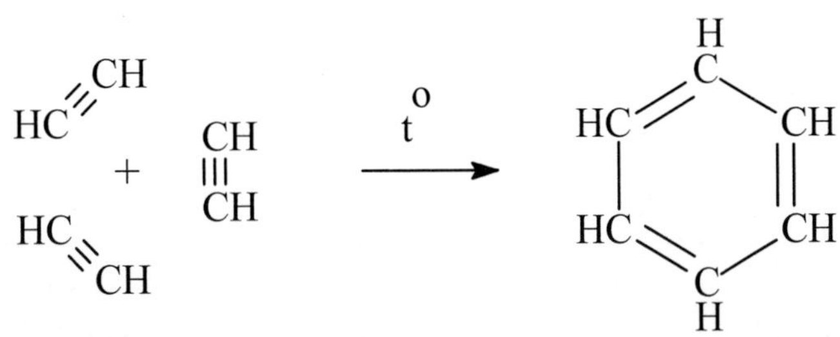

- Under laboratory conditions, benzene homologues are obtained by the Würz- Fittig method, wich is analogous to Wurz's method for preparing alkanes:

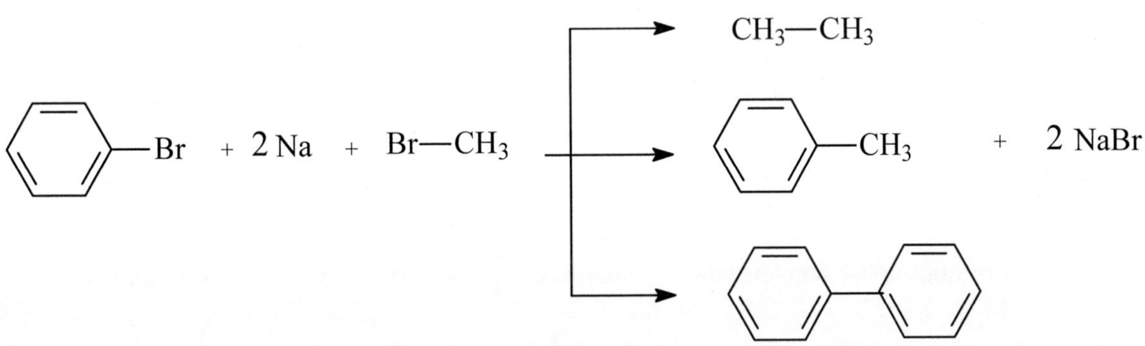

When the interaction takes place between halogen derivatives of different hydrocarbons a mixture of three products is obtained, in this example toluene, ethane and biphenyl.

- Another method of obtaining arenas is that of Friedel-Crafts:

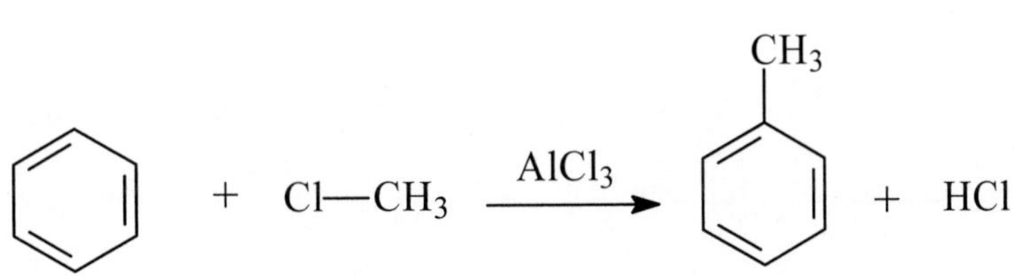

- Benzene homologues are most commonly produced by dehydrocyclization and aromatization:

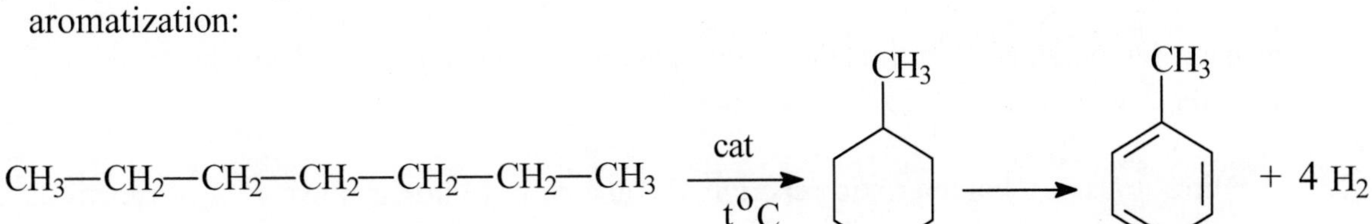

n-heptane methylcyclohexane methylbenzene (toluene)

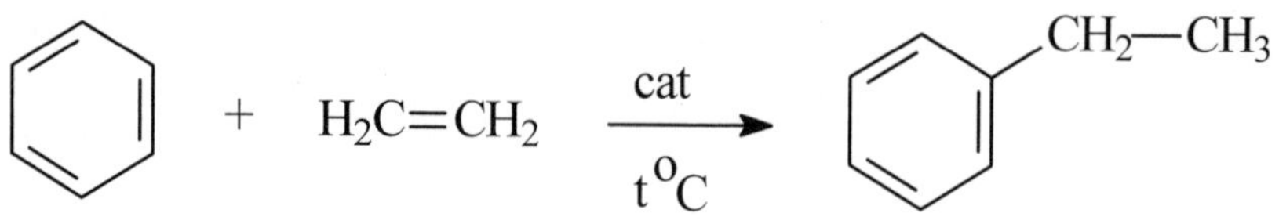

ethylbenzene

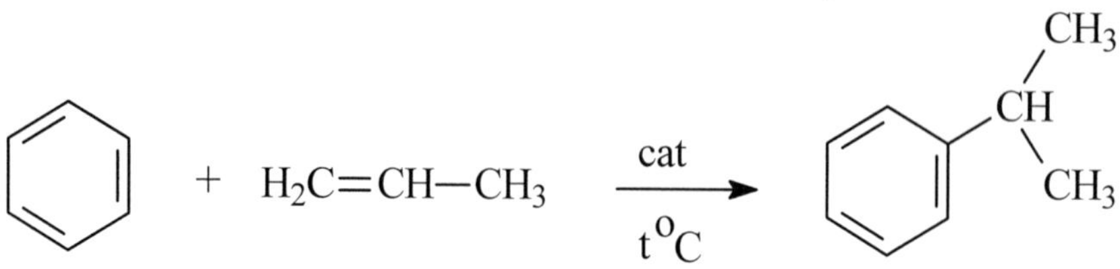

isopropylbenzene

Benzene is an essential feedstock in organic synthesis. It is used to produce pharmaceuticals, dyes, nitrobenzenes and other organic compounds. It is also used as a solvent.

Toluene is found in coal tar. It is a colorless liquid, lighter than water. It is used as a solvent, and in the chemical industry to make explosives.

Chemical properties of aromatic hydrocarbons

Aromatic hydrocarbons are characterized by substitution reactions that proceed by breaking carbon-hydrogen bonds. This is due to their electronic structure, the presence of a common delocalized π-electron system, which is more difficult to polarize and break than the double and triple bond in alkenes and alkynes.

Substitution reactions are characteristic of alkanes and arenes.

1. Halogenation of benzene takes place in the presence of an iron halide catalyst:

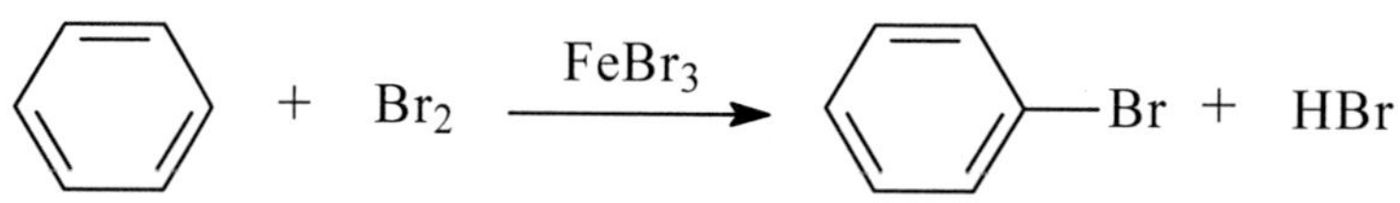

The reaction is substitutive:

$$FeBr_3 + Br_2 \longrightarrow [FeBr_4]^- \ Br^+$$

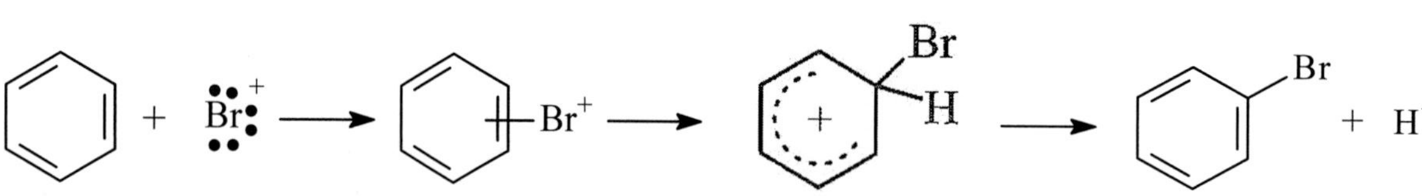

π-complex σ-complex bromobenzene

At ordinary temperature bromine does not react with benzene, i.e., benzene is treated as a saturated rather than an unsaturated hydrocarbon.

2. Nitration is a characteristic reaction for aromatic hydrocarbons. If a nitric mixture (a mixture of nitric acid and sulfuric acid) is added to benzene and slightly heated, a yellow substance with the smell of bitter almonds is formed - nitrobenzene. The reaction proceeds as an electrophilic substitution:

$$HNO_3 + 2\,H_2SO_4 \longrightarrow H_3O^+ + NO_2^+ + 2\,HSO_4^-$$

nitronium cation

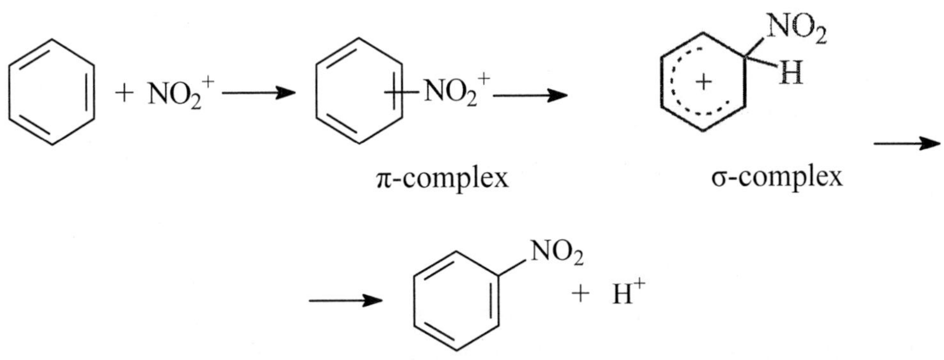

nitrobenzene

3. Arenas are easily sulfonated:

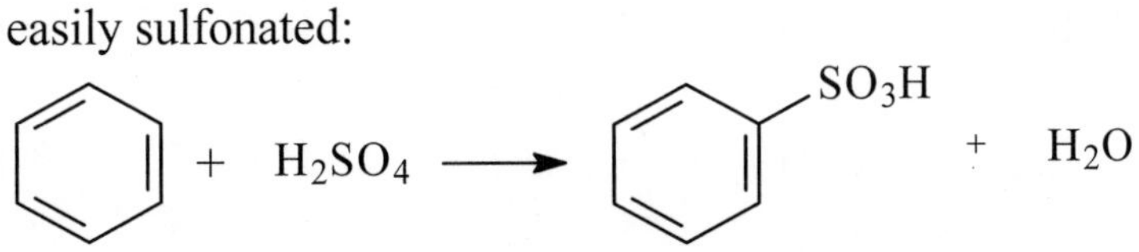

benzenesulfonic acid

4. The halogenation, nitration and sulfonation reactions proceed at a higher rate for benzene homologues. The increased reactivity in these reactions is due to the influence of alkyl substituents. The alkyl groups have a positive induction effect (+I), i.e., they supply electrons to the aromatic nucleus and increase the electron density at the o- and p-site with respect to the alkyl group. As a result, these sites have increased reactivity.

4a. For example, the nitration of methylbenzene (toluene) produces the symmetrical trinitrotoluene (known as explosive TNT), and bromination produces 2,4,6-tribromotoluene:

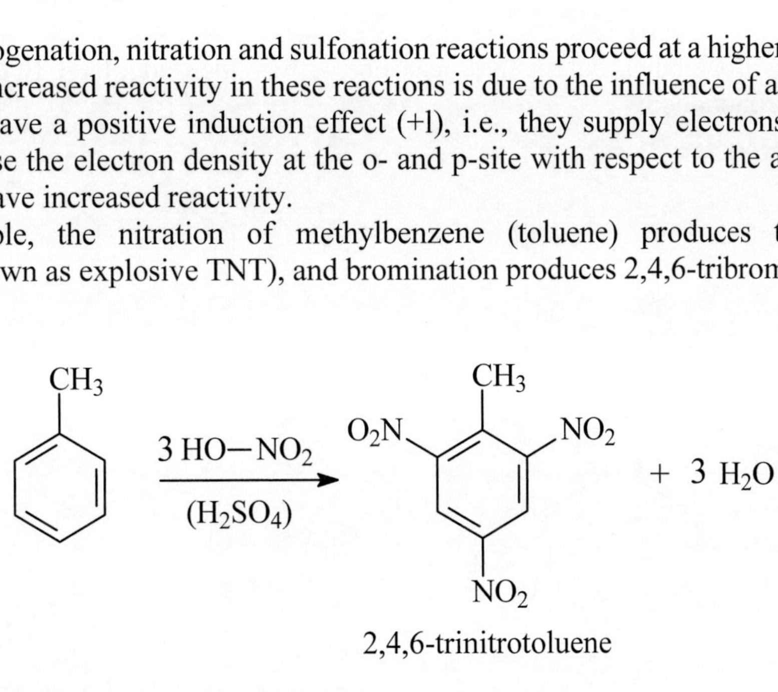

2,4,6-trinitrotoluene

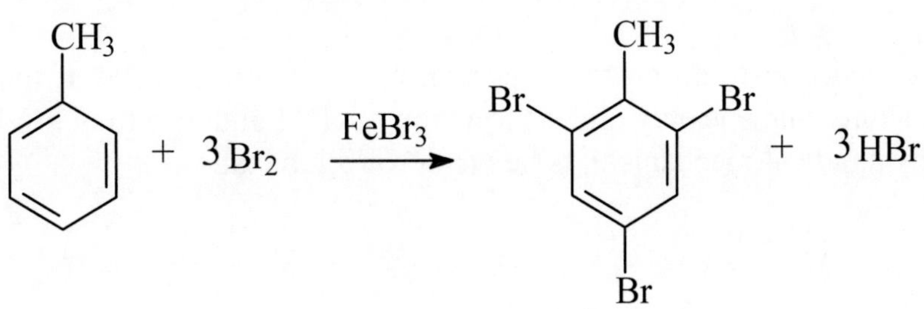

2,4,6-tribromotoluene

The influence of substituents on the reactivity of certain hydrogen atoms in the benzene nucleus so that they are most easily substituted is called the orienting influence.

4b. Upon irradiation with UV light, the reaction proceeds by a chain-radical mechanism and a hydrogen atom is substituted into the side group. Benzyl bromide is produced.

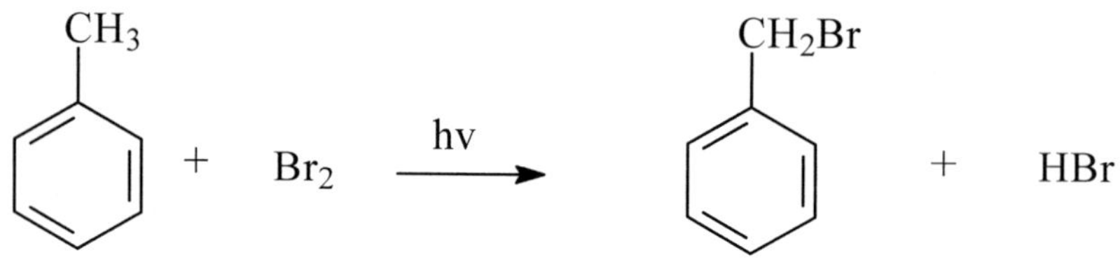

benzyl bromide

5. Arenes, although more difficult, can also participate in addition reactions - mainly hydrogenation and halogenation.

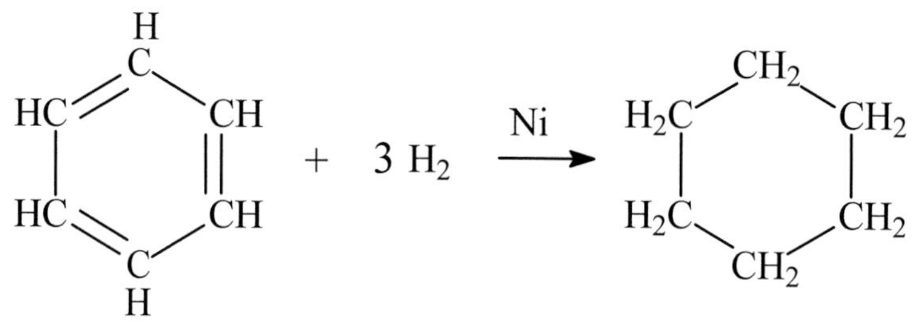

cyclohexane

It is easier to hydrogenate aromatic hydrocarbons with condensed nuclei. In these the aromatic character is less pronounced:

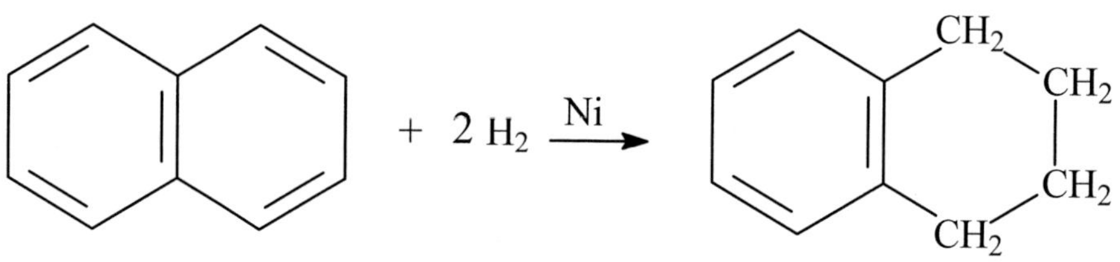

naphthalene tetrahydronaphthalene (tetralin)

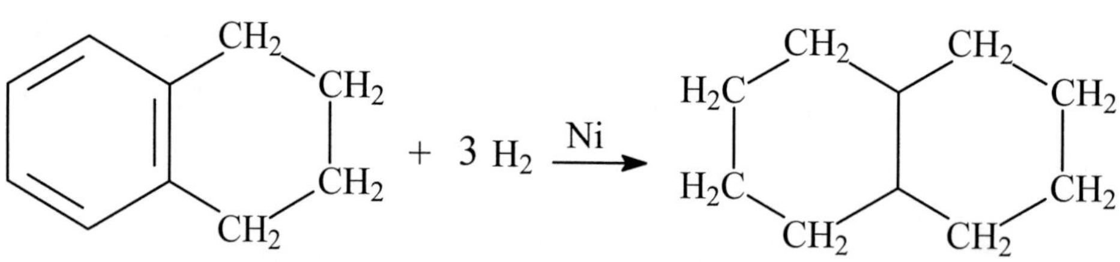

decahydronaphthalene (decalin)

The addition of halogen to an aromatic system proceeds more difficult than for alkenes. For example, benzene joins chlorine only under the action of UV light. The final product of the reaction, which proceeds by a radical mechanism, is hexachlorocyclohexane:

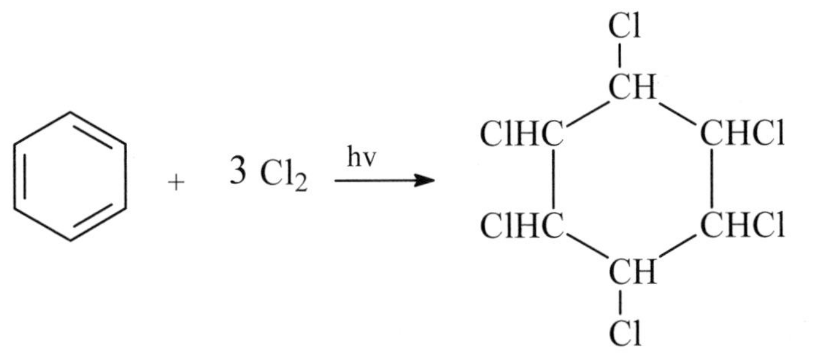

hexachlorocyclohexane ($C_6H_6Cl_6$)

Hexachlorocyclohexane ($C_6H_6Cl_6$) has been used as an insecticide.

6. In an oxygen atmosphere, arenas burn down to carbon dioxide and water.

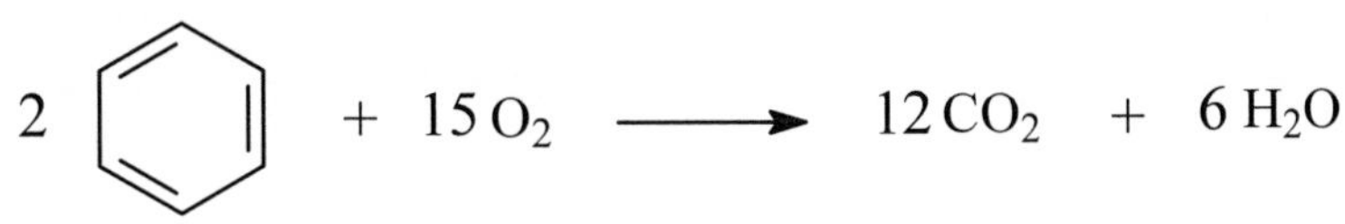

$$6\,C^{-} - 6\times 5\,e^{-} \longrightarrow 6\,C^{+4} \quad \left|\; 30 \;\right| \; 60 \; \left|\; 2 \right.$$

$$O_2 \longrightarrow 2\,O^{0} + 2\times 2\,e^{-} \longrightarrow 2\,O^{-2} \quad \left|\; 4 \;\right| \quad \left|\; 15 \right.$$

Under moderate oxidation, alkanes and benzene are not oxidized, but benzene homologues, regardless of side chain length, are oxidized to benzoic acid:

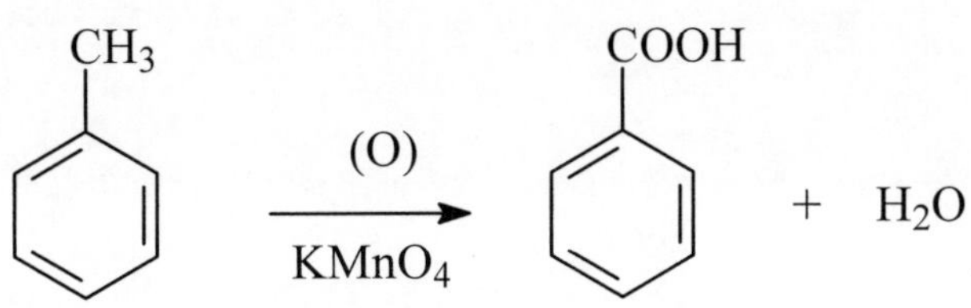

When the side chain contains more than one carbon atom, a mixture of carboxylic acids results from the oxidation-reduction process.

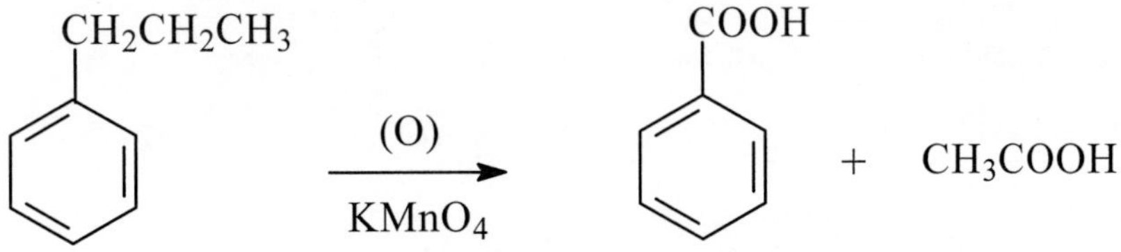

Sample problems and solutions

Task 1.

Give the structural formula of a hydrocarbon of composition C_9H_{12}, the oxidation of which yields benzenetricarboxylic acid, and the bromination of which in a 1:1 molar ratio in the presence of catalyst $FeBr_3$ yields only one product.

What is the starting hydrocarbon?

Write the formulas and names of its isomers.

One of the isomers is a raw material for the industrial production of phenol and acetone. What is the isomer?

Express with a chemical equation the oxidation process of the substance from item 3.

What are the oxidation products and what qualitative reactions will you use to prove them?

Exemplary solution

The starting hydrocarbon is alkylbenzene. To produce benzenetricarboxylic acid in oxidation means it must have three alkyl groups. The mutual arrangement of the substituents is symmetrical, i.e., the hydrocarbon is 1,3,5-trimethylbenzene.

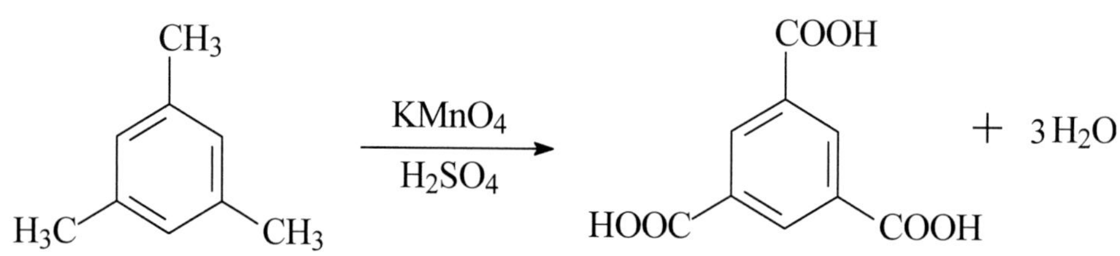

1,3,5-benzenetricarboxylic acid

Since the alkyl substituents are activating o- and p-orients in the bromination of this alkylbenzene in a 1:1 molar ratio with the participation of a catalyst only one product is possible, namely 1-bromo-2,4,6-trimethylbenzene.

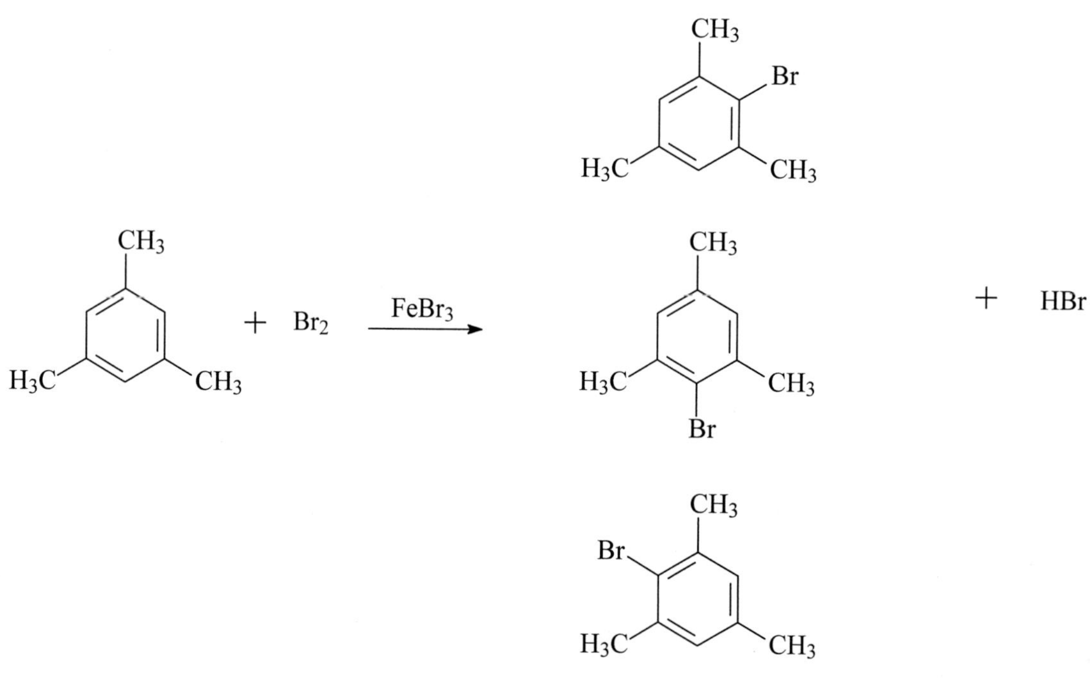

The isomers of 1,3,5-trimethylbenzene are as follows:

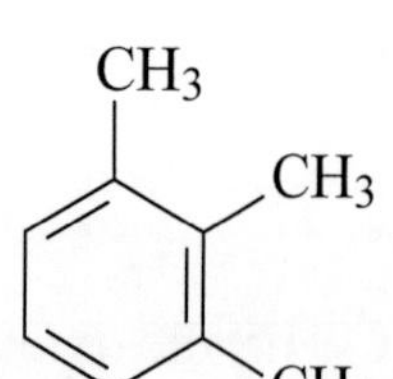

1,2,3-trimethylbenzene

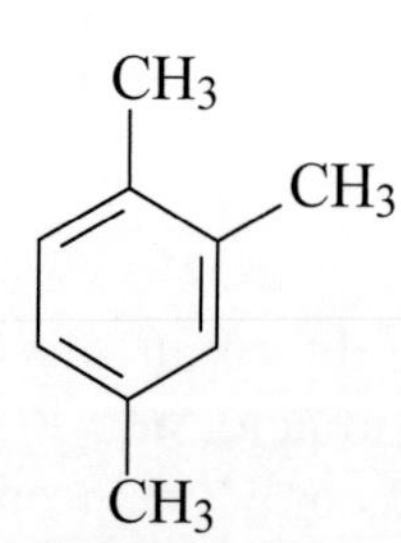

1,2,4-trimethylbenzene

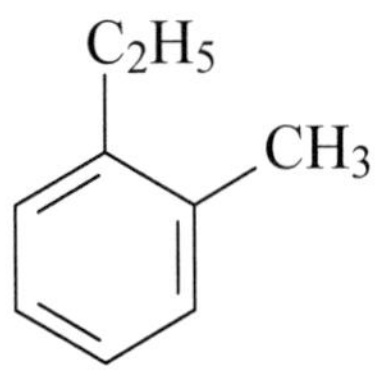

2-methylethylbenzene

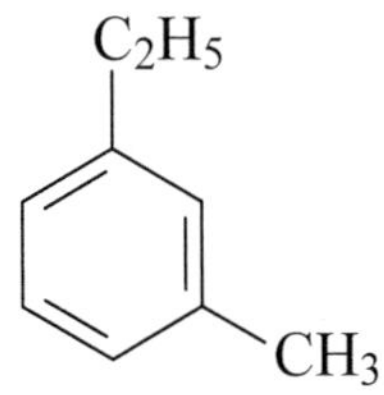

3-methylethylbenzene

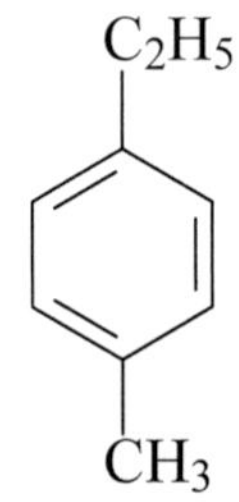

4-methylethylbenzene

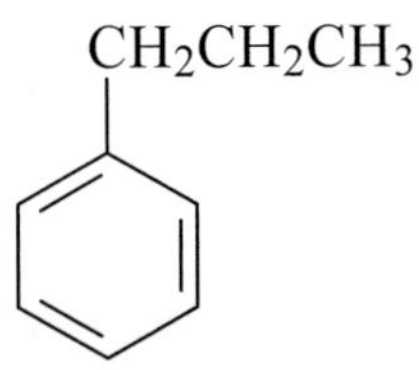

propylbenzene

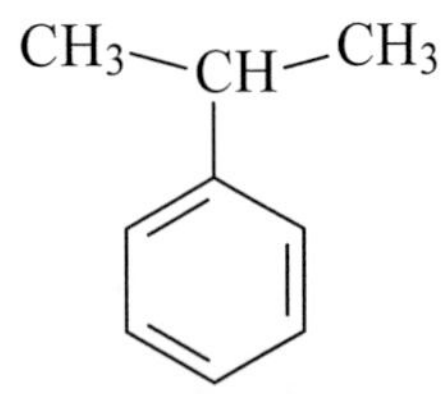

isopropylbenzene (cumene, coumol)

Cumene is known to be a feedstock for industrial production of phenol and acetone. The process takes place during oxidation of cumene in an alkaline medium with bond breaking between the isopropyl residue and the aromatic nucleus, which can be represented as follows:

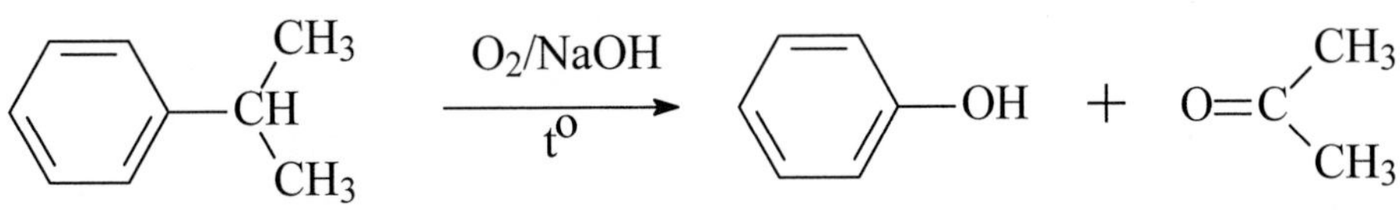

phenol acetone

With ferric trichloride, phenol forms a characteristic violet coloration. The reaction is sensitive and is used for the detection and qualitative demonstration of phenol.

Ketones are oxidized only by strong oxidants ($KMnO_4$, $K_2Cr_2O_7$). Oxidation breaks the carbon chain adjacent to the carbonyl group. A mixture of carboxylic acids is obtained.

Task 2.

Hydrocarbon **A** has gross formula C_9H_{10}. The compound decolors bromine water, on oxidation with a concentrated solution of $KMnO_4$ in the presence of H_2SO_4 forms benzoic acid, and on oxidation with a 2% solution of $KMnO_4$ yields 3-phenyl-1,2-propanediol.

Write the structural formula of the hydrocarbon. Express the processes with chemical equations. Suggest the names of all compounds.

What type of isomerism is possible in compounds?

In which of the compounds is the indicated type of isomerism present?

Approximate solution

Hydrocarbon **A** is 3-phenylpropene. The properties given in the problem statement - decolorisation in bromine water and oxidation with a 2% solution of $KMnO_4$ to give a diol - indicate the presence of a double bond in the compound. Energetic oxidation leads to benzoic acid - this is an indication that hydrocarbon **A** is alkenylbenzene.

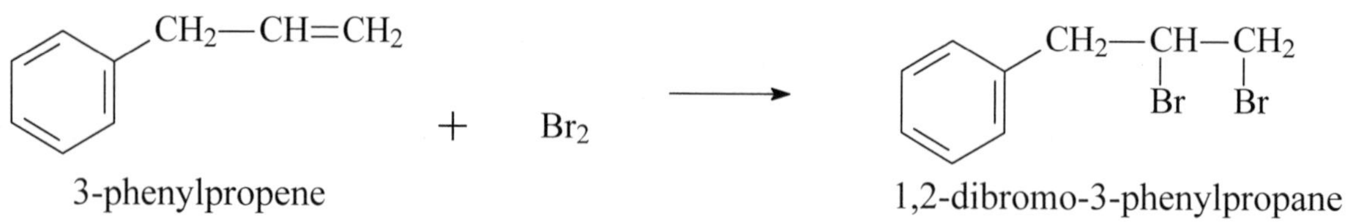

Decolorization of bromine water is a qualitative reaction to prove a multiple (double or triple) bond in the compound molecule.

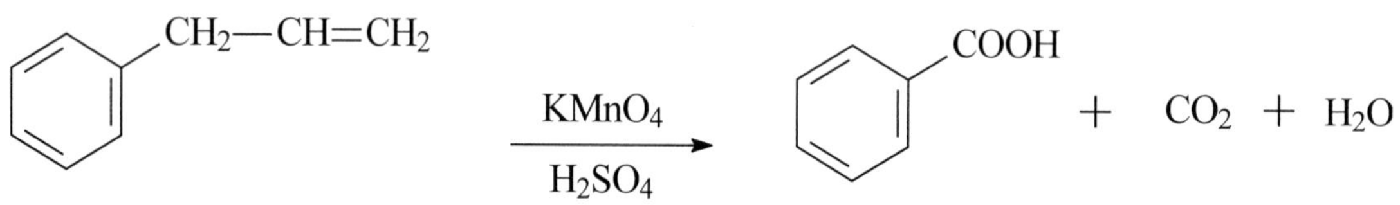

The energetic oxidation of alkyl or alkenyl arenes leads to benzoic acid. Regardless of the side chain length, the oxidation produces benzoic acid - the carbon atom immediately bonded to the aromatic nucleus is oxidized.

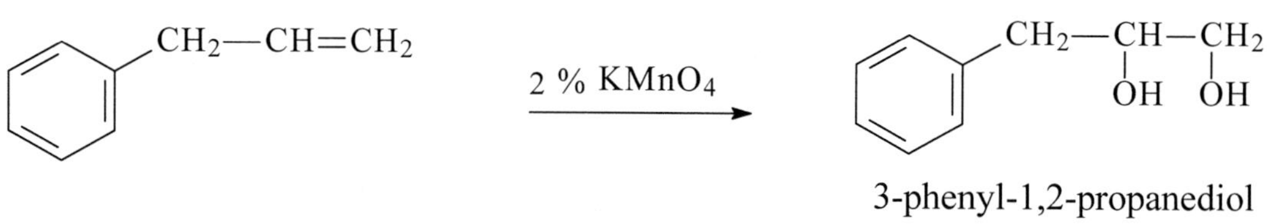

The moderate oxidation of hydrocarbon **A**, and especially the oxidation product, is unequivocal evidence of a double bond in the molecule.

The molecules of the compounds 1,2-dibromo-3-phenylpropane and 3-phenyl-1,2- propanediol have asymmetric (stereogenic) carbon atoms. Stereogenic (asymmetric) refers to a carbon atom bonded to four different substituents. In this case, symmetry elements (center, axis, and symmetry plane) are missing and the molecules of the compound relate to each other as an objcct and its mirror image, such as the left and right arms. Isomerism is called optical, and compounds exhibiting optical activity are designated as optical isomers.

24. HALOGEN DERIVATIVES OF HYDROCARBONS

DEFINITION, TYPES, NAMES, ISOMERISM, PHYSICAL AND CHEMICAL PROPERTIES. APPLICATION.

Halogenated derivatives of alkanes are viewed as derived from saturated hydrocarbons by substitution of hydrogen atoms with atoms of halogen elements.

Halogenated derivatives of alkanes are subdivided on the basis of several features:

- The **type of halogen atoms involved** are fluorine, chlorine, bromine, iodine and mixed:

C_2H_5F, CH_3Cl, C_2H_5Br, C_2H_5J, CCl_2F_2;

- By the **number of halogen atoms** in the molecule - mono-, di-, tri- and polyhalogen derivatives:

CH_3Cl, CH_2Cl_2, $CHCl_3$, CCl_4;

- According to **whether the halogen atom is bonded to a primary, secondary or tertiary carbon atom:**

$CH_3CH_2CH_2CH_2Cl$

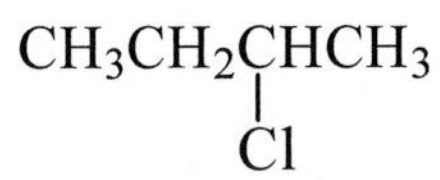

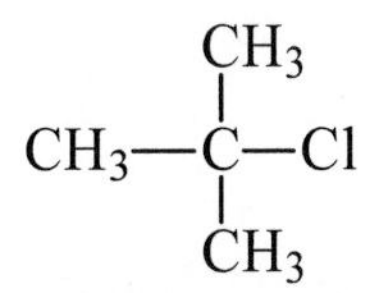

- According to the **type of hydrocarbon residue**, they are derivatives of saturated, unsaturated and aromatic hydrocarbons:

CH_3CH_2Cl

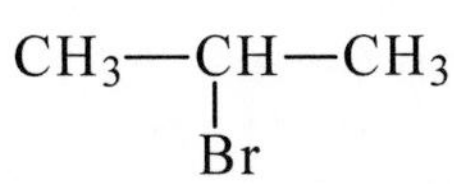

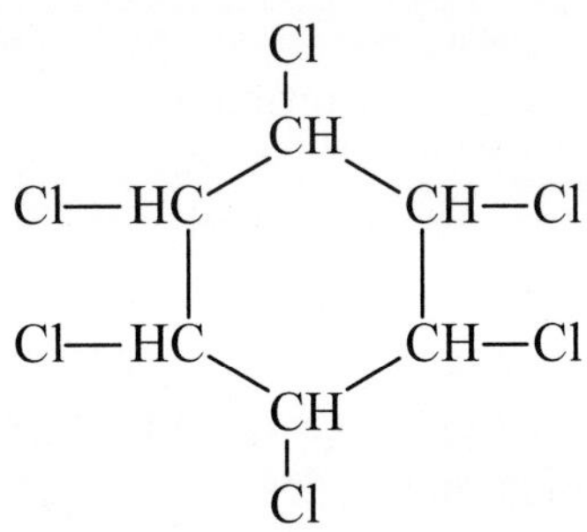

$CH_2{=}CHCl$ $CH_2{=}CHCH_2Cl$

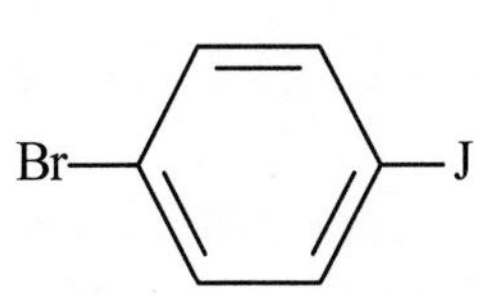

According to IUPAC, the **name of the hydrocarbon is prefixed with the name of the corresponding halogen atom. A number** in front of the name indicates the position of the halogen atom in the chain.

CH_3CH_2Cl

chloroethane

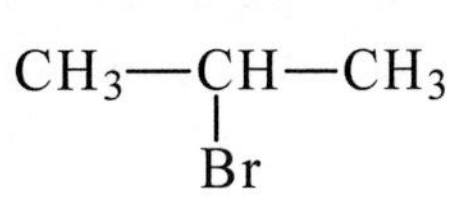

2-bromopropane

The other way to form the names of halogen derivatives is by using the **names of the corresponding hydrocarbon residues and the word chloride, bromide, iodide, fluoride**.

CH_3CH_2Cl

ethyl chloride

$CH_3CH_2CH_2CH_2Br$

butyl bromide

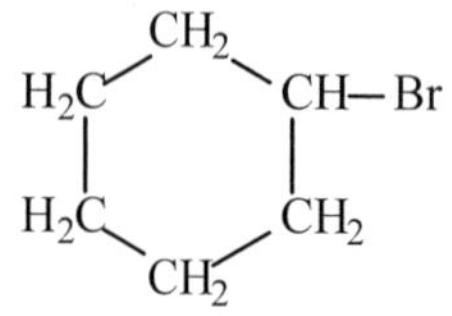

cyclohexyl bromide

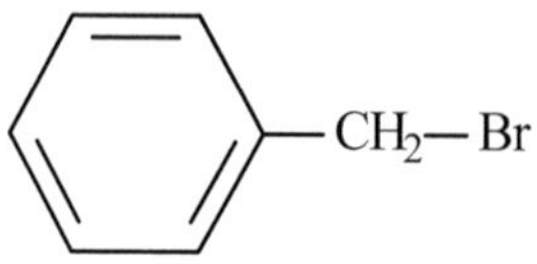

benzyl bromide

The following **types of isomerism occur** in halogenated derivatives of hydrocarbons:

- **Chain isomerism** - it is determined by the way and sequence of bonding of carbon atoms in the chain;

$CH_3CH_2CH_2CH_2Cl$

1-chlorobutane

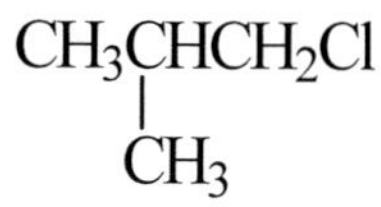

2-methyl-1-chloropropane
(isobutyl chloride)

- **Positional isomerism** - determined by the position of the substituent (functional group) in the carbon chain;

$CH_3CH_2CH_2CH_2Cl$

1-chlorobutane

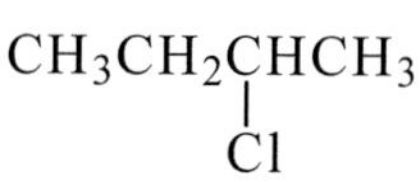

2-chlorobutane

- **Optical isomerism** - when the substituents at the carbon atom are different, optical isomerism is possible. The carbon atom is denoted as asymmetric (chiral) and two different arrangements of substituents in space (two configurations) are possible. They are conventionally denoted as D (from Latin dexter - right) and L (from Latin laevo - left). These isomers are called enantiomers, optical antipodes.

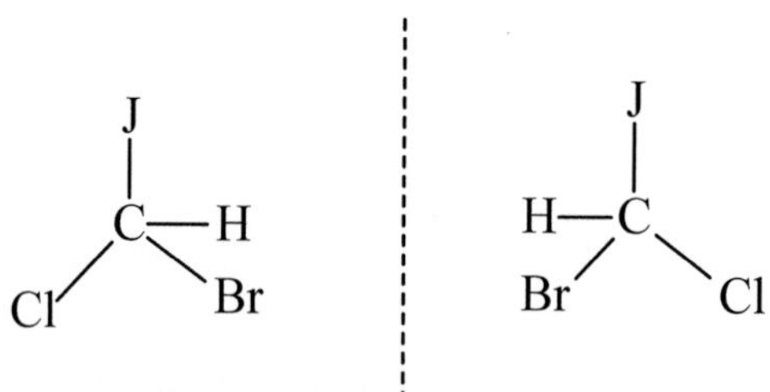

D- L-

The chemical bond C - X is polar due to the greater electronegativity of the halogen atoms relative to the carbon atom. They have a negative induction effect - they draw electron density away from the C - X bond, thus polarizing the neighboring chemical bonds in the molecule.

The energy of the C - X bonds decrease in the order F > Cl > Br > J.

The fluorine and chlorine derivatives of the lowest alkanes are colorless gases, the others are oily liquids. The alkyl halides boil at a higher temperature than the corresponding hydrocarbons, but at lower temperatures than the corresponding alcohols. Bromides boil at a higher temperature than chlorides and iodides boil at a higher temperature than bromides:

$$T_b(\text{R-F}) < T_b(\text{R-Cl}) < T_b(\text{R-Br}) < T_b(\text{R-J})$$

Their solubility in water is negligible, they dissolve well in organic solvents.

Chemical properties of halogenated derivatives of hydrocarbons

Due to the presence of a highly polar C - X bond in their molecule, halogen derivatives are reactive compounds and enter into a variety of chemical reactions. They interact with polar and ionic compounds. As a result, the halogen atoms are substituted with different atoms and atomic groups. New classes of organic compounds are produced.

1. For example, the reaction of bromoethane with water produces a primary alcohol, ethanol.

$$CH_3CH_2Br + H{-}OH \xrightleftharpoons{t^o} CH_3CH_2OH + HBr$$

Treatment of 2-bromopropane with water produces a secondary alcohol.

$$\underset{\displaystyle Br}{CH_3\overset{}{C}HCH_3} + H{-}OH \xrightleftharpoons{t^o} \underset{\displaystyle OH}{CH_3\overset{}{C}HCH_3} + HBr$$

2. Alcohols can also be produced by interaction with an aqueous solution of an alkali base, for example:

$$CH_3CH_2Br + NaOH\ (H_2O) \xrightarrow{t^o} CH_3CH_2OH + NaBr$$

3. The interaction of bromoethane with ammonia leads to aminoethane. This is one of the methods for obtaining amines:

$$CH_3CH_2Br + NH_3 \longrightarrow \underset{\text{ethylammonium bromide}}{[CH_3CH_2\overset{+}{N}H_3]\overset{-}{Br}} \xrightarrow{NH_3} \underset{\text{aminoethane (ethylamine)}}{CH_3CH_2NH_2 + NH_4Br}$$

4. The halogen atom can also be substituted with a cyano group. Nitriles are obtained. This is one of the methods for their preparation, allowing the carbon chain to be extended.

$$CH_3CH_2Br + NaCN \longrightarrow \underset{\text{propionitrile}}{CH_3CH_2CN} + NaBr$$

5. Halogenated derivatives are involved in an elimination process with a potassium based - alcohol solution. Atoms or atomic groups are separated from the starting molecules. Double bond, triple bond, or cyclic molecules are formed, for example:

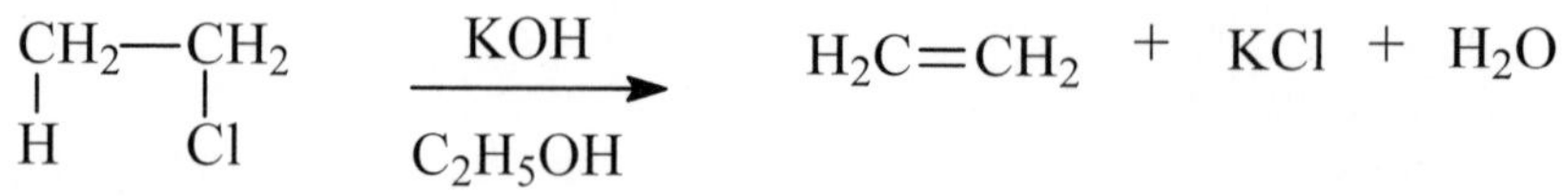

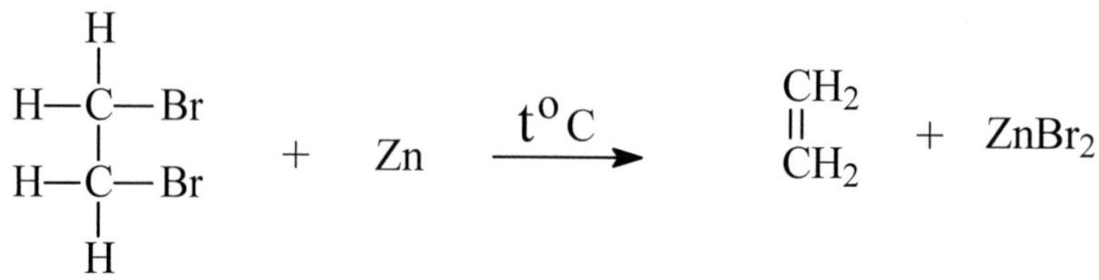

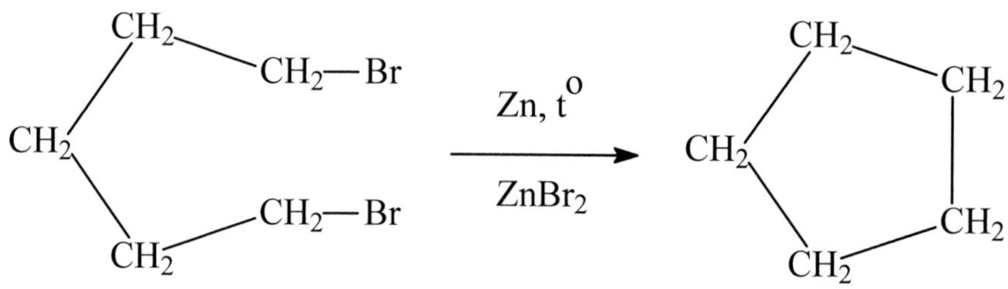

6. Under ordinary conditions, the interaction of halogenated derivatives with alkali metals results in the production of higher hydrocarbons. This is the Wurz method and is used for the laboratory preparation of alkanes.

$$CH_3J + 2Na + JCH_3 \longrightarrow CH_3CH_3 + 2NaJ$$

7. On strong cooling, the reaction may stop and alkali-metal derivatives of the hydrocarbons are produced.

$$CH_3Br + 2\,Li \longrightarrow CH_3Li + 2\,LiBr$$

Many metal derivatives have been obtained in this way. These are called organometallic compounds. Metal derivatives of alkynes are also of this class of organic compounds.

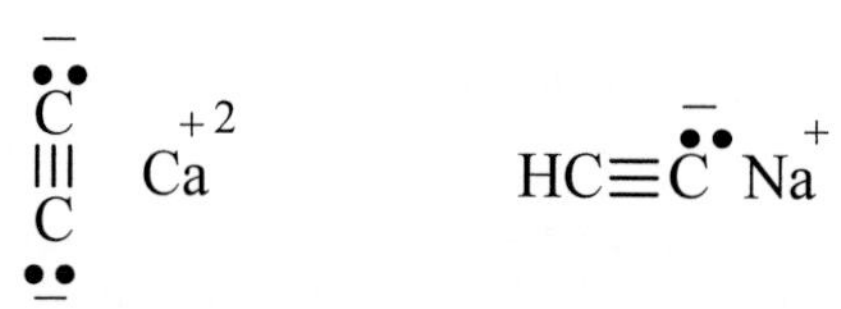

Carbon-metal chemical bonds are ionic or highly polar.

8. Alkyl halides burn hard as the flame contours turn green. Halogenated hydrogen is liberated on combustion and iodine derivatives can yield iodine.

9. With silver nitrate solution, no precipitate is immediately separated, but only after hydrolysis of the halogen derivative.

Methyl iodide in an alcoholic solution of silver nitrate decomposes rapidly to form silver iodide:

$$CH_3J + AgNO_3 + CH_3OH \longrightarrow AgJ\downarrow + HNO_3 + CH_3OCH_3$$

Due to the variety of chemical transformations in which halogen derivatives are involved, they are important raw materials for organic synthesis. Many of them are used as solvents ($CHCl_3$, CCl_4) and others are used to produce polymers (vinyl chloride - polyvinyl chloride).

10. High temperature, high pressure and catalyst salts of monovalent copper are required to replace the chlorine atom in the chlorobenzene molecule.

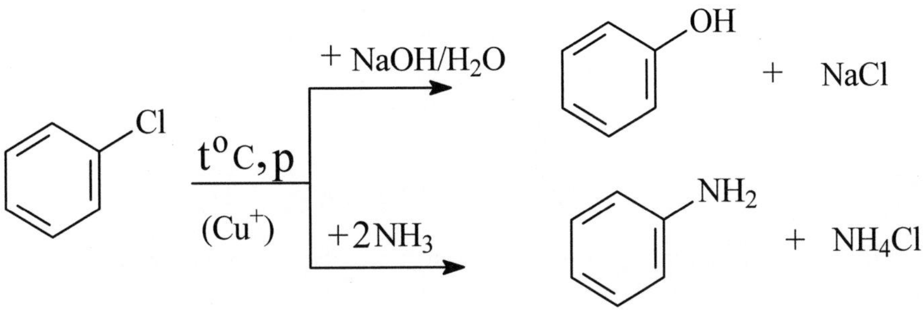

11. **To** the substitution reactions of halogenated derivatives can also be attributed the Würz, Würz-Fittig and Friedel-Crafts syntheses:

$$2\,R{-}C{\equiv}CH \xrightarrow{Na} 2\,R{-}C{\equiv}CNa + H_2$$

$$R{-}C{\equiv}CNa + I{-}CH_3 \longrightarrow R{-}C{\equiv}C{-}CH_3 + NaI$$

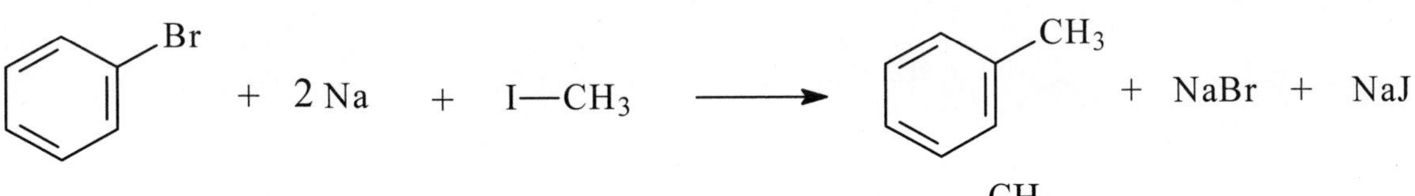

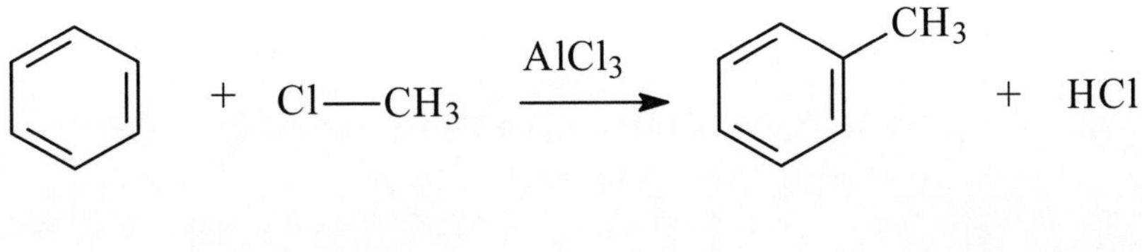

Preparation of halogenated derivatives of hydrocarbons

Halogenated derivatives are obtained from the corresponding hydrocarbons by halogenation or hydrogenation reactions. This produces mainly chloro- and bromo-organic compounds. Iodo- and fluoro-organic compounds are obtained by indirect methods. The direct interaction of iodine proceeds very slowly and the process is equilibrium. Direct fluorination of hydrocarbons leads to their charring. Fluorine derivatives are obtained by substitution reactions:

$$CCl_4 + 2\,HF \xrightarrow{t^oC,\,p} CCl_2F_2 + 2\,HCl$$

dichlorodifluoromethane (Freon 12)

Some of them are used for the destruction of insects, i.e., as insecticides, and others - in medicine, in refrigeration, for extinguishing fires.

Sample problems and solutions

Task 1.

What is the structural formula of a substance of composition C_8H_9Cl if it is known that on interaction with an aqueous solution of an alkaline base it forms an alcohol, and with an alcoholic solution of a base it is converted to a compound of formula C_8H_8, which readily polymerizes. Describe the possible isomerism types of the resulting alcohol - oxidation of which isomer of the alcohol produces acetophenone (methylphenyl ketone)? Express the above interactions with chemical equations and the names of the compounds.

Exemplary solution

The structural formula of a substance of composition C_8H_9Cl, can be represented by two possible isomers:

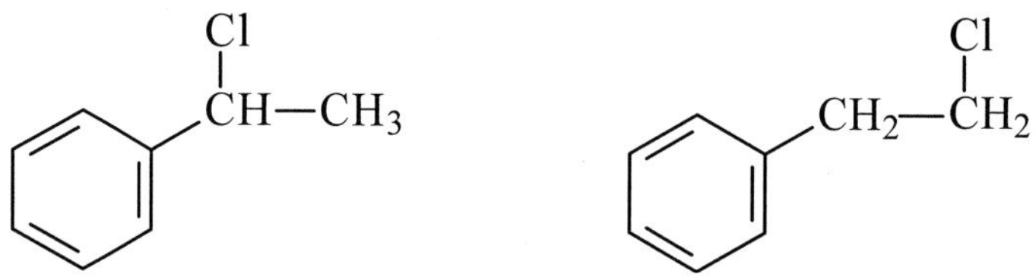

(1-chloroethyl)benzene
1-phenyl-1-chloroethane

(2-chloroethyl)benzene
2-phenyl-1-chloroethane

The basis for such a structure is given by the reaction of elimination of hydrogen halide with an alcoholic solution of an alkali base.

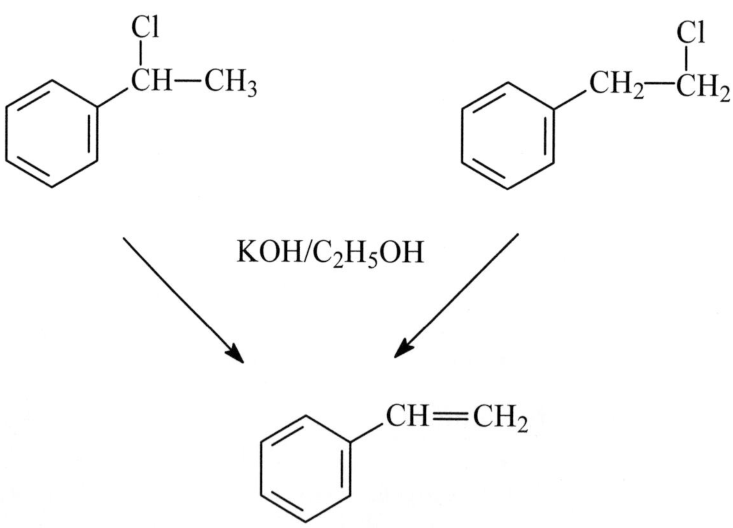

(C_8H_8)

Ethenylbenzene (styrene, vinylbenzene) is the most important representative of alkenylbenzenes. Due to the presence of an unsaturated hydrocarbon residue in the ethylbenzene molecule, it readily polymerizes:

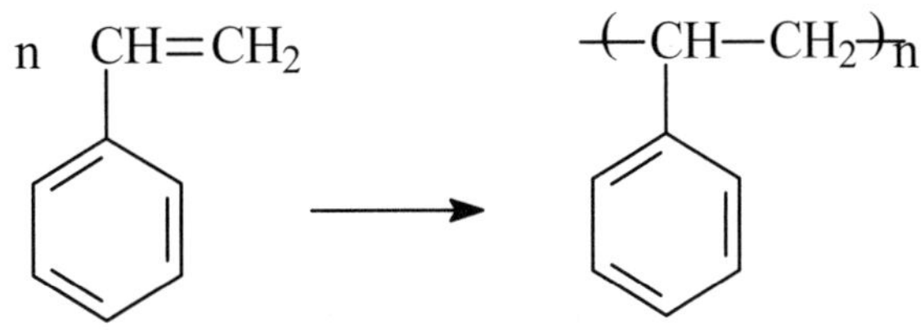

The polymer is used for the production of synthetic rubber and polystyrene products. If the starting halogen derivative is treated with an aqueous solution of an alkali base, an alcohol is obtained.

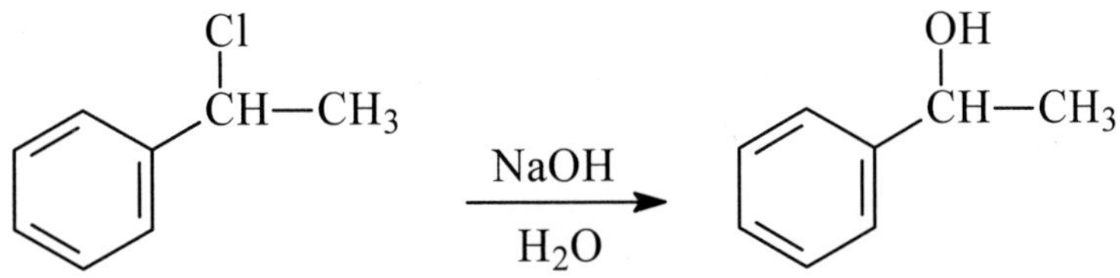

1-phenyl-1-ethanol

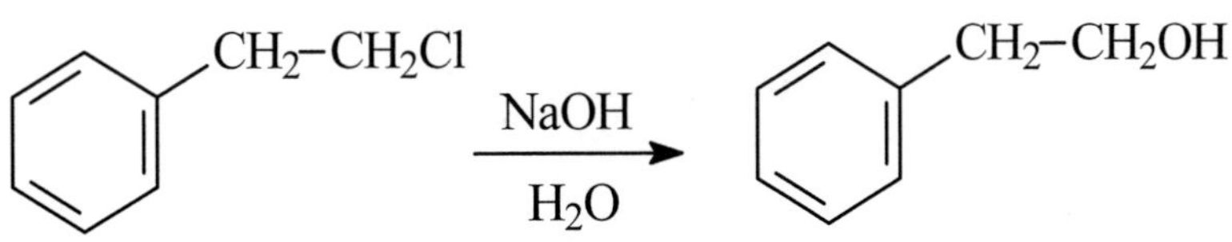

2-phenyl-1-ethanol

The isomerism in the resulting alcohol is positional. It is determined by the location of the hydroxyl group in the chain. The hydroxyl group can also be directly bonded to the aromatic nucleus, but then the compounds are isomers of the phenol, *not* the alcohol.

In the molecule of 1-phenyl-1-ethanol, there is an asymmetric carbon atom (bonded to four different substituents), which determines the existence of optical isomers - a pair of enantiomers referring to each other as the subject and its mirror image:

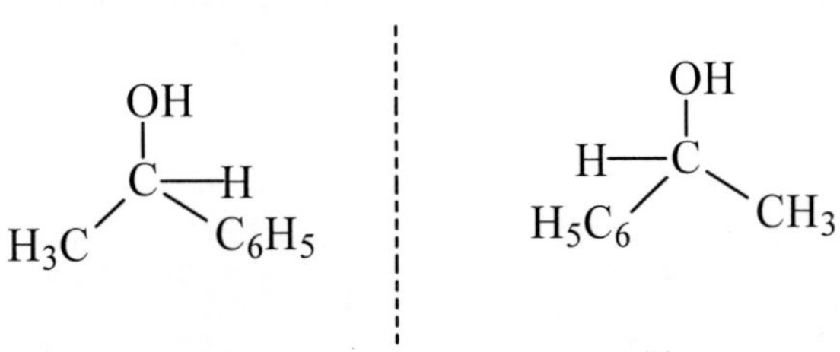

D-　　　　L-

Upon oxidation of the alcohol is produced:

(a) The primary alcohol is oxidized to an aldehyde. Aldehydes are also sensitive to weak oxidants and will yield carboxylic acid according to the scheme:

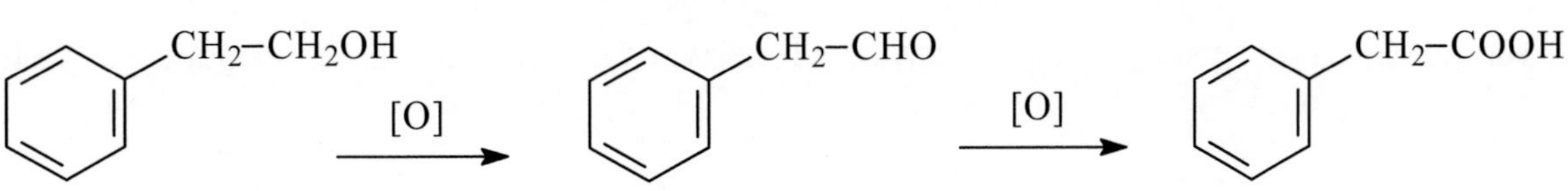

2-phenyl-1-ethanol　　2-phenylethanal　　2-phenylethanoic acid

b) Oxidation of the secondary alcohol will yield ketone:

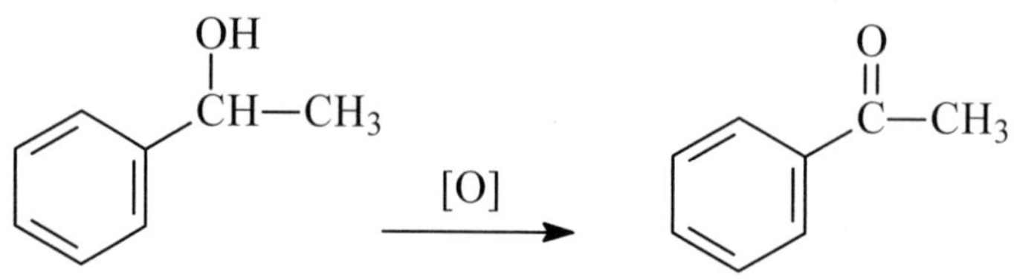

acetophenone (methylphenylketone)

Task 2.

Propose a three-step method for obtaining 4-bromobenzaldehyde from toluene without resorting to side-chain oxidation. Explain and motivate the individual steps of the conversion. Represent the processes with chemical equations.

Exemplary solution

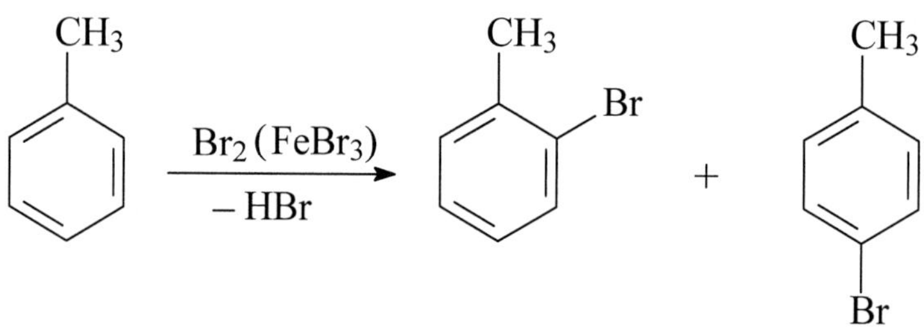

The halogenation of benzene and its homologues is carried out with the participation of a catalyst. An electrophilic substitution reaction takes place in the aromatic core. The available substituent, the methyl group, exhibits a positive induction effect and increases the electron density in the aromatic core at the o- and p-site. The attacking particle is directed to the sites of increased electron density. This is the reason why in benzene homologues a mixture of two isomers, in this case 2-bromotoluene (o-bromotoluene) and 4-bromotoluene (p-bromotoluene) is usually obtained. The reaction mixture is subjected to an appropriate purification and 2- bromotoluene and 4-bromotoluene are isolated.

If the bromination is carried out under irradiation of the reaction mixture with ultraviolet light or at high temperature (about 500° C), substitution of hydrogen atoms in the side chain (methyl group) will take place. 4-Bromotoluene undergoes bromination under irradiation with ultraviolet light:

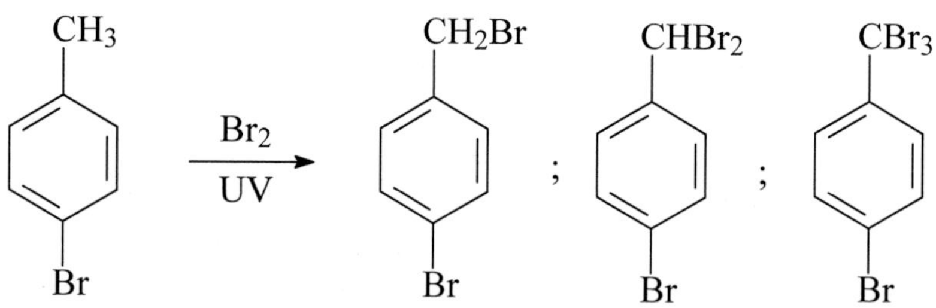

a **b** **c**

a - 4-bromobenzyl bromide; **b** - 4-bromobenzylidene bromide; **c** - 4-bromobenzotribromide

Halogenated hydrocarbon derivatives are known to hydrolyze in alkaline and acidic environments. Hydrolysis results in hydroxyl derivatives. Hydrolysis of product **a** yield 4-bromobenzyl alcohol:

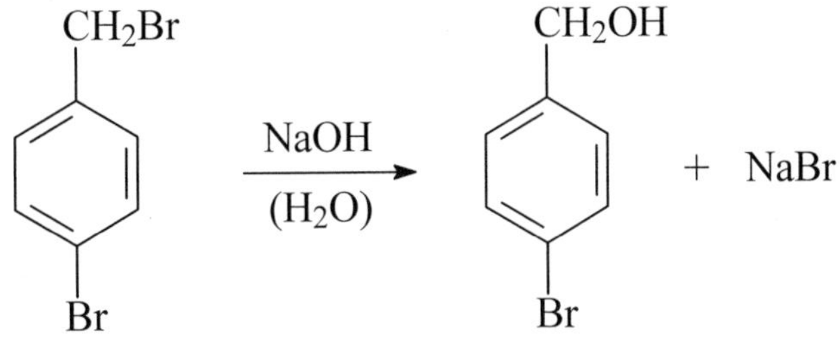

4-bromobenzyl alcohol

Hydrolysis of **b** produces a divalent bidentate alcohol. When the two hydroxyl groups are bonded to the same carbon atom the compound is perishable and spontaneously dehydrates. The final product of this process is an aldehyde:

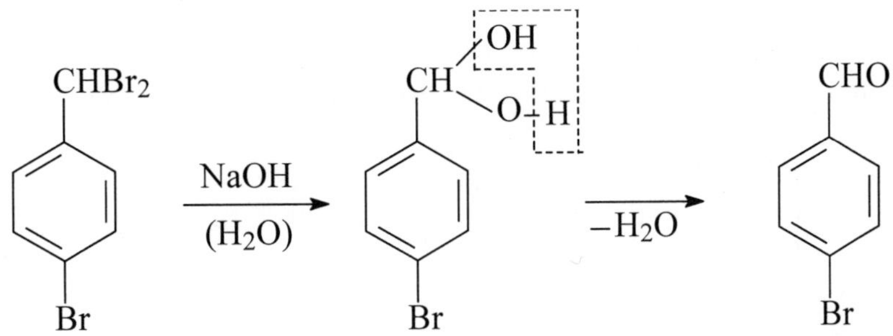

4-bromobenzaldehyde

Hydrolysis of **c** leads to a neutral trivalent alcohol which spontaneously dehydrates to carboxylic acid.

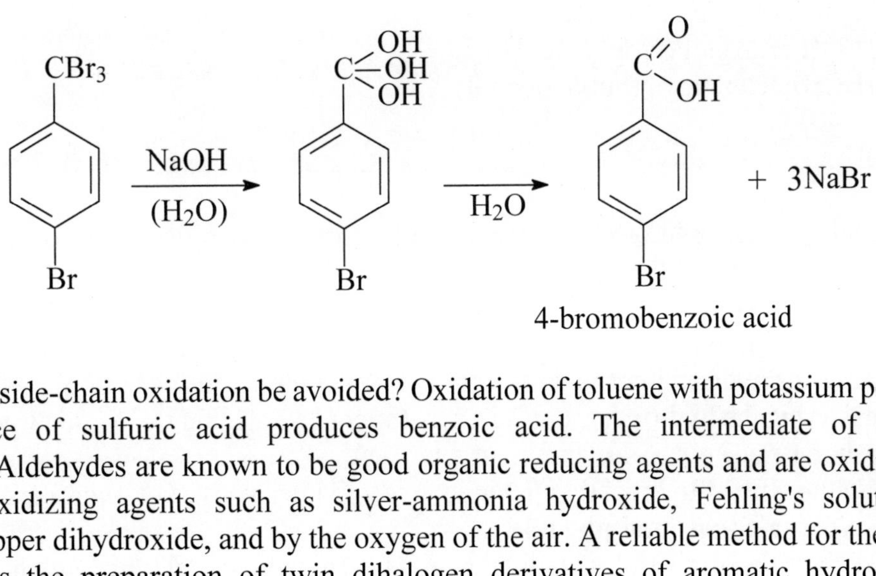

4-bromobenzoic acid

Why should side-chain oxidation be avoided? Oxidation of toluene with potassium permanganate in the presence of sulfuric acid produces benzoic acid. The intermediate of oxidation is benzaldehyde. Aldehydes are known to be good organic reducing agents and are oxidized by even the weakest oxidizing agents such as silver-ammonia hydroxide, Fehling's solution, freshly precipitated copper dihydroxide, and by the oxygen of the air. A reliable method for the preparation of aldehydes is the preparation of twin dihalogen derivatives of aromatic hydrocarbons and subsequent hydrolysis.

25. HYDROXYL DERIVATIVES OF HYDROCARBONS

TYPES AND NAMES. CLASSIFICATION ACCORDING TO THE NUMBER OF HYDROXYL GROUPS AND ACCORDING TO THE HYDROCARBON RESIDUE. HOMOLOGOUS ORDER. ISOMERISM - CHAIN, POSITIONAL AND OPTICAL. PHYSICAL PROPERTIES. COMPARATIVE CONSIDERATION OF CHEMICAL PROPERTIES OF ALCOHOLS AND PHENOLS DEPENDING ON ELECTRONIC STRUCTURE. APPLICATION

Types and names

Hydroxyl derivatives of hydrocarbons are compounds, which are considered to be derived from hydrocarbons, in the molecules of which one or more hydrogen atoms are substituted by hydroxyl groups.

Depending on the number of hydroxyl groups these compounds are – with one hydroxyl group – monofunctional (monovalent), with two hydroxyl groups - bifunctional (divalent), with three hydroxyl groups - trifunctional (trivalent) and so on polyfunctional (polyvalent) hydroxyl derivatives of hydrocarbons.

Monofunctional (monovalent) alcohols contain one hydroxyl group bonded to a hydrocarbon residue in their molecule. When the residue is alkyl, the alcohols are saturated monovalent. They form a homologous series:

CH_3OH methanol (methyl alcohol)
CH_3CH_2OH ethanol (ethyl alcohol)
$CH_3CH_2CH_2OH$ propanol (propyl alcohol)
$CH_3CH_2CH_2CH_2OH$ butanol (butyl alcohol)

$C_nH_{2n+1}OH$ ....R- OH .... alkanols

The nomenclature of hydroxyl derivatives of hydrocarbons is in accordance with their classification. The names of the hydroxyl derivatives with predominantly a lower hydrocarbon residue are formed (according to IUPAC) by adding the suffixes **-ol** (one hydroxyl group), **-diol** (two hydroxyl groups), **-triol** (three hydroxyl groups), etc. to the name of the hydrocarbon concerned. Names of hydroxyl derivatives with a more complex hydrocarbon residue can also be formed by considering the hydroxyl group as a substituent and placing the name **"hydroxy"** in front of the hydrocarbon name.

According to the type of hydrocarbon residue, hydroxyl derivatives of hydrocarbons are divided into two main types: **alcohols and phenols**.

Alcohols are hydroxyl derivatives of alkanes, alkenes, alkadienes, alkynes and the corresponding cyclic hydrocarbons. The common name alcohol was originally given to ethanol, the ethyl alcohol contained in wine.

Phenols are hydroxyl derivatives of aromatic hydrocarbons (arenes) in which the hydroxyl group is bonded to a carbon atom of the aromatic nucleus.

The basic classification of hydroxyl derivatives of hydrocarbons can be schematically represented as follows:

Hydroxyl derivatives of hydrocarbons

Alcohols (R-OH) Phenols (Ar-OH)

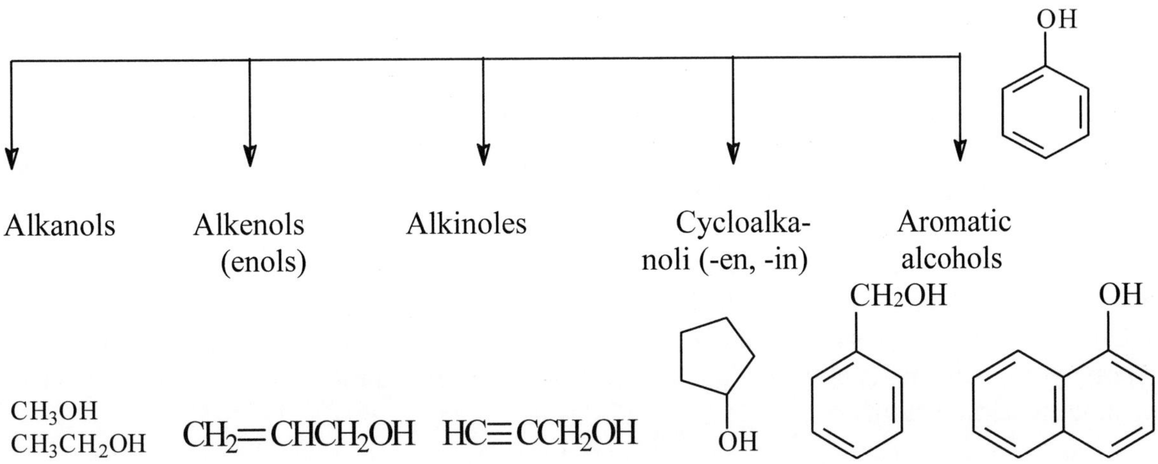

Isomerism

The carbon chains of saturated monovalent alcohols can be straight and branched. This determines the presence of chain isomers:

$CH_3-CH_2-CH_2-CH_2-CH_2-OH$

1-pentanol

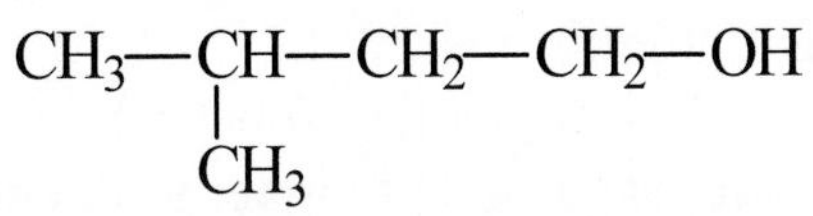

3-methyl-1-butanol

In alcohols, isomerism is also possible, which is determined by the location of the hydroxyl group in the chain - positional.

$CH_3-CH_2-CH_2-OH$

1-propanol

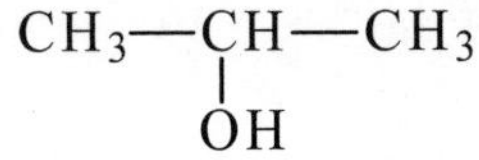

2-propanol

Monovalent alcohols come in three types depending on which carbon atom the hydroxyl group is bonded to:

$CH_3-CH_2-CH_2-CH_2-OH$

1-butanol
primary butyl alcohol

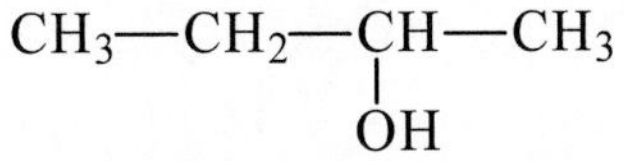

2-butanol
secondary butyl alcohol

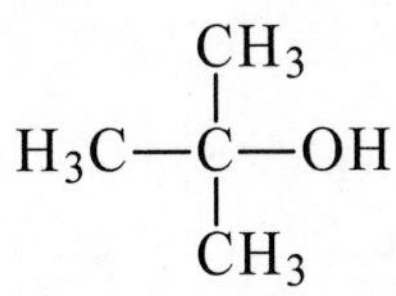

2-methyl-2-propanol
tertiary butyl alcohol

In alcohols, there is a type of spatial isomerism called optical isomerism. It is present when there is an asymmetric carbon atom in the compound molecule. An asymmetric carbon atom is one that is bonded to four different substituent atoms or atomic groups, for example, 2- butanol. Its molecule is not symmetrical: there is no plane, axis or center of symmetry. A molecule with an asymmetric carbon atom exists in the form of two optical antipodes that relate to each other as an object and its mirror image:

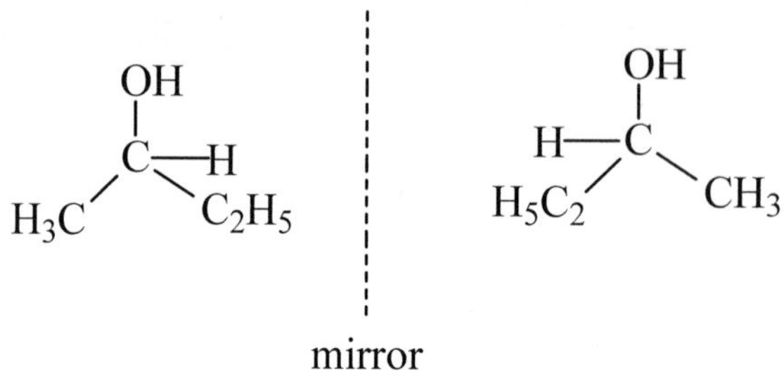

In nature, there are actually asymmetric molecules that relate to each other as an object and its mirror image. Such compounds are optical isomers. They have the same qualitative and quantitative composition of their molecules, but differ only in the spatial arrangement of certain atoms or atomic groups. The physical properties of optical isomers (m.p., m.b., solubility, etc.) as well as their basic chemical properties are the same. Optical isomers have a different rotation angle of polarized light.

In polyhydroxyalkanes (polyvalent alcohols), the number of enantiomers depends on the number of asymmetric carbon atoms.

Physical properties. Electronic structure

In the homologous order of alkanols, the first members are liquids and the higher ones are solids. The lower alkanols are well soluble in water, and the higher ones are less soluble as the relative proportion of hydrocarbon residue in them increases and the "similarity" to water decreases.

In the homologous order of alkanols, with increasing molecular mass of the homologues, a regular increase in boiling point is observed.

Phenol is a colorless crystalline substance with a characteristic odor. At ordinary temperature it is slightly soluble in water, but at higher temperature (~ 66 ° C) it dissolves indefinitely.

Alcohols and phenols have higher m.p. and b.p. than the corresponding hydrocarbons with the same number of carbon atoms. This is due to the formation of intermolecular hydrogen bonds (association):

Alcohols and phenols are significantly more soluble in water than the corresponding hydrocarbons. Water solubility decreases with increasing hydrocarbon residue and increases with increasing number of hydroxyl groups in the molecule. The reason for this is the formation of intermolecular hydrogen bonds with water molecules:

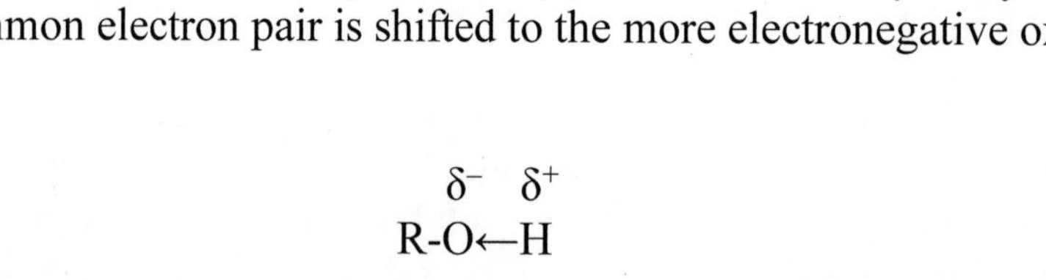

The chemical bond between the hydrogen and oxygen atoms in the hydroxyl group of alkanols is polar because the common electron pair is shifted to the more electronegative oxygen atom.

$$R\text{-}\overset{\delta^-}{O}\leftarrow\overset{\delta^+}{H}$$

As a result, the oxygen atom acquires a partial negative charge (δ^-) and hydrogen has a partial positive charge (δ^+). In molecules containing hydroxyl groups (H_2O, ROH), two dissimilar poles arise which determine the mutual attraction of the different molecules. They are oriented so that the divergent poles are located as follows:

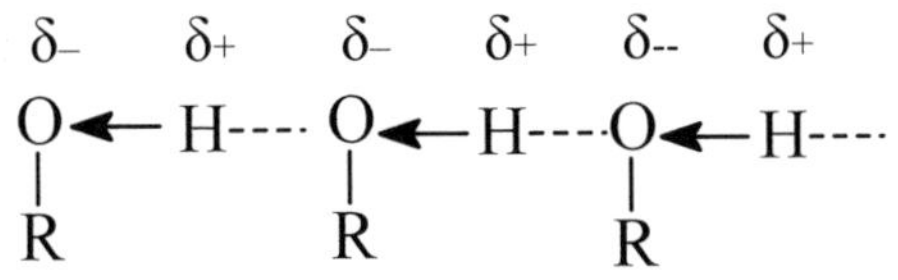

The bonds between molecules that result from this attraction are called hydrogen bonds. They are indicated by dotted lines. In both water and alcohols, the hydrogen bonds formed give rise to relatively stable groupings of several molecules:

$(H_2O)_x$

$x = 2, 3, 4, ...$

$(CH_3OH)_x$

The presence of these groupings determines the relatively high boiling point of water and alcohols.

Chemical properties of alcohols

The chemical properties of alcohols are determined by their involvement in the following types of reactions:

1. Reactions with O-H bond breaking;
2. Reactions with C - O bond breaking;
3. Reactions involving the hydrocarbon residue (R- or Ar-).

Reactions of the first type are possible because the hydrogen atom carries a partial positive charge. It can be replaced by an alkali metal. The resulting compounds are called alcoholates and have an ionic structure:

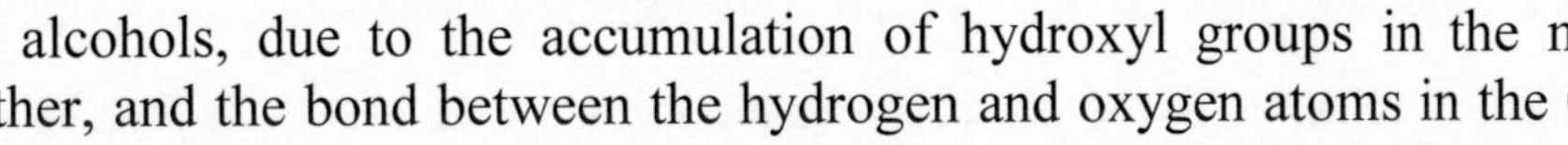

sodium ethylate (ethoxide)

The interaction of alcohols with sodium and with other alkali metals is not as violent as the reaction of sodium with water. This indicates that the polarity of the O - H bond in R - OH is less pronounced than in water. The reason for this is the influence of the alkyl residue.

In polyvalent alcohols, due to the accumulation of hydroxyl groups in the molecule, they influence each other, and the bond between the hydrogen and oxygen atoms in the O-H groups is more polar compared to the same bond in monovalent alcohols. This enhanced polarity also determines the acidic properties of polyvalent alcohols. Therefore, they interact with copper dihydroxide in addition to alkali metals:

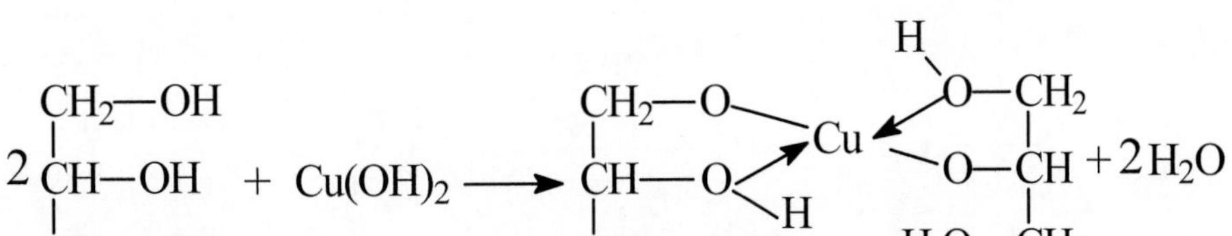

copper glycerate

The resulting copper glycerate is a dark blue, water-soluble complex compound. The reaction is used to detect polyvalent alcohols that contain several hydroxyl groups bonded to adjacent carbon atoms in their molecule.

The iodoform (haloform) reaction is used to prove ethyl alcohol as well as to determine the structure of alcohols containing the group:

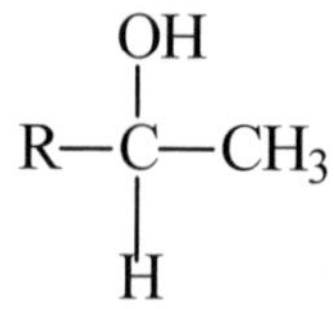

;

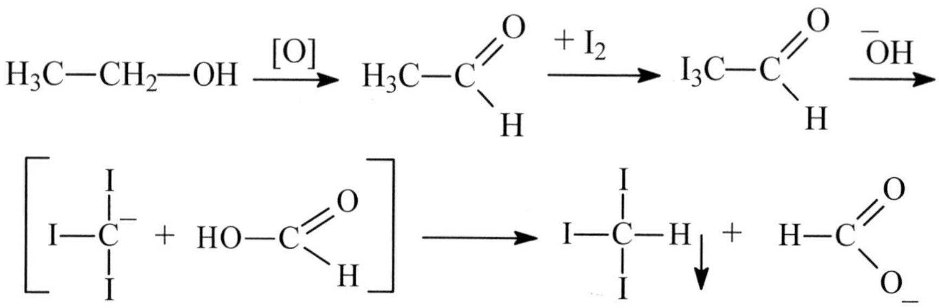

yellow sludge

The reaction is carried out with alkaline hypohalogenates in an alkaline medium. When sodium hypoiodite is used, the reaction is called iodoform.

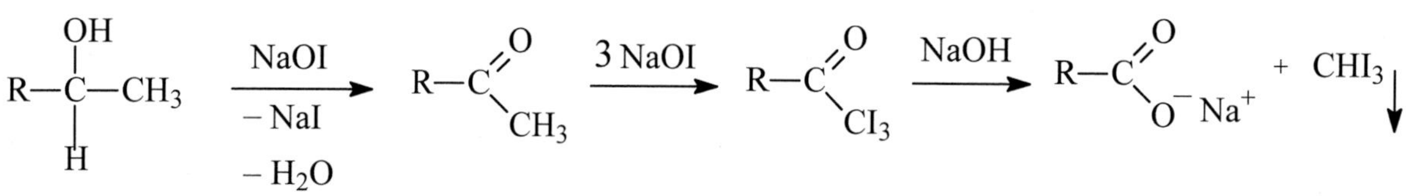

yellow sludge

The iodoform reaction is qualitative and, in addition to secondary alcohols and methyl ketones, it is also used to prove ethanol and ethanal. The gross reaction equation for ethanol is:

$$C_2H_5OH + 4\,I_2 + 6\,NaOH \longrightarrow CHI_3\downarrow + HCOONa + 5\,NaI + 5\,H_2O$$

Alcohols interact with acids. The resulting compounds are called esters, and the process, esterification. For example, the interaction of ethanol with acetic acid can be expressed as follows:

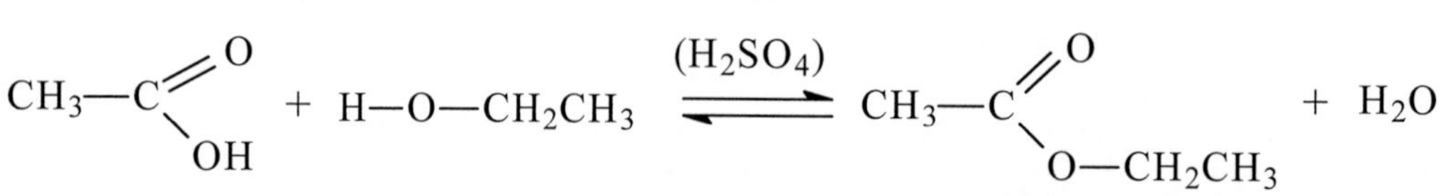

acetic acid ethyl ester
(ethyl acetate, ethyl ethanoate)

Esterification is a slow, molecular, reversible process. It can be shifted to the right if it takes place in the presence of concentrated sulfuric acid, which acts as a catalyst and water- releasing agent.

Glycerol is involved in the esterification process with nitric acid by the equation:

nitroglycerol (trinitroglycerin)

Nitroglycerol is a colorless oily liquid with a pleasant odor which explodes on impact or heating, decomposing instantly with the evolution of gases - a powerful explosive:

$$4C_3H_5(ONO_2)_3 \longrightarrow 12\,CO_2 + 6\,N_2 + 10\,H_2O + O_2 + Q$$

Mixed with inert substances, it is used under the name dynamite. It was discovered by Albert Nobel. Nitroglycerol also finds application in medicine in the treatment of cardiovascular diseases.

Alcohols are involved in chemical reactions in which the **entire hydroxyl group is affected**. When interacting with halogenated hydrocarbons (e.g., HBr) the following reaction takes place:

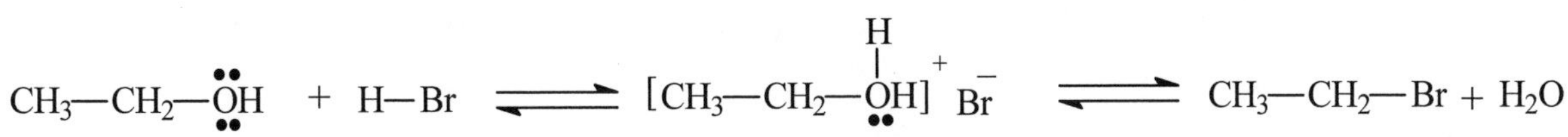

ethyloxonium bromide

The electronic mechanism of the reaction shows that the hydroxyl group is separated from the alkyl residue with the electron pair and forms a new covalent bond with the hydrogen cation from the hydrogen bromide in the separating water molecule (ethyloxonium bromide is formed intermediately).

A reaction between the molecules of the alcohols themselves is possible. For example, when ethanol is heated in the presence of catalytic amounts of concentrated sulfuric acid to ~ 140° C intermolecular dehydration takes place and ether is produced:

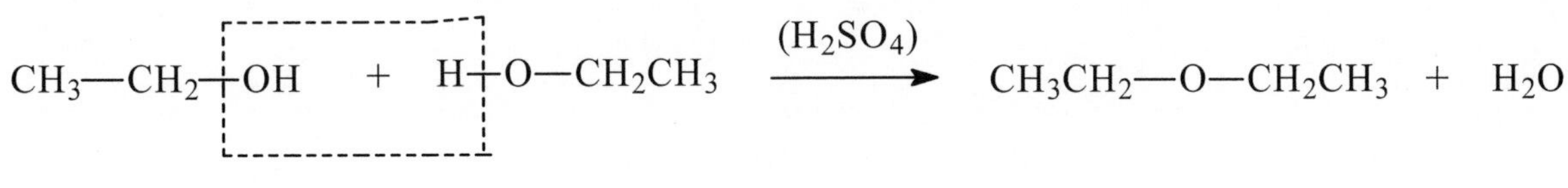

And when the same mixture is heated at t° > 140° C, another chemical reaction takes place to produce ethylene (an intramolecular dehydration process takes place):

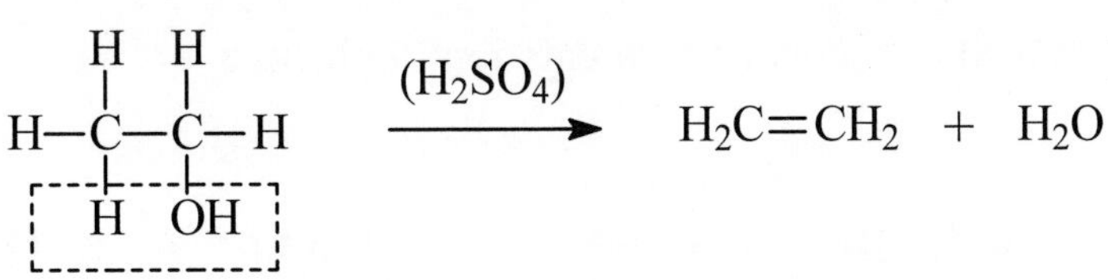

Alcohols burn:

$$2\,CH_3OH + 3O_2 \longrightarrow 2CO_2 + 4H_2O + Q$$

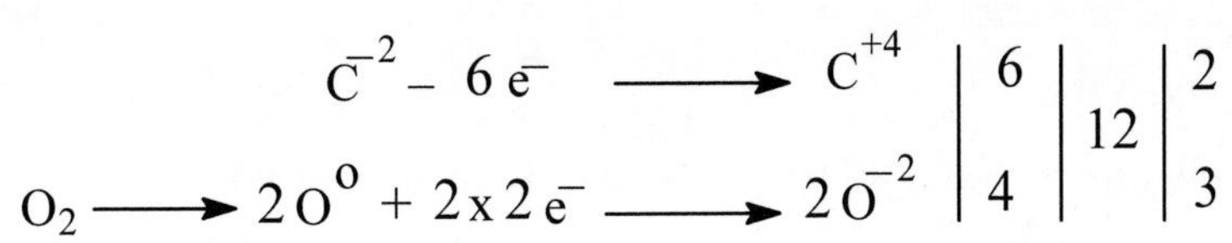

The energetic oxidation of ethanol leads directly to the production of acetic (ethanoic) acid.

$$CH_3CH_2OH \xrightarrow[H_2SO_4]{KMnO_4} CH_3COOH$$

$$\begin{array}{rcl|c|c|c} & \overset{-}{C} - 4\,e^- \longrightarrow & \overset{+3}{C} & 4 & \multirow{2}{*}{4} & 1 \\ O_2 \longrightarrow & 2O^{0} + 2\times2\,e^- \longrightarrow & 2O^{-2} & 4 & & 1 \end{array}$$

Obtaining alcohols

- Hydrolysis of halogen derivatives:

$$CH_3{-}CH_2{-}Br + H_2O \rightleftharpoons [CH_3{-}CH_2{-}\overset{H}{\overset{|}{\ddot{O}}}H]^{+}\,\overset{-}{Br} \rightleftharpoons$$

$$\rightleftharpoons CH_3{-}CH_2{-}\ddot{\underset{\cdot\cdot}{O}}H + H{-}Br$$

The process is reversible and has mainly theoretical significance.

- The method of alkaline hydrolysis of fatty and aromatic halogen derivatives of hydrocarbons finds practical application.

$$CH_3{-}CH_2{-}Br + \underbrace{Na^+ + \overset{-}{O}H} \longrightarrow CH_3{-}CH_2{-}OH + \underbrace{Na^+ + \overset{-}{Br}}$$

- Hydration of alkenes:

$$H_2C{=}CH_2 + H_2O \xrightarrow[(H^+)]{t^o,\ p} CH_3CH_2OH$$

Industrial **methanol** is produced by catalytic hydrogenation of carbon monoxide in a 1:2 ratio (so-called "synthesis gas"):

$$CO + 2\,H_2 \xrightarrow{cat.} CH_3OH$$

Ethanol is also produced by alcoholic fermentation of glucose:

$$C_6H_{12}O_6 \xrightarrow{cimasa} 2C_2H_5OH + 2CO_2$$

Glycerol is produced by hydrolysis of fats, a classical industrial method:

$$\begin{array}{l} CH_2{-}OCOR \\ | \\ CH{-}OCOR_1 \\ | \\ CH_2{-}OCOR_2 \end{array} + 3\,H_2O \xrightarrow{3\,NaOH} \begin{array}{l} CH_2{-}OH \\ | \\ CH{-}OH \\ | \\ CH_2{-}OH \end{array} + \begin{array}{l} RCOONa \\ R_1COONa \\ R_2COONa \end{array}$$

Chemical properties of phenols

The chemical properties of phenols are determined by their involvement in the following types of reactions:

1. Reactions with O-H bond breaking;
2. Reactions with C - O bond breaking;
3. Reactions involving the hydrocarbon residue (R- or Ar). Phenol reacts with sodium:

sodium phenolate (phenoxide)

Unlike alcohols, phenol also reacts with alkaline bases, which also produces phenolate:

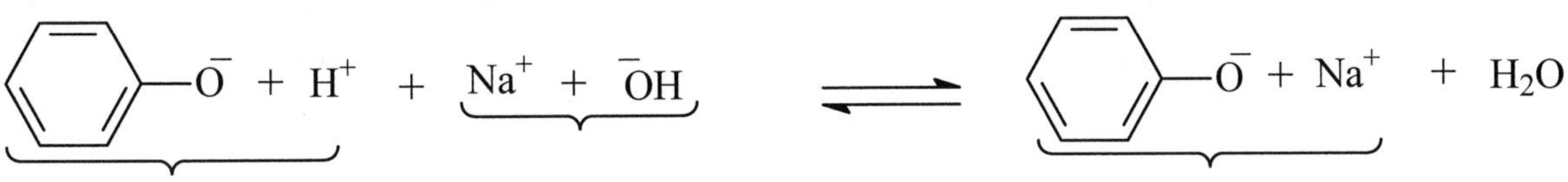

sodium phenolate

Phenol manifests as acid. Hence one of its old names - carbolic acid. In aqueous solution it dissociates:

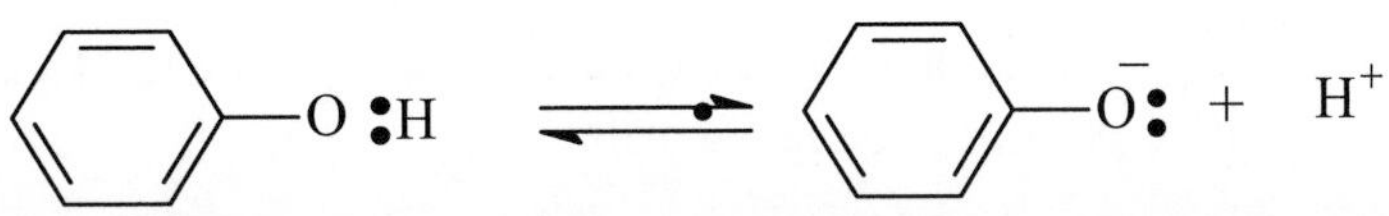

phenolate (phenoxide) anion

Phenol is a weak acid. The degree of electrolytic dissociation is greater than that of water, and less than that of carbonic acid. The increased acidity of phenol is explained as follows:

The difference between phenol and alcohols is in the hydrocarbon residues. The essence of the influence of the benzene nucleus on the hydroxyl group is the electronic interaction between the π-sextet of the aromatic nucleus and the free electron pair of the oxygen atom of the hydroxyl group. As a result of this interaction, the electron density of the oxygen atom decreases and it attracts the electron pair more strongly from its bond with hydrogen:

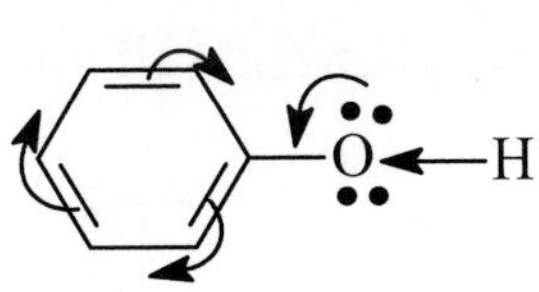

Because of this, the polarity of the oxygen-hydrogen bond is enhanced to the point that, in aqueous solution, phenol releases hydrogen ions relatively easily (H^+).

A qualitative reaction for phenols is the interaction with ferric trichloride, which produces a complex compound with a blue-violet coloration:

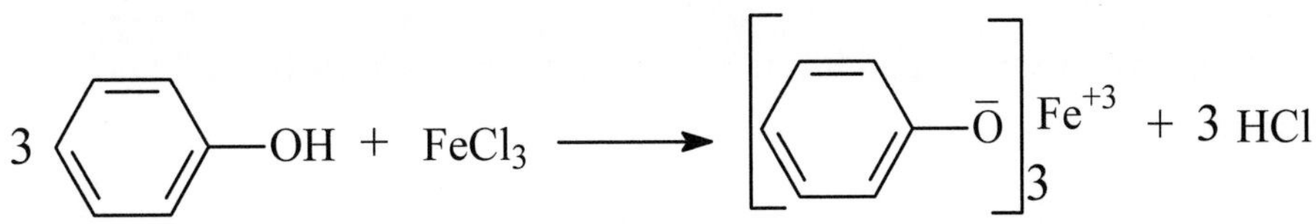

Phenols form esters, but not by direct interaction with acids, but mainly with their derivatives:

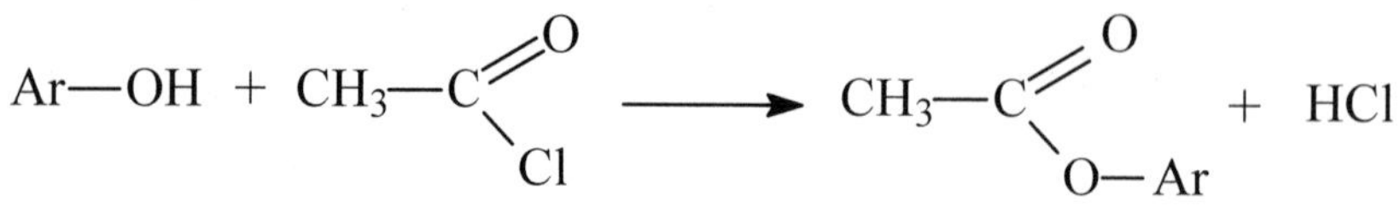

The presence of a phenyl residue in the phenol molecule determines its aromatic character (presence of a π-sextet).

While benzene reacts with bromine in the presence of a catalyst, for phenol this substitution reaction proceeds without a catalyst, i.e., more easily. When aqueous solutions of bromine and phenol are mixed, a white precipitate of 2,4,6-tribromophenol is released:

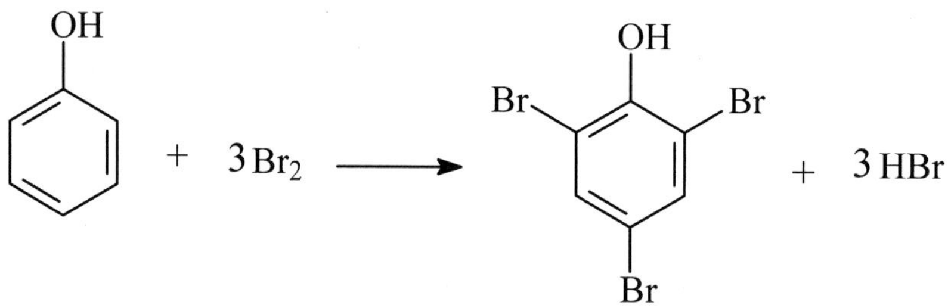

This indicates that the hydroxyl group, like the methyl group in the toluene molecule activates, and, to an even greater extent, the hydrogen atoms in the aromatic core and orients the substituents at the o- and p-sites. The hydroxyl group is bound directly to the aromatic nucleus exhibits -I (negative induction effect) and +M (positive mesomeric or conjugation effect).

Sulfonation and nitration of phenol proceed similarly:

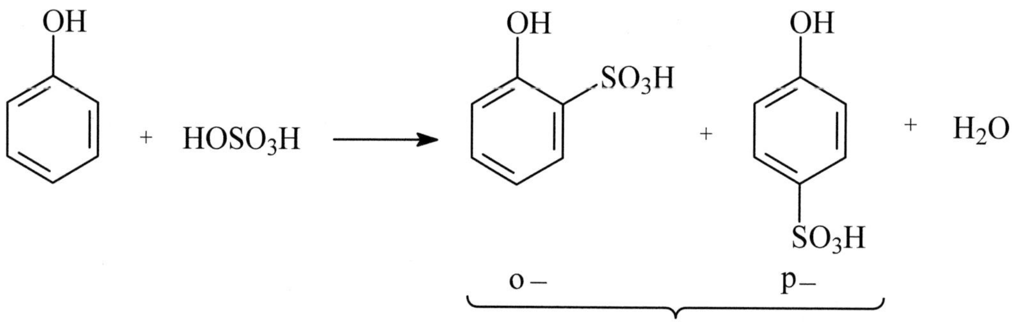

A mixture of 2-hydroxy- and 4-hydroxybenzenesulfonic acid is obtained.

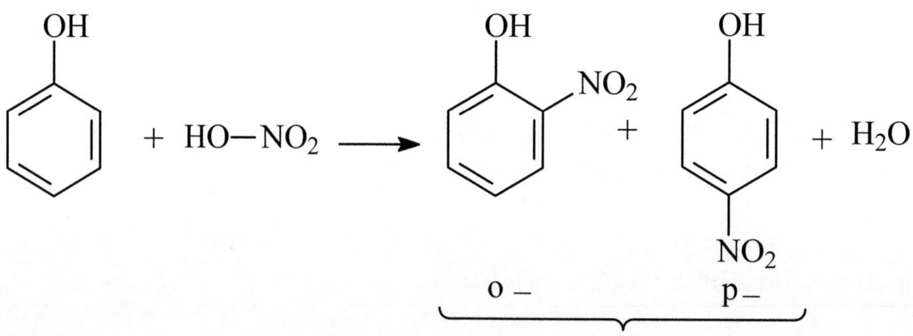

Depending on the conditions, in addition to 2-nitrophenol, 4-nitrophenol and

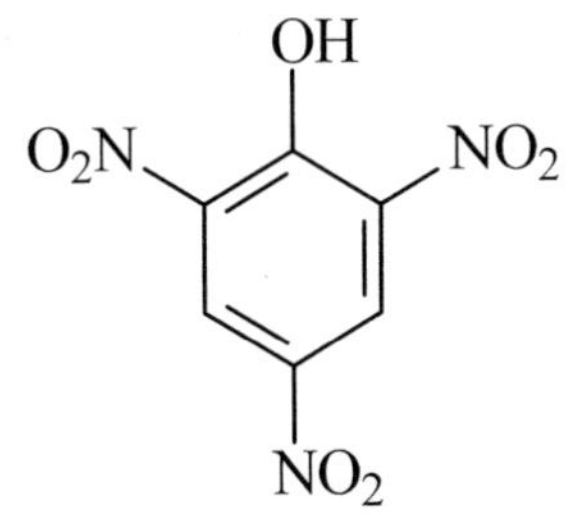

2,4,6-trinitrophenol (picric acid).

When heated in the presence of a catalyst, phenol reacts with formaldehyde to produce polyphenol formaldehyde (a synthetic polymer) with water evolution:

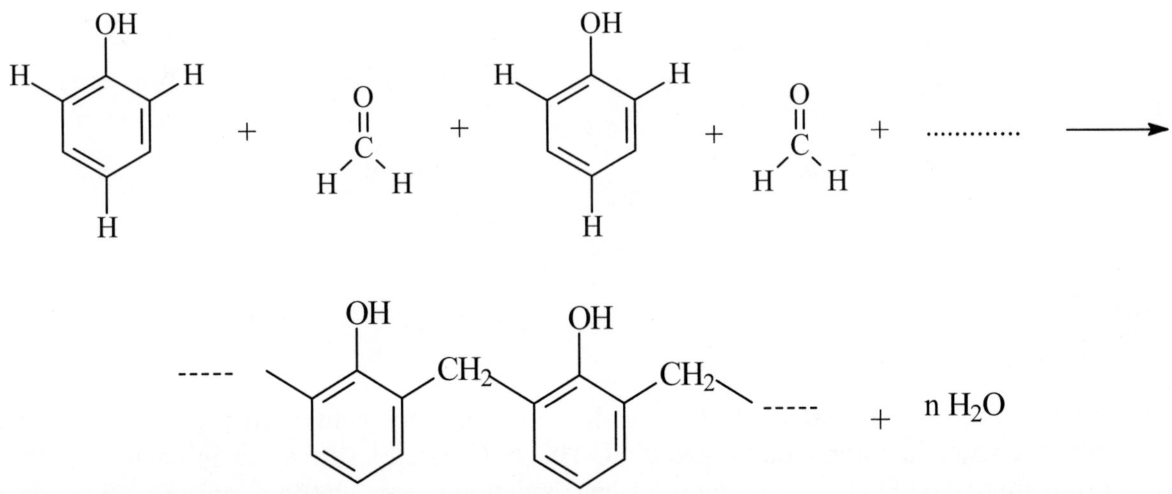

The process is called polycondensation.

Phenol is oxidized by oxygen in the air and yields products that are complex in composition.

Preparation of phenols

- Hydrolysis of halogen derivatives:

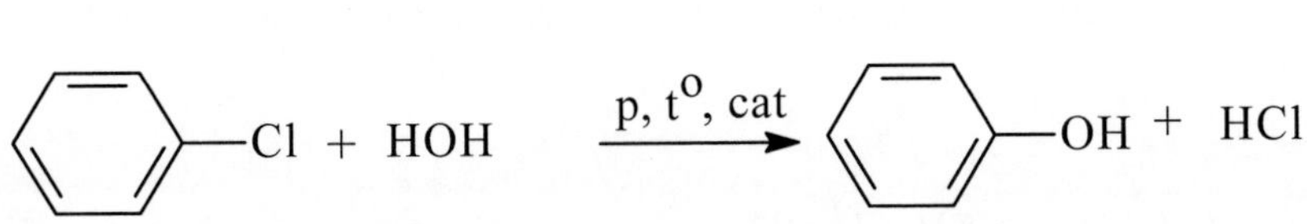

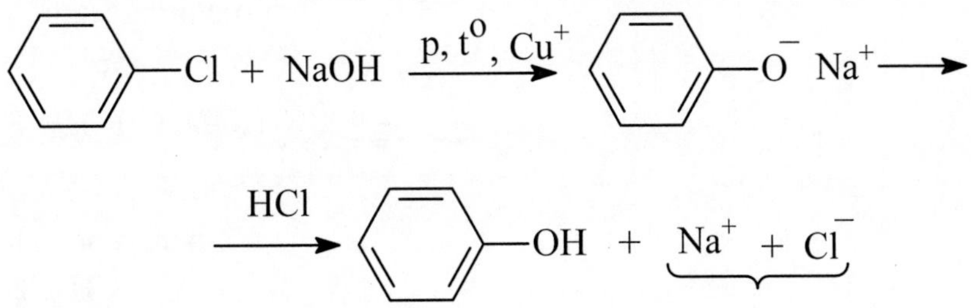

- **Alkaline fusion** of arylsulfonates:

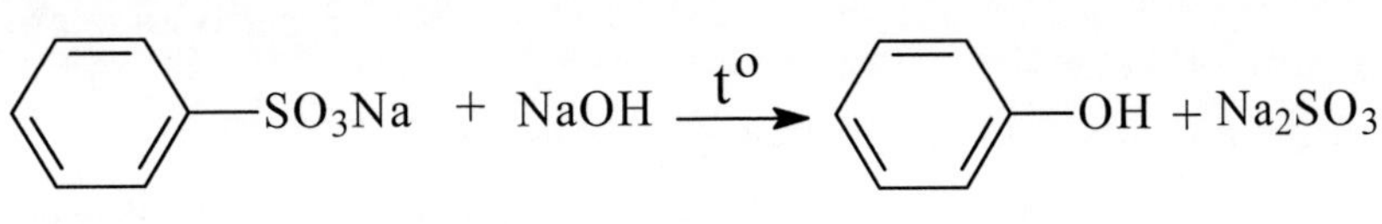

- **Kumol method**: in the first step, alkylation of benzene with propene is carried out to produce cumene, cumol (isopropylbenzene or 2-phenylpropane):

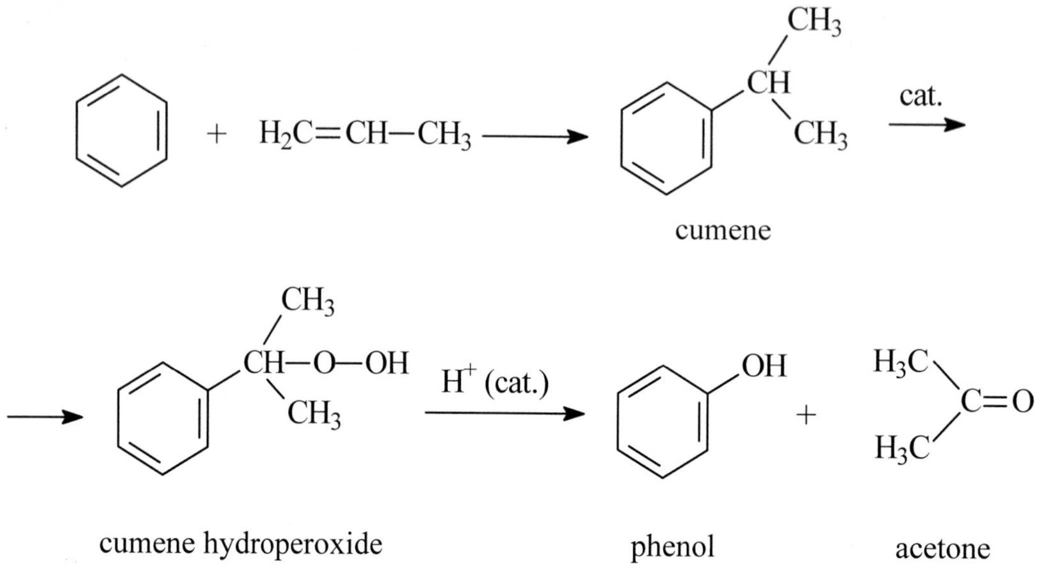

Sample problems and solutions

Task 1.

A substance of gross formula $C_5H_{12}O$ when heated with acetic acid in the presence of a sulfuric acid catalyst produces a compound of formula $C_7H_{14}O_2$. Oxidation of the starting material yields a product of composition $C_5H_{10}O$, which on further oxidation is converted to a mixture of acetic and propionic acids. Determine the nature of the starting substance and express the above conversions by chemical equations. Write the names of all the compounds.

Exemplary solution

The formula given in the condition corresponds to a saturated fatty alcohol - pentanol.

There are three possible isomers: 1-pentanol, 2-pentanol and 3-pentanol.

When alcohols react with carboxylic acids in the presence of a catalytic amount of sulfuric acid, a slow, molecular, reversible process called esterification takes place:

$$CH_3COOH + CH_3CH_2CH_2CH_2CH_2OH \overset{H^+}{\rightleftharpoons} CH_3COOCH_2CH_2CH_2CH_2CH_3 + H_2O$$

n-pentyl acetate

$$CH_3COOH + CH_3\underset{\displaystyle OH}{\underset{|}{C}}HCH_2CH_2CH_3 \overset{H^+}{\rightleftharpoons} CH_3COO\overset{\displaystyle CH_3}{\overset{|}{C}}HCH_2CH_2CH_3 + H_2O$$

2-pentyl acetate

$$CH_3COOH + CH_3CH_2\underset{\displaystyle OH}{\underset{|}{C}}HCH_2CH_3 \overset{H^+}{\rightleftharpoons} CH_3COO\overset{\displaystyle CH_2CH_3}{\overset{|}{C}}HCH_2CH_3 + H_2O$$

3-pentyl acetate

The isomeric esters correspond to a product of gross formula $C_7H_{14}O_2$.

Under moderate oxidation, alcohols are converted to carbonyl compounds - aldehydes and ketones, depending on the location of the hydroxyl group in the chain. Oxidation of primary alcohols produces aldehydes and oxidation of secondary alcohols produces ketones.

A crucial condition for the identification of the starting compound is the oxidation end products, acetic and propionic acids:

If the starting alcohol is 1-pentanol on moderate oxidation will give pentanal, and as a final product - pentanoic acid:

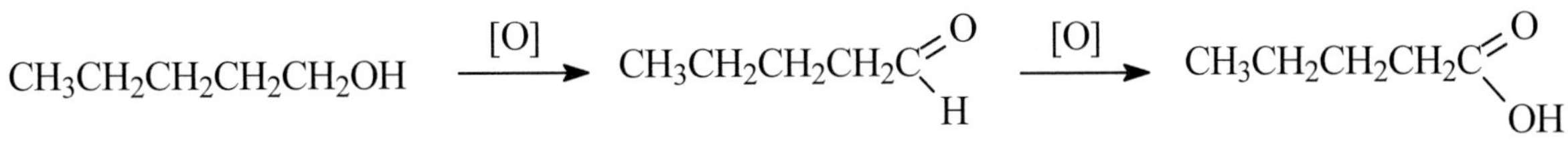

Oxidation of 2-pentanol will yield 2-pentanone:

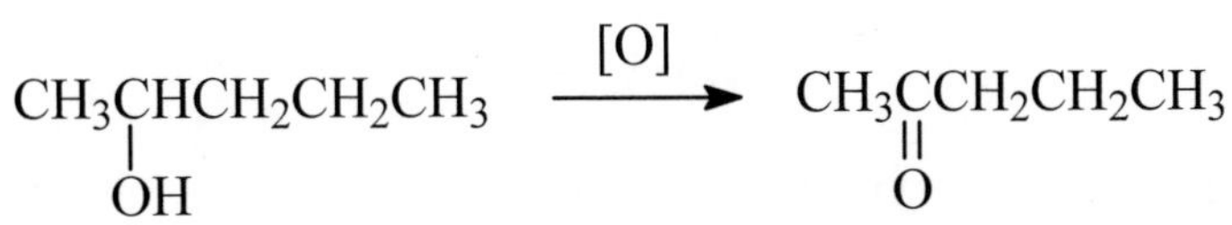

A vigorous oxidation of ketone will produce a mixture of formic, acetic, propionic and butyric acids:

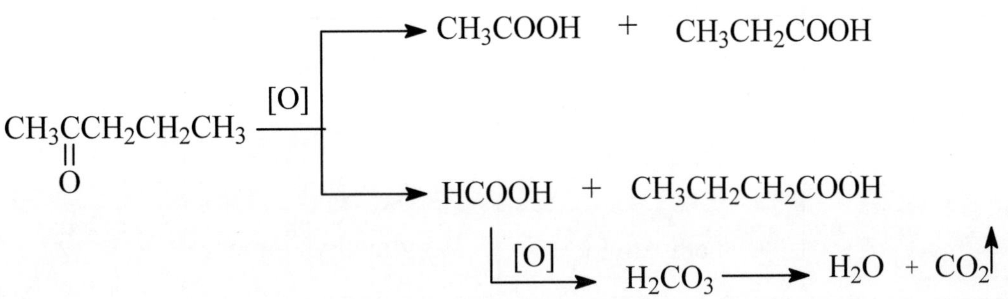

The mixture of carboxylic acids is produced because in the vigorous oxidation of ketones the carbon chain is broken adjacent to the carbonyl group, and this means the mixture of carboxylic acids indicated in equation. Under conditions of vigorous oxidation of ketones, only formic acid undergoes further oxidation (it has the properties of a reductant due to the presence of an aldehyde group in its structure) and is converted to the weak and perishable carbonic acid. It decomposes into carbon dioxide and water.

Oxidation of 3-pentanol produces 3-pentanone, and vigorous oxidation of the latter produces a mixture of two acids, acetic and propionic:

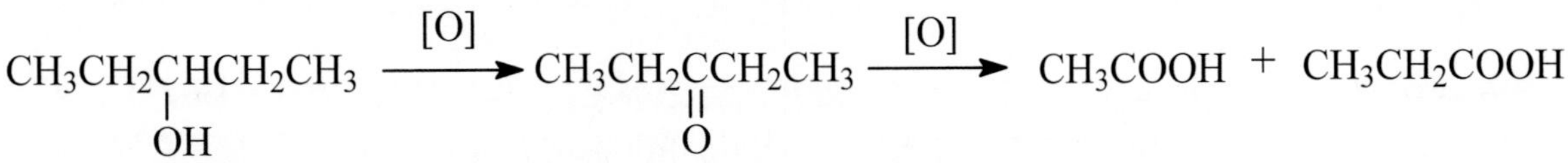

Therefore, the substance of composition $C_5H_{12}O$ is **3-pentanol**.

26. Nitrogen-containing Organic Compounds

Amines. Classification. Homologous order. Distribution. Structure and properties of amines. Basic character of fatty and aromatic amines. Application.

Amines are derivatives of hydrocarbons in which one or more hydrogen atoms in the molecule are substituted by an amino group or as derivatives of ammonia in which one, two or three hydrogen atoms in the molecule are substituted by hydrocarbon residues.

Classification

According to the type of hydrocarbon residues, amines are divided into fatty, aromatic and mixed; and according to the number of substituted hydrogen atoms in the ammonia molecule - into primary, secondary and tertiary.

	Primary	Secondary	Treatments
Fat amines	$CH_3—NH_2$ methylamine	CH_3 \ NH / CH_3CH_2 ethylmethylamine	CH_3 \ CH_3—N / CH_3 trimethylamine
Aromatic amines	$C_6H_5NH_2$ phenylamine (aniline)	C_6H_5 \ NH / C_6H_5 diphenylamine	C_6H_5 \ C_6H_5—N / C_6H_5 triphenylamine
Mixed	–	CH_3 \ NH / C_6H_5 methylphenylamine	CH_3 \ CH_3CH_2—N / C_6H_5 ethylmethylphenyl-amine

Quaternary ammonium salts and bases in which the nitrogen atom is saturated with hydrocarbon residues are also known:

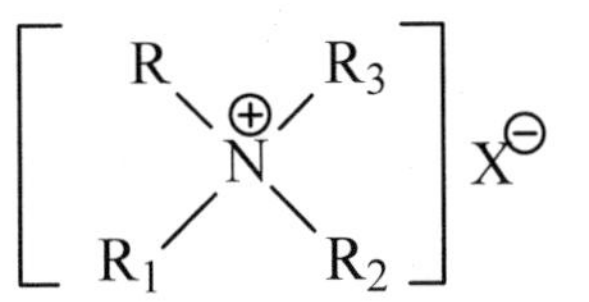

quaternary ammonium salt

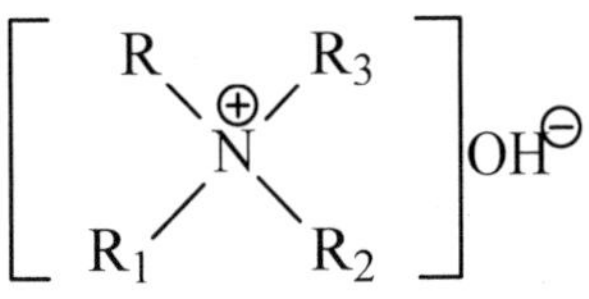

quaternary ammonium base

The names of amines are formed by placing the word amine after the names of the hydrocarbon residues. Under the old system, the names are obtained by alphabetizing the organic residues (hydrocarbon residues) related to nitrogen and adding the suffix -amine. The amino group is regarded as a substituent. Some amines also have trivial names.

$CH_3{-}NH_2$
methylamine
(aminomethane)

$C_6H_5NH_2$
phenylamine
(aminobenzene, aniline)

The monovalent residue of ammonia $-NH_2$ is called the amino group, and the divalent $>NH$ - imino group.

Each of the given types of amines form a homologous order. For example, primary amines, which are derivatives of saturated hydrocarbons, form a homologous series with the general formula:

$CH_3{-}NH_2$ - methylamine

$C_2H_5{-}NH_2$ - ethylamine

$C_3H_7{-}NH_2$ - propylamine

⋮

$C_nH_{2n+1}NH_2$ - alkylamine

According to the number of amino groups, amines are divided into monoamines, diamines, triamines:

CH_3NH_2	$H_2NCH_2CH_2NH_2$	$H_2NCH_2CH_2NHCH_2CH_2NH_2$
methylamine	ethylenediamine	diethylenetriamine

Distribution

Fatty amines occur in nature. Some of them are products of the vital activity of plants. Others are formed when protein substances break down. Because of their unpleasant odor and poisonous action, they make stale foods – meat, fish, etc. – unsafe to eat.

Aniline is contained in small amounts in coal tar.

Structure and properties of amines

The lower fatty amines are gases with an odor similar to ammonia, the middle ones are liquids with a fishy smell, and the higher ones are solids with no odor. The lower ones, like ammonia, are readily soluble in water because they form hydrogen bonds with water. As the molecular mass increases, solubility in water decreases and in organic solvents increases. Aromatic amines are insoluble in water.

Compared to hydrocarbons, lower primary and secondary alkylamines with the same molecular mass have higher boiling points and lower boiling points compared to alcohols. The reason for this is that the intermolecular hydrogen bonds in amines are weaker than for alcohols.

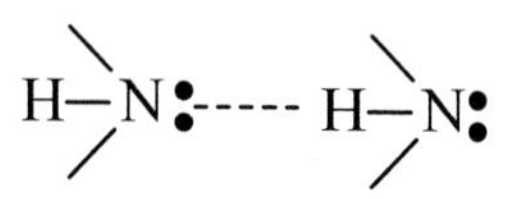

This is due to the lower electronegativity of the nitrogen atom (VA group) compared to the oxygen atom (VIA group). Intermolecular hydrogen bond formation is not possible with tertiary amines and their boiling points are low.

Amines contain the functional amino group that determines their properties:

The similarity of the chemical properties of ammonia and amines is explained by the similarity in their electronic structure. Like ammonia, amines contain an unshared electron pair at the nitrogen atom:

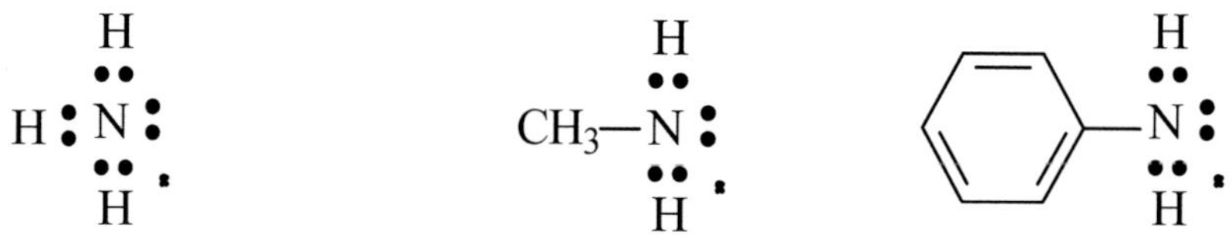

The unenriched electron pair determines the basic character of these compounds as it attracts the hydrogen cations from water and acids, forming the corresponding hydroxides and salts. Because of their basic nature, amines are also called organic bases.

In alkylamine molecules, the nitrogen atom is in a trivalent sp^3-hybrid state. These compounds have a pyramidal structure:

The valence angles C - N - H and H - N - H are ≈ 107°, i.e., smaller than the tetrahedral one, which is due to the repulsion between the uncharged electron pair of the nitrogen atom and the electron pairs of the N - C and N - H bonds.

It has been found that in aromatic amines there is an n,π-strain – between the free electron pair (n) at the nitrogen atom and the π-sequence of the aromatic nucleus. This implies a transition of the nitrogen atom into a sp^2 -hybrid state (similar to Csp^2), determining the planar arrangement of σ-bonds in the molecule.

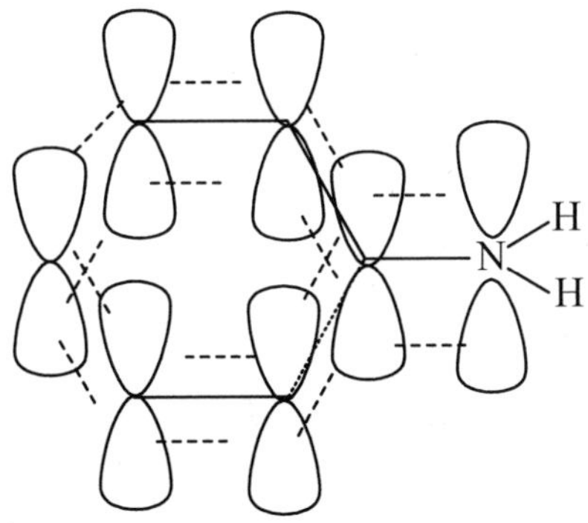

Chemical properties of amines

The interaction between ammonia and amines with water can be expressed as follows:

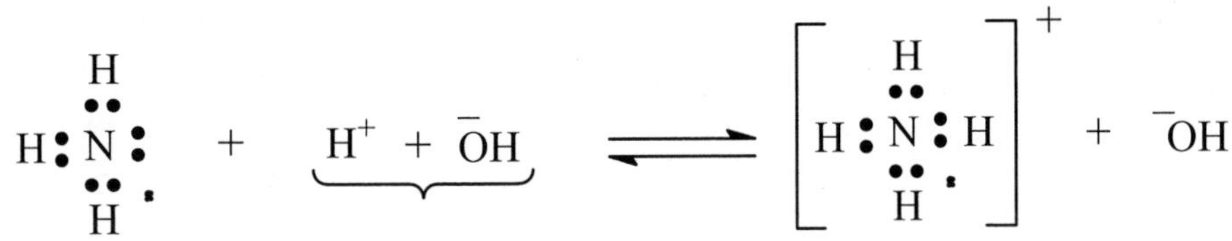

ammonium hydroxide

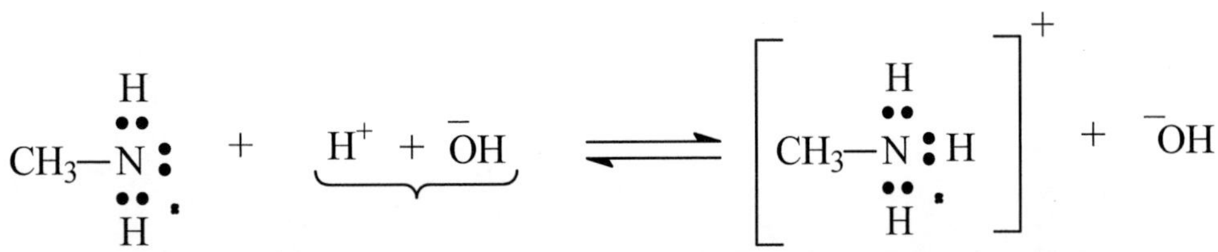

methylammonium hydroxide

The process is equilibrium. In the case of aromatic amines, the equilibrium is shifted to a greater extent towards the starting substances. This means that they are weaker bases than fatty amines and ammonia.

Fatty and aromatic amines differ in the type of hydrocarbon residues to which the amino group is bonded. The different influence exerted by the hydrocarbon residues on the amino group determines the difference in the strength of the basic character of these compounds.

In fatty amines, under the influence of the positive induction effect (+I) of the alkyl residue, the electron pair from the C - N bond is pulled towards the nitrogen atom. The electron density of the nitrogen atom increases and its ability to bond to hydrogen ions increases. The basic character of fatty amines is therefore more pronounced than that of ammonia. In some cases, the spatial effect of these substituents becomes the reason why tertiary amines have less reactivity than secondary ones.

In aromatic amines, the free electron pair at the nitrogen atom participates in conjugation with the π-sextet of the aromatic nucleus and is delocalized (+M - positive conjugation effect, mesomeric effect). Because of this, the electron density at the nitrogen atom decreases thereby decreasing its ability to bond hydrogen ions. The basic character of aniline is therefore less pronounced than that of ammonia.

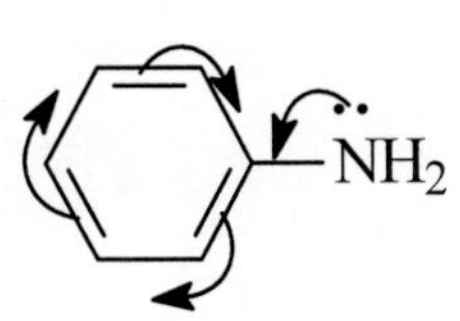

Therefore, ammonia, methylamine and phenylamine are arranged in the following order of decreasing basicity:

$$CH_3NH_2 > NH_3 > C_6H_5NH_2$$

The basicity of aromatic amines decreases in the order:

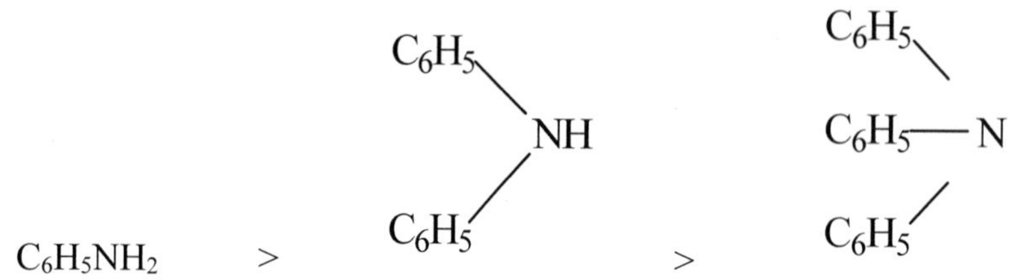

When interacting with acids, amines, like ammonia, form salts:

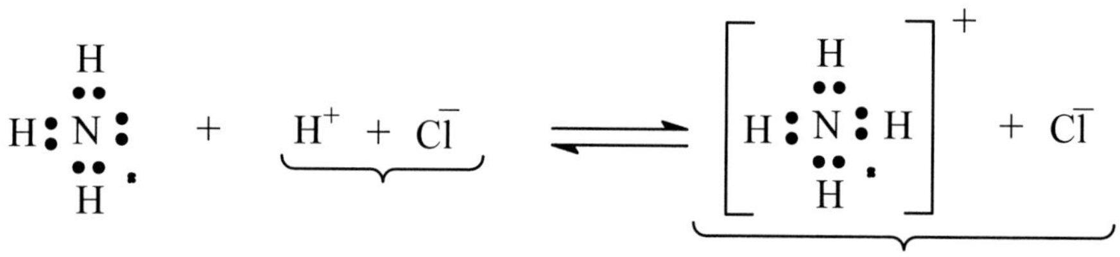

ammonium chloride

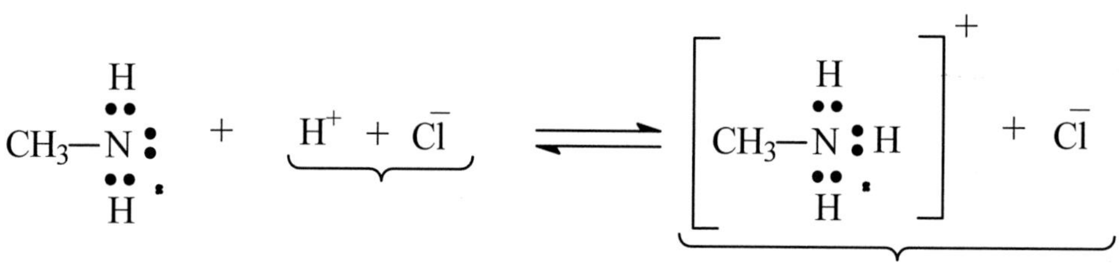

methylammonium chloride

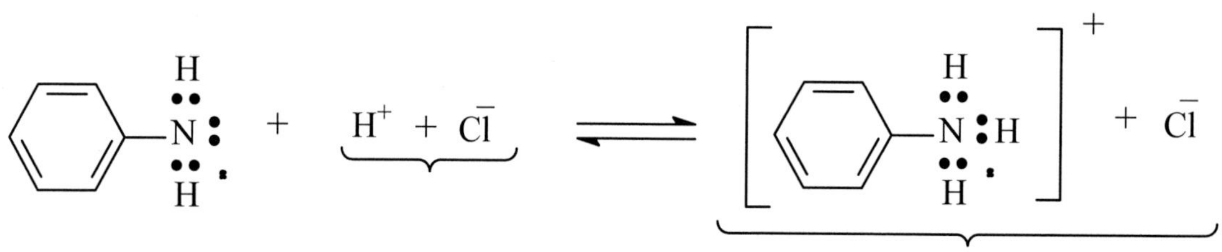

phenylammonium chloride

Oxidation of amines results in different products according to the type of amine:

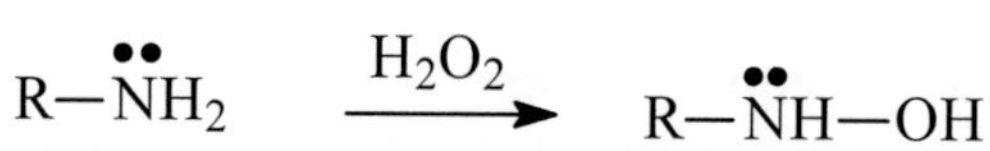

N-alkylhydroxylamine

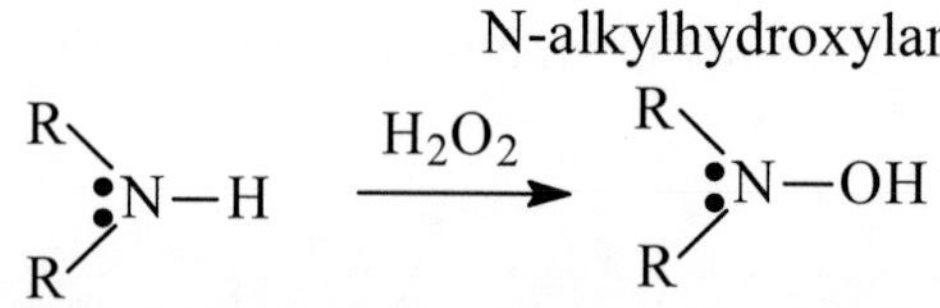

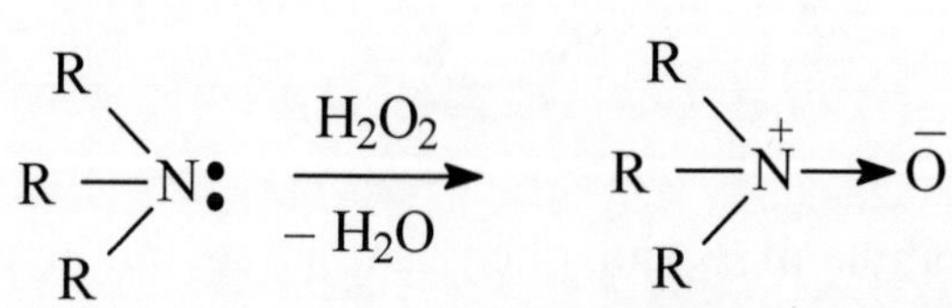

aminooxide

The oxidation of aromatic amines is complex, various products are formed:

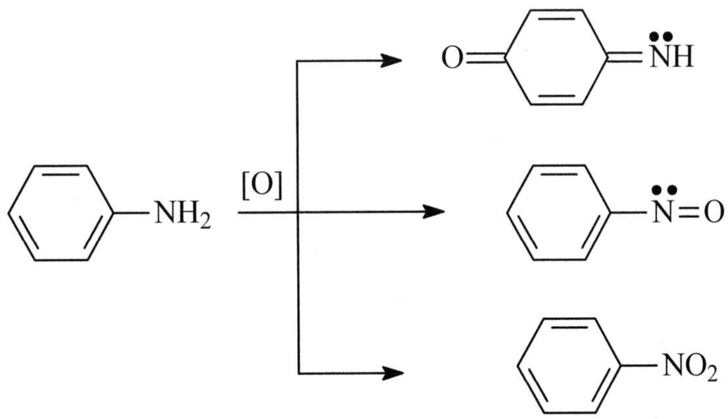

The oxidation of aniline to nitrobenzene also occurs under biological conditions under the catalytic action of enzymes.

Primary arylamines react with nitrous acid in acidic media to form diazonium salts:

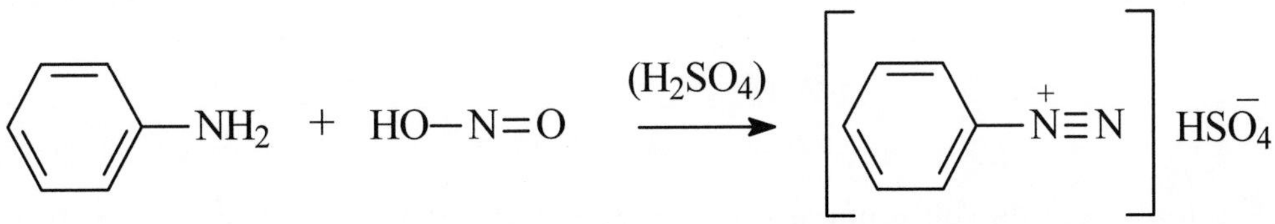

phenyldiazonium hydrogensulphate

Primary fatty amines form very perishable diazonium salts, which immediately decompose to alcohol (ROH) with nitrogen evolution.

The reactions involving the hydrocarbon residue in amines refers mainly to arylamines.

Compared to benzene, aniline is more easily halogenated, nitrated and sulfonated. This is due to the influence of the amino group, which facilitates substitution reactions and orients the substituents at the o- and p-site in the aromatic core (+M).

If bromine water is added to an aqueous solution of aniline, a white precipitate of 2,4,6-tribromoaniline is obtained:

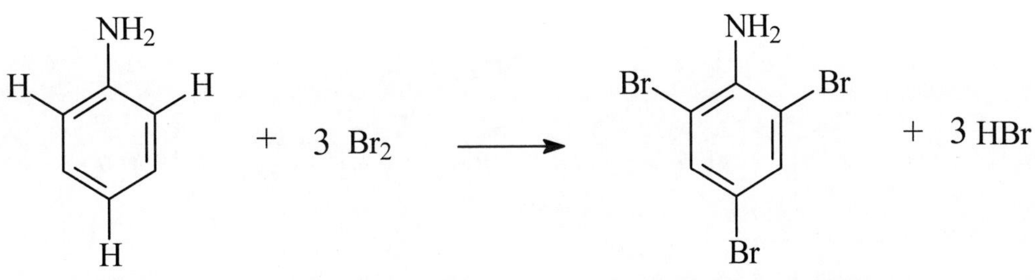

2,4,6-tribromoaniline

When aniline is reacted with an equimolecular amount of concentrated sulfuric acid, anilinesulfate is initially formed, which, on heating, is converted to p-aminobenzenesulfonic acid (sulfanilic acid):

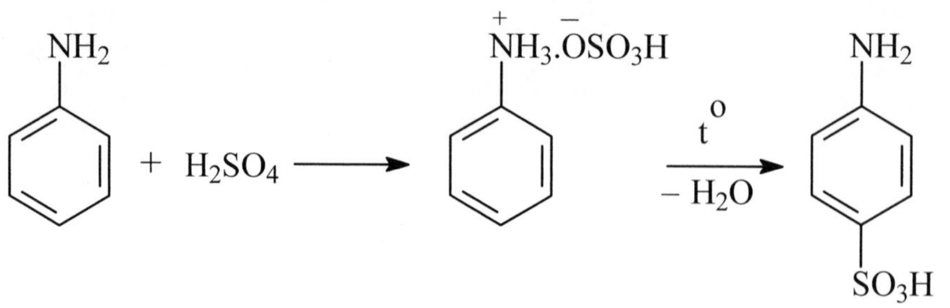

anilinesulfate (anilinium hydrogensulfate)

p-aminobenzenesulfonic acid (sulfanilic acid)

Sulfanilic acid has an internal salt structure (betaine structure):

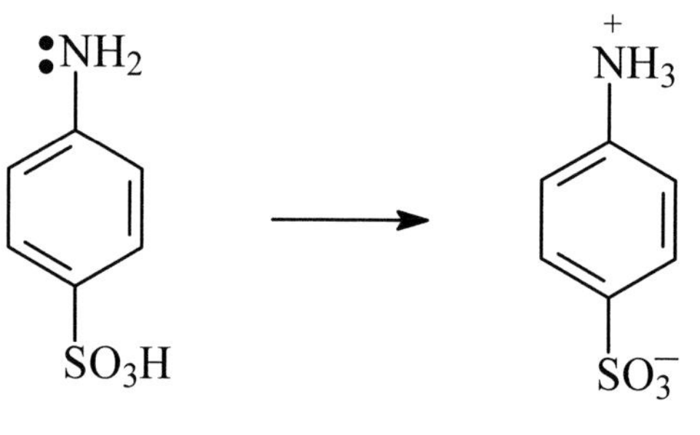

In substitution reactions, the amino group, like the hydroxyl group, activates the aromatic nucleus (+M-effect) and facilitates electrophilic substitution reactions, which proceed more readily compared to benzene.

Aniline is involved in polycondensation processes:

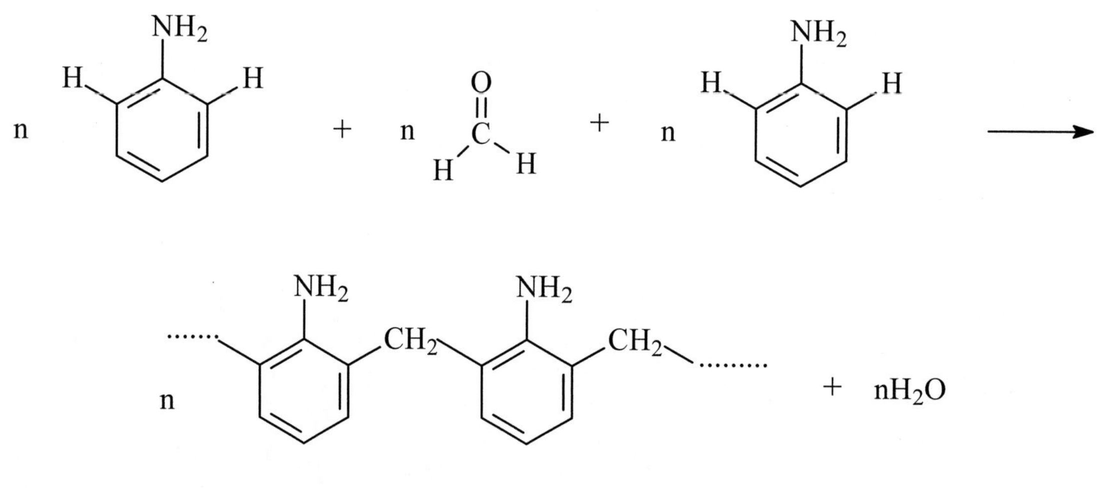

polyaniline formaldehyde

A qualitative reaction for aniline is its oxidation by chlor lime. The resulting product is red-violet in color.

Methods for the preparation of amines

1. Amines are produced when halogen derivatives of hydrocarbons react with ammonia:

$$CH_3I + 2NH_3 \xrightarrow{cat} CH_3NH_2 + NH_4I$$

$$CH_3NH_2 + CH_3I + NH_3 \longrightarrow (CH_3)_2NH + NH_4I$$

$$(CH_3)_2NH + CH_3I + NH_3 \longrightarrow (CH_3)_3N + NH_4I$$

$$(CH_3)_3N + CH_3I \longrightarrow$$

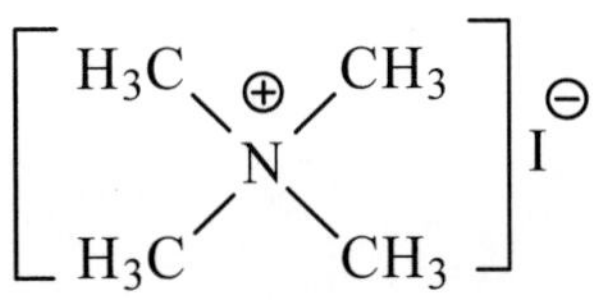

tetramethylammonium iodide

2. Amines are also produced by the reaction of alcohols with ammonia on heating in the presence of a catalyst:

$$R{-}OH + NH_3 \xrightarrow[300^{o}C]{Al_2O_3} R{-}NH_2 + H_2O$$

3. Aniline was obtained by Zinin in 1842 by reduction of nitrobenzene:

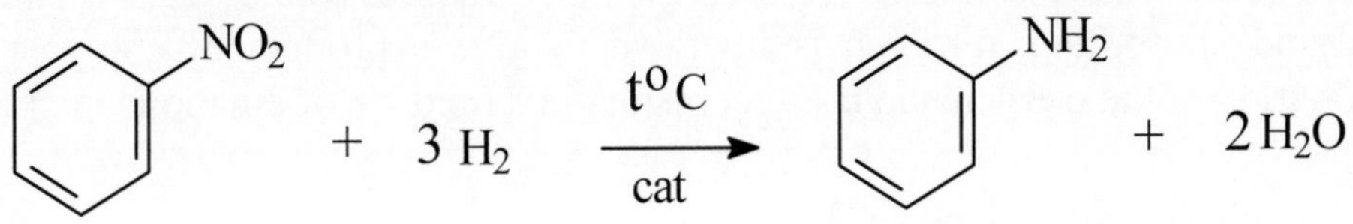

Due to its high reactivity, aniline is used in organic synthesis.

Isomerism in amines

Isomerism is possible with amines:

- Positional:

$CH_3CH_2CH_2NH_2$	$CH_3CH(NH_2)CH_3$
1-aminopropane (1-propanamine)	2-aminopropane (isopropylamine) (2-propanamine)

- Chain isomerism - depends on the bonding sequence of the carbon atoms in the chain:

$CH_3CH_2CH_2CH_2CH_2NH_2$	$CH_3CH(CH_3)CH_2CH_2NH_2$	$CH_3CH_2CH(CH_3)CH_2NH_2$
1-aminopentane (1-pentanamine)	1-amino-3-methylbutane (3-methyl-1-butanamine)	1-amino-2-methylbutane (2-methyl-1-butanamine)

- For tertiary amines with three different substituents, enantiomerism can be expected due to the chirality of the molecule (the electron pair is considered as the fourth substituent):

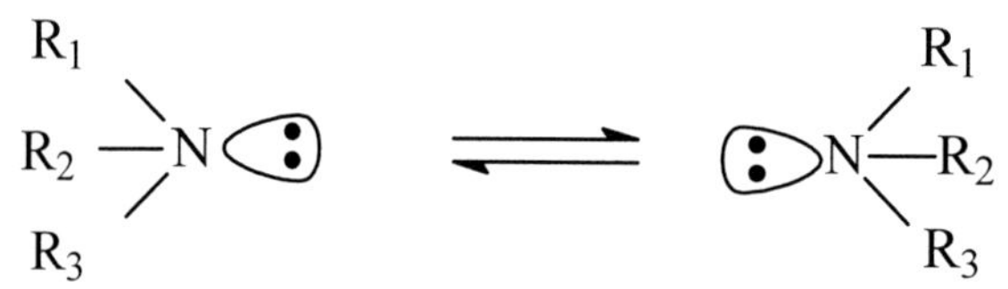

In quaternary ammonium salts with four different substituents, the tetrahedral molecules are also asymmetric and have enantiomers that have been found experimentally in a number of cases.

Aniline is highly poisonous. Its action occurs both by inhalation of its vapors and through the skin. Aniline reduces the oxygen-carrying capacity of the blood and acts on the nervous system.

Sample problems and solutions

Task 1.

A substance of the gross formula $C_5H_{13}N$ (**A**) with hydrochloric acid forms a salt. When reacted with nitrous acid on cooling, nitrogen is liberated to give a product of composition $C_5H_{12}O$. Energetic oxidation of the oxygenated product results in a mixture of ethanoic and propanoic acids.

1. Determine the structure of substance **A**;
2. Represent with a chemical equation the interaction of **A** with hydrochloric acid;
3. Express the preparation of the oxygenated product involving nitrous acid;
4. Comment on the conditions and possibilities for obtaining a mixture of ethanoic and propanoic acids by vigorous oxidation of the substance of gross formula $C_5H_{12}O$.

__Exemplary solution__

- The substance of gross formula $C_5H_{13}N$ is a fatty amine. The amino group is bonded to the third carbon atom. The compound is 3-aminopentane.

$$CH_3CH_2CH(NH_2)CH_2CH_3$$

- Fatty amines interact with mineral acids. This is due to the presence of an uncharged electron pair at the nitrogen atom of the amino group. The proton from the acid binds to the uncharged electron pair and an ammonium salt is formed.

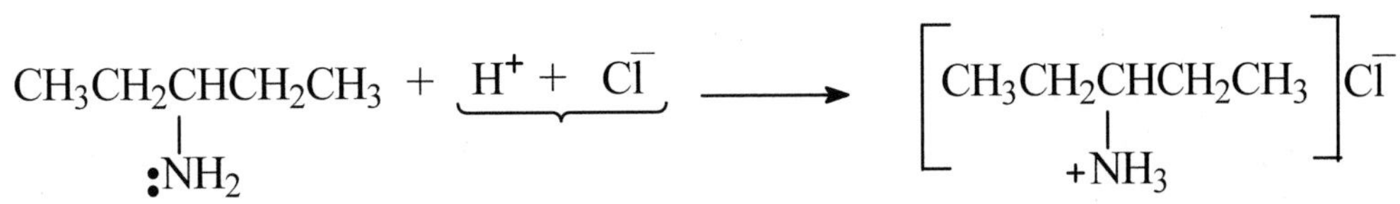

3-pentylammonium chloride.

- Low temperature treatment of fatty amines with nitric acid (about 0° C) leads to the formation of perishable diazonium salts, which decompose with the release of nitrogen and oxygen-containing products:

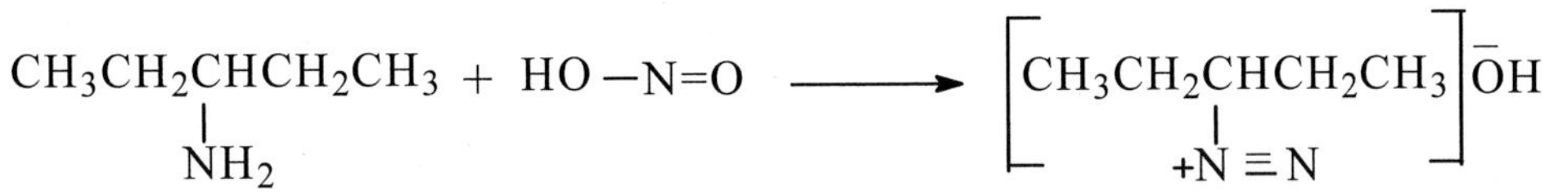

3-pentyldiazonium hydroxide

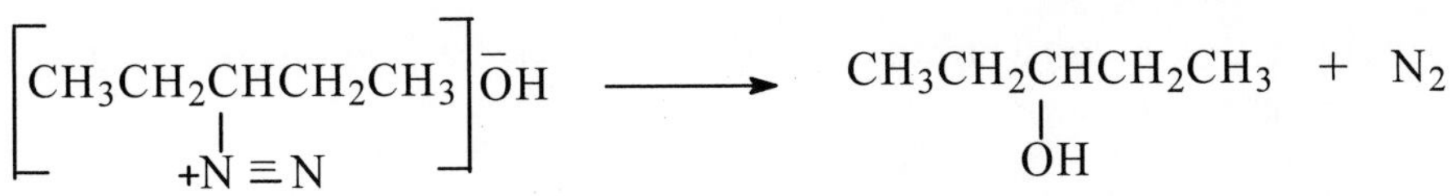

3-pentanol

- Oxidation or dehydrogenation of secondary alcohols leads to the formation of ketones. The product of the moderate oxidation of 3-pentanol is 3-pentanone:

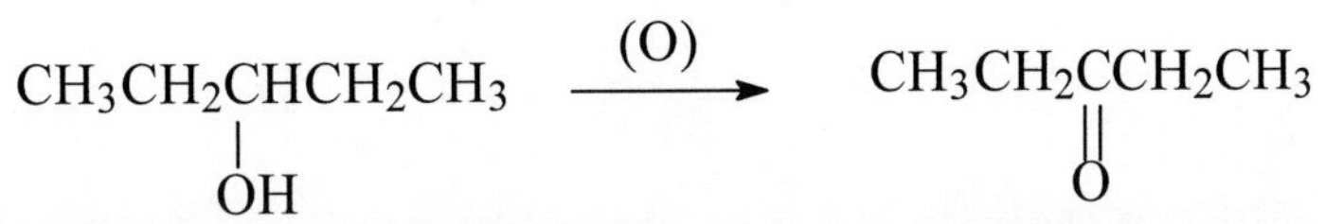

Ketones, unlike aldehydes, are oxidized only by strong oxidants. The carbon chain breaks adjacent to the carbonyl group and a mixture of carboxylic acids is produced.

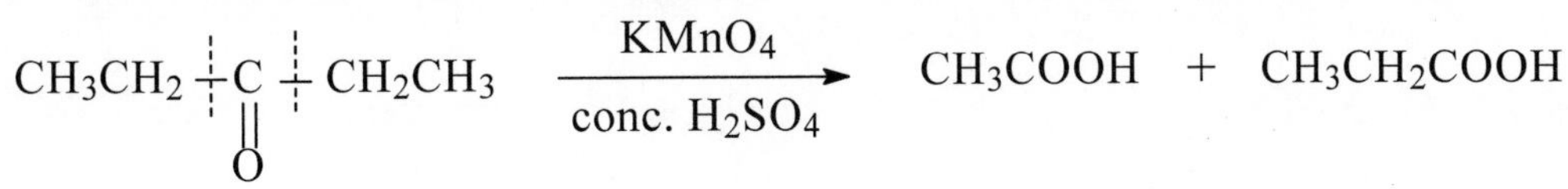

27. Carbonyl Derivatives of Hydrocarbons

Definition, Types. Names. Isomerism. Physical Properties. Molecular Structure. Comparative Consideration of the Chemical Properties of Aldehydes and Ketones in Relation to their Structure. Application.

Carbonyl compounds are derivatives of hydrocarbons whose molecules contain the functional group called carbonyl.

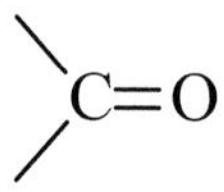

Depending on the type of hydrocarbon substituents in the carbonyl group, carbonyl compounds are divided into two main groups:

aldehydes and ketones

R = alkyl-, alkenyl-, etc. or aryl-.

In **aldehyde** molecules, one substituent at the carbonyl group is a hydrogen atom and the other is a hydrogen atom or hydrocarbon residue. Their functional group is called an aldehyde.

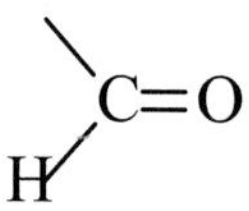

In **ketone** molecules, both substituents at the carbonyl group are hydrocarbon residues.

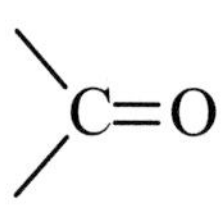

Names

For aldehydes and ketones, names according to the IUPAC nomenclature are used, as well as some older rational names or trivial names.

IUPAC names for aldehydes are formed from the name of the corresponding hydrocarbon with the same number of carbon atoms and the suffix **-al**, and for ketones from the name of the hydrocarbon with the same number of carbon atoms and the suffix **-on.**

A carbonyl group is denoted by the prefix **-oxo** when it is defined as a substituent in more structurally complex organic compounds.

The classification of aldehydes and ketones is built according to the type of hydrocarbon residues in their molecules and has the same appearance as that of hydrocarbons. It can also be extended according to the number of functional groups in these compounds.

Each individual species of aldehydes and ketones may have its own homologous order.

Construction

The structure of carbonyl compounds is mainly determined by their functional group structure. In the carbonyl group, the carbon atom is in the sp^2-hybrid state – there are three sp^2-orbital and one unhybridized 2p-orbital. The valence bonding of the carbon and oxygen atoms is via σ- and π-bonds. At the oxygen atom there are two free electron pairs. With the other two sp^2 -orbitals, the carbon atom forms σ-bonds with the substituents at the carbonyl group - R-, Ar- or H-.

In this orbital model of the carbonyl group, the σ-skeleton lies in one plane, the angle between the σ- linkages is ~120°, and the π-linkage plane is perpendicular to the σ-skeleton plane of the functional group (similar to alkenes).

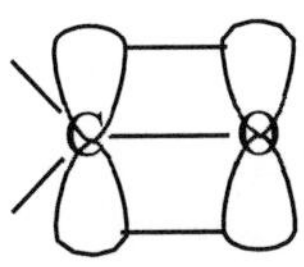

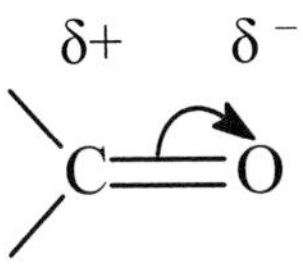

Due to the different electronegativity of the carbon and oxygen atoms, the bond in the carbonyl group is polar. The major contribution to its polarity is made by the more mobile π- bond.

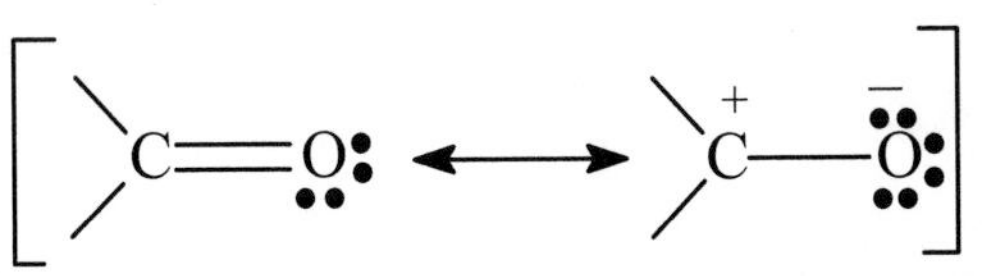

or

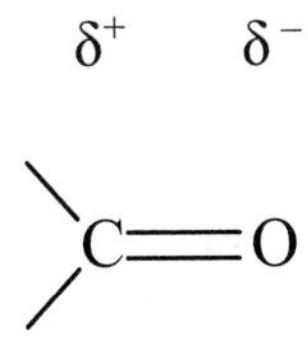

Physical properties

The lower aldehydes and ketones are liquids (a HCHO is a gas) and the higher ones are solids. The carbonyl group makes these compounds polar. Because of this, aldehydes and ketones have a higher b.p. (by ~50 - 80° C) compared to corresponding less polar organic compounds with the same molecular composition and the same molecular mass e.g., ethers.

Lower aldehydes and ketones have some solubility in water due to hydrogen bond formation with its molecules. For higher aldehydes and ketones, the solubility in water decreases. Carbonyl compounds are soluble in organic solvents of medium polarity.

Reaction ability

The chemical properties of aldehydes and ketones are determined by the reactivity of the carbonyl group, its associated substituents, and the combined action of these structural parts of the molecule.

Unlike the double bond in alkenes, due to the greater electronegativity of the oxygen atom, the bond in the carbonyl group is highly polar.

The reactivity of carbonyl compounds depends on the magnitude of the partial positive charge at the carbonyl carbon atom. This, in turn, depends on whether the substituents are electron-donating or electron-acceptor.

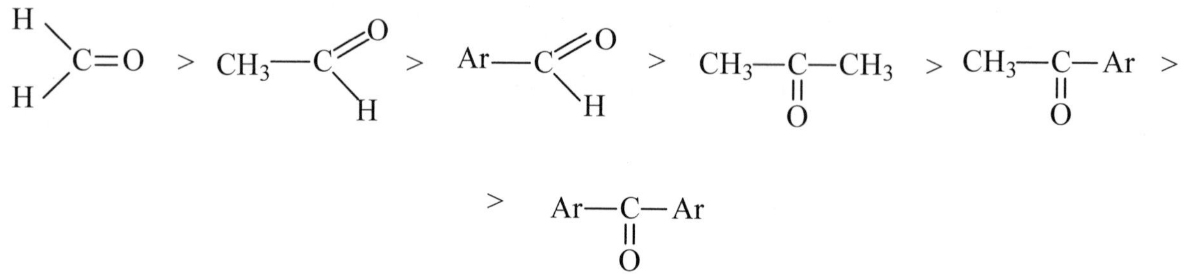

The activity of the carbonyl group is also influenced by the spatial effect of the substituents - bulky substituents make it more difficult for the nucleophile to attack:

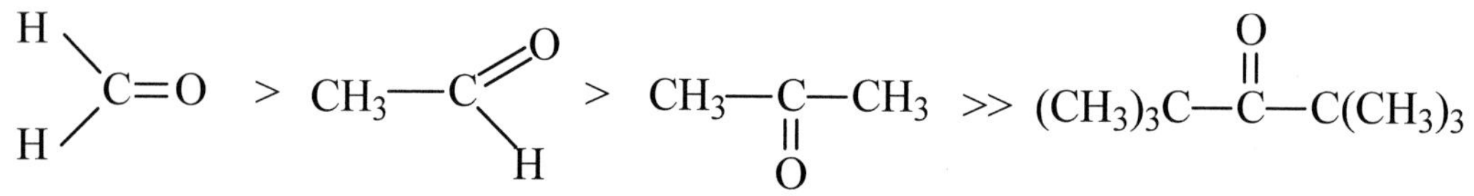

Aromatic aldehydes are characterized by electrophilic substitution reactions in the aromatic nucleus: halogenation, nitration, sulfonation. The aldehyde group is a deactivating m- orientant.

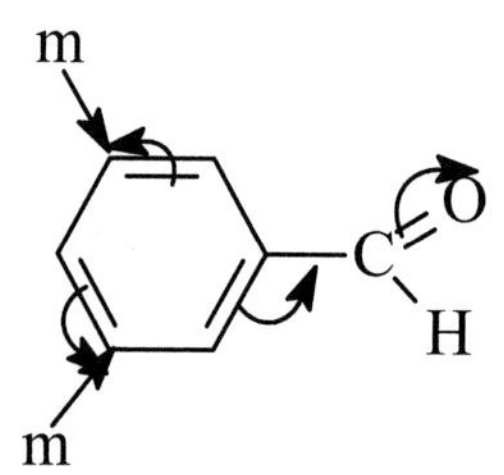

Due to electronic interaction, the carbon atoms of the aromatic nucleus have a reduced electron density. The lowest decrease is in m- carbon atoms relative to the functional group. Newly occurring substituents bind at the m-site relative to the carbonyl group:

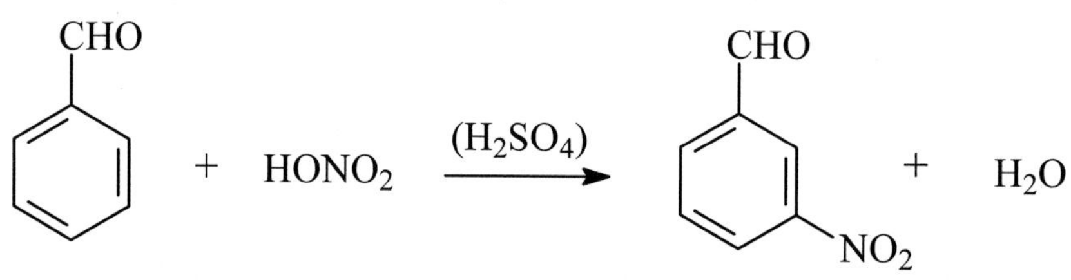

m-nitrobenzaldehyde

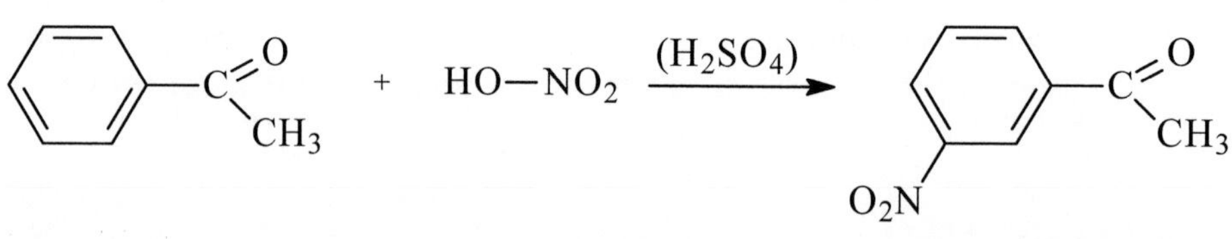

m-nitroacetophenone

Formaldehyde is contained in smoke; it has a preservative effect on organs and tissues in anatomical practice; it can be found in the leather industry; and in decontamination of greenhouse areas in vegetable production. Large quantities of formaldehyde are also used to make plastics.

Classification of aldehydes

The classification of aldehydes is built according to the type of hydrocarbon residues in their molecules and has the same appearance as that of hydrocarbons. It can also be extended according to the number of functional groups in these compounds.

Monoaldehydes

(with one aldehyde group)

Alkanali

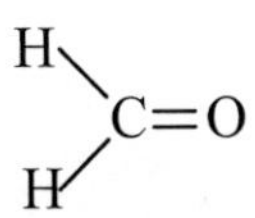

methanal
(formic aldehyde, formaldehyde)

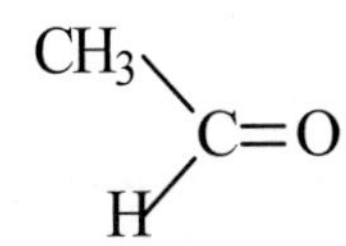

ethanal
(acetaldehyde)

Alkenals

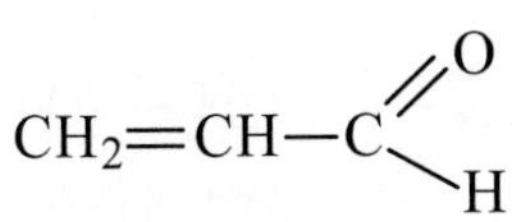

2-propenal
(acrolein)

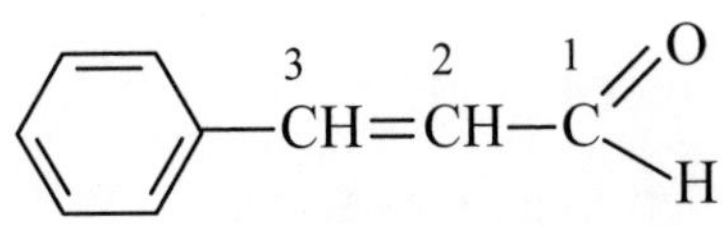

3-phenyl-2-propenal
(cinnamaldehyde)

Alkinali

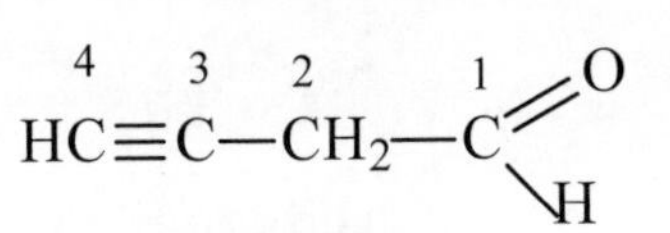

3-butinal

Aromatic aldehydes

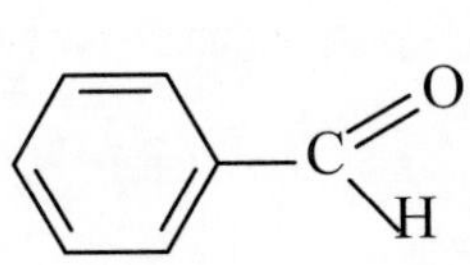

benzaldehyde (benzenecarbaldehyde)

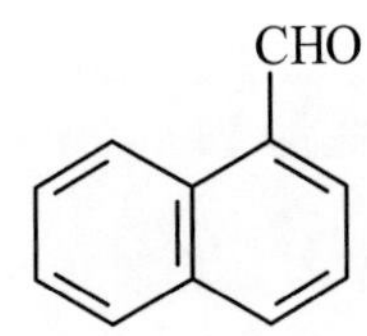

1-naphthalenecarbaldehyde

Dialdehydes

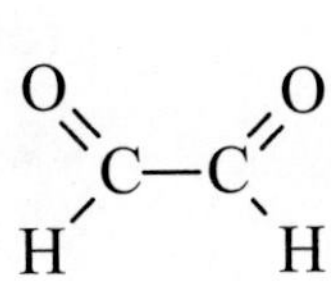

etanedial (glyoxal)

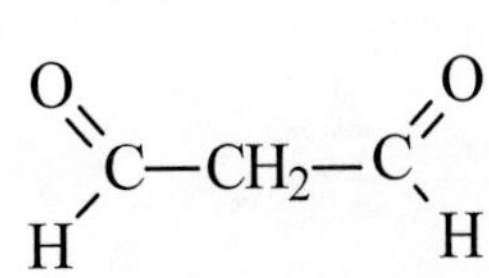

propanedial

Carbonyl compounds include two groups of organic derivatives also containing a carbonyl functional group – ketenes.

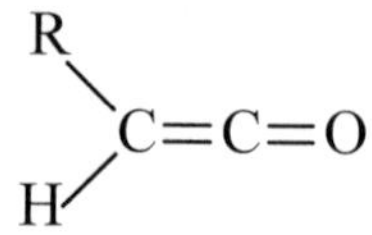

ketene

Classification of ketones

The classification of ketones is built according to the type of hydrocarbon residues in their molecules and has the same appearance as that of hydrocarbons. It can also be extended according to the number of functional groups in these compounds.

Monoketons
(with one keto group)

Alkanones

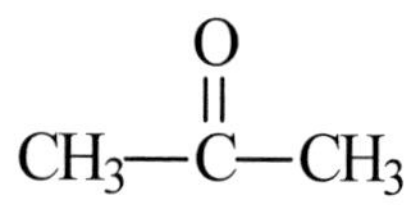

propanone (acetone)

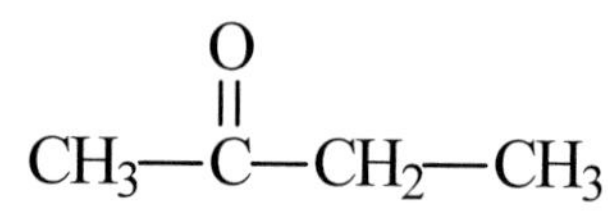

butanone (ethylmethylketone)

Alkenones

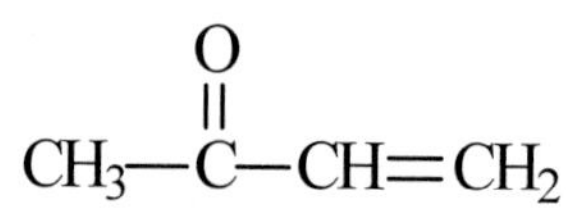

butenone

Alkinones

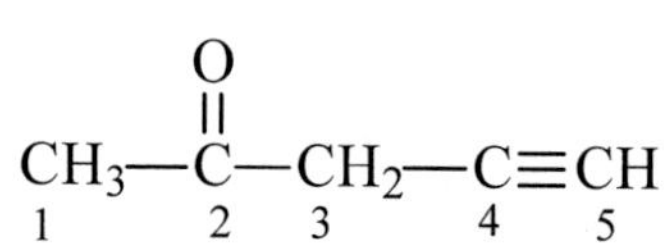

4-pentyn-2-one

Aromatic ketones

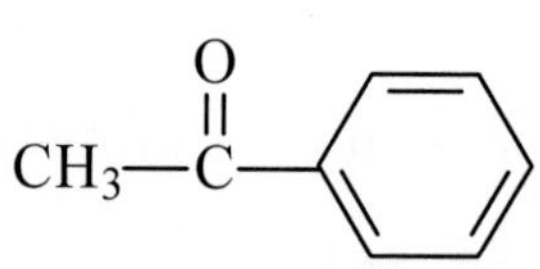

methylphenylketone
(acetophenone)

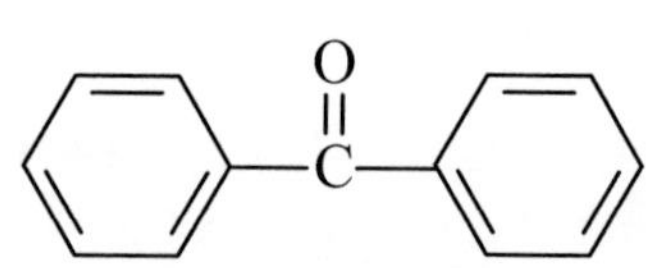

benzophenone
(diphenylketone)

Dicketons

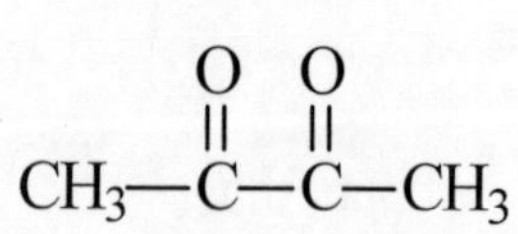

butanedione (diacetyl)

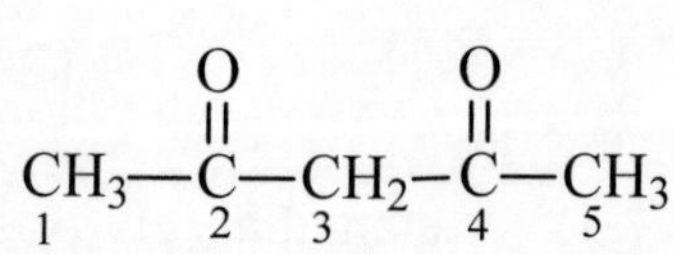

2,4-pentanedione (acetylacetone)

Quinones

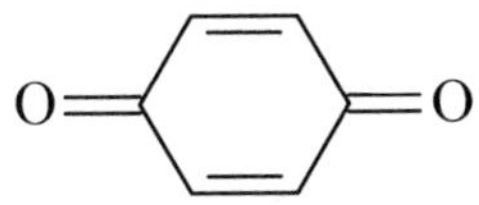

p-quinone

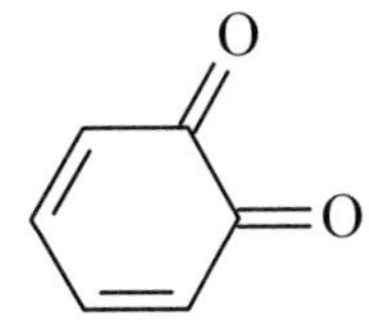

o-quinone

Isomerism in aldehydes

In aldehydes, there is a chain isomerism defined by the different bonding sequence of the carbon atoms in the chain.

$CH_3CH_2CH_2CHO$

butanal

$\overset{3}{C}H_3\underset{\underset{CH_3}{|}}{\overset{2}{C}}H\overset{1}{C}HO$

2-methylpropanal

In aromatic aldehydes, positional isomerism exists:

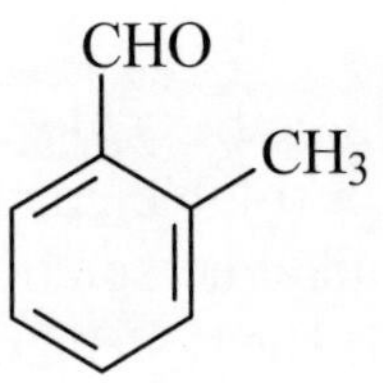

o-

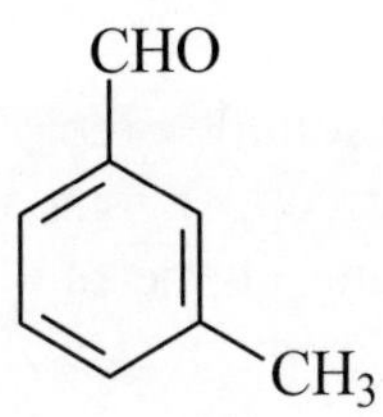

m-

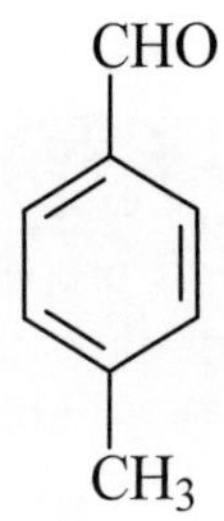

p-

Isomerism in ketones

Positional isomerism (according to the position of the carbonyl group in the molecule) is possible for mono- and diketones.

$CH_3CH_2CH_2-\underset{\underset{O}{\|}}{C}-CH_3$

2-pentanone

$CH_3CH_2-\underset{\underset{O}{\|}}{C}-CH_2CH_3$

3-pentanone

$\underset{1}{CH_3}-\underset{2}{\overset{\overset{O}{\|}}{C}}-\underset{3}{\overset{\overset{O}{\|}}{C}}-\underset{4}{CH_2}-\underset{5}{CH_3}$

2,3-pentanedione

$\underset{1}{CH_3}-\underset{2}{\overset{\overset{O}{\|}}{C}}-\underset{3}{CH_2}-\underset{4}{\overset{\overset{O}{\|}}{C}}-\underset{5}{CH_3}$

2,4-pentanedione

Chemical properties of aldehydes

The chemical properties of aldehydes are determined by the reactivity of the carbonyl group, its associated substituents, and the combined action of these structural parts of the molecule. The polarity of the double bond in the carbonyl group causes further polarization of the carbon-hydrogen bond in the aldehyde group, making this bond more labile.

1. Redox processes

The increased reactivity of the hydrogen atom of the aldehyde group determines the reducing action of aldehydes. Specific oxidizing agents for the aldehyde group (including monosaccharides) are silver ammonia hydroxide solution or freshly precipitated copper dihydroxide and Fehling's solution. These are used for proof of the aldehyde group.

If aldehyde is added to a silver ammonia hydroxide solution and the mixture is heated, a shiny silver is given off on the walls of the vessel (silver mirror reaction):

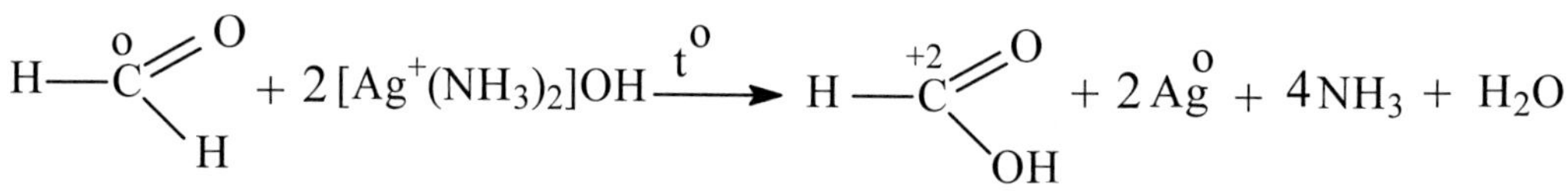

$$\begin{array}{lcl|c|c|c} C^{0} - 2\bar{e} & \longrightarrow & C^{+2} & 2 & & 1 \\ & & & & 2 & \\ Ag^{+} + e^{-} & \longrightarrow & Ag^{0} & 1 & & 2 \end{array}$$

Copper dihydroxide is a sparingly soluble precipitate. For a smoother redox process, a Fehling solution is used whereby the divalent copper ions are bound in a soluble complex compound. A Fehling solution is prepared by mixing copper sulfate and an alkaline solution of a segnet salt (potassium sodium tartrate, potassium sodium salt of tartaric acid. The systematic name of the acid is 2,3-dihydroxy-1,4-butanedicarboxylic acid). The result is a water-soluble complex of divalent copper ions with the two hydroxyl groups of tartaric acid. The oxidation-reduction process proceeds by heating a Fehling solution with aldehyde, which produces a tile-red precipitate of dicopper oxide:

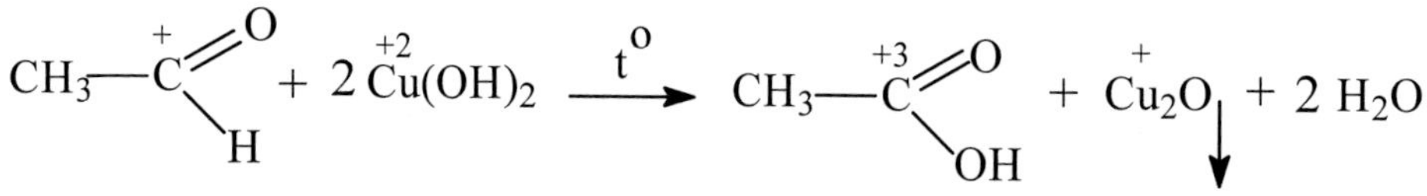

$$\begin{array}{lcl|c|c|c} C^{+} - 2\bar{e} & \longrightarrow & C^{+3} & 2 & & 1 \\ & & & & 2 & \\ Cu^{+2} + e^{-} & \longrightarrow & Cu^{+} & 1 & & 2 \end{array}$$

2. Adjunctive reactions

The double bond in the carbonyl group (aldehyde or keto group) determines the ability of the compounds to enter into addition reactions. The hydrogenation of aldehydes yields primary alcohols and that of ketones yields secondary alcohols:

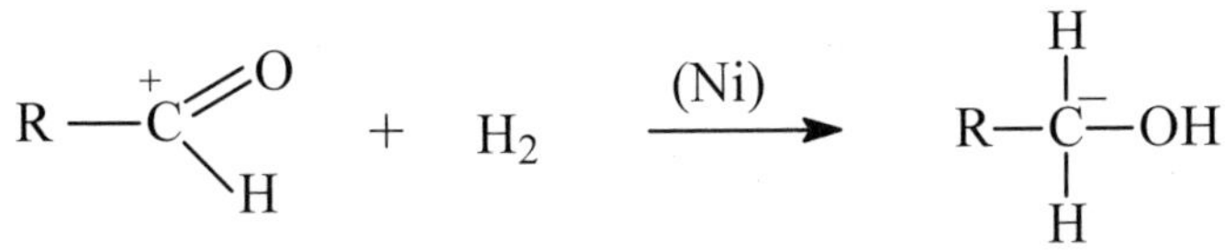

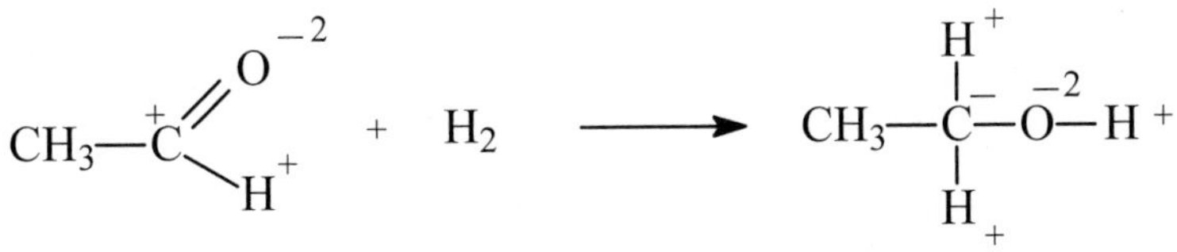

$$C^{+} + 2e^{-} \longrightarrow C^{-} \quad \Big| \; 2 \; \Big| \; 2 \; \Big| \; 1$$

$$H_2 \longrightarrow 2H^{0} - 2 \times 1\,e^{-} \longrightarrow 2H^{+} \quad \Big| \; 2 \; \Big| \; \Big| \; 1$$

Aldehydes interact with ammonia and amines, hydrogen cyanide, water:

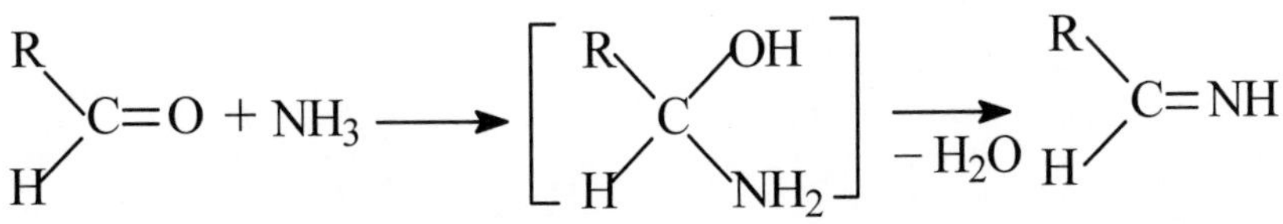

aldimine

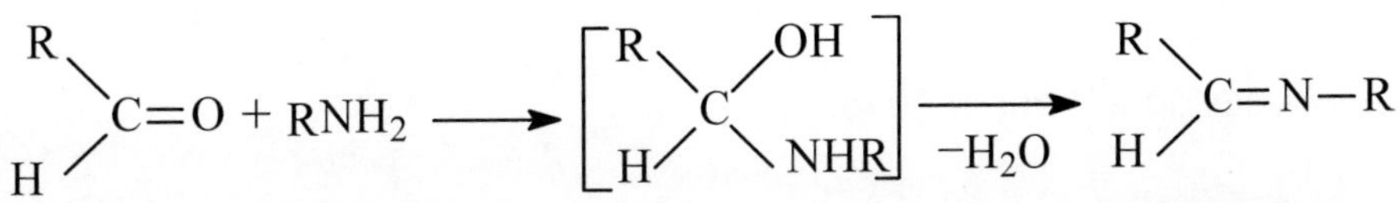

alkylimine

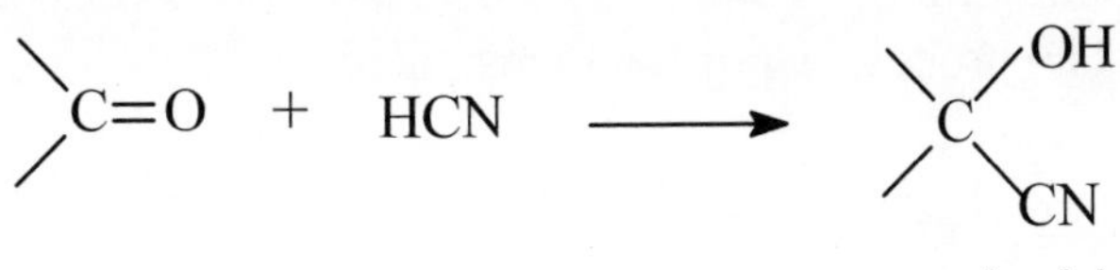

cyanhydrin

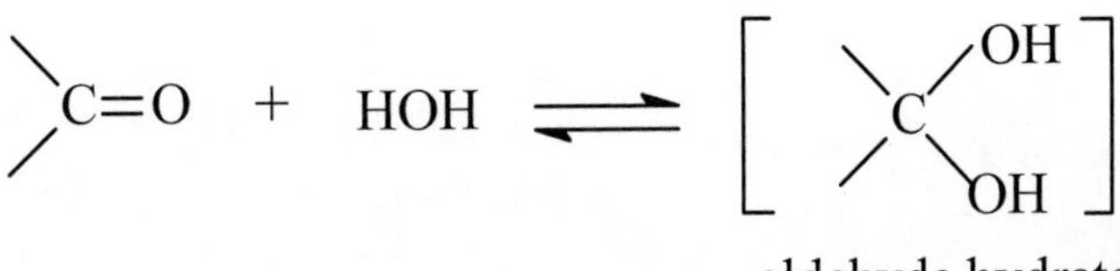

aldehyde hydrate

3. Polymerization

Some aldehydes participate in cyclic and linear polymerization reactions with the carbonyl group. Lower alkanes polymerize under the catalytic action of protons under ordinary conditions. Formaldehyde forms stable trioxymethylene and paraformaldehyde:

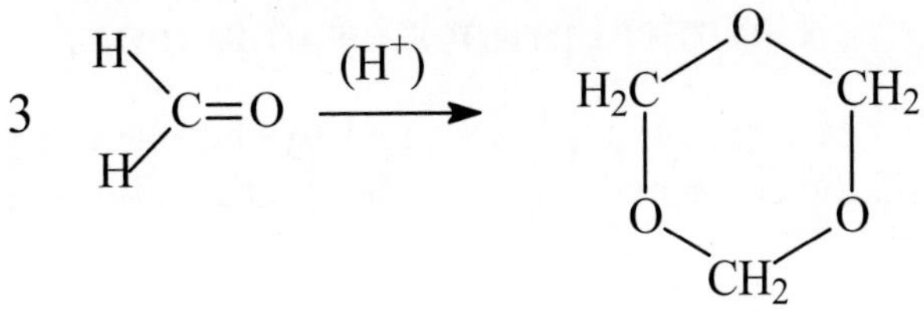

trioxymethylene

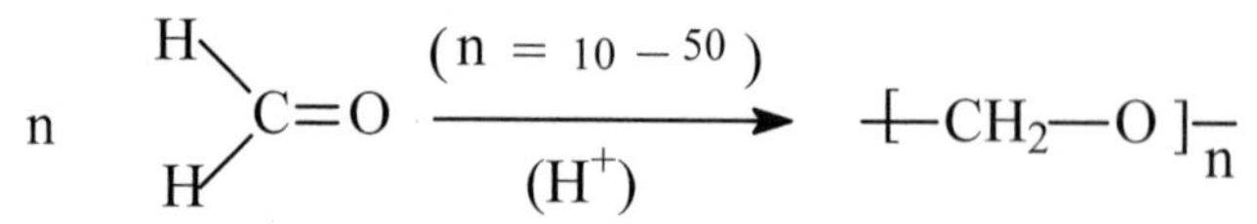

polyformaldehyde Acetaldehyde forms a trimer - paraldehyde and a tetramer - metaldehyde:

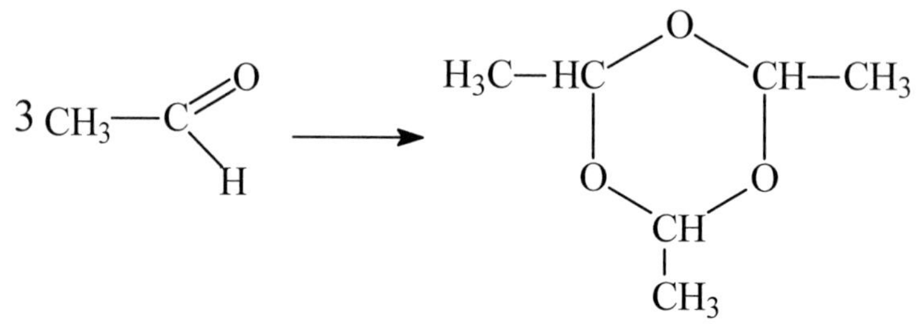

Paraldehyde

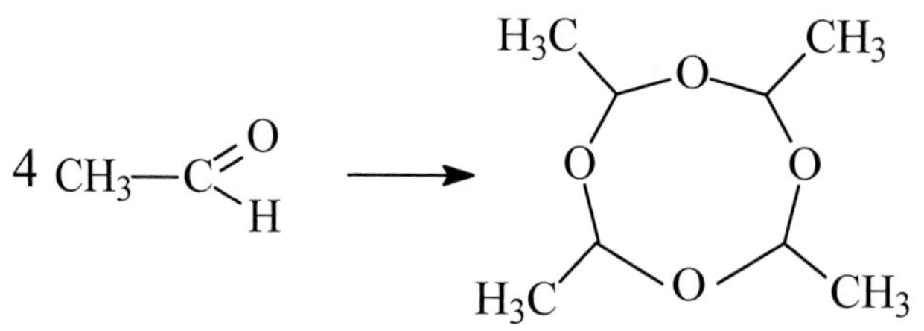

metaldehyde

Aromatic aldehydes do not polymerize.

4. Polycondensation

Formaldehyde with phenol, in the presence of a catalyst, forms polyphenol formaldehyde with water evolution. The process is polycondensation:

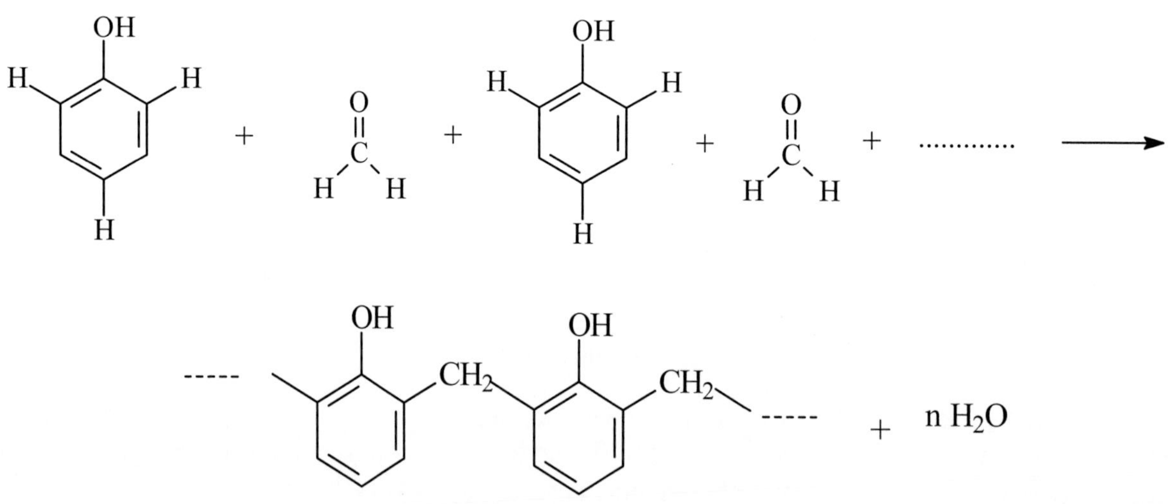

Chemical properties of ketones

The chemical properties of ketones are determined by the reactivity of the carbonyl group, its associated substituents, and the combined action of these structural parts of the molecule.

1. Oxidation

Ketones are much more difficult to oxidize than aldehydes because they do not contain a hydrogen atom directly bonded to the carbon atom of the carbonyl group. They do not reduce Fehling's solution and a silver ammonia hydroxide solution (Tolens reagent).

Ketones are oxidized by strong oxidants (potassium permanganate solution in the presence of sulfuric acid solution). The carbon chain is broken adjacent to the carbonyl group and a mixture of carboxylic acids is obtained.

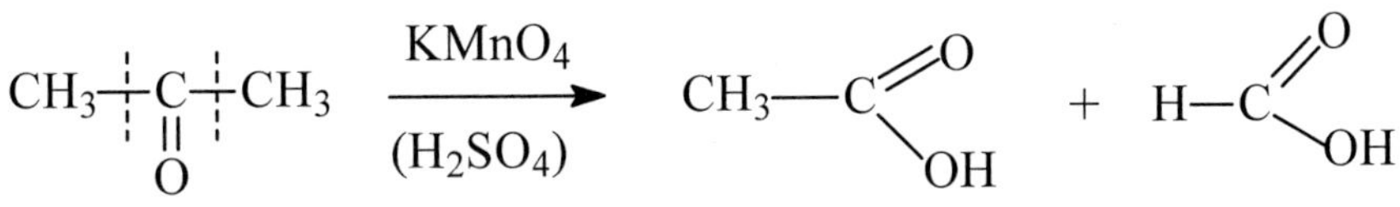

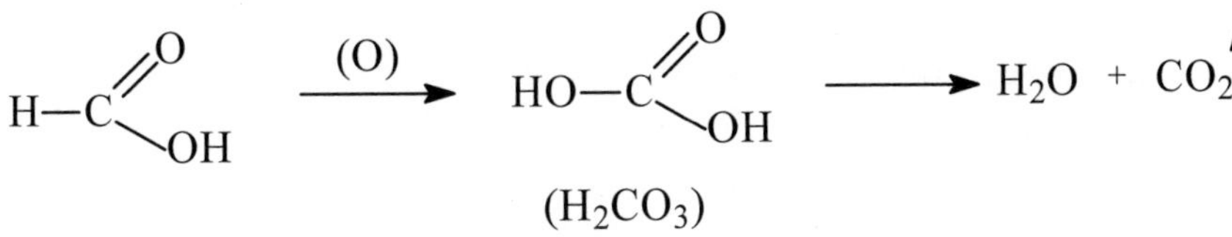

Formic acid is converted into the weak and perishable carbonic acid, which decomposes.

2. Adjunctive reactions

The double bond in the carbonyl group (keto group) determines the ability of the compounds to enter into addition reactions:

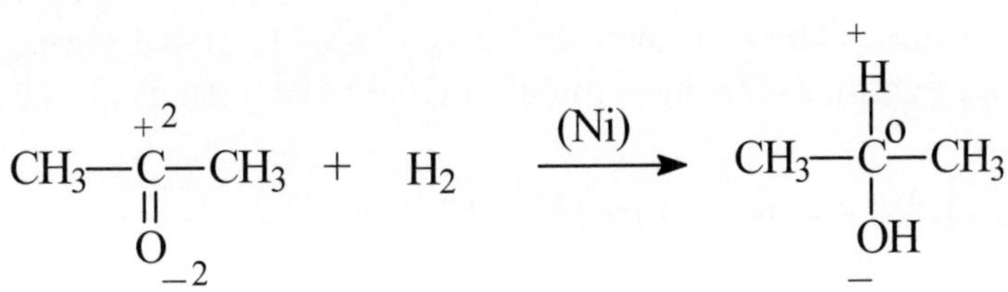

$$C^{+2} + 2e^- \longrightarrow C^0 \quad | \; 2 \; | \; 2 \; | \; 1$$

$$H_2 \longrightarrow 2H^0 - 2 \times 1e^- \longrightarrow 2H^+ \quad | \; 2 \; | \; 2 \; | \; 1$$

Ketones interact with ammonia, amines, hydrogen cyanide, water:

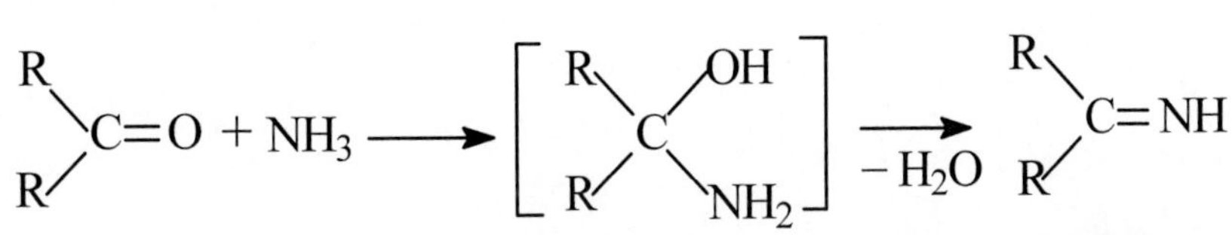

ketimine

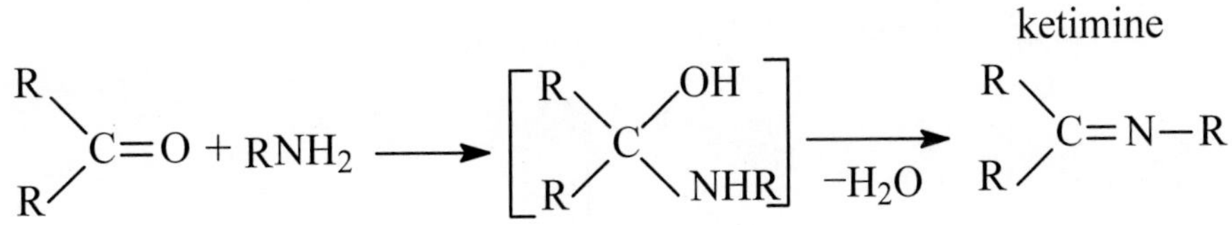

alkylimine

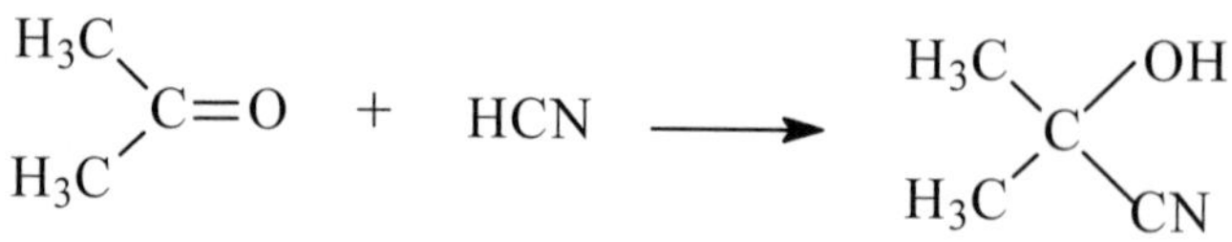

2-methyl-2-hydroxypropanenitrile
(cyanhydrin)

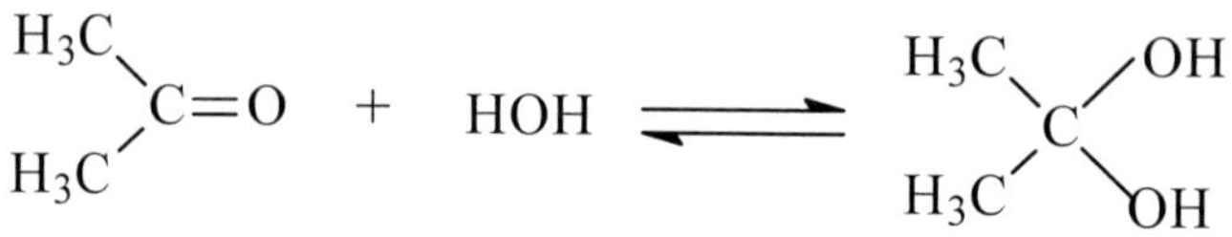

2,2-propanediol

3. Substitute reactions

Substitution reactions take place in the alkyl residues of aldehydes and ketones. A typical example is the substitution of a hydrogen atom in the methyl group of acetones by a halogen. In acidic media, only one hydrogen atom in the ketone molecule is replaced by a bromine atom:

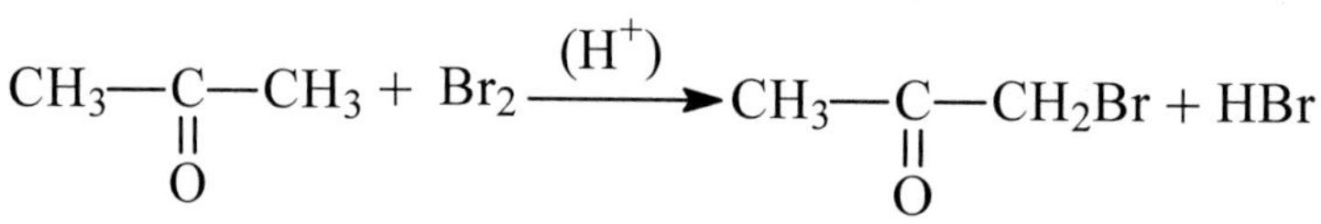

monobromoacetone (monobromopropanone)

In an alkaline medium, bromination occurs when the three hydrogen atoms in the methyl group of the acetone are substituted. The result is tribromoacetone (1,1,1-tribromopropanone).

Halogenated ketones exhibit a tear-inducing effect and have been used as incendiaries in wartime.

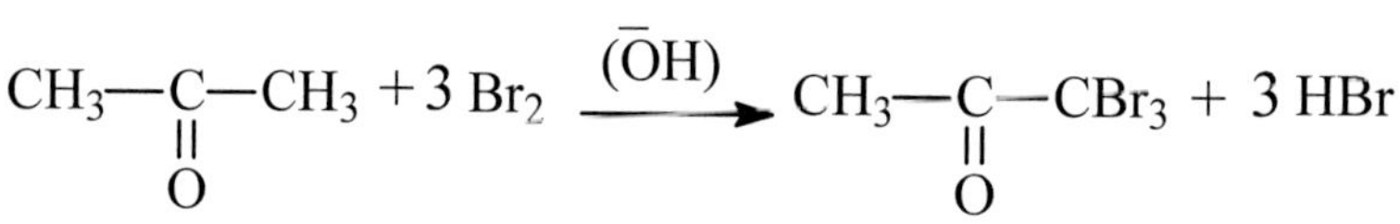

Preparation of aldehydes

1. Aldehydes are produced by moderate oxidation of primary alcohols. On heating, in the presence of a copper (Cu) catalyst, the alcohols undergo dehydrogenation:

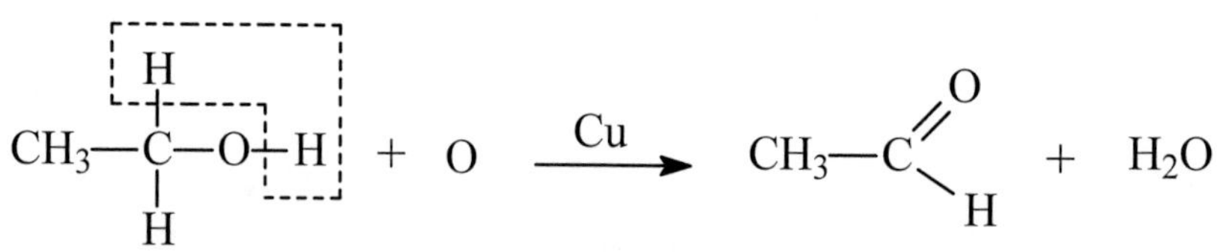

2. Acetaldehyde is produced by hydration of ethyne in the presence of divalent mercuric ions (Hg^{+2}) in a sulfuric acid medium - Kucherov's method:

$$HC{\equiv}CH \;+\; H{-}OH \xrightarrow[(H_2SO_4)]{Hg^{+2}} [H_2C{=}CHOH] \rightleftharpoons CH_3CHO$$

3. Industrially, some carbonyl compounds are produced by the oxidation of hydrocarbons.

$$CH_4 + O_2 \xrightarrow{cat} HCHO + H_2O$$

$$H_2C{=}CH_2 + O_2 \xrightarrow[CuCl_2]{PdCl_2} CH_3CHO$$

The direct oxidation of toluene in the presence of divanadium pentoxide results in benzaldehyde.

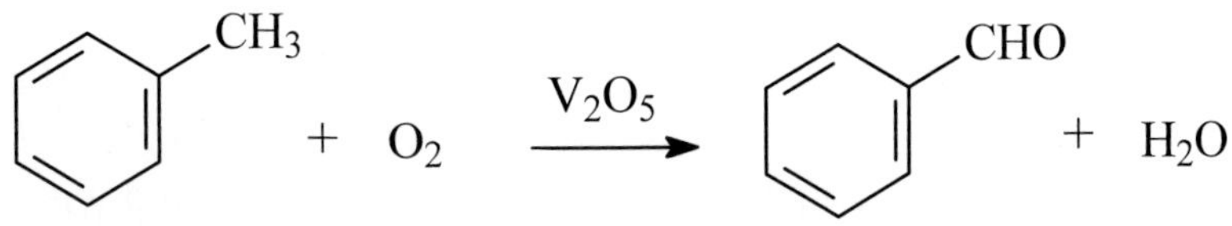

Preparation of ketones

1. Ketones are produced by moderate oxidation of secondary alcohols:

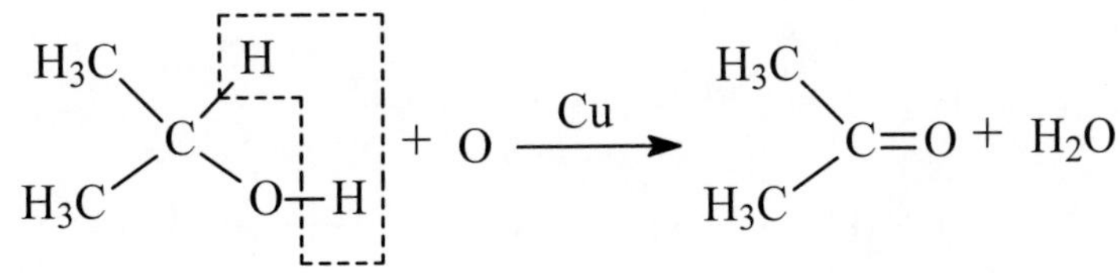

2. Ethyne homologues are converted to ketones on addition of water:

$$HC{\equiv}C{-}CH_2{-}CH_3 + H_2O \xrightarrow[(H_2SO_4)]{Hg^{+2}} \left[H_2C{=}\underset{OH}{\underset{|}{C}}{-}CH_2{-}CH_3\right] \rightleftarrows CH_3{-}\underset{O}{\underset{||}{C}}{-}CH_2{-}CH_3$$

3. Acetone is produced by the oxidation of propene produced in the cracking process of petroleum in the presence of a catalyst.

An industrial method for producing acetone is the cumol method. This method produces two valuable products for industry (phenol and acetone):

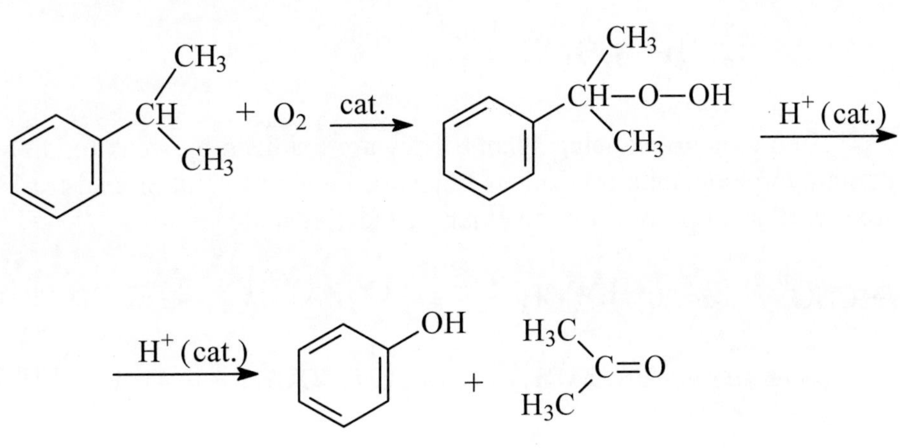

Sample problems and solutions

Task 1.

Express with chemical equations the preparation from carbon and inorganic reagents of a compound of gross formula C_2H_4O known to exhibit reducing properties, polymerize, and participate in addition reactions. From the above compound, obtain the ethyl ester of alanine.

Exemplary solution

The interaction between carbon and hydrogen at high temperatures produces methane. The reaction has a rather theoretical significance because it allows the bond between inorganic and organic substances. The methane undergoes halogenation when irradiated with UV light to produce chloromethane. The Würz reaction of chloromethane with metallic sodium leads to ethane and sodium chloride is released.

$$C + 2\,H_2 \xrightarrow{t^o} CH_4$$

$$CH_4 + Cl_2 \xrightarrow{hv} CH_3Cl + HCl$$

$$CH_3Cl + 2\,Na + ClCH_3 \longrightarrow CH_3{-}CH_3 + 2\,NaCl$$

The resulting ethane is halogenated under illumination of the reaction mixture to give chloroethane.

$$CH_3{-}CH_3 + Cl_2 \xrightarrow{hv} CH_3{-}CH_2Cl + HCl$$

Chloroethane under the action of an aqueous solution of an alkaline base is converted to ethanol.

$$CH_3{-}CH_2Cl + NaOH \xrightarrow{H_2O} CH_3{-}CH_2OH + NaCl$$

Ethanol is converted to ethanal (acetaldehyde) by moderate oxidation involving a copper catalyst and at low temperature.

$$2\,CH_3{-}CH_2OH \xrightarrow{Cu,\,t^o} 2\,CH_3{-}CHO + H_2$$

Aldehydes are good organic reducing agents. They are oxidized by weak oxidizing agents such as silver-ammonia hydroxide solution (Tolens reagent), Fehling's solution and freshly precipitated copper dihydroxide. Ethanal is converted to ethanoic (acetic) acid.

$$CH_3{-}CHO + 2[Ag(NH_3)_2]OH \xrightarrow{t^o} CH_3COOH + 2\,Ag + 4\,NH_3 + H_2O$$

$$CH_3{-}CHO + 2\,Cu(OH)_2 \xrightarrow{t^o} CH_3COOH + Cu_2O\downarrow + 2\,H_2O$$

Ethanal readily polymerizes. It is converted into cyclic trimer and cyclic tetramer.

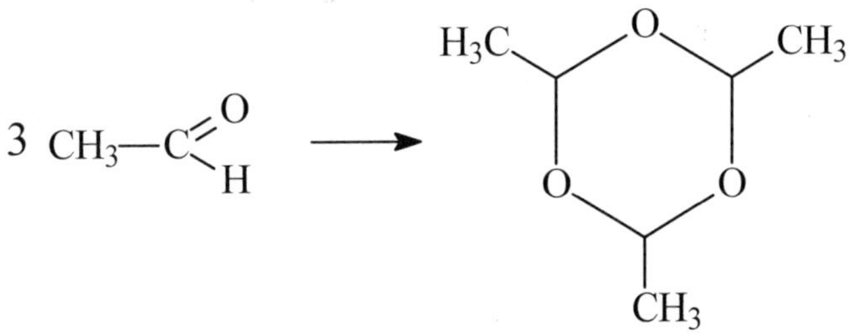

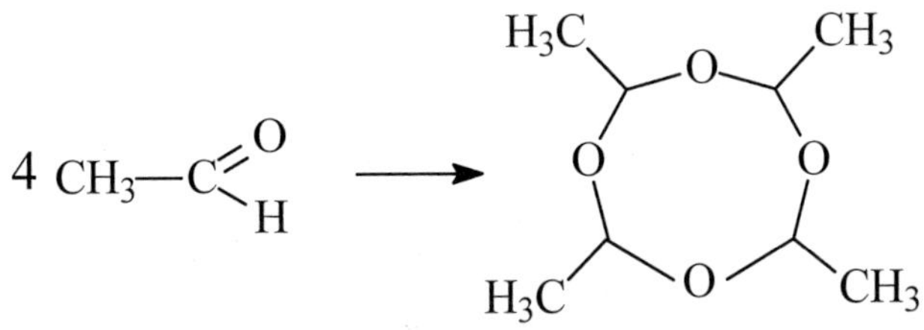

Aldehydes contain a double bond in their functional group and are therefore involved in addition reactions. Hydrogenation of aldehydes yields primary alcohols. Hydrogen addition occurs with the participation of a Ni catalyst and upon heating.

$$CH_3{-}CHO + H_2 \xrightarrow{Ni,\ t^{o}} CH_3{-}CH_2OH$$

Aldehydes also join polar hydrogen-containing molecules. When reacted with hydrogen cyanide, hydroxynitriles are produced and the carbon chain is increased by one carbon atom.

$$CH_3{-}C(=O){-}H + HCN \longrightarrow CH_3{-}CH(OH){-}CN$$

2-hydroxypropanenitrile

When hydrolysis of hydroxynitrile is carried out, hydroxy acid is obtained and an ammonium salt is released. If sulfuric acid is used in the hydrolysis process, diammonium sulfate is released; if the acid is hydrochloric, ammonium chloride is released.

$$CH_3{-}CH(OH){-}CN \xrightarrow{3\ H_2O/H^+} CH_3{-}CH(OH){-}COOH + NH_4^+$$

2-hydroxypropanoic acid

Hydroxy acids react with ammonia and are converted to amino acids. In this case, 2-hydroxypropanoic acid is converted to 2-aminopropanoic acid (alanine).

$$CH_3{-}CH(OH){-}COOH + NH_3 \longrightarrow CH_3{-}CH(NH_2){-}COOH + H_2O$$

Alanine is involved in the esterification process with ethanol. Esterification is a slow, molecular, reversible process catalyzed by mineral acids.

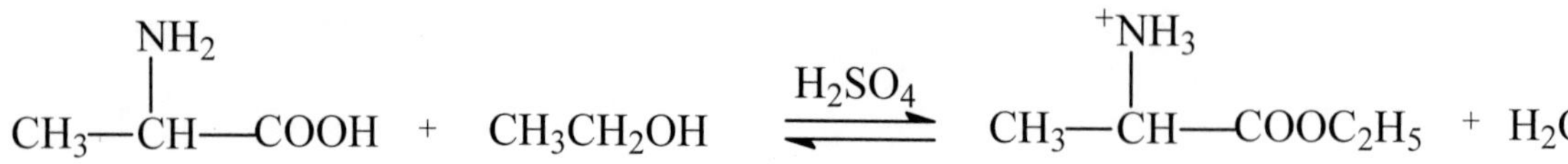

When sulfuric acid is used as a catalyst, the amino group of alanine is converted to an ammonium cation. As a result of the interaction with the alcohol, an ethyl ester of alanine is obtained.

28. CARBOXYLIC ACIDS

CLASSIFICATION, HOMOLOGOUS ORDER AND NAMES OF SATURATED MONOBASIC CARBOXYLIC ACIDS. ISOMERISM. MOLECULAR STRUCTURE AND PROPERTIES OF FATTY AND AROMATIC ACIDS - COMPARATIVE CONSIDERATION. USES.

Derivatives of hydrocarbons containing a functional carboxyl group in their molecule are called carboxylic acids.

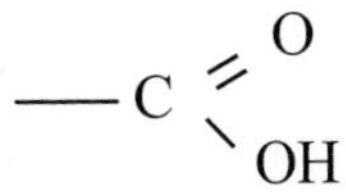

= **A carboxyl group** is composed of two simpler functional groups: a carbonyl

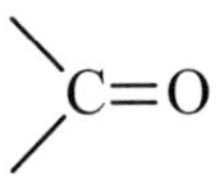

and hydroxyl (– OH), hence its name.

Classification, homologous order and names of saturated monobasic carboxylic acids

The classification of carboxylic acids is based on two principles:

According to the number of carboxylic groups: mono-, di-, etc. carboxylic acids;

According to the type of hydrocarbon residue bonded to the carboxyl group (-COOH) and the presence of other functional groups (substituents) in it: alkanoic, alkenoic, alkynoic, cycloalkanoic, aromatic, and other carboxylic acids.

The alkanoic monocarboxylic (monobasic) acids form the following homologous order:

HCOOH - methanoic (formic) acid
CH_3COOH - ethanoic (acetic) acid
CH_3CH_2COOH - propanoic (propionic) acid
....
$C_{17}H_{35}COOH$ - octadecanoic (stearic) acid

having the general formula $C_nH_{2n+1}COOH$ (R - COOH)

According to the IUPAC nomenclature, their names are formed by adding the suffix **-oic** and the word **acid** to the name of the corresponding alkane with the same number of carbon atoms.

In the case of a branched alkyl chain, a main (straight) chain is chosen, at one end of which is the carboxyl group. This chain is numbered by giving the carbon atom of the carboxyl group the number one (1). The names of the substituents are listed in alphabetical order or according to their complexity, from simpler to more complex, in front of the basic name, and their positions are indicated by numbers.

$$\begin{array}{ccccc} 5 & 4 & 3 & 2 & 1 \\ CH_3— & CH— & CH_2— & CH— & COOH \\ & | & & | & \\ & CH_3 & & C_2H_5 & \end{array}$$

2-ethyl-4-methylpentanoic acid

Some of the carboxylic acids have trivial names.

Carboxylic acids containing a double and a triple bond in the hydrocarbon residue are known:

$CH_2{=}CH{-}COOH$	$CH_3{-}CH{=}CH{-}COOH$	$HC{\equiv}C{-}COOH$
propenoic acid	2-butenoic acid	propynoic acid

Their names are formed similarly to the alkanoic acids by indicating with a number the site of the complex bond.

Aromatic carboxylic acids may be regarded as derivatives of aromatic hydrocarbons in which one or more hydrogen atoms are substituted by carboxyl groups. The carboxyl group is bonded immediately to the aromatic nucleus:

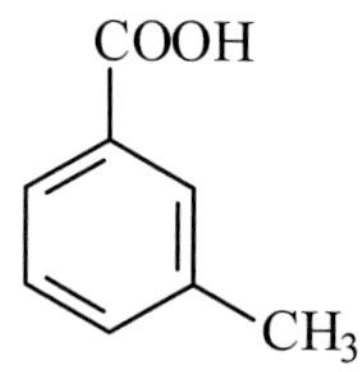

3-methylbenzoic acid

Acids in the molecule of which the carboxylic group is linked to the side chain are considered as derivatives of alkanoic acids, with which they are similar in properties.

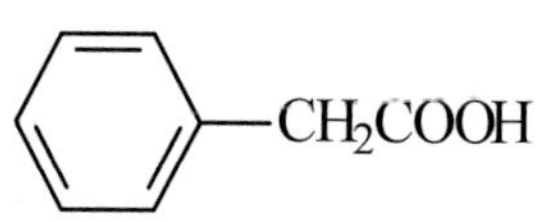

phenylacetic acid

Acids in which the carboxylic group is bonded immediately to the aromatic nucleus are called aromatic or arenic acids.

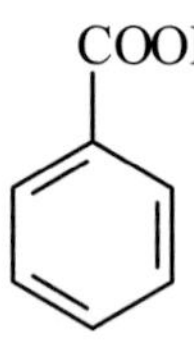

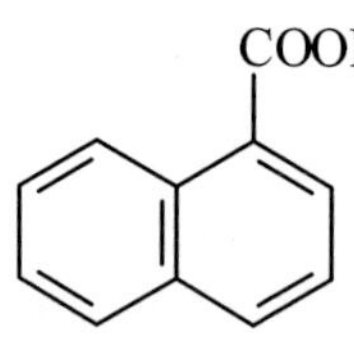

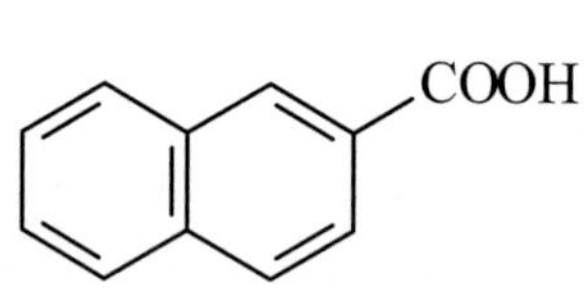

benzenecarboxylic acid; (benzoic acid) 1-naphthalenecarboxylic acid; 2-naphthalenecarboxylic acid

Their names are formed by adding "carboxylic acid" to the name of the arene. The acid residue is called acyl (R is alkyl-, alkenyl-, aryl-, etc.), e.g.:

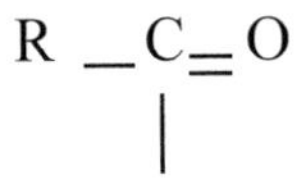

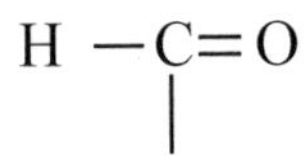
- formyl

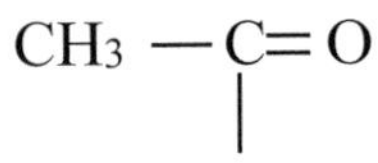
- acetyl

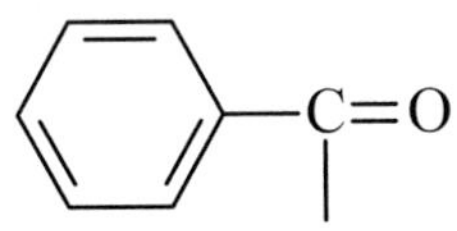
- benzoyl

The groups of carboxylic acids considered have their homologous series.

Isomerism. Types of isomerism

Different types of isomerism occur in carboxylic acids – chain, positional, π- diastereomerism depending on the properties of the hydrocarbon residue present.

Chain:

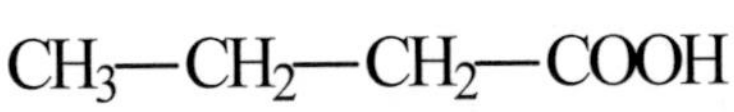
n-butanoic acid

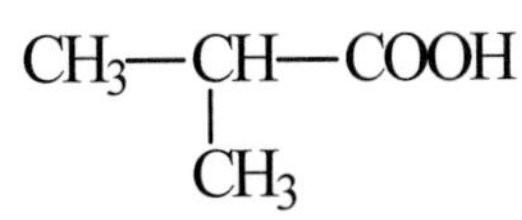
isobutanoic acid
(2-methylpropanoic acid)

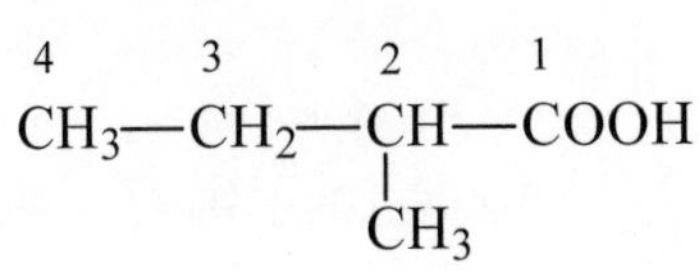
2-methylbutanoic acid

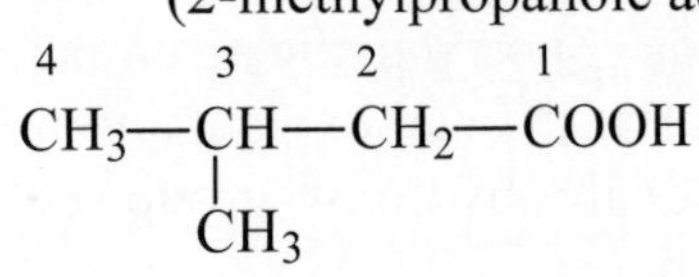
3-methylbutanoic acid

Positional:

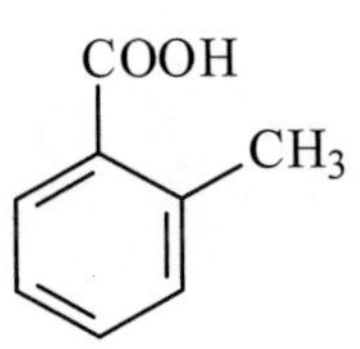
o-methylbenzoic acid

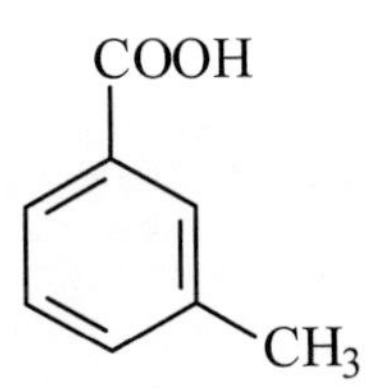
m-methylbenzoic acid

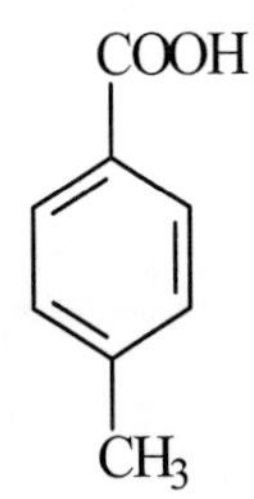
p-methylbenzoic acid

Geometric isomerism (π-diastereomerism):

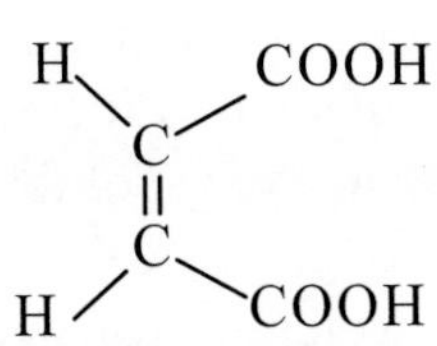
cis-2-butenedioic acid

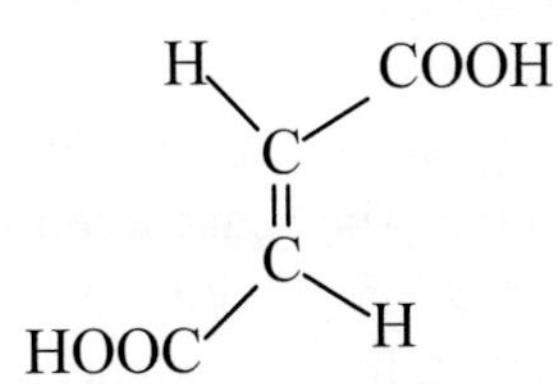
trans-2-butenedioic acid

The carboxyl group is a complex functional group. The mutual influence of the carbonyl group and the hydroxyl group is so strong that they are treated as a new functional group possessing qualitatively new properties.

In alcohols, due to the high electronegativity of oxygen, the bond between the hydrogen and oxygen atoms in the hydroxyl group is highly polar – the pair of electrons bonding the atoms is pulled towards the oxygen atom.

The double bond in the carbonyl group is also polar, as a result, the carbon atom acquires a partial positive charge.

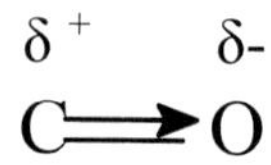

To compensate for this charge, it draws towards itself the electron pair that binds it to the oxygen atom of the hydroxyl group. As a result, the polarisation of the bond in the hydroxyl group is also strengthened, i.e., the partial positive charge of the hydrogen atom increases even more.

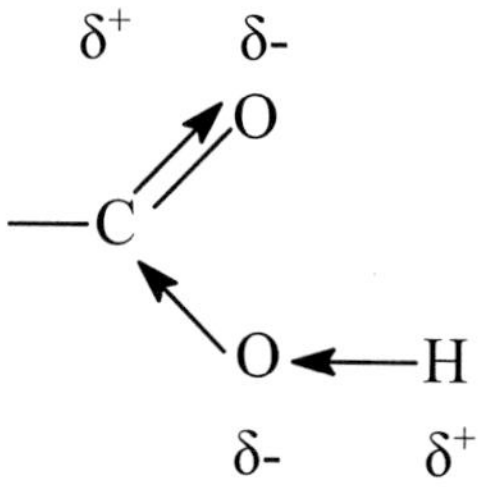

The greater degree of dissociation of carboxylic acids is due to the better stabilization of the carboxylate anion as opposed to the alcoholate anion.

The stabilization of the carboxylate anion is due to the interaction between the electron pair of the oxygen atom of the hydroxyl group, determining the anionic (negative) charge, and the π-electrons of the carbonyl group. As a result of this delocalization effect, the negative charge is evenly distributed between the two oxygen atoms in the carboxylate anion:

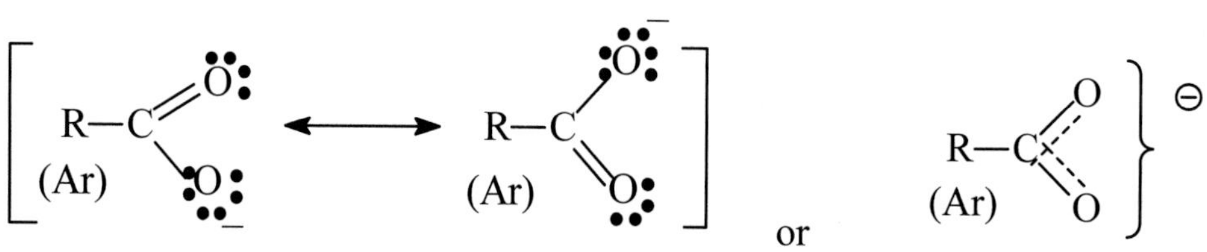

X-ray structural studies show that the lengths of the two carbon-oxygen bonds in the carboxylate anion are the same in contrast to those in the undissociated acid.

The first representatives of the homologous order of alkanoic acids (methanoic, ethanoic, propanoic) are easily mobile liquids with a pungent odor; the next are oily liquids with an unpleasant odor (butyric, valerian, capric); and the higher acids are solids without odor. Under ordinary conditions there are no gaseous carboxylic acids. As the molecular mass increases, their physical properties change properly. The first three members of the series are very soluble in water, the next less so, and the higher ones are practically insoluble in water.

Aromatic acids are crystalline substances, slightly soluble in water at ordinary temperature.

The reason for the relatively high boiling point is the occurrence of intermolecular hydrogen bonds that bind the acid molecules into relatively stable groupings.

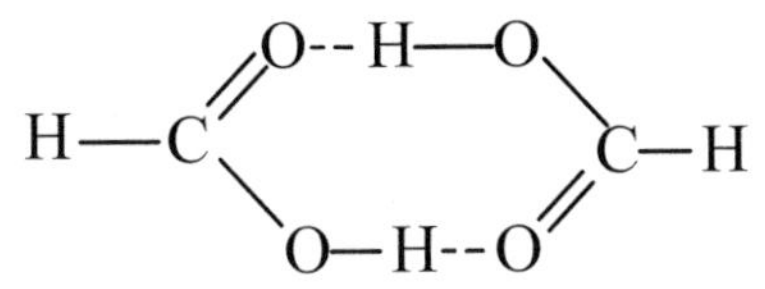

Carboxylic acids have important implications for practice. Formic acid causes inflammation when it gets on the skin.

Acetic acid is used as a seasoning and preservative. Acetic acid finds application in organic synthesis for production of dyes, medicines, artificial silk etc.

Acetates find a variety of uses - in the dye industry, for pest control in agriculture. Ethyl acetate is used as a solvent for resins, varnishes, and some types of rubber.

Oleic acid, together with palmitic and stearic acids, is a constituent of fat. It is unsaturated in nature and, under ordinary conditions, is a liquid.

Benzoic acid has a pronounced antiseptic effect and is used as a preservative, and its salts and esters are used in medicine as antirheumatic agents.

Salicylic acid is an aromatic hydroxycarboxylic acid. It contains carboxyl and hydroxyl groups and combines the properties of acid and phenol.

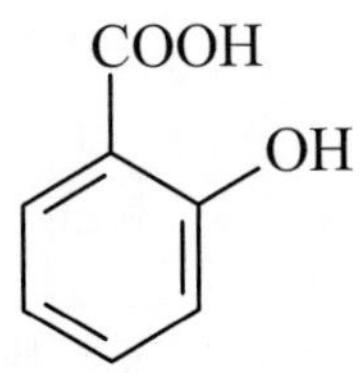

,

It is used to produce pharmaceutical preparations. Aspirin is acetylsalicylic acid, i.e., an ester of salicylic and acetic acids.

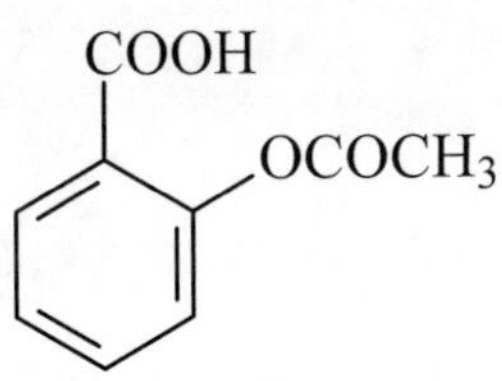

It is one of the most popular drugs with multifaceted action and almost no side effects.

Chemical properties of carboxylic acids

1. Carboxylic acids in aqueous solution are dissociated by hydrogen ion evolution i.e., the carboxylic group determines acidic properties.

$$R{-}COOH + H_2O \rightleftharpoons R{-}COO^- + H_3O^+$$

Aqueous solutions of the acids change the blue litmus to red, indicating that they dissociate electrolytically:

$$CH_3{-}COOH \rightleftharpoons CH_3{-}COO^- + H^+$$

The dissociation of monocarboxylic acids in solution proceeds to varying degrees depending on the type of hydrocarbon residue bound to the carboxylic group.

Electron-donor substituents increase the electron density at the oxygen atom from the hydroxyl group and hinder the dissociation of the carboxyl group. These acids are weaker:

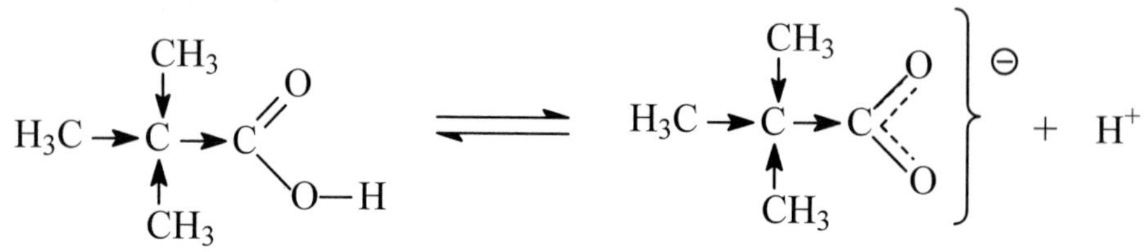

Electron acceptor substituents inversely affect the dissociation of the carboxyl group and the stabilization of the carboxylate anion by increasing the strength of the acids:

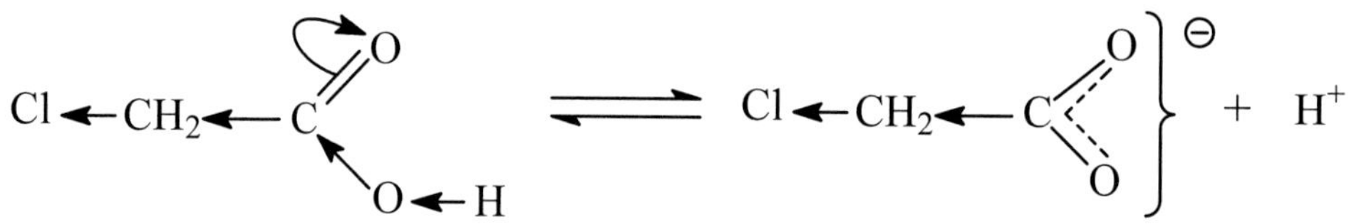

$$ClCH_2COOH < Cl_2CHCOOH < Cl_3CCOOH$$

The strength of halogenated acids increases with an increasing number of chlorine atoms in the example of chloro-, dichloro- and trichloroacetic acids.

In aqueous solution, acids with electron acceptor substituents dissociate to a greater extent, i.e., they are stronger.

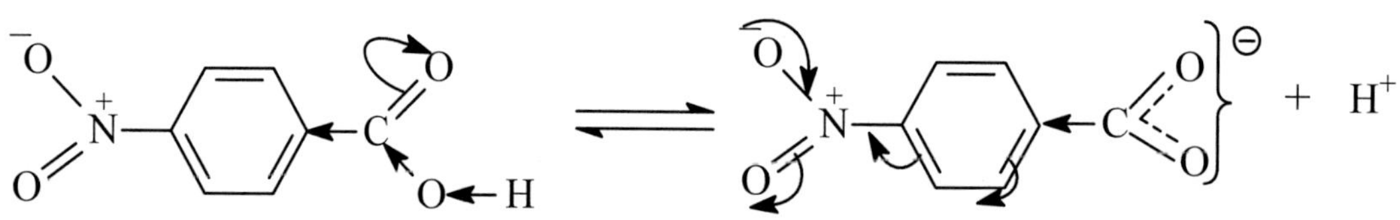

2. Metals that are ahead of hydrogen in the order of the relative activity of metals (ORAM) react with acids, releasing hydrogen and producing salts.

$$2\,\underbrace{CH_3—COO^- + H^+} + 2\,Na \longrightarrow 2\,\underbrace{CH_3—COO^- + Na^+} + H_2$$

sodium ethanoate (acetate)

$$Na^0 - e^- \longrightarrow Na^+ \quad | \; 1 \; | \; 2 \; | \; 2$$

$$2\,H^+ + 2 \times 1\bar{e} \longrightarrow 2\,H^0 \longrightarrow H_2 \quad | \; 2 \; | \; 2 \; | \; 1$$

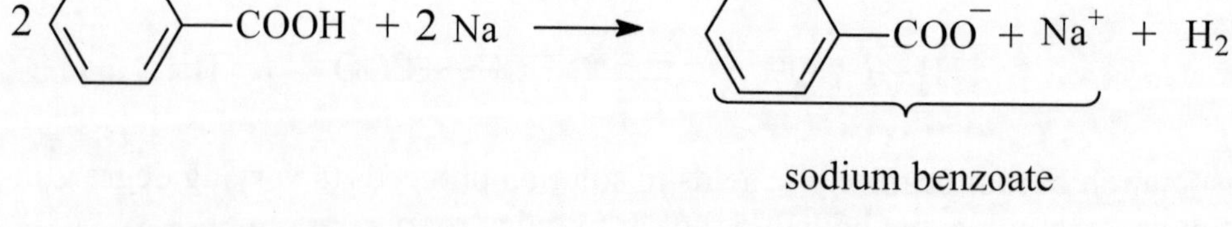

sodium benzoate

$$\underbrace{2CH_3{-}COO^- + 2H^+} + Zn \longrightarrow (CH_3COO)_2Zn + H_2$$

zinc diacetate

3. When interacting with bases, neutralization is performed:

$$\underbrace{CH_3{-}COO^- + H^+} + \underbrace{Na^+ + OH^-} \rightleftharpoons \underbrace{CH_3{-}COO^- + Na^+} + H_2O$$

The process is reversible because the resulting sodium acetate is a salt of a strong base and a weak acid and hydrolyzes.

4. Salts are also produced when acids react with ammonia.

$$CH_3COOH + NH_3 \longrightarrow CH_3COONH_4$$

ammonium acetate

5. On heating, ammonium salts give off water and amides are produced.

$$CH_3COONH_4 \xrightarrow{t^o} CH_3CONH_2 + H_2O$$

Acetamide

Further dehydration of the amides yields nitriles.

$$CH_3CONH_2 \xrightarrow{t^o} CH_3CN + H_2O$$

6. Alkanoic and arenoic acids are esterified with alcohols.

$$R(Ar){-}COOH + R_1OH \overset{(H^+)}{\rightleftharpoons} R(Ar){-}COOR_1 + H_2O$$

$$CH_3{-}C(=O){-}OH + HOC_2H_5 \overset{(H_2SO_4)}{\rightleftharpoons} CH_3{-}C(=O){-}O{-}C_2H_5 + H_2O$$

ethyl ethanoate (ethyl acetate)
acetic acid ethyl ester

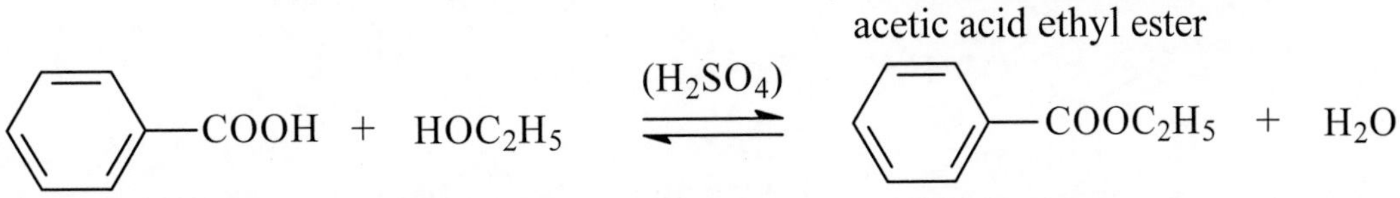

ethylbenzoate

Esterification is a slow, molecular and reversible process catalyzed by mineral acids. The equilibrium can be shifted to the right (towards the products of the reaction) if it proceeds in the presence of concentrated sulfuric acid, which acts as a water-removing agent and as a catalyst.

The mechanism of the reaction shows that a hydroxyl group is separated from the carboxyl group, and a hydrogen atom is separated from the hydroxyl group of the alcohol.

7. Carboxylic acids interact with thionyl chloride ($SOCl_2$), sulfuryl chloride (SO_2Cl_2) phosphorus trichloride (PCl_3), phosphorus pentachloride (PCl_5), phosphorus oxychloride ($POCl_3$):

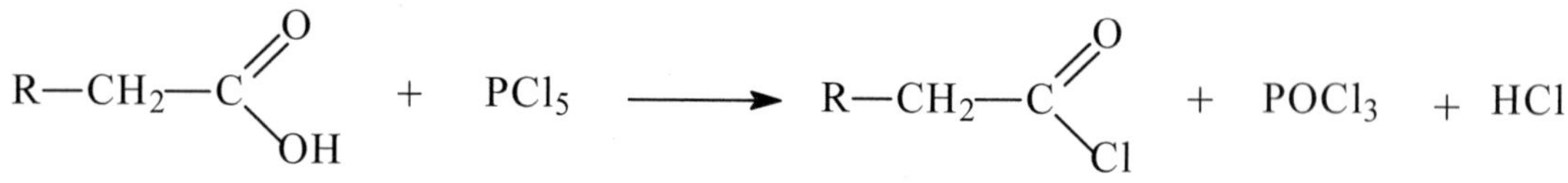

8. In alkanoic acid molecules, substitution of hydrogen atoms at the alkyl residue by an atom of a halogen element is possible when the acids react with halogen in the presence of phosphorus:

$$CH_3COOH + Cl_2 \xrightarrow{P} ClCH_2COOH + HCl$$

The reaction is substitutive. Usually, a hydrogen atom is substituted at the α-site relative to the carboxyl group.

$$CH_3CH_2COOH + Cl_2 \xrightarrow{P} CH_3\underset{\substack{|\\ Cl}}{C}HCOOH + HCl$$

2-chloropropanoic acid

9. Unsaturated fatty acids are involved in the characteristic π-bond addition reactions such as hydrogenation, HX (HCl) addition, X_2 (Cl_2, Br_2), etc.

$$CH_3(CH_2)_7CH{=}CH(CH_2)_7COOH + H_2 \xrightarrow{Ni} CH_3(CH_2)_{16}COOH$$

9-octadecenoic acid (oleic acid) → octadecanoic acid (stearic acid)

10. On heating, the acids are dehydrated to form anhydrides:

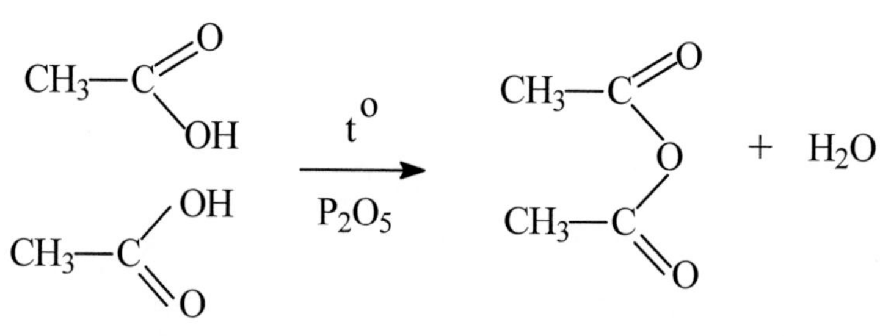

acetic acid anhydride

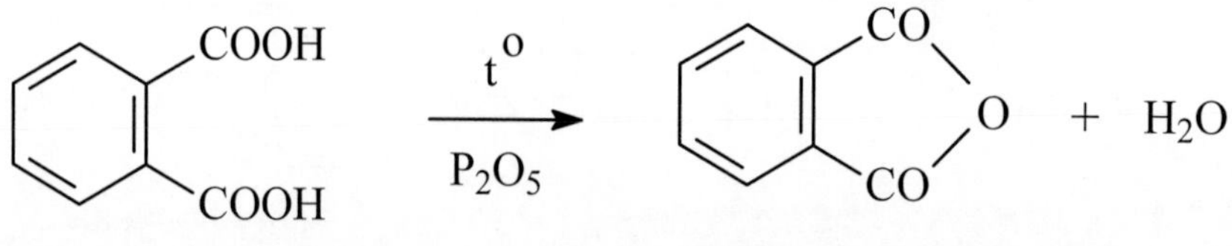

phthalic anhydride

11. Due to the presence of an aromatic nucleus in its molecule, benzoic acid also enters into all the characteristic reactions for aromatic compounds. Unlike the hydroxyl group in phenol, the carboxyl group orients the substituents at the m-site in the benzene nucleus. For example, nitration of benzoic acid yields m-nitrobenzoic acid:

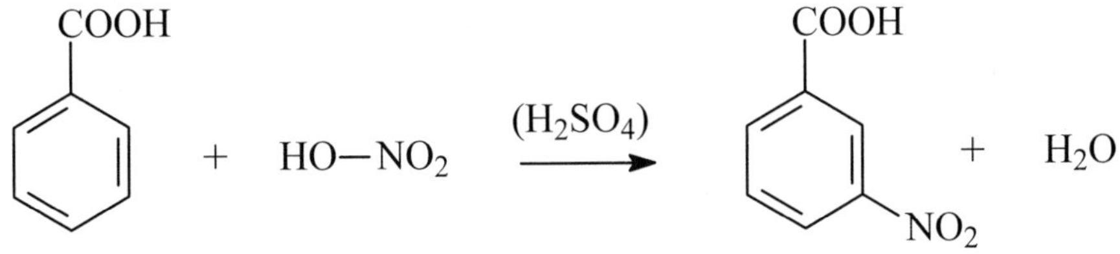

The carboxyl group is a deactivating m-orientant and therefore nitration, sulfonation, halogenation reactions proceed more difficult than in unsubstituted arenes.

12. Carboxylic acids burn to produce carbon dioxide and water.

$$CH_3COOH + 2O_2 \longrightarrow 2\,CO_2 + 2\,H_2O$$

$$\begin{array}{lcl|c|c|c} C^{-3} - 7e^- & \longrightarrow & C^{+4} & \multirow{2}{*}{8} & & 1 \\ C^{+3} - e^- & \longrightarrow & C^{+4} & & 8 & \\ O_2 \longrightarrow 2O^0 + 2\times 2e^- & \longrightarrow & 2O^{-2} & 4 & & 2 \end{array}$$

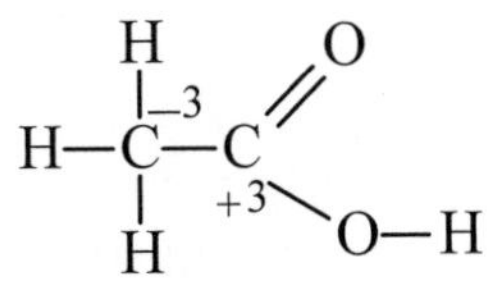

Sample problems and solutions

Task 1.

Write the structural formula of a substance of composition $C_8H_8O_2$, the chlorination of which with excess chlorine on heating (or scattered sunlight) yields a substance of composition $C_8H_5Cl_3O_2$.

On hydrolysis with water in an acid medium, the chlorinated product is converted to terephthalic acid (1,4-benzenedioic acid). Condensation of terephthalic acid with ethylene glycol (1,2-ethanediol) yields a product which is used to produce the fibre terylene. Introduce the mechanism of chlorination. What products are obtained if the chlorination of the starting substance takes place with the participation of a catalyst? Describe with chemical equations the above processes and write the names of the products.

Exemplary solution

Terephthalic acid is 1,4-benzenedicarboxylic acid

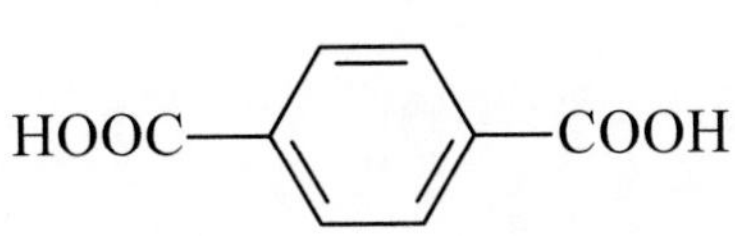

Its production can be carried out by oxidation of 1,4-dimethylbenzene (p-xylene) or from 4-methylbenzoic acid

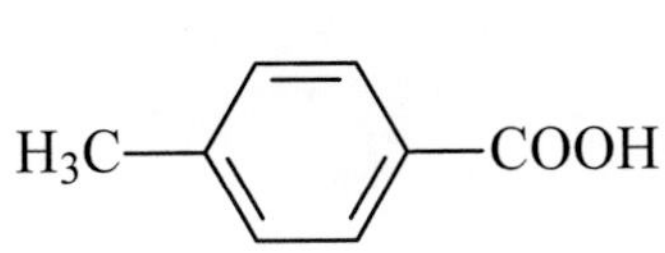

The acid has the composition $C_8H_8O_2$. The isomers 3-methylbenzoic acid and 2- methylbenzoic acid also correspond to this gross formula.

In the halogenation of aromatic hydrocarbons, depending on the reaction conditions, a hydrogen atom substitution can take place in the aromatic core or in the side chain. In the problem condition, it is stated that chlorination takes place with excess chlorine upon heating or irradiation of the reaction mixture with diffused sunlight. The process can be represented by the following chemical equation:

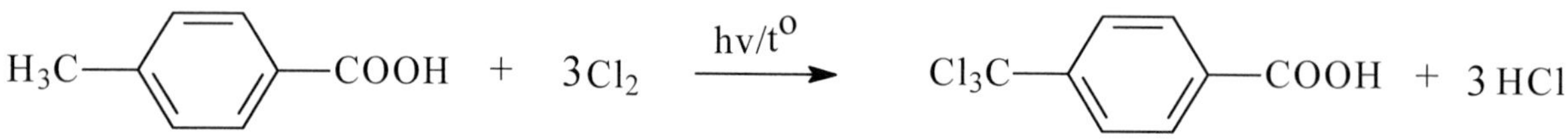

4-methylbenzoic acid 4-trichloromethylbenzoic acid

Halogenation occurs by a chain-radical mechanism. At high temperature or scattered sunlight, the chlorine molecule breaks apart homolytically:

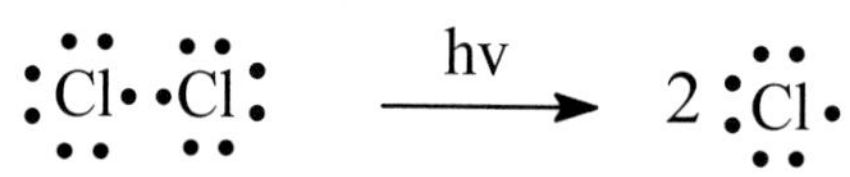

The resulting chlorine atoms have a single electron in the valence electron layer and are therefore of significant reactivity. The chlorine atom attacks the methyl group of 4- methylbenzoic acid. Hydrogen chloride and a hydrocarbon radical are released:

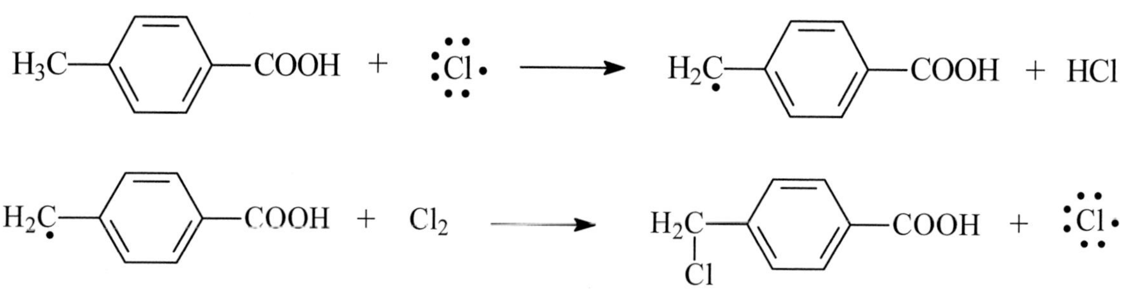

4-chloromethylbenzoic acid

The chlorination process continues until the three hydrogen atoms in the methyl group are completely substituted. 4-Trichloromethylbenzoic acid has the gross formula $C_8H_5Cl_3O_2$.

On hydrolysis in acidic medium, the chlorinated product is converted to terephthalic acid according to the scheme:

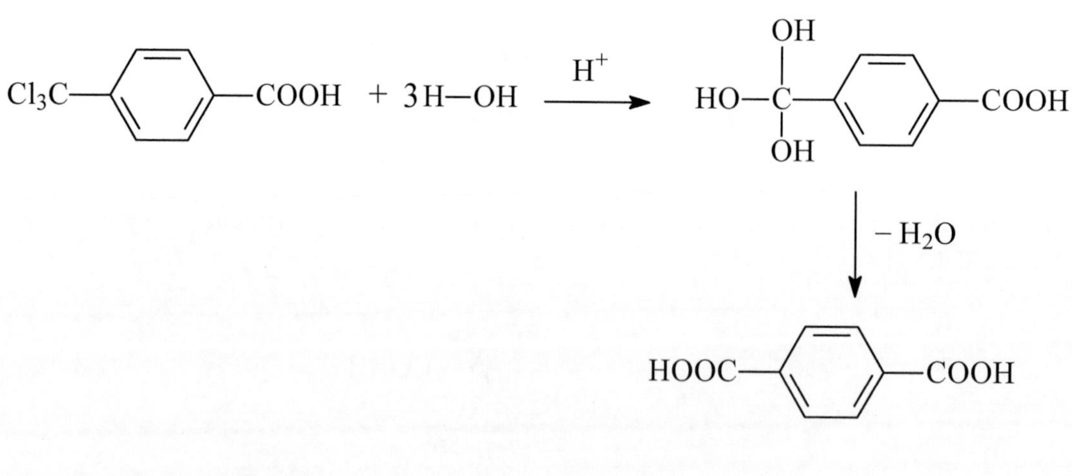

Hydrolysis results in a bidentate triol, a perishable compound that spontaneously dehydrates to the stable terephthalic acid.

The isomers of 4-methylbenzoic acid do not satisfy the condition of the problem. These conversions yield phthalic (1,2-benzenedicarboxylic) and isophthalic (1,3-benzenedicarboxylic) acids, respectively.

Terephthalic acid is a feedstock for the production of the polyester fibre terylene (lavsan). It is produced by the reaction of terephthalic acid with ethylene glycol or dimethyl terephthalate (the dimethyl ester of 1,4-benzenedicarboxylic acid) with ethanediol:

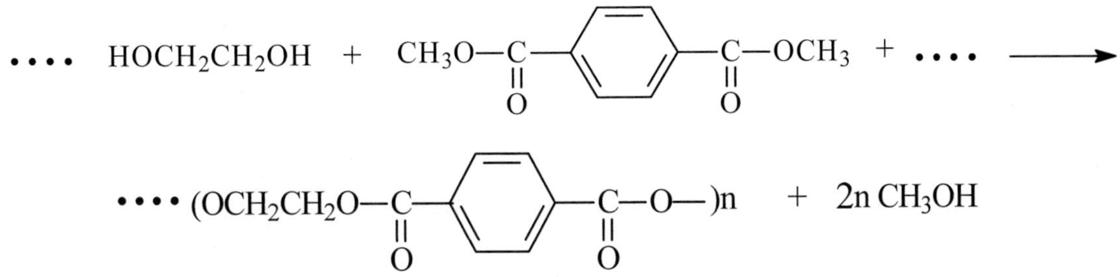

The product is called polyethylene terephthalate or polyethylene glycol terephthalate.

The feedstocks are products of the chemical processing of petroleum.

If the chlorination of 4-methylbenzoic acid (the starting substance) is carried out with the participation of a catalyst, 4-methyl-3-chlorobenzoic acid, a product of electrophilic substitution in the aromatic nucleus will be obtained. The presence of two substituents in the aromatic core uniquely determines the location of the third. The carboxyl group is a deactivating m-orientant (-I, -M-effect) and the methyl group is an activating o- and p-orientant (+I-effect). The orientational influence of the two substituents will result in a single product:

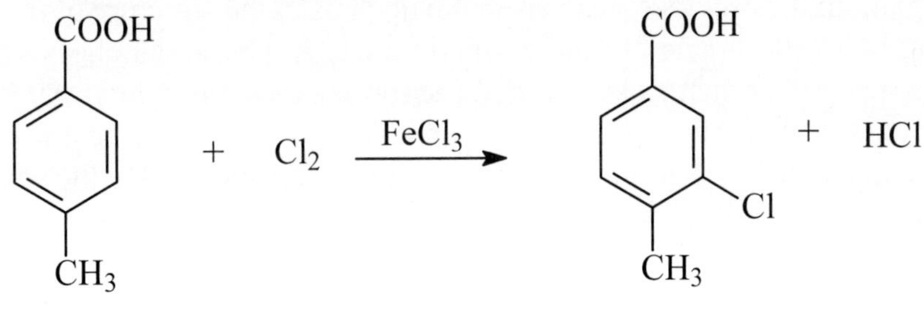

Task 2.

Determine the structure of a substance of composition $C_8H_8O_3$, if it is known that it dissolves well in water and aqueous solutions of alkaline bases, forms esters with acids and alcohols, and does not form a colored compound with ferric trichloride. It exists in two enantiomeric forms. Express with chemical equations the processes involved and the names of all the compounds.

Exemplary solution

The substance of composition $C_8N_8O_3$, satisfying the requirements specified in the problem statement, is:

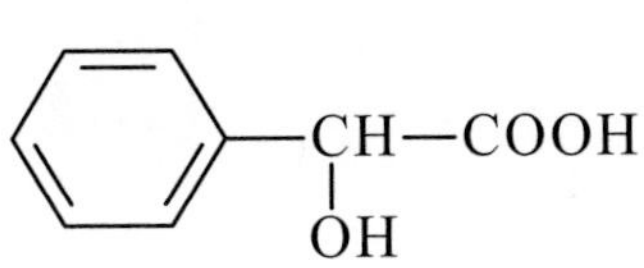

2-phenyl-2-hydroxyethanoic (almond) acid

The presence of a carboxyl and the hydroxyl group attached to the adjacent carbon atom ensure the good solubility of the substance in water and aqueous solutions of alkaline bases. Solubility in water is due to the formation of intermolecular hydrogen bonds of the functional groups of the substance with the water molecules. Aqueous solutions of alkaline bases convert the acid to an alkaline salt. The ionic crystal structure of salt is a prerequisite for its good solubility in water.

The substance forms esters with acids, in which the alcohol hydroxyl group of the hydroxy acid is involved in the process:

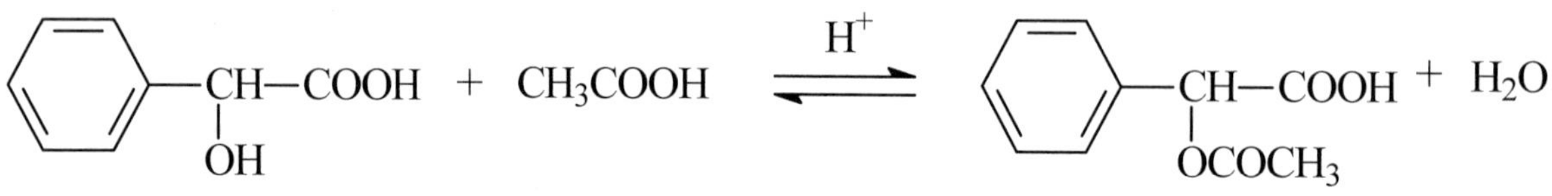

2-acethyloxy-2-phenylacetic acid

With alcohols it forms esters. The carboxyl group is involved in this process:

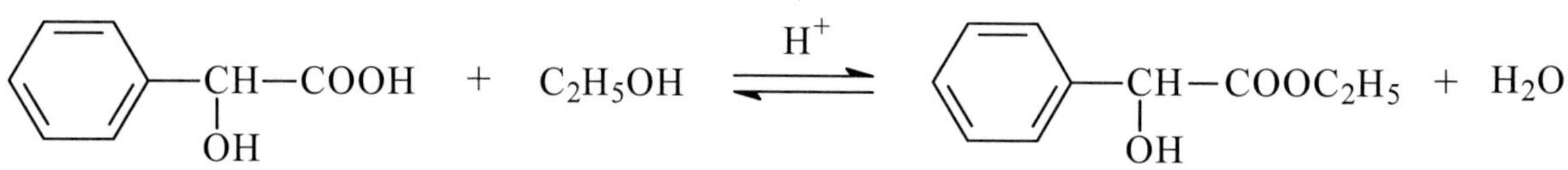

ethyl 2-phenyl-2-hydroxyacetate

Esterification is a slow, molecular, reversible process catalyzed by mineral acids, which produces ester and water.

For the correct solution to the problem, the statement that the substance does not form a colored compound with ferric trichloride is important. The aqueous solution of ferric trichloride is a reagent for proving a phenolic hydroxyl group. The mention of ferric trichloride carries useful and suggestive information: the presence of an aromatic nucleus in the structure of the compound and of a hydroxyl group indirectly bonded to the aromatic nucleus, i.e., an alcohol hydroxyl group. It is esterified with an acid rather than an acid anhydride (unlike a phenolic hydroxyl group).

The proposed compound 2-phenyl-2-hydroxyacetic (almond) acid contains an asymmetric carbon atom (a carbon atom bonded to four different substituents). This means that the substance is optically active and exists in two enantiomeric forms, which relate to each other as the unsymmetrical object and its mirror image.

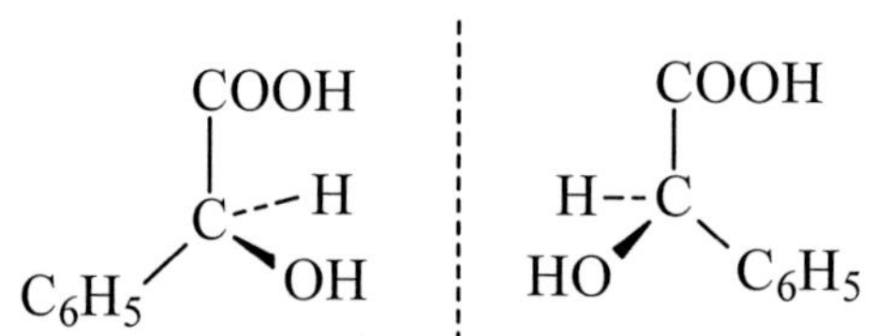

D− L−

29. FATS

COMPOSITION AND STRUCTURE. NAMES. PROPERTIES OF FATS. USES. DISTRIBUTION. SOAPS. COMPOSITION, PROPERTIES AND USES. SYNTHETIC CLEANING AGENTS – WASHING ACTION.

Composition and construction

Fats are a mixture of esters of glycerol with higher fatty saturated or unsaturated carboxylic acids. They have a general formula

$$\begin{array}{l} CH_2\text{—}OCOR \\ | \\ CH\text{—}OCOR' \\ | \\ CH_2\text{—}OCOR'' \end{array}$$

where R-, R'-, R"- are the same or different hydrocarbon residues. Fats are also called lipids. Their composition and structure have been proved by Chevreul. He found that when fats are heated in the presence of alkaline bases they are hydrolyzed to glycerol and a mixture of carboxylic acids, chiefly higher ones. Berthelot first carried out the synthesis of fats (1854) by heating glycerol with higher fatty acids. The process is reversible.

$$\begin{array}{l} CH_2OH \\ | \\ CHOH \\ | \\ CH_2OH \end{array} + \begin{array}{l} RCOOH \\ \\ R'COOH \\ \\ R''COOH \end{array} \xrightleftharpoons{(H^+)} \begin{array}{l} CH_2\text{—}OCOR \\ | \\ CH\text{—}OCOR' \\ | \\ CH_2\text{—}OCOR'' \end{array} + 3\,H_2O$$

In the composition of fats are involved acids of different lengths of carbon chain - from three to twenty-four carbon atoms, with a predominance of higher acids containing from sixteen to twenty carbon atoms in their chain. Only the propanoic and pentanoic acids have an odd number of carbon atoms. All the others in the composition of fats are of unbranched chain and of an even number of carbon atoms.

Fats are actually triglycerides and their correct name is triacylglycerols.

The names of glycerides are formed by replacing the acid residue ending "il" with "in" and denoting the number of acid residues.

$$\begin{array}{l} CH_2\text{—}OCOC_{17}H_{35} \\ | \\ CH\text{—}OCOC_{17}H_{35} \\ | \\ CH_2\text{—}OCOC_{17}H_{35} \end{array}$$

tristearin

Acid residues of unsaturated acids may also be involved in the composition of fats.

$$CH_3(CH_2)_7CH{=}CH(CH_2)_7COOH$$

9-octadecenoic acid (oleic acid)

$$CH_3CH_2CH{=}CHCH_2CH{=}CHCH_2CH{=}CH(CH_2)_7COOH$$

9,12,15-octadecentryoic acid (linolenic acid)

They enter into the composition of mixed glycerides.

$$\begin{array}{l} CH_2{-}OCOC_{17}H_{33} \\ | \\ CH{-}OCOC_{15}H_{31} \\ | \\ CH_2{-}OCOC_3H_7 \end{array}$$

oleo-palmito-butyrin

Oleo-palmito-butyrin is a constituent of cow butter.

The composition and structure of fats has also been demonstrated by synthesis of mixed glycerides.

Fats are widely distributed in nature. They are found in the animal and plant world. Fats also contain small amounts of vitamins such as A, D and E, free carboxylic acids, dyes, aromatics, etc. Along with carbohydrates and proteins, fats are a staple food for the body and are a very rich source of energy.

Properties of fats

Fats are: **solid** (also called fats). They are mainly of animal origin; **semi-solid** and **liquid** (called oils). They are of vegetable origin, too. Liquid fats of animal origin (e.g. fish oil) are rare. **Only saturated fatty acids** are involved in the composition of **solid fats**, and both saturated and unsaturated fatty acids are involved in the composition of other fats. The melting point of a fat is higher when the carbon content of the saturated acids is also higher.

Fats are **lighter than water** and practically **do not dissolve in it**. They dissolve well in less polar organic solvents such as chloroform, ether, benzene, gasoline, etc. Due to this solubility, they are extracted from natural sources. Fats have poor thermal conductivity, which is of great importance for the thermal regulation of animal organisms.

Fats are characterized by appearance, color, odor, melting point for fats, refractive index for oils, and also by some characteristic numbers:

Iodine number - measured by the amount of halogen, e.g. iodine, that joins 100 g of fat. The iodine value is a measure of the unsaturated acid residue content of glycerides.

Acid number - measured in milligrams of potassium base used to neutralize free acids in 100 g of fat. It is a measure of the free acid content of the fat composition.

Saponification number - milligrams of potassium base required to neutralize free and to saponify ester-bound acids in 100 g of fat.

Chemical properties

Hydrolysis of fats - this is one of the most important properties of fats as they are esters:

$$\begin{array}{l} CH_2{-}OCOR \\ | \\ CH{-}OCOR' \\ | \\ CH_2{-}OCOR'' \end{array} + 3\,H_2O \overset{(H^+)}{\rightleftharpoons} \begin{array}{l} CH_2OH \\ | \\ CHOH \\ | \\ CH_2OH \end{array} + \begin{array}{l} RCOOH \\ \\ R'COOH \\ \\ R''COOH \end{array}$$

1,2,3-propanetriol

The process is reversible. If hydrolysis is carried out in an alkaline medium the process is irreversible. In this case, not free carboxylic acids but their salts are obtained.

$C_{17}H_{35}COONa$
(sodium stearate)

The biological oxidation of fat releases twice as much heat energy as the oxidation of the same amount of carbohydrate or protein.

Fat hydrogenation - Liquid fats that contain unsaturated acid residues are hydrogenated. The reaction is carried out in an autoclave under heating and in the presence of an Ni catalyst (finely dispersed in the fat).

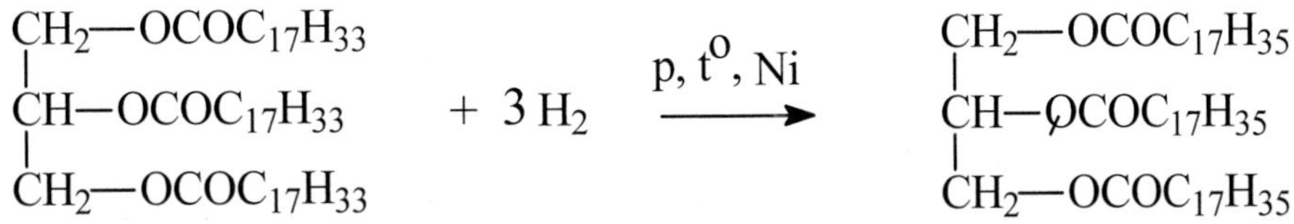

By hydrogenation, the oils are hardened. The reaction has great economic importance for the use of available and cheap liquid fats (vegetable oils, fish oil, etc.). The solid fats obtained by hydrogenation are used in the soap industry, margarine production, etc.

Drying of fats - Glycerides of unsaturated acids on standing in the air gradually harden due to polymerization processes. They are converted into high molecular weight resins. Such fats are drying. Linseed, hemp, poppy, etc. oils are volatile.

Sunflower, cottonseed and other oils are semi-drying oils, which partially polymerize and thicken in air. Saturated and slightly unsaturated glycerides are non-drying (olive, almond, peanut, etc.).

Rancidity of fats - When left to stand for a long time, fats acquire an unpleasant odor due to biochemical changes. The process is called rancidity. Under the action of certain bacteria, fats are partially hydrolysed and the resulting free acids undergo ß-oxidation and decarboxylation to ketones with an unpleasant odor.

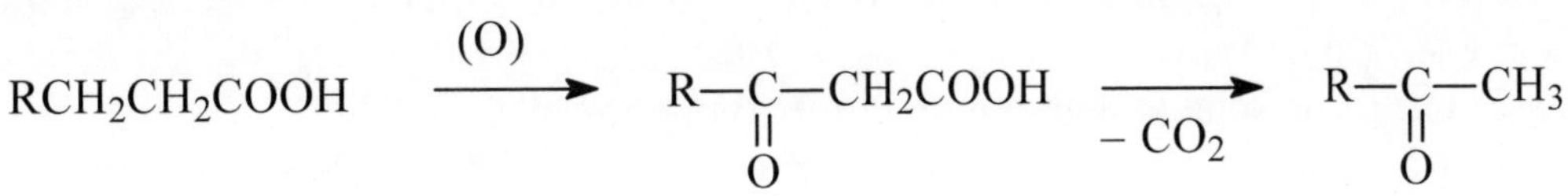

ß-ketoacid

Distribution and use of fats

Fats are involved in the composition of animal and plant cells along with carbohydrates and proteins. Liquid plant fats are a constituent of protoplasm and are nutrients in storage.

About half of the fat produced is used as high-calorie food. The rest is used for technical purposes such as soap production, in medicine and cosmetics. The dried oils are used to make stains, swamp paints, oil varnishes, paints for painting, etc.

Soaps. Composition and properties. Washing action.

Soaps are salts of higher fatty acids with up to 22 carbon atoms. Alkaline soaps in aqueous solution form colloidal solutions and have a washing action. The alkaline earth salts of the higher fatty acids are insoluble in water and have no washing action.

Sodium soaps are solid and potassium soaps are liquid:

Claims:
$C_{17}H_{35}COONa$ sodium stearate
$C_{17}H_{33}COONa$ sodium oleate
$C_{15}H_{31}COONa$ sodium palmitate

Liquid:
$C_{17}H_{35}COOK$ potassium stearate

Because of their washing action, alkaline soaps have been used by humans since more than 3000 years ago.

Washing action is a physicochemical process that consists of the following:

The alkaline salts of the higher fatty acids in aqueous solution dissociate electrolytically:

$$C_{17}H_{35}COO^-Na^+ \rightleftharpoons C_{17}H_{35}COO^- + Na^+$$

The resulting anion has a strongly electronegative and polar carboxylate group and a long non-polar carbon chain containing 15 to 17 carbon atoms. Through the polar group, the soap dissolves in water. However, the non-polar portion makes this dissolution difficult and tends to displace the soap molecules from the interior of the water to its surface. The anions thus accumulate on the surface of the water and greatly lower the surface tension. When the soap solution is agitated, air bubbles enter the soap solution forming a lasting foam. The soap solution wets the impurities and penetrates between them and the surface of the object being washed. The foam, through its large surface area, draws out and retains the impurities in the solution.

The orientation of the soap molecule can be schematically represented as follows:

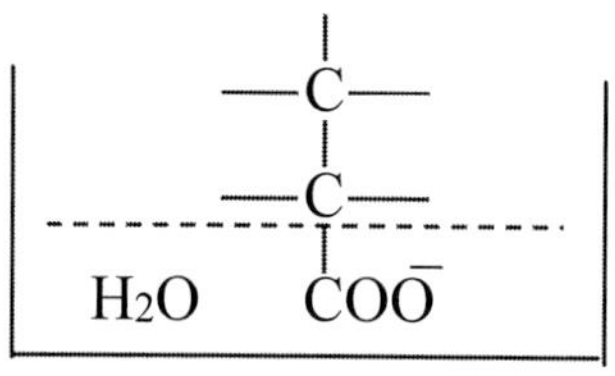

The soap anions surround the particles of impurity (e.g., grease) and form the micelle. The nonpolar residue (R-), denoted by traces, is dissolved in the oil, and the polar end is strongly attracted by water molecules. The oil particle is thus detached from the surface of the object and passes into the soap solution, emulsifying.

Soaps are strong surfactants with a high emulsifying power.

The cleansing action of alkaline soaps is also enhanced by the alkaline base obtained from the hydrolysis of the soap. It changes the composition of some of the impurities and thus facilitates their extraction:

$$\underbrace{C_{17}H_{35}COO^- + Na^+} + \underbrace{H^+ + {}^-OH}_{\uparrow\downarrow \; H_2O} \rightleftharpoons C_{17}H_{35}COOH + \underbrace{Na^+ + {}^-OH}$$

Hydrogen ions (H+) bind in the weak electrolyte stearic acid and an excess of hydroxide ions remain in solution, which determine the alkaline character of the medium. Therefore, alkaline soaps are not suitable for washing wool and silk fabrics.

A characteristic feature of soaps is that in acidic media (H+), or in hard waters containing calcium or magnesium salts, they crosslink due to the formation of insoluble salts.

$$\underbrace{2C_{17}H_{35}COO^- + 2Na^+} + \underbrace{Ca^{2+} + 2Cl^-} \longrightarrow \underbrace{2Na^+ + 2Cl^-} + (C_{17}H_{35}COO^-)_2Ca\downarrow$$

When the **soaps are crossed**, their washing action is reduced.

Use and production of soap

Soaps are mainly used in households. They are also used in engineering: for wetting fabrics before dyeing, for preventing rubber products from sticking to moulds, for sealing paper, in the production of grease, lubricating oils, etc. They have not only a washing but also an antiseptic effect.

Soap is obtained by two methods

Degreasing with an alkaline base:

$$C_3H_5(OCOC_{17}H_{35})_3 \quad + \quad 3\,NaOH \quad \longrightarrow \quad C_3H_5(OH)_3 \quad + \quad 3\,C_{17}H_{35}CO\bar{O}Na^+$$

sodium stearate

Hydrolysis of fats with water vapor and neutralisation of the resulting acids with disodium carbonate:

$$C_3H_5(OCOC_{17}H_{35})_3 \quad + 3\,H_2O \quad \rightleftharpoons \quad C_3H_5(OH)_3 \quad + \quad 3C_{17}H_{35}COOH$$

$$2C_{17}H_{35}COOH \; + \; Na_2CO_3 \; \longrightarrow \; 2C_{17}H_{35}COONa \quad + \quad H_2CO_3$$

$$H_2CO_3 \rightleftharpoons H_2O \; + \; CO_2\uparrow$$

The second method is economically more advantageous due to the cheaper disodium carbonate compared to sodium base, and in addition the glycerol obtained is purer.

The soap forms a concentrated lyophilic colloidal solution from which it is isolated by crosslinking (coagulation) under the action of an electrolyte - sodium chloride. The process is called desalting.

Construction and washing action of synthetic targeting agents (detergents)

Not only do the alkali salts of the higher carboxylic acids have an emulsifying effect, but so do any molecules that contains a polar group and a long nonpolar moiety. Soaps and synthetic detergents are similar in this respect, although they differ in chemical structure. The main practical difference between soaps and detergents is the unequal solubility of their various salts. While the calcium and magnesium salts of the higher fatty acids are insoluble in water, the same salts of detergents are soluble and can be used in hard waters.

It is a prerequisite that synthetic detergents readily break down under the action of microorganisms.

Synthetic detergents can be anions, cations or neutral molecules. The most commonly used detergents are saturated alkyl sulfonates or alkyl sulfonates:

alkylsulfonate

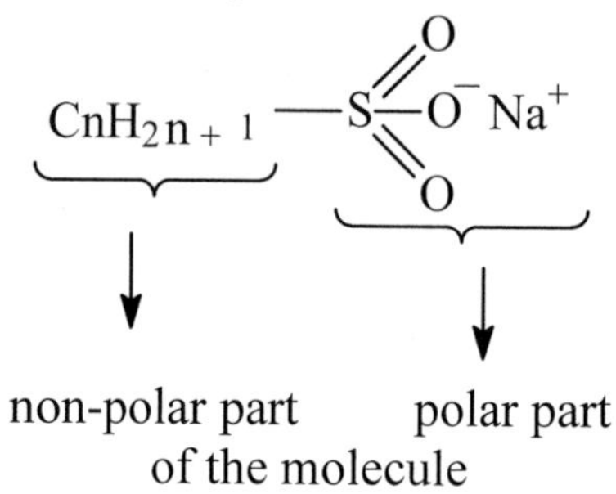

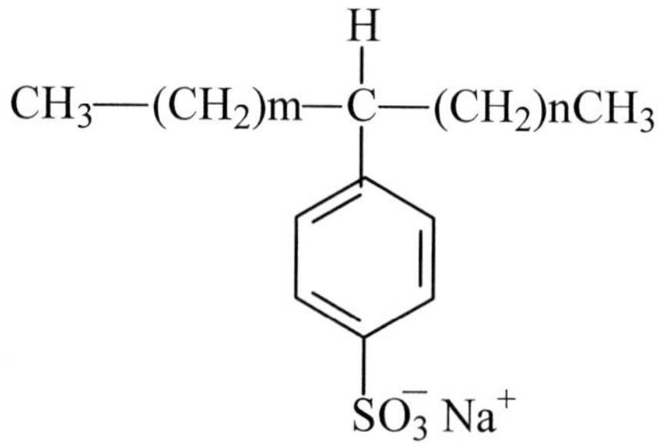

$$(m + n) \approx 12,$$

and sodium salts of sulfuric esters of higher alcohols

$$C_{17}H_{35}CH_2OSO_2\bar{O}\,\overset{+}{Na}$$

They are anionic surfactants, as is ordinary soap.

Quaternary ammonium salts, which are cationic surfactants, have good washing, emulsifying and foaming ability.

$$\left[R_4\overset{+}{N}\right]\bar{Br}$$

They are also called invert soaps. Invert soaps have bactericidal action and are important antiseptics in medical practice.

Detergents include various additives such as salts of inorganic acids, which impart anticorrosive action, soften water and increase washing action; salts of organic acids, which have bleaching, disinfecting and flavoring action.

Detergents, like soaps, are surfactants and the mechanism of their washing action is the same. Detergents have some advantages:

- **They do not hydrolyse** and their solutions do not have an alkaline reaction, therefore they do not damage tissues;
- They also have **better washing action** at lower temperature (~30° C) than soap.

Therefore, they do not cause changes in the strength and color of the tissues;

- **They do not lose their washing action** in hard water or seawater because their ions either do not interact with calcium and magnesium ions or form soluble salts with them.

The disadvantages of synthetic detergents are:

- **They have a strong degreasing effect on the skin**, and in some cases allergic diseases occur;
- **There is a danger of environmental pollution if they are** not easily destroyed in natural conditions.

Use and receipt

Detergents are used in addition to household detergents as:
- **Emulsifiers** in the production of rubber, plastics, chocolate, margarine, etc.;
- **Wetting agents** - in the flotation of ores, in the metal and textile industries, in the production of photographic materials, etc.;
- **Stabilisers** - in the manufacture of agricultural pest control products, etc.

The production of detergents consists in mixing, homogenizing the starting materials and obtaining stable solutions or mixtures. The main raw materials are petroleum refining products.

30. BIFUNCTIONAL NITROGEN- AND OXYGEN-CONTAINING DERIVATIVES OF HYDROCARBONS

AMINO ACIDS. CLASSIFICATION OF AMINO ACIDS. ISOMERISM. NAMES. CHEMICAL PROPERTIES OF AMINO ACIDS. METHODS OF PREPARATION OF AMINO ACIDS. POLYPEPTIDES. PROTEINS. COMPOSITION, PRIMARY AND SECONDARY STRUCTURE. PROPERTIES.

Amino acids

Amino acids are derivatives of carboxylic acids, in the molecules of which one or more hydrogen atoms of the hydrocarbon residue are replaced by amino groups.

Amino acids are solids with a high m.p. (> 230° C) and are readily soluble in water, but difficult to dissolve in less polar organic solvents (ether, benzene, etc.). They have a relatively large dipole moment. These properties are an indication of the ionic (salt-like) nature of amino acids.

Amino acids occupy a central place in the nitrogen metabolism of living organisms. They are sources for the formation of substances necessary for vital activity - proteins, peptides, enzymes, hormones, etc. Man and animals obtain amino acids through protein food. Many amino acids are used in medicine in the treatment of various diseases.

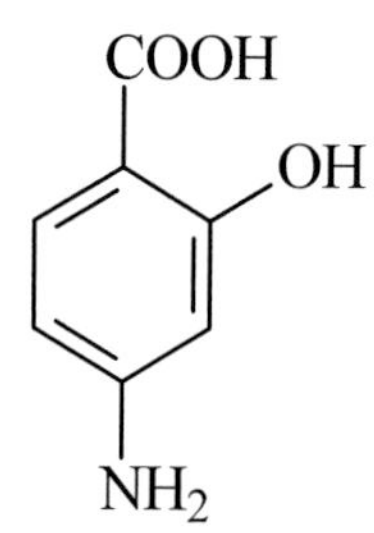

One of the most powerful chemotherapeutics for the treatment of tuberculosis e p-aminosalicylic acid (PASA).

Protein substances. Primary and secondary structure. Properties

The most important **biological functions** of protein substances are:

- structural;
- catalytic (enzymes are protein substances);
- regulatory;
- transport (hemoglobin);
- protective: mechanical (keratin), immune biological (antibodies);
- redundancy.

Protein substances find applications as food products, serums, vaccines, hormones, etc. Proteins are used in plastic, textile, leather industries, etc.

Amino acids. Classification of amino acids

I. **Depending on the number of amino and carboxylic groups,** amino acids are: monoamino-carboxylic, diaminonocarboxylic, monoamino-dicarboxylic, etc.

$$H_2N-CH_2COOH$$

aminoacetic acid

$$H_2N-CH_2-\underset{\substack{|\\ NH_2}}{CH}-CH_2-CH_2COOH$$

4,5-diaminopentanoic acid

$$H_2N-CH_2-CH_2-\underset{\substack{|\\ COOH}}{CH}-CH_2-CH_2COOH$$

2-(2-aminoethyl)-1,5-pentanedioic acid

II. **Depending on the position of the amino group relative to the carboxyl group,** amino acids are: 2- or α-; 3- or β-; 4- or γ-, etc.

$$R-CH_2-\overset{2}{\underset{\substack{|\\ NH_2}}{CH}}-\overset{1}{COOH}$$

2-aminoalkanoic acid (α-amino acid)

$$R-\overset{3}{\underset{\substack{|\\ NH_2}}{CH}}-\overset{2}{CH_2}-\overset{1}{COOH}$$

3- aminoalkanoic acid) (β-amino acid)

III. **According to the type of hydrocarbon residue:**

fatty amino acids

$$R-CH_2-\underset{\substack{|\\ NH_2}}{CH}-COOH$$

aromatic amino acids

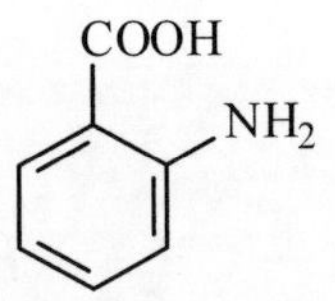

2-aminobenzoic acid
(anthranilic acid)

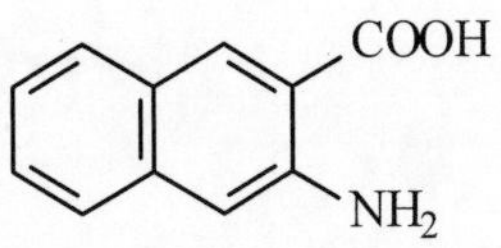

3-amino-2-naphthalene carboxylic acid

Isomerism

For α-, β-, γ-, δ-, etc. amino acids with the same number of carbon atoms there is **chain and positional** isomerism:

(a) **positional isomerism**

$$CH_3CH_2\underset{\substack{|\\ NH_2}}{CH}COOH$$

2-aminobutanoic acid

$$CH_3\underset{\substack{|\\ NH_2}}{CH}CH_2COOH$$

3-aminobutanoic acid

$$\underset{\substack{|\\ NH_2}}{CH_2}CH_2CH_2COOH$$

4-aminobutanoic acid

b) **chain isomerism**

$$CH_3\overset{\substack{CH_3\\ |}}{CH}CH_2\underset{\substack{|\\ NH_2}}{CH}COOH$$

2-amino-4-methylpentanoic acid

$$CH_3CH_2\overset{\substack{CH_3\\ |}}{CH}\underset{\substack{|\\ NH_2}}{CH}COOH$$

2-amino-3-methylpentanoic acid

All α-aminocarboxylic acids except glycine (aminoacetic acid) have an **asymmetric carbon atom** and possess **two enantiomers - D and L**.

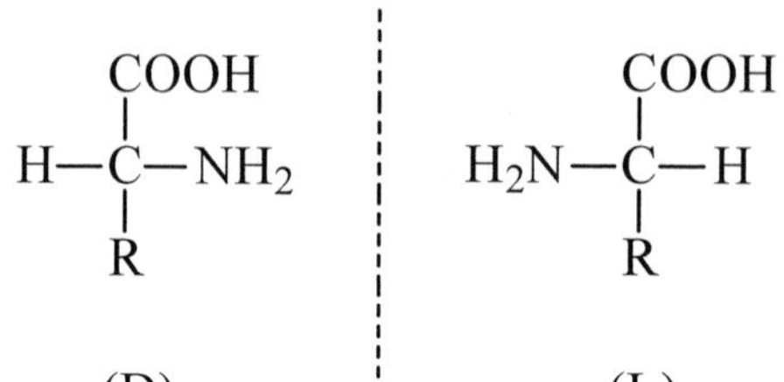

(D) (L)

In nature, mainly L-enantiomers are distributed.

Names

The names of amino acids are formed from those of carboxylic acids plus the prefix - amino. A number or Latin letter marks the position of the amino group. IUPAC rules shall be followed.

Many of the amino acids have trivial names.

H_2N-CH_2COOH

aminoacetic acid (aminoethanoic acid) (glycine)

$CH_3-CH(NH_2)-COOH$

2-aminopropanoic acid (alanine)

$H_2N-CH_2-(CH_2)_4-COOH$

6-aminohexanoic acid (ε-aminocaproic acid)

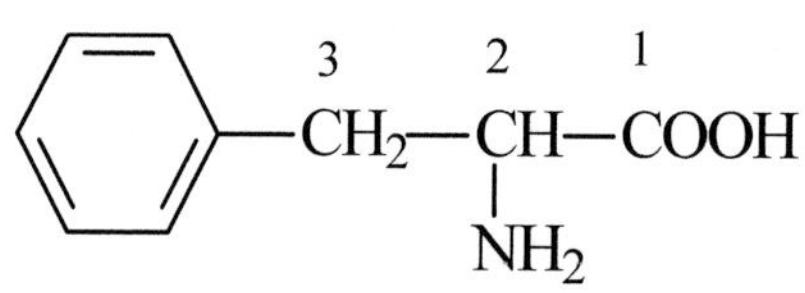

2-amino-3-phenylpropanoic acid (phenylalanine)

Chemical properties of amino acids

α-Aminocarboxylic acids are **ampholytes** - they possess acidic COOH and basic (- NH_2) functional groups. In aqueous solutions, an acid-base equilibrium exists involving bipolar ions (intramolecular salts).

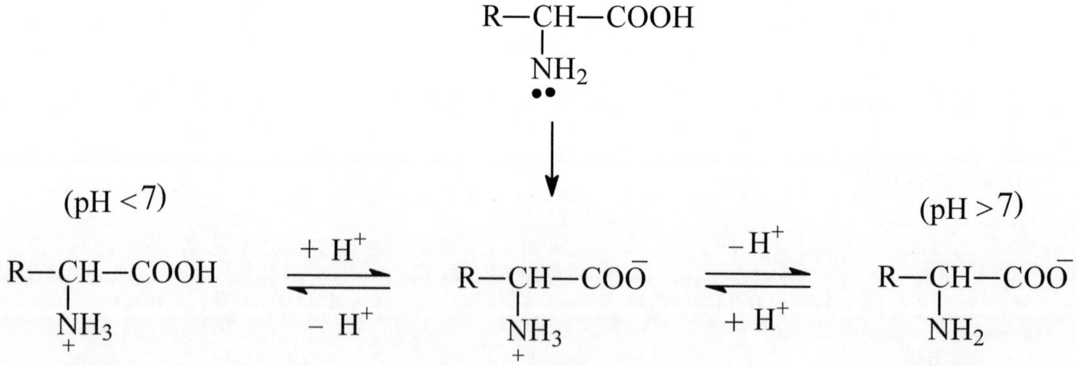

ammonium salt (acid) bipolar ion aminocarboxylate (base)

In an acidic medium (pH < 7), amino acids are referred to as acids, and in an alkaline medium (pH > 7) as bases. At a given pH value, each amino acid exists as a bipolar ion i.e., it is neutral. This pH value is called the **isoelectric point** and is a characteristic value for each amino acid.

The presence of two functional groups opposite in nature determines the amphoteric character of the compounds. They interact with both bases and acids to form the corresponding salts:

$$H_2N{-}CH_2COOH \;+\; \underbrace{Na^+ + \bar{O}H} \;\longrightarrow\; \underbrace{H_2N{-}CH_2COO^- \;+\; Na^+} \;+\; H_2O$$

$$H_2N{-}CH_2COOH \;+\; \underbrace{H^+ + \bar{Cl}} \;\longrightarrow\; H_3\overset{+}{N}{-}CH_2COOH \;+\; \bar{Cl}$$

By **mutual neutralization of** the amino and carboxyl groups, an internal salt is formed.

$$H_2N{-}CH_2COOH \;\rightleftharpoons\; H_3\overset{+}{N}{-}CH_2CO\bar{O}$$

Amino acids form **stable chelate complexes** with transition metals ($CuCl_2$, AgCl, $(CH_3COO)_2Cu$, etc.:

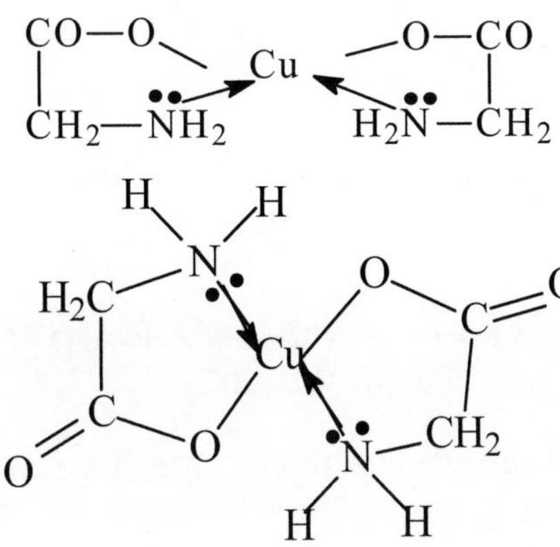

These complexes are dark blue in color and are used in analytical practice and for synthetic purposes.

Amino acids are **esterified**.

$$R{-}\underset{NH_2}{\underset{|}{C}H}{-}COOH \;+\; C_2H_5OH \;\underset{t^o}{\overset{H^+}{\rightleftharpoons}}\; R{-}\underset{NH_2}{\underset{|}{C}H}{-}COOC_2H_5 \;+\; H_2O$$

When **amino acids are heated,** the following products are obtained depending on the position of the amino group:

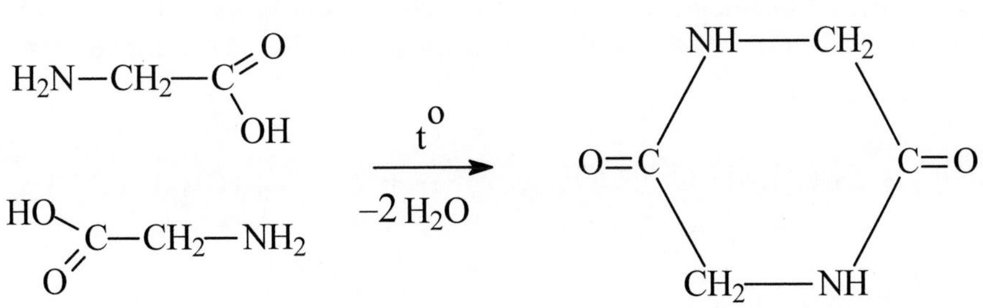

diketopiperazine

ß-Amino acids on heating give off ammonia and form α,ß-unsaturated carboxylic acids:

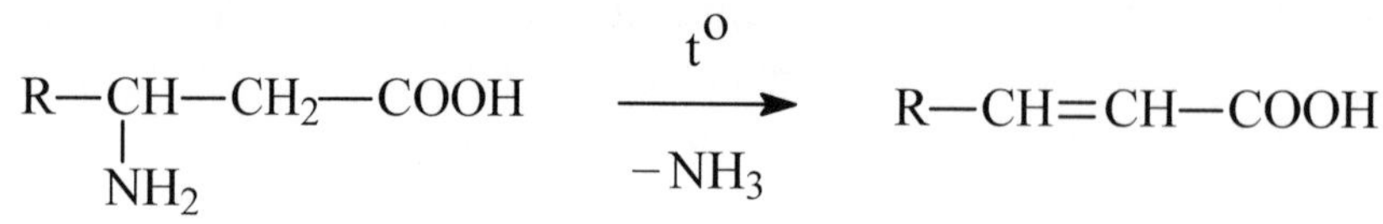

γ-, δ- and ε-Amino acids as a result of **intramolecular dehydration form** ring compounds - lactams:

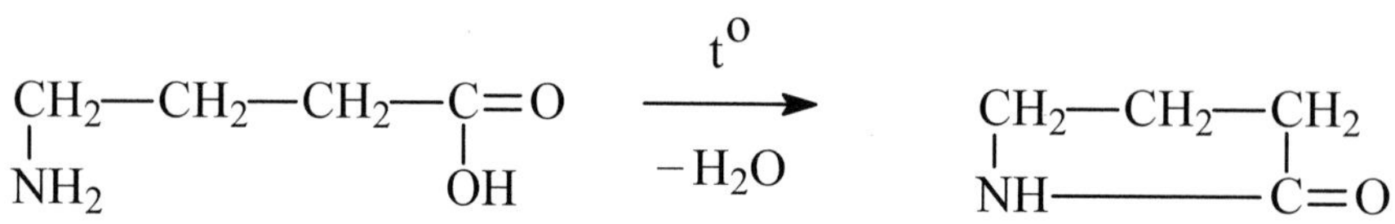

lactam

The dehydration of ε-aminocaproic acid has great practical application. The resulting ε-caprolactam is a feedstock for the polyamide fibre capron:

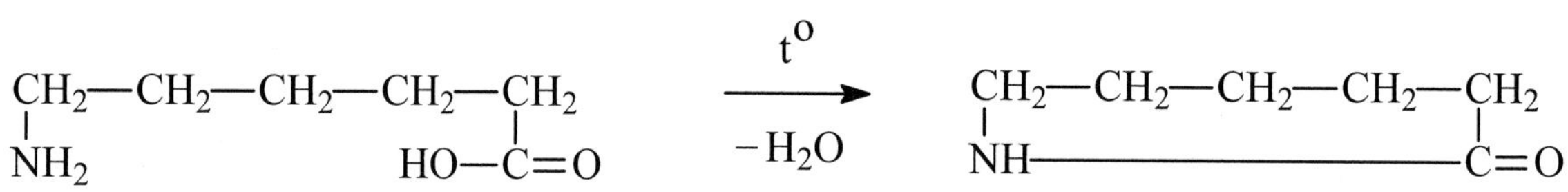

ε-Aminocaproic acid forms **high molecular weight linear amides** and by polycondensation gives the following chain:

$$n\ H_2N(CH_2)_5COOH \longrightarrow [-HN(CH_2)_5CO-]n.....$$

Amino acids interact with **aldehydes** whereby the amino group is blocked. The resulting compounds are Schiff bases:

$$H_2C{=}O + H_2NCH_2COOH \xrightarrow[-H_2O]{t^o} H_2C{=}NCH_2COOH$$

Amino acids react with **nitrous acid**, releasing nitrogen and producing hydroxy acid. One mole of nitrogen is released per amino group, which is why the reaction is used as a method for quantifying amino acids (Slyke method):

$$HO{-}N{=}O + H_2N{-}CH_2COOH \longrightarrow H_2O + N_2 + HO{-}CH_2COOH$$

When **heated in high boiling solvents,** amino acids decompose with the release of carbon dioxide and are converted to amines. Similar decomposition also takes place under the action of bacteria:

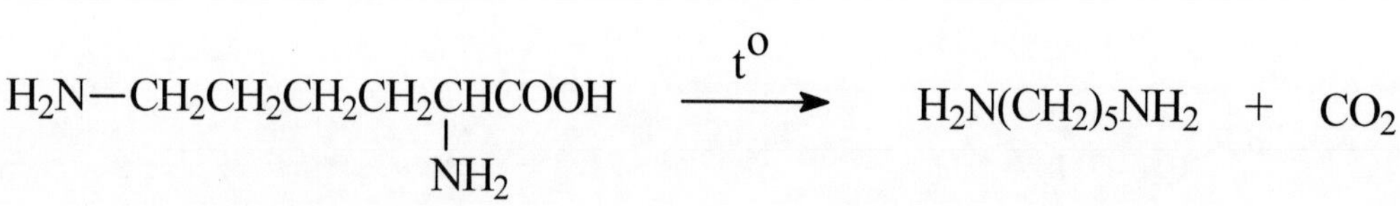

2,6-diaminohexanoic acid (lysine) 1,5-diaminopentane (cadaverine)

The amphoteric nature of amino acids determines the possibility of interaction of two or more identical or different amino acid molecules with each other. In this, one molecule participates in the reaction with its carboxyl group and the other with its amino group.

$$H_2N{-}CH_2CO\boxed{OH + H}{-}NH{-}CH_2COOH \longrightarrow H_2N{-}CH_2CO{-}HN{-}CH_2COOH + H_2O$$

$$H_2N{-}CH_2CO{-}HN{-}CH_2COOH + H{-}NH{-}CH_2COOH \xrightarrow[-\,H_2O]{}$$

$$H_2N{-}CH_2CO{-}HN{-}CH_2CO{-}NH{-}CH_2COOH \longrightarrow \cdots\cdots\cdot$$

polypeptide

Since the newly formed molecule contains both an amino group and a carboxyl group it can react with a third amino acid molecule and so on. According to the number of bonded amino acids, peptides are dipeptides, tripeptides, polypeptides.

The first synthesis of a polypeptide was carried out by Emile Fischer at the beginning of the last century (1907).

A typical qualitative reaction for peptide bond detection is the **biuret reaction**. In an alkaline solution of a peptide under the action of copper sulfate, depending on the size of the peptide molecule, a characteristic color is produced - blue-violet to red-violet.

In aromatic amino acids, due to the conjugation of the uncharged electron pair of the nitrogen atom from the amino group with the aromatic nucleus (p,π-conjugation),

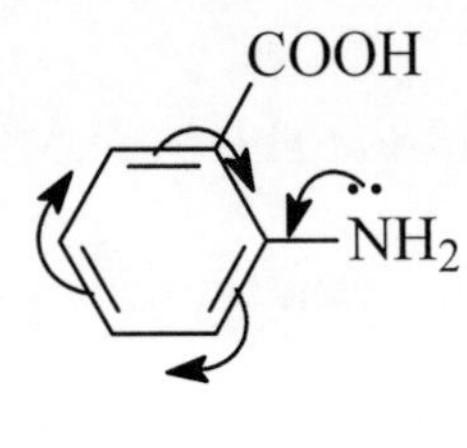

the basicity is lowered and hence the less pronounced saline character. These acids have a lower melting point compared to the melting points of fatty amino acids. Electrophilic reactions are facilitated. The amino group is o-, p-oriented.

Methods of obtaining amino acids

Amino acids are produced by several methods:

1. **α-Halogenated acids are acted** on **with ammonia:**

$$CH_3CH_2\underset{\displaystyle Cl}{\underset{|}{C}}HCOOH + 2NH_3 \longrightarrow CH_3CH_2\underset{\displaystyle NH_2}{\underset{|}{C}}HCOOH + NH_4Cl$$

2. ß-Amino acids are derived from **α, ß-unsaturated acids and ammonia**:

$$CH_2{=}CHCOOH + NH_3 \longrightarrow \underset{\displaystyle NH_2}{\underset{|}{C}}H_2CH_2COOH$$

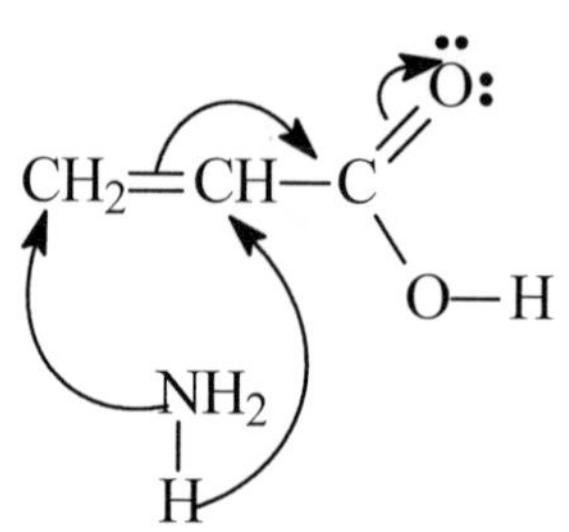

The addition of polar molecules (HX, NH_3, H_2O) proceeds against the Markovnikov's rule. This is due to the negative conjugation effect (mesomeric effect), which results in the π- electrons of the double bond participating in conjugation with the carboxyl group.

3. By **hydrolysis of protein substances** from animal and plant products (blood, horn, etc.).

Polypeptides. Protein substances. Composition and importance of protein substances

Proteinaceous substances (proteins) are natural high-molecular-weight biopolymers composed of α-amino acid residues linked in long polypeptide chains.

Proteins are one of the most important components of every cell. They are involved in the cytoplasm, cell nucleus, blood plasma and lymph. They are found in the motor and supporting tissues of the body (muscles, cartilage, bones, tendons), in the covering and protective formations of the skin (hairs, feathers, nails, etc.). Proteins are one of the body's energy sources. They are the basis of all life.

In the **hydrolysis of proteins**, 20 α-amino acids were detected. Nine of them are called essential because they cannot be synthesized in the human organism, but are obtained through food. Proteins are made up of the following elements:

C 50 - 55 % ; O 21 - 23 % ; S 0.3 - 2.5 % ; H 6.5 - 7.5 % ; N 15.0 - 17.6 %

Some proteins also contain phosphorus, iron (in haemoglobin), magnesium, halogen elements, etc.

The number of α-amino acid residues in proteins varies from 51 (in insulin) to hundreds and tens of thousands.

Structure of protein substances

Simple proteinaceous substances, proteins, are made of α-amino acid residues only. The mode of binding of α-amino acid residues is established by hydrolysis, which takes place in a weakly acidic medium or in the presence of enzymes.

Partial hydrolysis of protein substances produces polypeptides. This indicates that the α-amino acid residues are peptidically linked.

proteins + water $\longrightarrow$ polypeptides $\longrightarrow$ peptides $\longrightarrow$ α-amino acids

In addition to glycine, alanine and phenylalanine in the composition of proteins are involved:

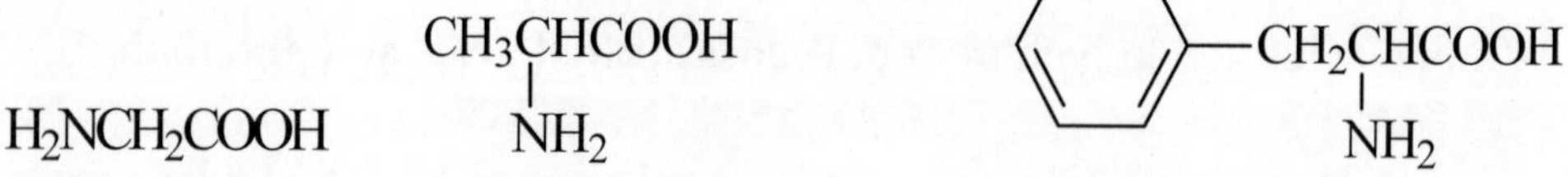

aminoacetic (glycine) | 2-aminopropanoic (α-alanine) | 2-amino-3-phenylpropanoic acid (phenylalanine)

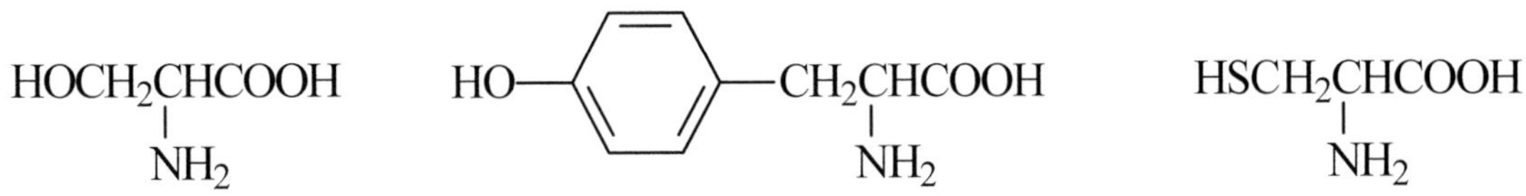

serine tyrosine cysteine

cystine aspartic acid

In 1903 E. Fischer established the peptide theory of protein structure, which has been proven in many ways.

This is the general formula of α-amino acids, and the formation of the polypeptide chain can be represented by the following scheme of polycondensation reaction:

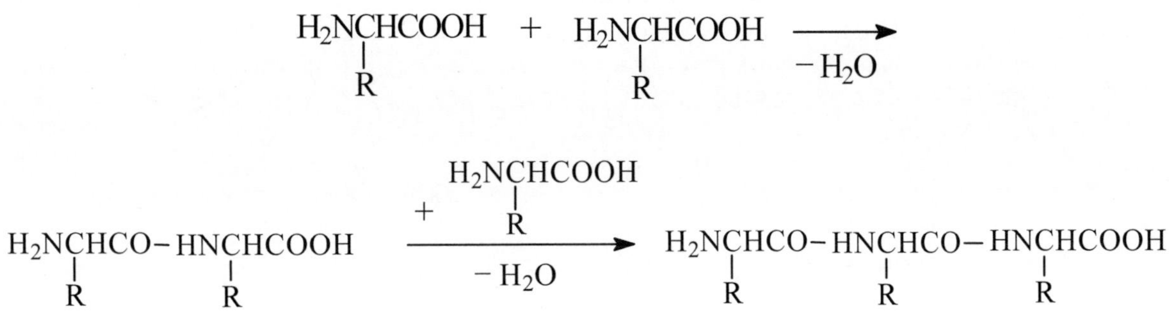

etc.

Structure of the polypeptide chain:

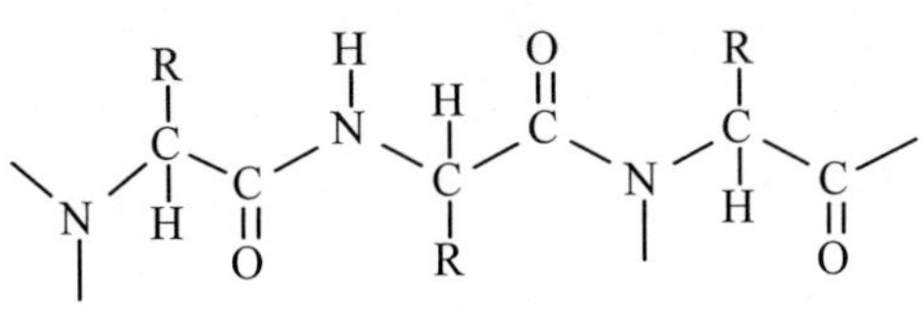

In most cases, polypeptide chains are open, but there are also cyclic polypeptides where both ends of the chain are linked.

The group - NH - CO - is called peptide. All atoms within it lie in one plane as the bonds - NH - and CO are directed opposite.

The R- residues of the side chain alternate on one side and the other side along the chain.

The peptide structure of the proteins was confirmed by X-ray structure analysis and by the first laboratory synthesis of ribonuclease in 1970. Proteins can be built from one to four polypeptide chains. For example, ribonuclease has one chain, insulin has two, haemoglobin has four.

The properties and functions of proteins depend on their primary, secondary, tertiary and quaternary structures.

A) **Primary structure of proteins** - determined by the number, type and sequence of α-amino acid residues in the protein macromolecule.

The differences in the proteins are primarily due to differences in the binding sequence of α-amino acid residues in the macromolecules. Of the three α-amino acids, six isomers are possible that result from different binding modes.

B) Secondary structure

One of the most important conditions in the formation of protein structure is the formation of a maximum number of hydrogen bonds between the carbonyl and amino groups of the main chain.

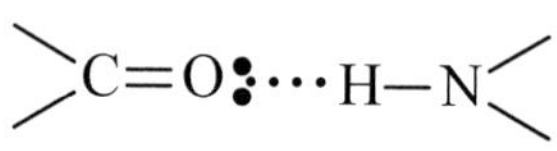

Hydrogen bonds can form between favorably positioned peptide groups in separate parts of one or more chains. Each peptide group -NH - CO - can participate in the formation of two hydrogen bonds through the uncharged electron pairs of the oxygen and nitrogen atoms. The formation of a large number of hydrogen bonds in a polypeptide chain causes it to fold into a helix (like a spring). This spatial organization of a protein macromolecule is called a secondary helical structure.

Another type of helical structure, called a corrugated or sheet-like or ß-form, is also possible. It is formed by maximally stretched polypeptide chains linked together by transverse hydrogen bonds. It is found in silk fibroin.

C) Tertiary structure

It is a higher degree of organization of protein macromolecules. It arises from the formation of additional hydrogen bonds, ionic, ester, disulfide, hydrophobic, and other interactions involving functional groups from R-residues in the general formula of amino acid residues:

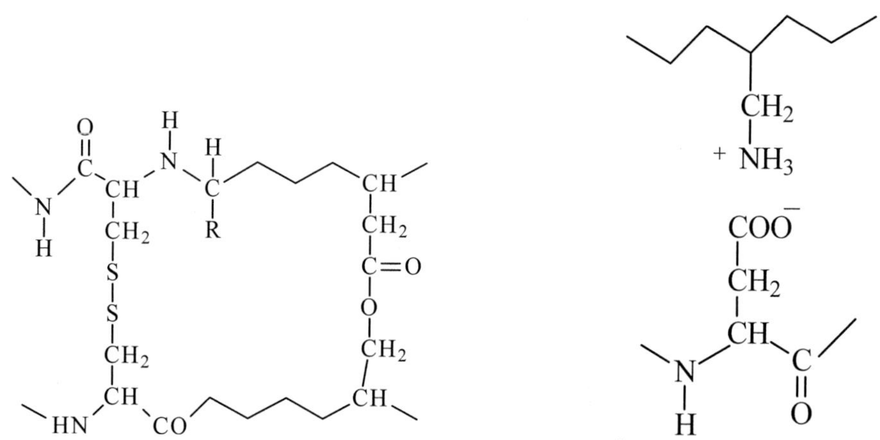

disulfide bond ester bond ionic bond

When occurring between more distant sites in the polypeptide chain, the bonds cause further folding and intertwining of the protein macromolecules. As a result, the molecules stabilize and acquire a more compact shape. The tertiary structure is absolutely necessary for the normal functions of proteins.

D) Quaternary structure

It is a supramolecular structure. It is possible in proteins that consist of several chains. The quaternary structure is formed by individual subunits (structural groupings). Weak hydrogen and hydrophobic bonds occur between them that are not engaged in the other structures (primary, secondary and tertiary). These bonds are weak, which is why proteins with a quaternary structure can reversibly dissociate.

Types of protein substances

1. **According to the composition and structure** of macromolecules, proteins are divided into two large groups: proteins and proteids.

A) **Proteins** (simple) - are made of α-amino acid residues only;

B) **Proteins** (complex) - are made up of proteins and a non-protein moiety called a prosthetic group.

Depending on the non-protein part, proteins can be:

- **Phosphoproteins** - contain a phosphoric acid residue;
- **Glycoproteins** - contain a carbohydrate residue;
- **Lipoproteins** -contain residual fat (lipids);
- **Chromoproteins** - contain a residue of colored substances;
- **Nucleoproteins** - are protein macromolecules linked to nucleic acids.

Proteinaceous substances can have different macromolecular shapes, which are mainly determined by the secondary and tertiary structure.

2. **According to the shape of macromolecules**, proteins are:

A) **fibrillar (filamentous)** - with elongated filamentous macromolecules. Such are, for example, collagen (of the skin), keratin (hair, wool), fibrinogen of silk, etc.

B) **globular (spherical).** They have a spherical or elliptical shape of macromolecules. For example, egg white, blood hemoglobin, blood plasma, etc.

Properties of protein substances

Proteinaceous substances are solid, mostly amorphous. Most of them are colorless or slightly yellow colored. The variety in the structure of proteins determines their diverse properties.

The solubility of protein substances depends on the arrangement of polar groups in their molecules. A small fraction of proteins are soluble in water (egg white) and most are soluble in electrolyte solutions. Therefore, many water-insoluble proteins in cells are in the dissolved state. Generally, globular proteins form colloidal solutions and fibrillar proteins are insoluble. Proteins do not dissolve in alcohol and other organic solvents.

A characteristic feature of protein substances is their relative instability. Under the action of various factors (heating, irradiation with various rays, the presence of acids, bases, salts of heavy metals, alkaloids, alcohols, etc.) changes in the structure of protein substances occur. The process is called **denaturation**. The changes that occur are mainly related to disruption of the secondary structure. In denaturation, the bonds that determine the secondary structure are broken. As a result, the specific shape of the protein macromolecules changes and they lose their biological action.

Reversible and irreversible coagulation of colloidal solutions of protein substances is possible. Reversible coagulation is caused, for example, by the action of acetone, sodium chloride, ammonium chloride, etc.

In irreversible coagulation, deeper changes occur in which peptide bonds are broken

—HN⋮CO—

and chemical interaction between the protein and the denaturing agent takes place. Irreversible coagulation occurs under the action of salts of heavy metals (lead, copper, mercury, etc.), at high temperature, etc.

When proteins are burned, volatiles with a discernible odor are emitted, which serves to detect proteins.

Chemically, proteins can participate in a variety of reactions, as they contain different functional groups: hydroxyl-, imino-, amino-, carboxyl-, aromatic core, sulfhydryl, disulfide groups, etc.

An important property of protein substances is their hydrolysis, which takes place in acidic or alkaline environments. In many cases it is preferred to carry out enzymatic hydrolysis under the action of enzymes called peptidases or proteases.

Typical qualitative reactions for protein detection are:

A) **Biuret reaction** - used to detect a peptide bond. The reaction is carried out in copper sulfate solution in alkaline medium. A complex compound with blue violet color is formed.

B) **Xanthoprotein reaction** - used to detect aromatic nuclei in the side chains of amino acid residues, e.g. in phenylalanine, tyrosine, etc. A golden yellow coloration is obtained. The reaction is carried out in nitric acid.

Protein substances are of great importance for the existence of living matter.

Sample problems and solutions

Task 1.

Suggest a scheme for obtaining the amino acid glycine from starch. With which of the following substances can glycine react: HCl, NaOH, C_2H_5OH, PCl_3, CaO, H_2NCH_2COOH. Express the possible reactions with chemical equations.

Exemplary solution

Starch is a natural high-molecular substance. It has been established that starch is not uniform. Its composition includes amylose and amylopectin. Their quantity depends on the origin of the starch, but usually the amylose content is 25 % and the amylopectin content is 75 %. Starch is composed of α-glucose residues and its structure can be expressed as follows:

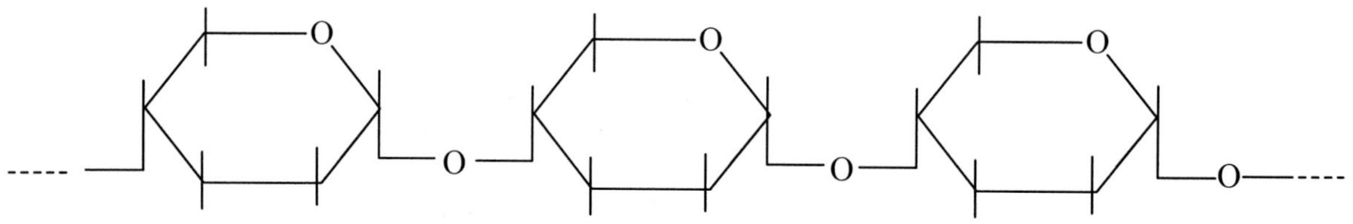

On heating in the presence of dilute inorganic acids or under the action of the enzyme amylase, starch is hydrolysed to glucose.

Hydrolysis is a stepwise process:

$$\underset{\text{starch}}{(C_6H_{10}O_5)_n} + nH_2O \xrightarrow[\text{(enzyme)}]{H^+} \underset{\text{glucose}}{nC_6H_{12}O_6}$$

Under certain conditions glucose undergoes alcoholic fermentation and ethanol is produced and carbon dioxide is released. The enzyme cymase causes alcoholic fermentation:

$$C_6H_{12}O_6 \longrightarrow 2\,C_2H_5OH + 2\,CO_2$$

Alcohol is converted to ethanal by dehydrogenation.

$$C_2H_5OH \xrightarrow[t^o]{Cu} CH_3CHO + H_2$$

Ethanal is readily oxidized to ethanoic (acetic) acid using freshly precipitated copper dihydroxide. The sky-blue color of copper dihydroxide changes to tile-red because dicopper oxide is released:

$$CH_3CHO + 2\,Cu(OH)_2 \xrightarrow{t^o} CH_3COOH + Cu_2O\downarrow + 2\,H_2O$$

The hydrogen atoms in the methyl group of ethanoic acid exhibit greater mobility due to electron interactions in the carboxyl group. The mobility allows substitution of a hydrogen atom in the methyl group by an atom of a halogen element. Depending on the quantitative ratios between the acid and the halogen, all three hydrogen atoms can be substituted. The process is catalysed by red phosphorus.

$$CH_3COOH + Cl_2 \xrightarrow{P} CH_2(Cl)COOH + HCl$$

chloroethanoic acid

Chloroethanoic acid is treated with a twofold excess of ammonia to give the final product, glycine (aminoethanoic acid):

$$CH_2(Cl)COOH + 2NH_3 \longrightarrow H_2NCH_2COOH + NH_4Cl$$

In their structure amino acids contain two functional groups of opposite chemical character - a carboxyl and an amino group. This implies the interaction between the two functional groups. Therefore, amino acid molecules exist in the form of bipolar ions. This is the reason why amino acids dissolve well in water and have high melting temperatures. In addition, amino acids react with both substances with acidic properties and those with basic properties.

$$H_2NCH_2COOH + HCl \longrightarrow [H_3\overset{+}{N}CH_2COOH]Cl^-$$

With hydrochloric acid reacts the amino group. The proton from the miteral acid bonds to the uncharged electron pair of the nitrogen atom and an ammonium salt is formed - glycylammonium chloride.

$$H_2NCH_2COOH + NaOH \rightleftharpoons H_2NCH_2COONa + H_2O$$

Glycine interacts with sodium base. A neutralisation process takes place. The process is reversible, since sodium glycinate is a salt of weak acid and strong base and easily hydrolyzes.

Ethanol undergoes esterification, a slow, molecular, reversible process catalyzed by mineral acids. Ester and water are separated.

$$H_2NCH_2COOH + C_2H_5OH \rightleftharpoons H_2NCH_2COOC_2H_5 + H_2O$$

The amino acid can react with phosphorus trichloride to produce the acid chloride glycinoyl chloride or aminoacetyl chloride.

$$3\ H_2NCH_2COOH + PCl_3 \longrightarrow 3\ H_2NCH_2COCl + H_3PO_3$$

The interaction of glycine with CaO is possible - the interaction of an acid with a basic oxide. Calcium glycinate is obtained.

$$2\ H_2NCH_2COOH + CaO \longrightarrow (H_2NCH_2COO)_2Ca + H_2O$$

Amino acids can interact with each other.

$$H_2NCH_2COOH + H_2NCH_2COOH \longrightarrow H_2NCH_2CONHCH_2COOH + H_2O$$

The dipeptide glycylglycine is produced. The process is known as polycondensation.

Task 2.

How will 1-iodobutane and inorganic reagents yield 2-aminopentanoic acid? Write the chemical equations for the interactions of 2-aminopentanoic acid with sodium base, with hydrochloric acid, and with ethanol in the presence of sulfuric acid.

Exemplary solution

1-Iodobutane reacts with potassium cyanide to give pentanenitrile.

$$CH_3CH_2CH_2CH_2I + KCN \longrightarrow CH_3CH_2CH_2CH_2CN + KI$$

The hydrolysis of pentanenitrile in an acidic medium results in the production of pentanoic acid.

$$2\,CH_3CH_2CH_2CH_2CN + 4\,H{-}OH \xrightarrow[t^o]{H_2SO_4} 2\,CH_3CH_2CH_2CH_2COOH + (NH_4)_2SO_4$$

Pentanoic acid undergoes halogenation in the presence of red phosphorus. The hydrogen atoms bonded to the carbon atom adjacent to the carboxyl group are mobile and this is a prerequisite for their participation in a substitution reaction.

$$CH_3CH_2CH_2CH_2COOH + Cl_2 \xrightarrow{P} CH_3CH_2CH_2\underset{\displaystyle Cl}{\underset{|}{C}}HCOOH + HCl$$

The 2-chloropentanoic acid is treated with an excess of ammonia to give 2-aminopentanoic acid.

$$CH_3CH_2CH_2\underset{\displaystyle Cl}{\underset{|}{C}}HCOOH + 2\,NH_3 \longrightarrow CH_3CH_2CH_2\underset{\displaystyle NH_2}{\underset{|}{C}}HCOOH + NH_4Cl$$

The amino acid can interact with sodium base. The process is neutralization.

$$CH_3CH_2CH_2\underset{NH_2}{\underset{|}{C}}HCOOH + NaOH \rightleftharpoons CH_3CH_2CH_2\underset{NH_2}{\underset{|}{C}}HCOONa + H_2O$$

The interaction with hydrochloric acid involves the amino group. An ammonium salt is obtained.

$$CH_3CH_2CH_2\underset{NH_2}{\underset{|}{C}}HCOOH + HCl \rightleftharpoons [CH_3CH_2CH_2\underset{\overset{+}{NH_3}}{\underset{|}{C}}HCOOH]\,Cl^-$$

The interaction of an acid with an alcohol in the presence of sulfuric acid is called esterification.

$$CH_3CH_2CH_2\underset{NH_2}{\underset{|}{C}}HCOOH + CH_3CH_2OH \overset{H_2SO_4}{\rightleftharpoons} CH_3CH_2CH_2\underset{NH_2}{\underset{|}{C}}HCOOCH_2CH_3 + H_2O$$

31. Oxygen-containing Bifunctional Derivatives

Carbohydrates. General classification of carbohydrates. Composition, molecular structure and comparative consideration of the properties of glucose, sucrose, starch and cellulose

The name carbohydrate arose to denote compounds that have the same ratio of oxygen and hydrogen atoms as water and that are compounds of carbon - $C_m(H_2O)_n$. Later, a number of compounds have been found which do not conform to this formula, but by other properties are referred to carbohydrates.

Classification of carbohydrates

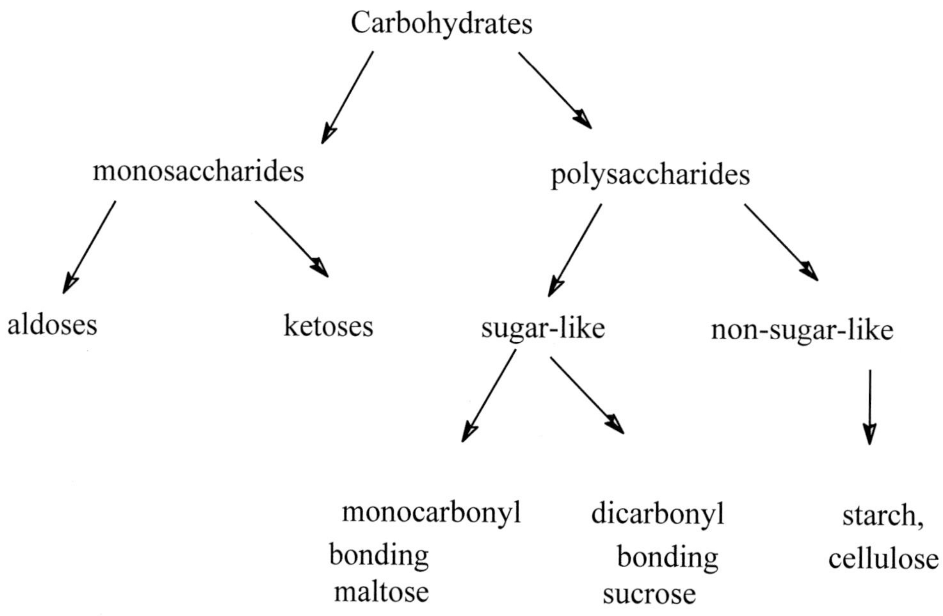

Monosaccharides are regarded as the products of multivalent alcohols in whose molecules on oxidation of one hydroxyl group at a primary or secondary carbon atom is converted to an aldehyde or ketone group, respectively.

Aldozis | **Ketozis**

$HO{-}CH_2{-}CHO$

aldobiosis

$HOCH_2{-}C(=O){-}CH_2OH$

ketotriosis

$HO{-}CH_2{-}CH(OH){-}CHO$

aldotriosis

$HO{-}CH_2{-}CH(OH){-}C(=O){-}CH_2OH$

ketotetrosis

⋮

$HO{-}CH_2{-}CH(OH){-}CH(OH){-}CH(OH){-}CH(OH){-}CHO$

aldohexosis

$HO{-}CH_2{-}CH(OH){-}CH(OH){-}CH(OH){-}C(=O){-}CH_2OH$

ketohexosis

Physical and chemical properties of monosaccharide

Monosaccharides are colorless, hard-crystallising substances that taste sweet and are soluble in water. When heated, they melt, caramelize and char with the evolution of water vapor. Glucose is a colorless crystalline substance which is well soluble in water, and with a sweet taste.

It occurs as a syrupy liquid because it is difficult to crystallize in the presence of impurities. When heated, it melts, caramelizes and chars with the evolution of water vapor. It therefore contains carbon, hydrogen and oxygen atoms.

Composition and construction

It has been established that the formula of glucose is $C_6H_{12}O_6$ and that the carbon atoms are linked in a straight chain.

Qualitative and quantitative analysis suggests the existence of carboxyl, carbonyl and hydroxyl groups.

Glucose does not change the color of blue litmus, therefore it does not contain a carboxyl group.

From the large number of oxygen atoms and the sweet taste, the presence of hydroxyl groups can be assumed. The identification of these groups, as in the case of glycerol, is possible by the interaction of glucose with freshly precipitated copper dihydroxide at room temperature. A dark blue colored, water-soluble complex compound called copper glucosate is obtained. This is a qualitative reaction for the detection of hydroxyl groups in polyvalent alcohols bonded to adjacent carbon atoms. The number of hydroxyl groups was established by esterification of glucose with acetic acid anhydride. One mole of glucose is esterified with five moles of acetic acid anhydride; therefore, it contains five hydroxyl groups.

Glucose reacts with silver ammonia hydroxide (Tolens reagent) and with Fehling's solution, therefore it contains an aldehyde group.

This allows the structure of glucose to be expressed as follows:

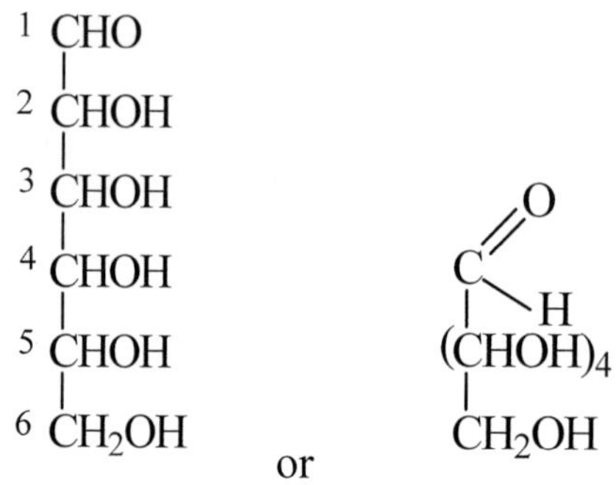

It has been found that, in glucose, a rearrangement of atoms is possible in which a hydrogen atom from the hydroxyl group at the fifth carbon atom jumps to the oxygen atom from the aldehyde group; the bonds released at the oxygen atom at the fifth site and the carbon atom from the aldehyde group saturate each other and a ring is formed. The ring form can also be expressed as follows:

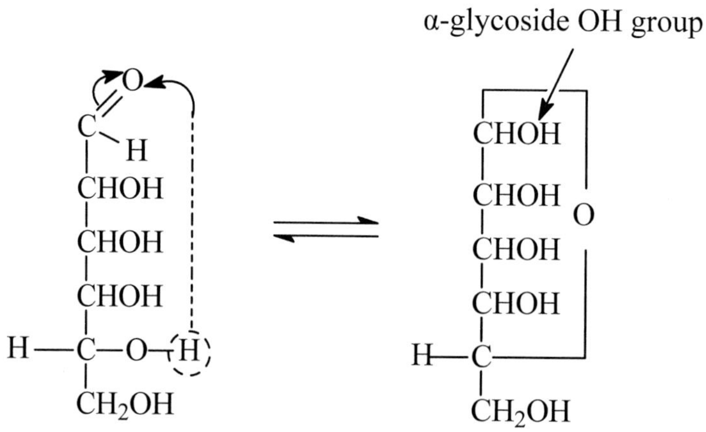

glucose (aldehyde form) glucose (ring form)

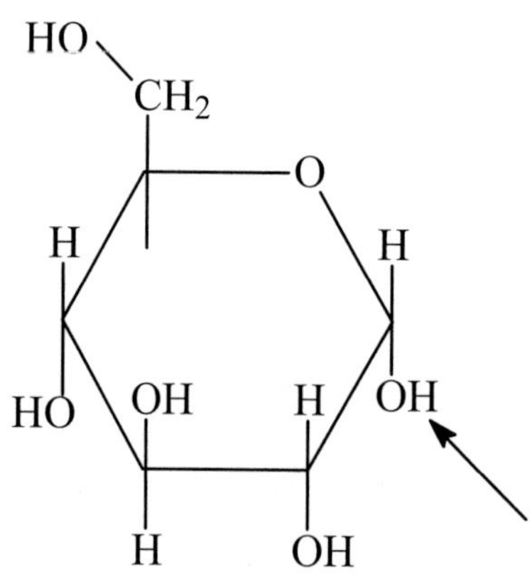

α- glycoside OH group

The newly formed hydroxyl group is called a glycosidic group. Prospective formulas of glucose:

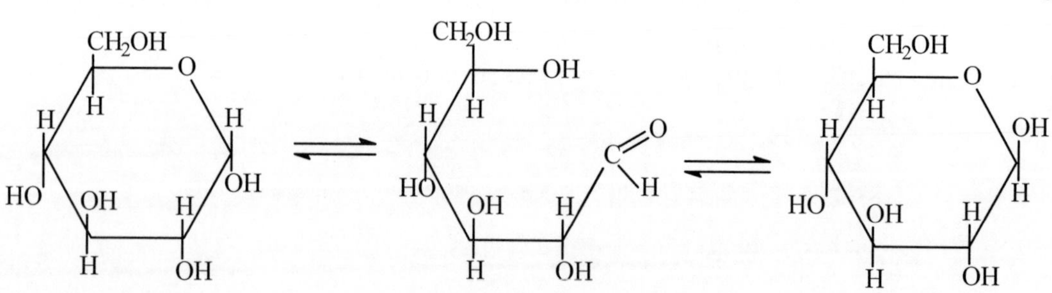

α-glucose aldehyde form β-glucose

In the solid state, glucose is in its ring form. In solution there is an equilibrium between the aldehyde and ring forms.

Other compounds are known with the same formula, which are isomers of glucose (16 in number). One of the most important is fructose. Glucose is found in grapes in the greatest quantity and is called grape sugar, while fructose is fruit sugar. Both are found in honey. Fructose differs from glucose in that it tastes sweeter and is also dissolved in alcohol.

Fructose has been found to contain a ketone group at the second carbon atom. It is also possible to regroup the atoms and form a ring.

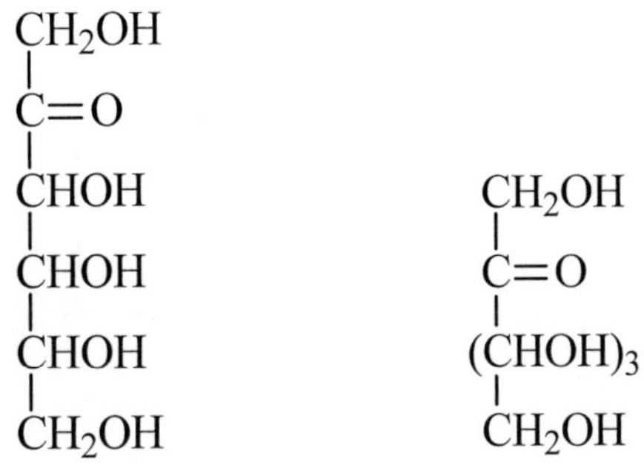

fructose

It can also be expressed as follows:

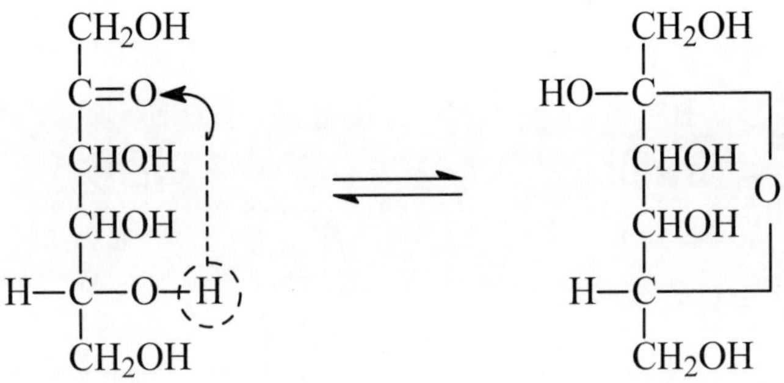

It can also be expressed as follows:

fructose (ring shape)

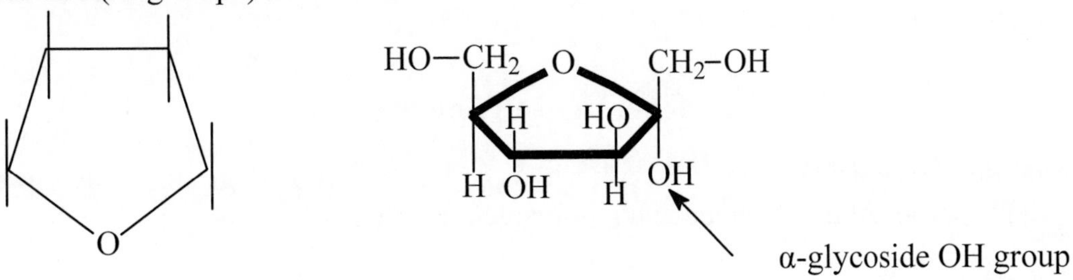

Due to the same qualitative and quantitative composition with glucose, the presence of hydroxyl and carbonyl groups, fructose, like glucose, interacts with copper dihydroxide, hydrogen, esterifies. Unlike ketones, fructose interacts with silver ammonia hydroxide and Fehling's solution as it isomerizes to glucose in alkaline medium. Fructose ferments under the action of various enzymes.

Of the pentoses, the most important is ribose

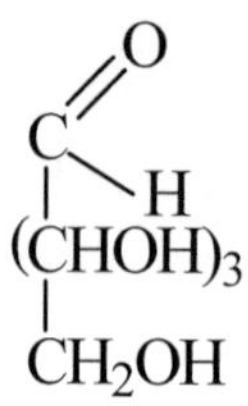

It also involves the regrouping of atoms.

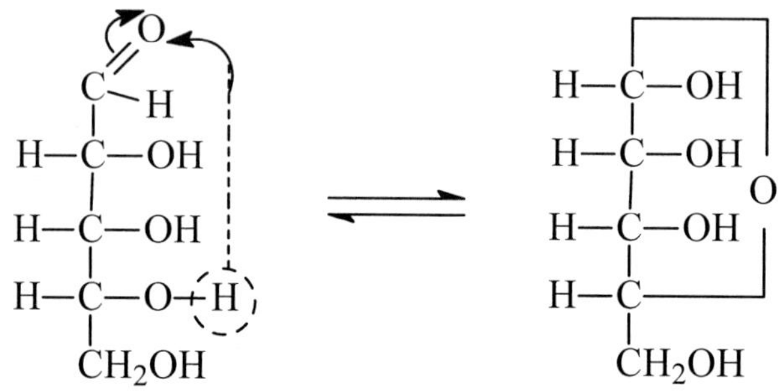

ribose (ring form)

Ribose is a constituent of nucleic acids.

Compounds similar in structure to monosaccharides also occur in nature. This is *L-ascorbic acid*, known under the name *vitamin C*. L-ascorbic acid is the lactone of 2-keto-L-gulonic acid in a stable enol form:

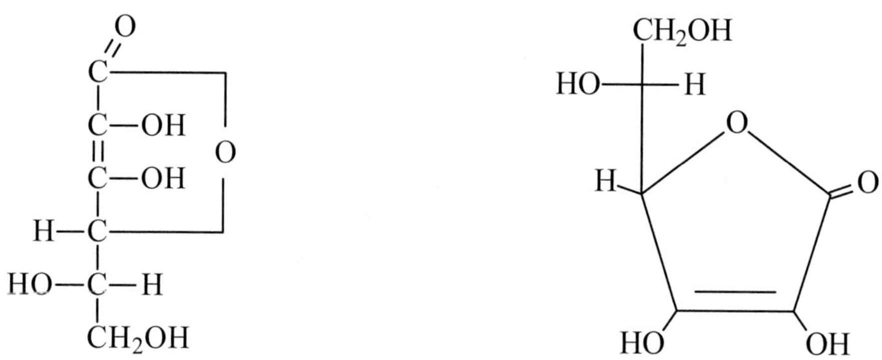

γ-lactone of 2-keto-L-gulonic acid

Under industrial conditions, vitamin C is obtained from D-glucose by a multi-step synthesis. It has acidic character and reducing properties. It is contained in some fruits and vegetables. It is one of the important vitamins. Deficiency of vitamin C in the body causes scurvy disease (bleeding gums and teeth falling out).

Chemical properties

The glucose molecule contains alcoholic hydroxyl and carbonyl groups. Therefore, glucose exhibits the properties of an aldehyde and a polyvalent alcohol.

- **Chemical reactions involving hydroxyl groups:**
- Interaction with **freshly precipitated copper dihydroxide.**

A dark blue, water-soluble complex compound is obtained, which is more accurately described as follows:

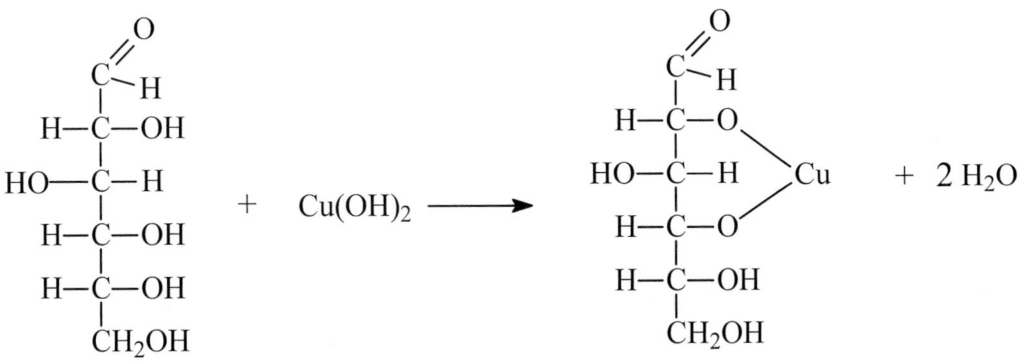

copper glucosate

It has been found that copper dihydroxide can also react with two molecules of glucose to give the following complex compound:

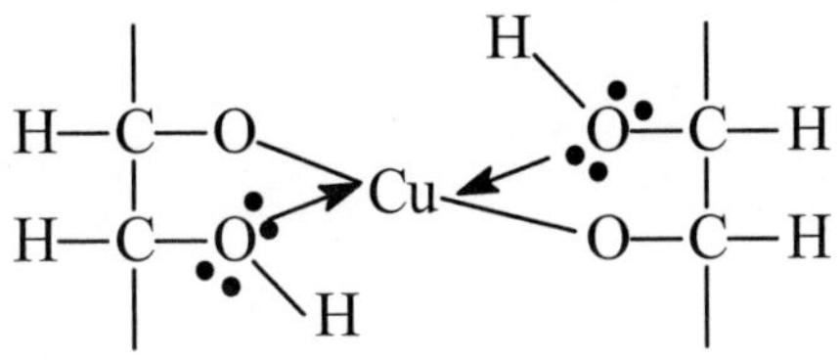

Interaction with **calcium dihydroxide:**

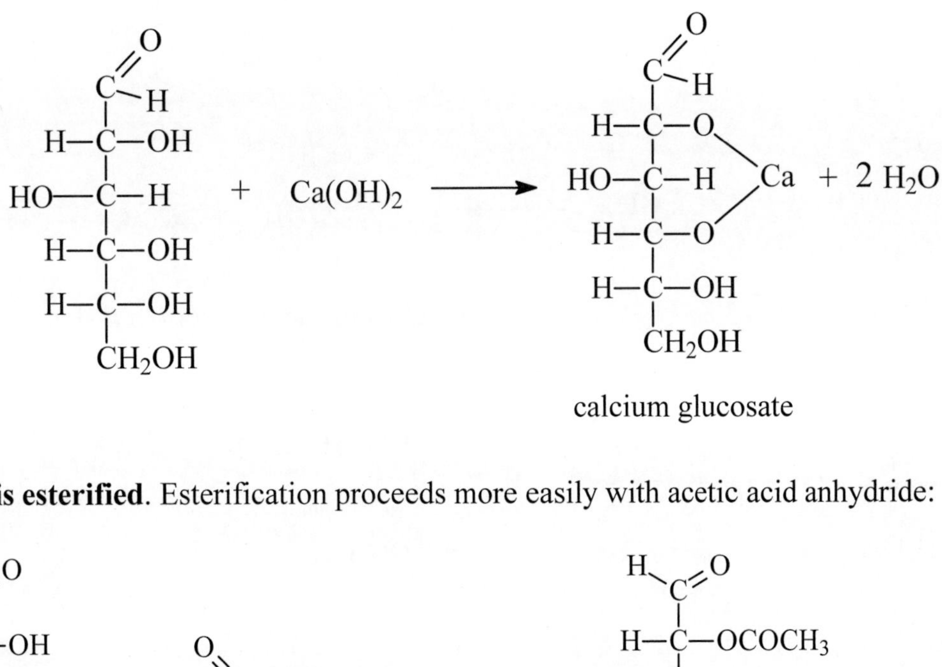

calcium glucosate

Glucose is esterified. Esterification proceeds more easily with acetic acid anhydride:

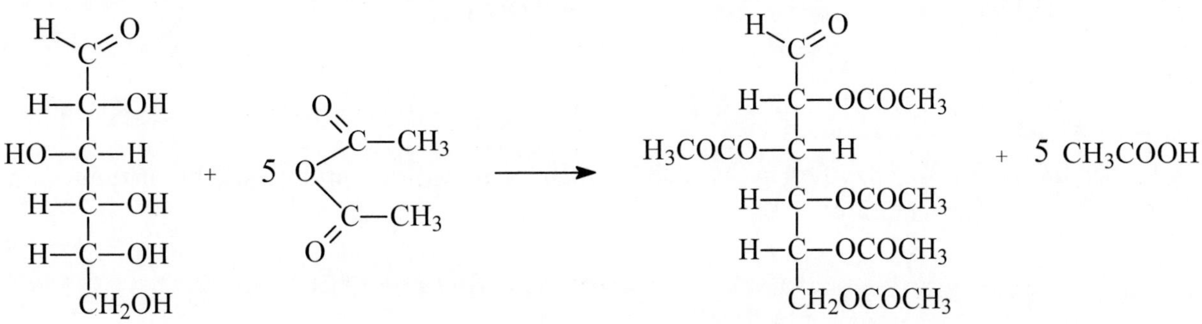

pentaacetylglucose

2. Chemical reactions affecting the aldehyde group:

Interaction with hydrogen **cyanide**

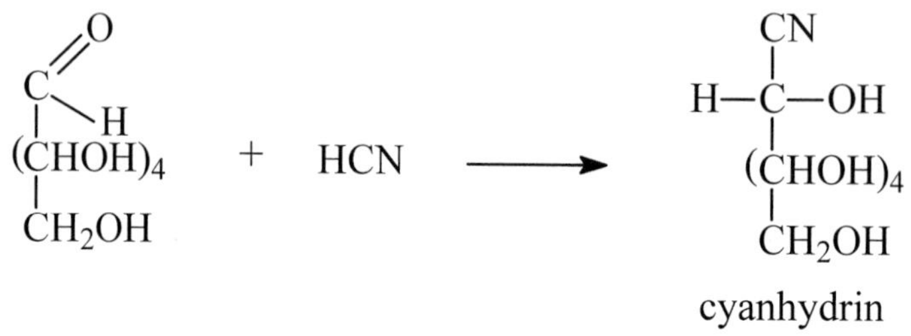

Interaction with hydrogen - hexavalent alcohol is obtained.

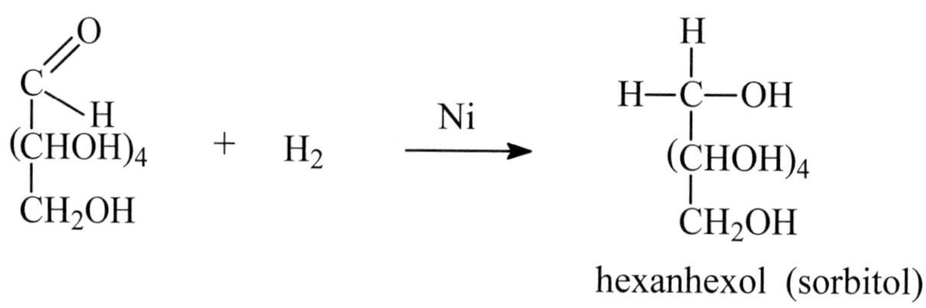

The presence of an **aldehyde group** is demonstrated by the interaction of glucose with **Fehling's solution** and with a diammoniacal solution of silver hydroxide **(Tolens reagent)** - the "silver mirror" reaction.

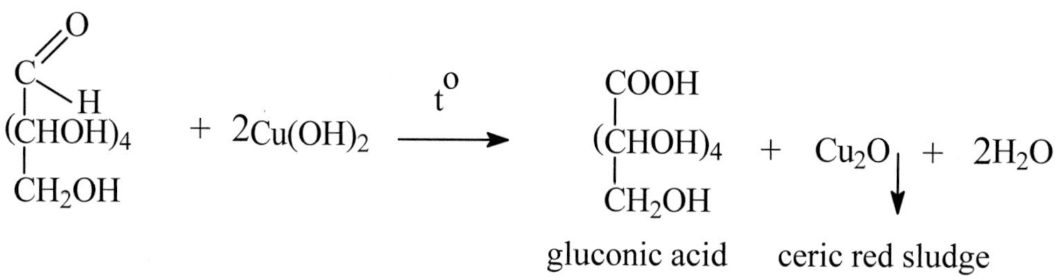

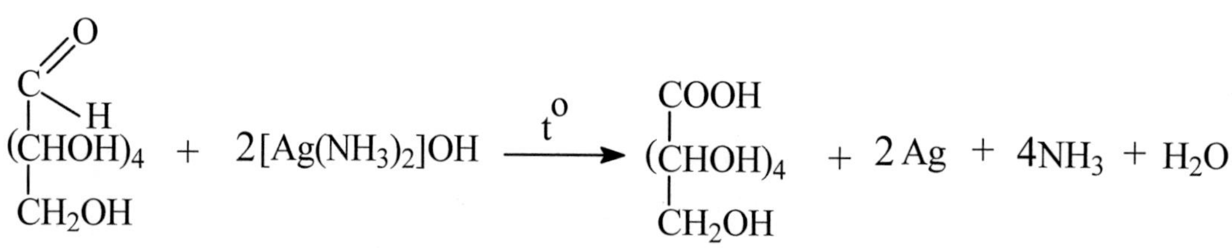

3. Reactions involving the whole molecule

Under the action of enzymes, glucose is fermented to various products. The enzyme cimase causes alcoholic fermentation:

$$C_6H_{12}O_6 \longrightarrow 2\,C_2H_5OH + 2\,CO_2$$

Glucose is easily absorbed by the body. Its oxidative breakdown in the body releases the required amount of energy. Substances processed with glucose are difficult to crystallize. Its reducing properties are used in tanning leather and making mirrors.

Disaccharides

Disaccharides can be viewed as derived from monosaccharides upon dehydration of two hydroxyl groups. There are two types of binding of monosaccharide molecules into disaccharide and two types of disaccharides, respectively.

Monocarbonyl bonding is accomplished by the oxygen atom produced by dehydration of one hydroxyl and one glycosidic (obtained by regrouping the atoms) groups. Such disaccharides have a potential free aldehyde group and can interact with silver ammonia hydroxide and Fehling's solution.

Dicarbonyl bonding is accomplished by the oxygen atom produced by dehydration of two glycoside groups. These disaccharides have no potential aldehyde group and cannot reduce silver ammonia hydroxide and Fehling's solution. They are called trehalose-type disaccharides.

Maltose is a monocarbonyl-linked disaccharide and can reduce silver ammonia hydroxide and Fehling's solution.

The dicarbonyl-linked disaccharide is sucrose. In practice it is known as common sugar. The sucrose is melted, caramelized, producing a liquid amorphous mass called candy sugar.

At a higher temperature, it chars. It has a sweet taste and is soluble in water.

It also consists of carbon, hydrogen and oxygen atoms. The similarity in properties with glucose suggests the presence of the same functional groups. The sweet taste and oxygen atoms suggest the presence of hydroxyl groups. Sucrose does not reduce silver ammonia hydroxide and Fehling's solution - it is a dicarbonyl-linked disaccharide.

The presence of hydroxyl groups can be demonstrated by the interaction with copper and calcium dihydroxides. With copper dihydroxide it forms a dark blue solution. If reacted with lime milk, a clarification of the solution is observed due to the formation of soluble calcium saccharate.

$$C_{12}H_{22}O_{11} + Cu(OH)_2 \longrightarrow \underset{\text{copper saccharate}}{C_{12}H_{20}O_{11}Cu} + 2H_2O$$

$$C_{12}H_{22}O_{11} + Ca(OH)_2 \longrightarrow \underset{\text{calcium saccharate}}{C_{12}H_{20}O_{11}Ca} + 2H_2O$$

The sucrose molecule has been found to have the following structure:

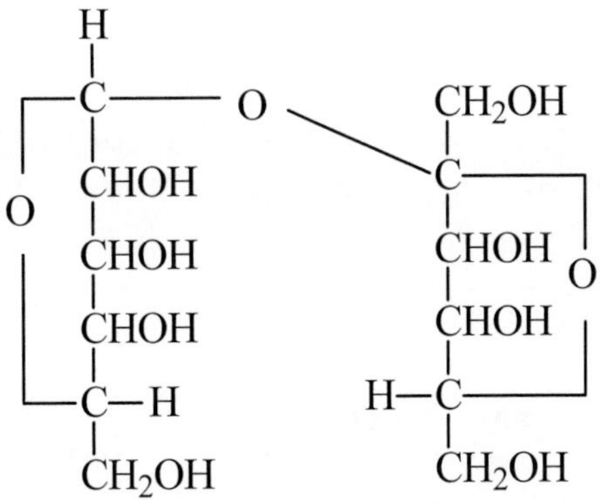

It is formed by the dehydration of two glycoside hydroxyl groups from glucose and fructose molecules. Glucose and fructose are involved with their ring forms. The structure of sucrose can also be expressed in the following way, which indicates that it contains α-D-glucose and ß-D-fructose:

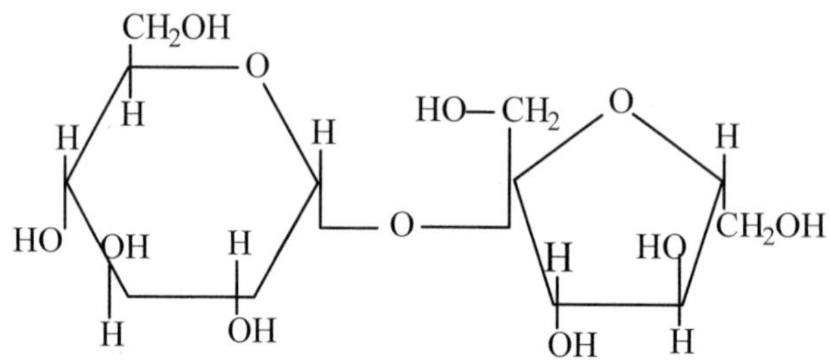

This structure of sucrose has been established by its hydrolysis. Glucose and fructose are obtained in equal amounts:

$$\underset{}{C_{12}H_{22}O_{11}} + H_2O \xrightarrow[\text{enzyme}]{} \underset{\text{glucose}}{C_6H_{12}O_6} + \underset{\text{fructose}}{C_6H_{12}O_6}$$

The resulting mixture of glucose and fructose is called invert sugar. Natural invert sugar is honey.

Polysaccharides

Polysaccharides are hydrolysed to monosaccharides. In turn, polysaccharides are sugar- like and non-sugar-like. The sugar-like ones are made up of a small number - two, three monosaccharide residues. They resemble monosaccharides in that they taste sweet and are well soluble in water.

Non-saccharide polysaccharides are composed of a large number of monosaccharide residues, they have no sweet taste and are slightly soluble or insoluble in water.

The most common polysaccharides are starch and cellulose.

Starch is a white, amorphous-looking substance. Electron microscopic studies show that it has a microcrystalline structure. It does not melt on heating but chars. It is hygroscopic.

With an alcoholic solution of iodine (iodine tincture) forms a compound with a blue color, which disappears on heating, and reappears on cooling. This is a **qualitative reaction for starch detection.**

When heated to 160-185° C, a specific mass is formed - the kóla. It is used for collaging the tissues.

Starch does not react with silver ammonia hydroxide and Fehling's solution. It reacts only after hydrolysis to give glucose.

This indicates that the starch is made up of glucose residues. Different starch molecules contain different numbers of glucose residues from 200 to 6000; the large number of glucose residues mean that starch molecules are macromolecules. Starch is a naturally high molecular weight substance.

On heating in the presence of dilute inorganic acids or under the action of the enzyme amylase, starch is hydrolysed to glucose.

Hydrolysis is a stepwise process:

$$\underset{\text{starch}}{(C_6H_{10}O_5)_n} + nH_2O \xrightarrow[\text{(enzyme)}]{H^+} \underset{\text{glucose}}{nC_6H_{12}O_6}$$

It has been established that starch is not a homogeneous substance. It consists of two substances: amylose and amylopectin. Their quantity depends on the origin of the starch, but usually the amylose content is 25 % and the amylopectin content 75 %. Starch is composed of α- glucose residues and its structure can be expressed as follows:

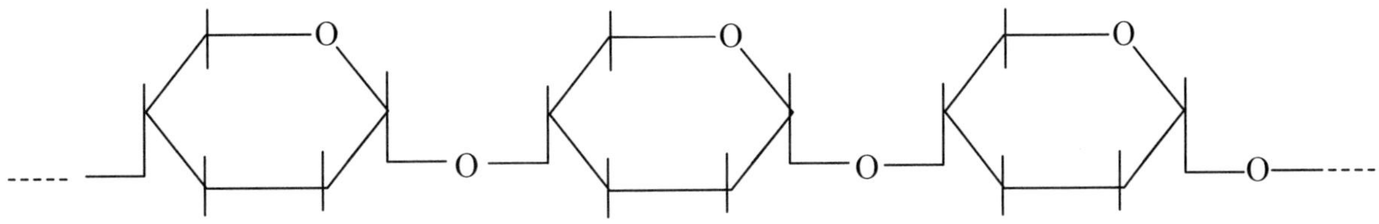

Cellulose is widely distributed in nature. It is contained in cell walls, in wood, in fibrous plants. Almost pure cellulose is cotton (98 %).

Cellulose, like starch, is a high-molecular-weight substance composed of glucose residues, but differs from starch in the type, structure and spatial arrangement of glucose residues in macromolecules. It is made up of ß-glucose residues that are linked in straight chains to form long fine strands that are very strong because of the numerous hydrogen bonds between the glucose residues.

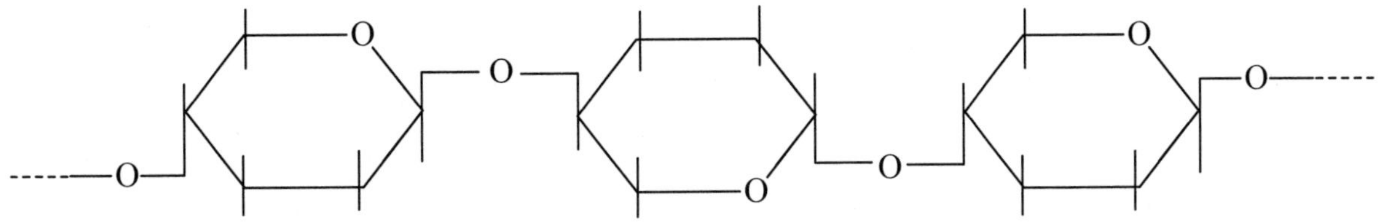

cellulose

Cellulose is insoluble in water. It dissolves in Schweitzer's reagent (ammonia solution of copper dihydroxide), in 40 % hydrochloric acid, tin dichloride, etc.

When heated, the cellulose does not melt but chars. It burns. Glucose residues in the cellulose macromolecule can be up to 10000.

Cellulose is hydrolysed in steps on heating in the presence of inorganic acids or under the action of enzymes - cellulases and glucose is obtained:

$$(C_6H_{10}O_5)_n + nH_2O \xrightarrow[(H^+)]{\text{cellulase}} nC_6H_{12}O_6$$

The structure of cellulose can be represented as follows:

$$\left[C_6H_7O_2\begin{matrix}\diagup OH\\ -OH\\ \diagdown OH\end{matrix}\right]_n$$

Chemical properties

The presence of hydroxyl groups determines the property of cellulose to participate in the esterification process.

Cellulosic nitrates have great application in practice. Different numbers of hydroxyl groups are esterified under different conditions. Esterification takes place when cellulose is treated with a nitric mixture (consisting of concentrated sulfuric and concentrated nitric acids).

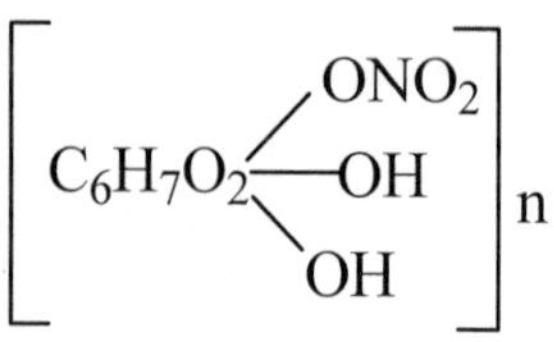

cellulose mononitrate (colloxiline)

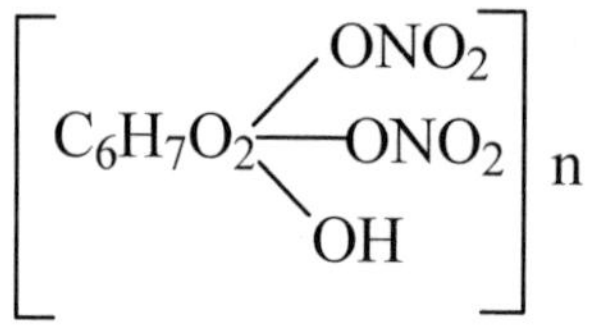

cellulose dinitrate (pyroxiline)

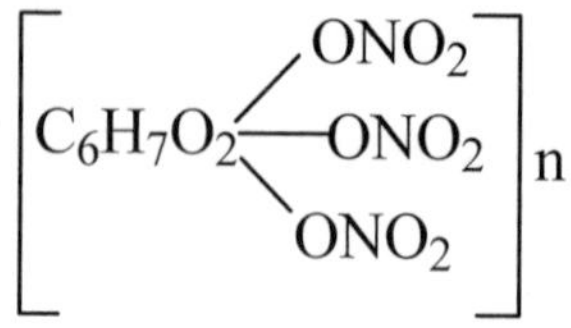

cellulose trinitrate

Cellulose esters with acetic acid - cellulose acetates - are also of great use.

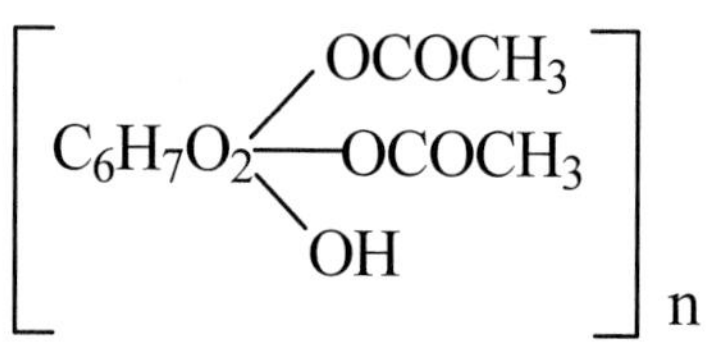

cellulose diacetate

Composition, molecular mass and structure of some carbohydrates

Composition (formula)	Molecular mass	Construction
Glucose $C_6H_{12}O_6$	Strictly defined, the same for all molecules; M = 180; low molecular weight compound	Pentahydroxyaldehyde, molecular crystal lattice
Sucrose $C_{12}H_{22}O_{11}$	Strictly defined, the same for all molecules; M = 342; low molecular weight compound	Polyhydroxyl derivative with carbonyl groups employed in the creation the oxygen bond between mono harridan residues; molecular crystal lattice
Starch $(C_6H_{10}O_5)_n$	Different for different groups of molecules; M = 30000 to 1000000; high molecular weight compound	Monosaccharide residues are bind- in highly branched and in straight chains; granular structure
Cellulose $(C_6H_{10}O_5)_n$	Different for different groups of molecules; M = 200000 to 5 - 6000000; high molecular weight compound	Monosaccharide residues are bind- in a straight chain; fibrous structure tour

The similarity in the properties of carbohydrates stems from the same qualitative composition and the same functional groups contained in them. The differences are due to the different structure, molecular mass and spatial arrangement of the carbohydrate molecules.

The similarities as well as the differences between the carbohydrates are shown in the following table:

Properties	Glucose	Sucrose	Starch	Cellulose
Physical constants	There are well-defined values of physical constants	There are well-defined values of physical constants	The values of the physical constants move within certain limits	The values of the physical constants move within certain limits
Solubility in water	Dissolves	Dissolves	A little soluble	Does not dissolve
Strength and robustness	Does not possess	Does not possess	Does not possess	Very large
Hydrolysis	Does not hydrolyse	Hydrolysed	Hydrolysed	Hydrolysed
Reduction properties	Strong reducer	Not a reducer	Not a reducer	Not a reducer
Esterification	Esterify	Esterify	Esterify	Esterify

When comparing the properties of carbohydrates, there is a change in properties from monosaccharides to polysaccharides.

Sample problems and solutions

Task 1.

Express with chemical equations the preparation from **ethane** and inorganic reagents of phenylacetate and acetic acid. Acetylate glucose. Name the resulting compound. Identify the asymmetric carbon atoms in **glucose**. What type of isomerism do they define and how many such isomers does the carbonyl form of glucose have?

Exemplary solution

The preparation of acetic acid from ethane can be represented as follows

- Ethane undergoes halogenation by a chain-radical mechanism (UV-light irradiation):

$$CH_3{-}CH_3 \ + \ Cl_2 \xrightarrow{hv} CH_3{-}CH_2Cl \ + \ HCl$$

The mechanism of the halogenation reaction is presented in detail in the topic "Alkanes". Chloroethane is treated with an aqueous solution of an alkaline base to give ethanol:

$$CH_3{-}CH_2Cl \ + \ NaOH \xrightarrow{H_2O} CH_3{-}CH_2OH \ + \ NaCl$$

If an alcoholic solution of an alkaline base is used, an elimination reaction takes place. In this case, instead of ethanol, ethene will be obtained.

Alcohol undergoes dehydrogenation using glowing copper wire. Ethanal (acetaldehyde) is obtained.

$$CH_3{-}CH_2OH \xrightarrow{Cu} CH_3{-}CHO + H_2$$

Aldehydes are good organic reducing agents and are oxidized by weak oxidizing agents (silver-ammonia hydroxide - Tolens reagent, Fehling's solution or freshly precipitated copper dihydroxide):

$$CH_3{-}CHO + 2\,[Ag(NH_3)_2]OH \xrightarrow{t^o} CH_3{-}COOH + 2\,Ag + 4\,NH_3 + H_2O$$

Acetic acid cannot be used to esterify phenols and polyvalent alcohols or aldoses and ketoses. These hydroxyl derivatives of hydrocarbons exhibit albeit weak acidic properties. This is the reason for carrying out the esterification with acetic acid anhydride.

Acetic acid anhydride is produced by intermolecular dehydration using diphosphoricpentoxide:

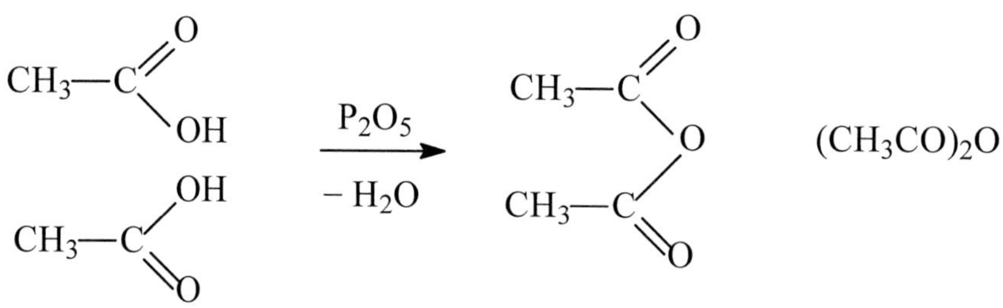

Acetic acid can also be obtained according to the scheme:

$$H_3C{-}CH_3 \underset{+\,H_2}{\overset{-\,H_2}{\rightleftharpoons}} H_2C{=}CH_2 \underset{+\,H_2}{\overset{-\,H_2}{\rightleftharpoons}} HC{\equiv}CH$$

$$HC{\equiv}CH \xrightarrow[Hg^{+2};\ H^+]{H_2O} [CH_2{=}CH{-}OH] \rightleftharpoons CH_3CHO$$

$$CH_3CHO \xrightarrow{2Cu(OH)_2} CH_3COOH + Cu_2O\downarrow + 2H_2O$$

The acetylene obtained by the above scheme can be used to obtain phenol. Initially, cyclic trimerisation takes place and benzene is obtained.

From benzene, phenol can be obtained by halogenation to chlorobenzene, and then the chlorine atom is replaced by a hydroxyl group.

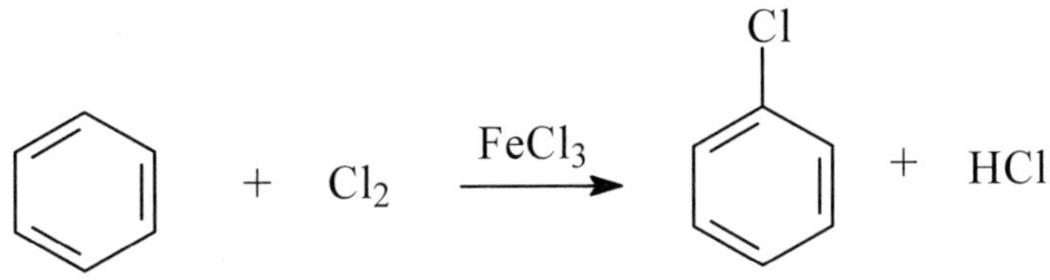

The carbon-halogen bond is partially double due to electron interactions between an uncharged electron pair from the chlorine atom with the π-sequence of the aromatic nucleus. This makes substitution of the chlorine atom more difficult compared to halogen in halogenoalkanes and requires more rigid reaction conditions. The reaction proceeds under elevated pressure, high temperature and catalyst salts of monovalent copper.

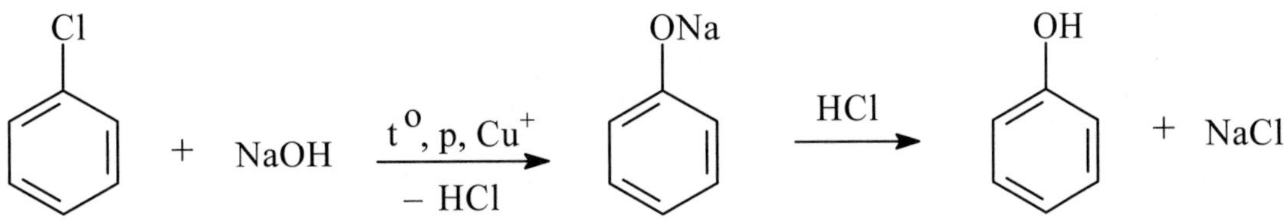

Phenol is subjected to esterification with acetic acid anhydride. The process is irreversible because the by-product is not water but acetic acid. The product is called **phenylacetate**.

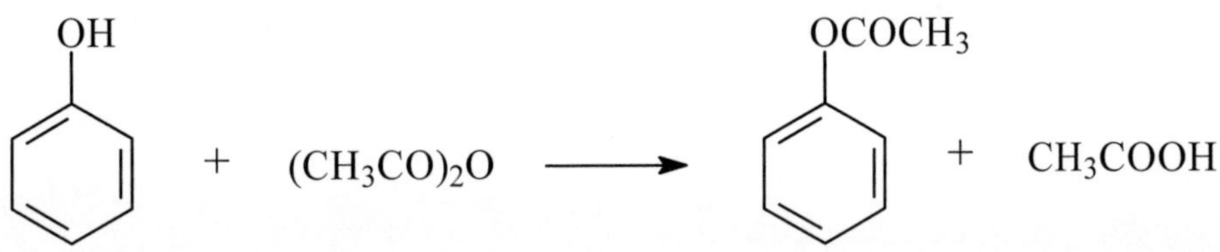

Glucose is esterified. Esterification proceeds more easily with acetic acid anhydride:

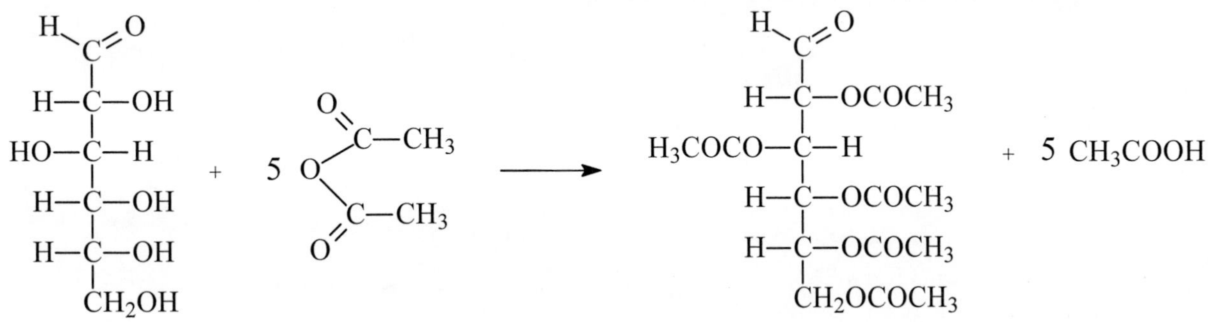

pentaacetylglucose

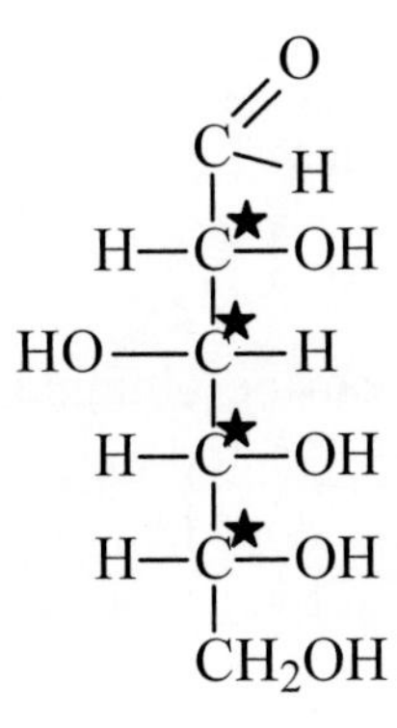

There are four asymmetric carbon atoms in the glucose molecule. A carbon atom bonded to four different substituents is called asymmetric. Its presence in an organic molecule is a prerequisite for optical isomerism. In an asymmetric carbon atom, there are two optical isomers that rotate the plane of polarized light in two opposite directions. The two isomers are optical antipodes. The number of optical isomers is determined by the formula 2^n. The value of **n** is determined by the number of asymmetric carbon atoms. In this case **n = 4**. The number of optical isomers of the carbonyl form of glucose is 16.

Task 2.

To obtain from cellulose ethanol, and from it suitable inorganic reagents, a compound of gross formula $C_7H_5NO_3$, which is known to exhibit reducing properties and to join hydrogen cyanide. What are the products of complete nitration and acetylation of cellulose?

Exemplary solution

The structure of cellulose can be represented by the following formula

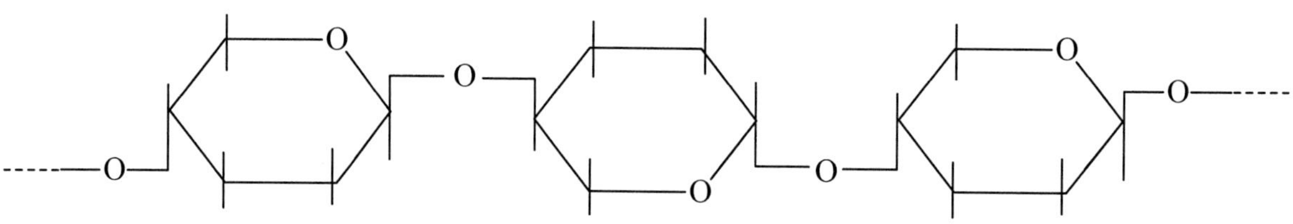

cellulose

and abbreviated:

$$\left[C_6H_7O_2(OH)_3\right]_n \quad \text{or} \quad (C_6H_{10}O_5)_n$$

Cellulose is hydrolysed in steps when heated in the presence of inorganic acids or under the action of enzymes - cellulases - to produce glucose:

$$(C_6H_{10}O_5)_n + nH_2O \xrightarrow[(H^+)]{\text{cellulase}} nC_6H_{12}O_6$$

Under the action of enzymes, glucose is fermented to various products. The enzyme cimase causes alcoholic fermentation:

$$C_6H_{12}O_6 \xrightarrow{\text{cimase}} 2\,C_2H_5OH + 2\,CO_2$$

From ethanol, by suitable reactions, can be obtained a substance of gross formula $C_7H_5NO_3$, which is known to exhibit reducing properties and to join hydrogen cyanide.

The manifestation of reducing properties indicates that a carbonyl group is contained in the structure of the compound. On the basis of the gross formula it can be assumed that the compound is:

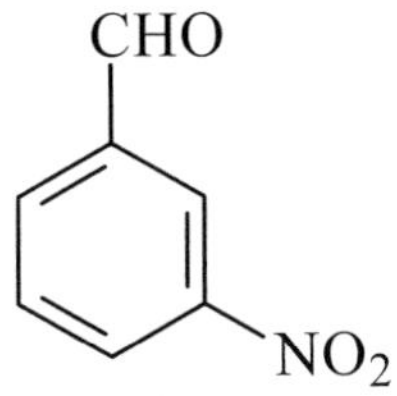

3-nitrobenzaldehyde

The alcohol may be subjected to dehydration and, depending on the temperature regime, ethene or diethyl ether may be obtained:

$$CH_3CH_2OH \xrightarrow{180^oC} H_2C{=}CH_2 + H_2O$$

$$2\,CH_3CH_2OH \xrightarrow{140^oC} CH_3CH_2{-}O{-}CH_2CH_3 + H_2O$$

The ethene is treated with bromine water, which decolours as a result of an addition reaction. This is a qualitative reaction to prove a multiple bond - double or triple.

$$H_2C{=}CH_2 + Br_2 \longrightarrow CH_2Br{-}CH_2Br$$

1,2-dibromoethane

Both dehydration and halogenation are redox processes involving a change in the hybrid state of carbon atoms.

$$CH_2Br{-}CH_2Br + 2\,KOH \xrightarrow{C_2H_5OH} HC{\equiv}CH + 2\,KBr + 2\,H_2O$$

When 1,2-dibromoethane is treated with an alcoholic solution of potassium base, an elimination reaction takes place and ethyne, potassium bromide and water are obtained.

Cyclic trimerization of ethyne at high temperature results in the production of benzene.

Option I:

The next step is the alkylation of benzene to methylbenzene by the Friedel-Crafts reaction. The reaction proceeds as an electrophilic substitution in the aromatic core.

The methyl group is an activating o- and p-orientant. The new substituent orients at the o- and p-site relative to the methyl group.

The nitration of toluene (methylbenzene) is a characteristic reaction for aromatic hydrocarbons and is carried out with a nitric mixture (a mixture of nitric acid and sulfuric acid concentrate), yielding two products, o-nitrotoluene and p-nitrotoluene. At the end of the process, p-nitrotoluene is isolated from the reaction mixture.

The reaction proceeds as an electrophilic substitution.

$$HO{-}NO_2 + 2\,HO{-}SO_3H \longrightarrow H_3O^+ + \overset{+}{N}O_2 + 2\,HSO_4^-$$

nitronium cation

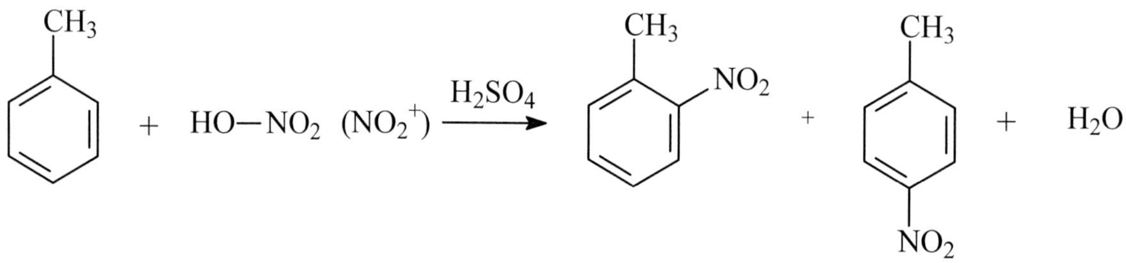

The resulting p-nitrotoluene is halogenated by a chain-radical mechanism (irradiation of the system with UV-light). The product is p-nitrobenzyl chloride.

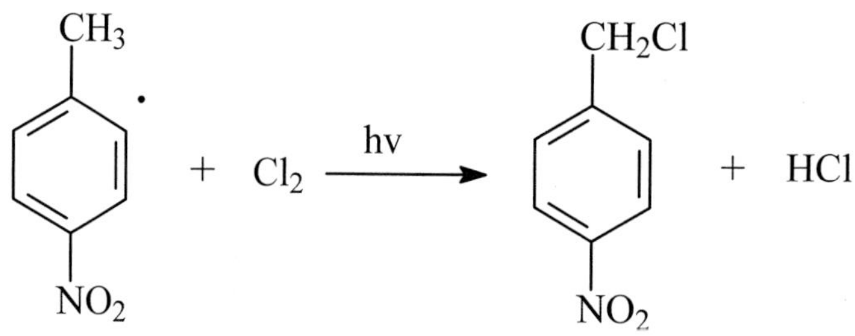

p-Nitrobenzyl chloride is treated with an aqueous solution of an alkaline base. A substitution reaction takes place and the product is p-nitrobenzyl alcohol.

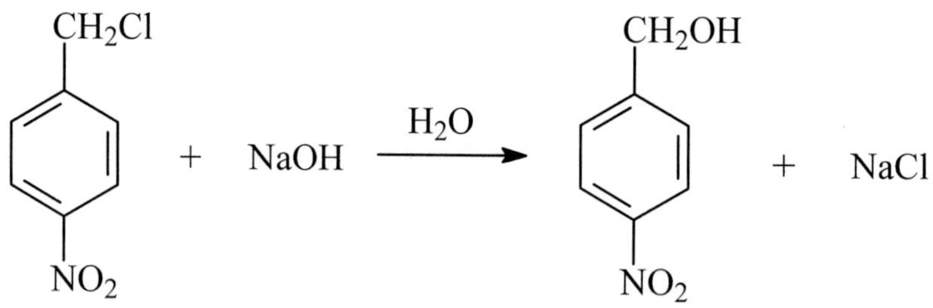

p-Nitrobenzyl alcohol undergoes dehydrogenation. The product of gross formula $C_7H_5NO_3$ is obtained.

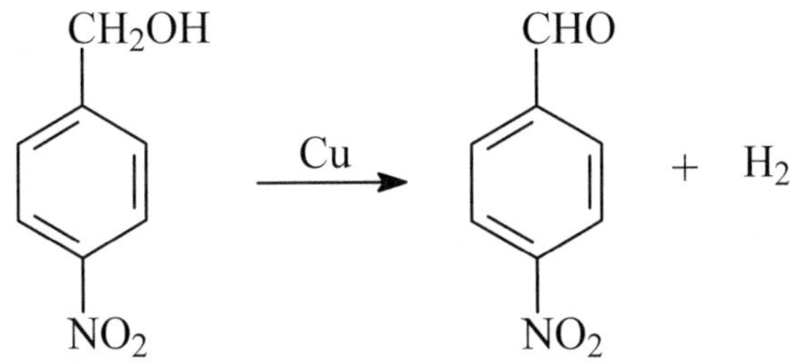

Aldehydes are good organic reducing agents and are oxidized by weak oxidizing agents such as silver-ammonia hydroxide (Tolens reagent), Fehling's solution, freshly precipitated copper dihydroxide, even by atmospheric oxygen.

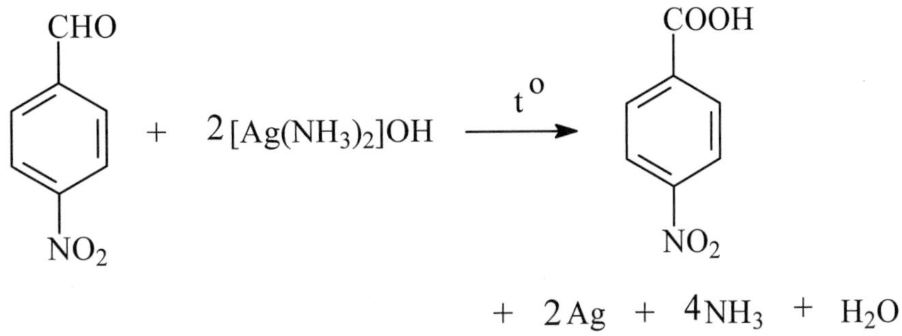

At the end of the redox process, 4-nitrobenzoic acid is produced, elemental silver, ammonia and water are released. For this reason, the reaction is called the ‘silver mirror reaction’.

Aldehydes join polar hydrogen-containing molecules such as hydrogen cyanide.

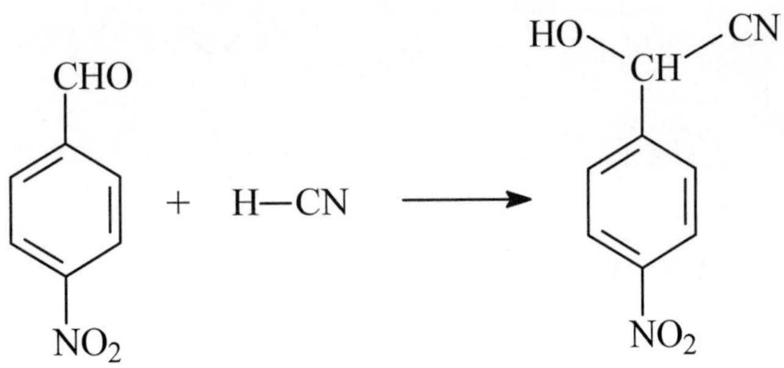

The product is called 2-(4-nitrophenyl)-2-hydroxyacetonitrile.

Option II:

The next step is the nitration of benzene to nitrobenzene: Nitration is a characteristic reaction for aromatic hydrocarbons. If a nitric mixture (a mixture of nitric acid and sulfuric acid) is added to benzene and heated slightly, a yellow substance with the smell of bitter almonds is formed - nitrobenzene. The reaction proceeds as an electrophilic substitution:

$$HNO_3 + 2H_2SO_4 \longrightarrow H_3O^+ + NO_2^+ + 2HSO_4^-$$

nitronium cation

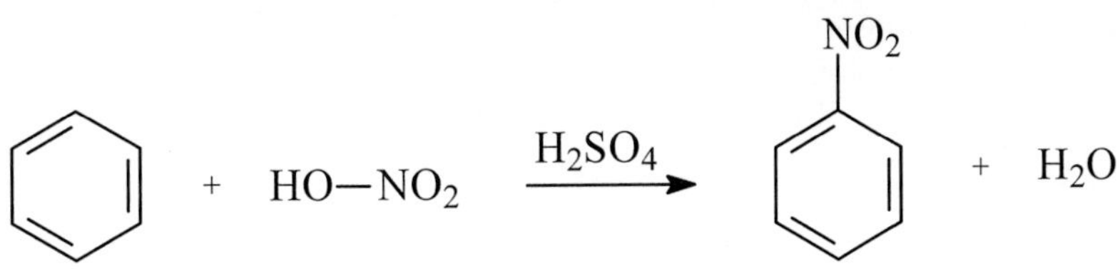

The nitro group is a deactivating meta-orientant. The newly arriving substituent orients at the m-site relative to the substituent present in the aromatic core.

Alkylation is carried out by the Friedel-Crafts reaction:

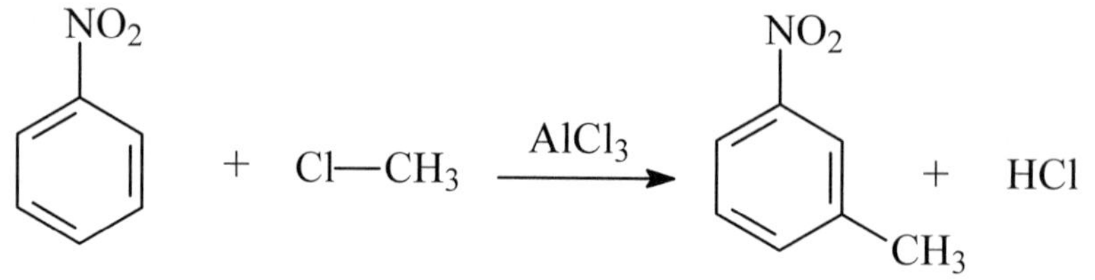

m-nitrotoluene

The resulting m-nitrotoluene is halogenated by a chain-radical mechanism (irradiation of the system with UV-light). The product is m-nitrobenzyl chloride.

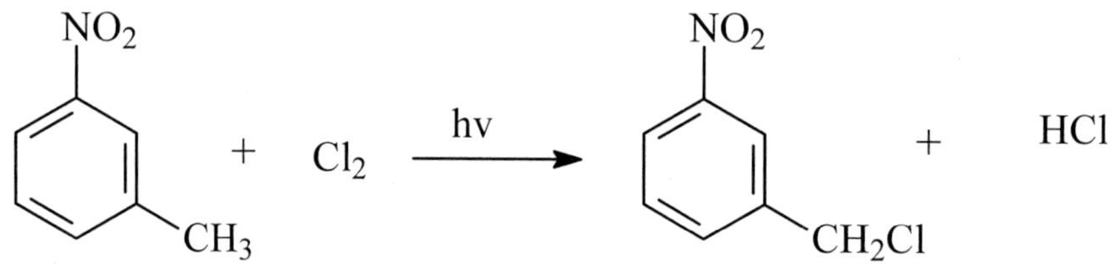

m-Nitrobenzyl chloride is treated with an aqueous solution of an alkaline base. A substitution reaction takes place and the product is m-nitrobenzyl alcohol.

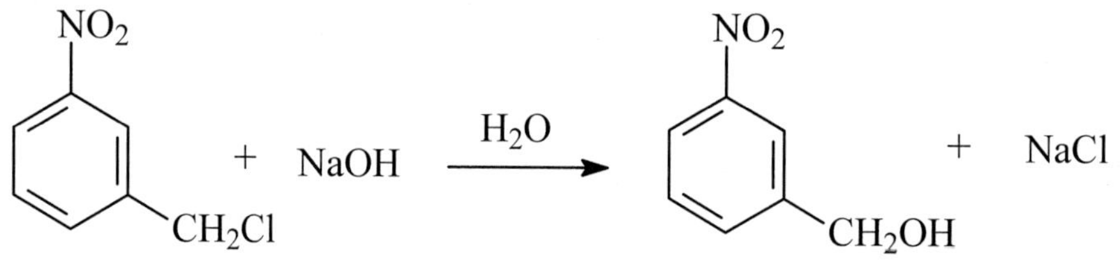

m-Nitrobenzyl alcohol undergoes dehydrogenation. The product of gross formula $C_7H_5NO_3$ is obtained.

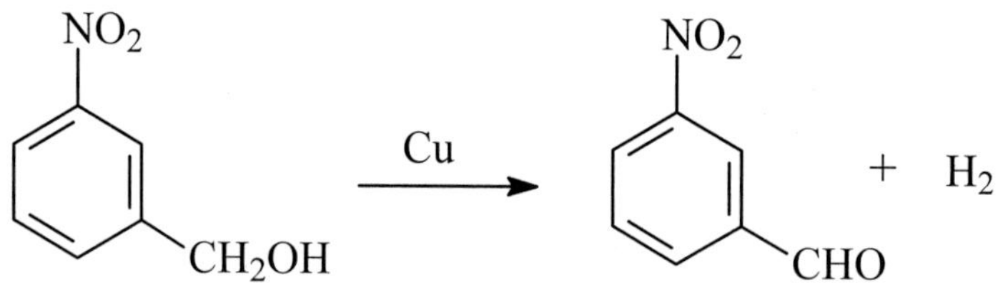

Aldehydes are good organic reducing agents and are oxidized by weak oxidizing agents such as silver-ammonia solution (Tolens reagent), Fehling's solution, freshly precipitated copper dihydroxide, even by atmospheric oxygen.

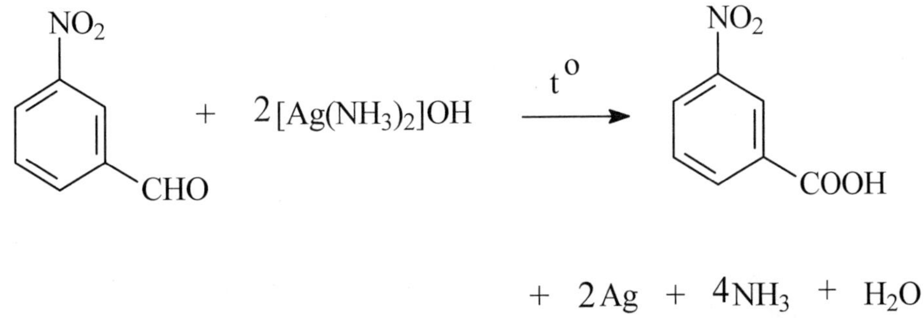

At the end of the redox process, 3-nitrobenzoic acid is produced, elemental silver, ammonia and water are released. For this reason, the reaction is called the 'silver mirror reaction'.

Aldehydes join polar hydrogen-containing molecules such as hydrogen cyanide.

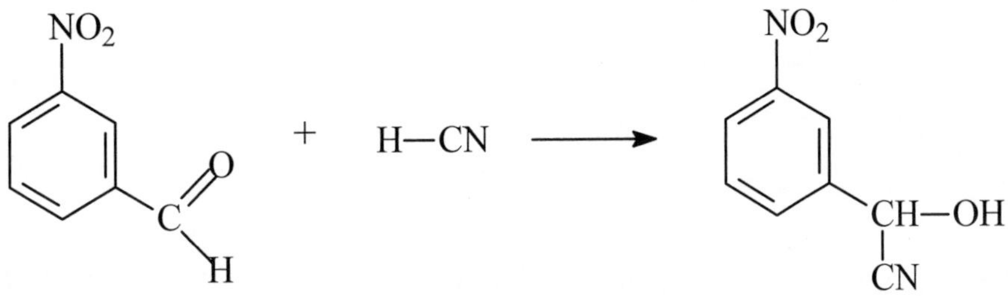

The product is called 2-(3-nitrophenyl)-2-hydroxyacetonitrile.

The presence of hydroxyl groups determines the property of cellulose to esterify.

Cellulosic nitrates have great application in practice. Different numbers of hydroxyl groups are esterified under different conditions. Esterification takes place when cellulose is treated with a nitric mixture (consisting of concentrated sulfuric and concentrated nitric acids).

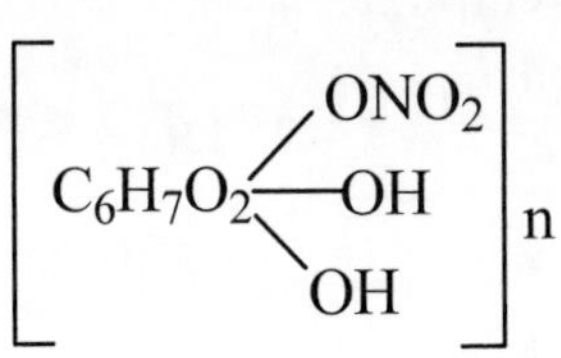

cellulose mononitrate

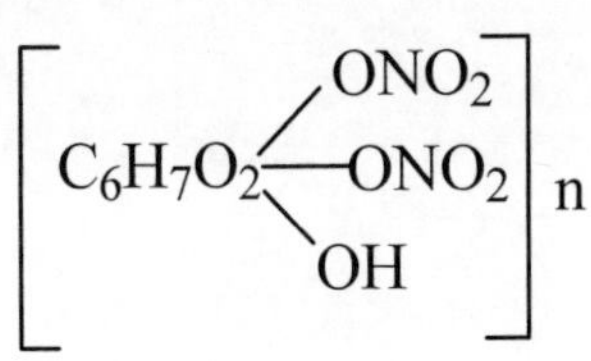

cellulose dinitrate

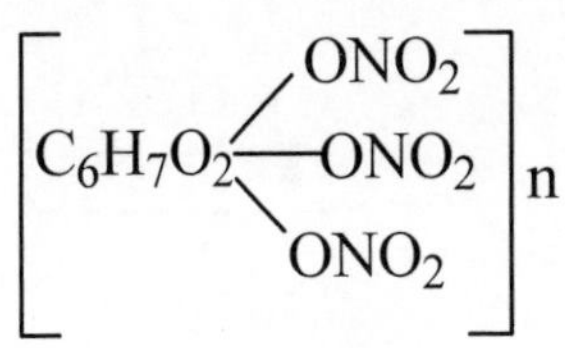

cellulose trinitrate

The product of complete nitration of cellulose is cellulose trinitrate, and in acetylation of cellulose - cellulose triacetate:

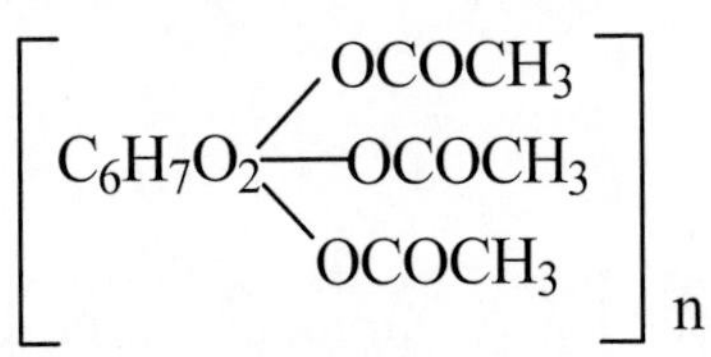

cellulose triacetate

32. QUALITATIVE REACTIONS

The elements of Group IA of the Periodic Table (the alkaline elements) do not form water-insoluble compounds. Therefore, the elements or their ions can be identified using an alcohol lamp. **Lithium** colors the flame **carmine red; sodium – yellow; potassium – violet.**

The volatile salts of the alkaline earth elements (Group IIA of the Periodic Table) characteristically color the flame of the spirit lamp. **Magnesium - white; calcium - tile red; strontium - bright red; barium - green.**

The reaction with **silver nitrate** is qualitative for the ions of the **halogen elements and the corresponding hydrogen halide acids.**

$$\underbrace{Ag^+ + NO_3^-} + \underbrace{K^+ + Cl^-} \longrightarrow AgCl\downarrow + \underbrace{K^+ + NO_3^-}$$

white precipitate

Hydrochloric acid and soluble chlorides precipitate a white, water-insoluble **and** acid- soluble **silver chloride. Silver chloride is colored violet by light (silver salts are photosensitive).**

Similarly, the silver cation is related to the anions of hydrobromic and hydroiodic acids and their soluble salts.

$$\underbrace{Ag^+ + NO_3^-} + \underbrace{K^+ + Br^-} \longrightarrow AgBr\downarrow + \underbrace{K^+ + NO_3^-}$$

pale yellow precipitate

$$\underbrace{Ag^+ + NO_3^-} + \underbrace{K^+ + I^-} \longrightarrow AgI\downarrow + \underbrace{K^+ + NO_3^-}$$

yellow sludge

The reaction to prove the silver ion is the easy solubility of AgCl in ammonia. Due to the formation of a complex ion:

$$AgCl\downarrow + 2\,NH_3 \longrightarrow [Ag(NH_3)_2]^+\,Cl^-$$

silver ammonia chloride

On heating the complex compound silver ammonia chloride, the ammonia is separated and silver chloride is precipitated again.

Sulfuric acid and soluble sulfates are demonstrated with alkaline earth metal ions (Group IIA of the Periodic Table), mainly barium ions. Insoluble alkaline earth sulfates are formed, e.g.

$$\underbrace{2\,Na^+ + SO_4^{-2}} + \underbrace{Ba^{+2} + 2\,Cl^-} \longrightarrow BaSO_4\downarrow + \underbrace{2\,Na^+ + 2\,Cl^-}$$

$$\underbrace{2\,K^+ + SO_4^{-2}} + \underbrace{Ca^{+2} + 2\,Cl^-} \longrightarrow \underset{\text{gypsum}}{CaSO_4\downarrow} + \underbrace{2\,K^+ + 2\,Cl^-}$$

The ions of group IIA elements precipitate white crystalline sulfates.

Nitric acid can be proved by one of the products resulting from the interaction of the acid with metals located after hydrogen in the Order of Relative Activity of Metals (ORAM).

Concentrated **HNO_3** oxidizes the metals Hg and Cu. A reddish-brown fume of nitrogen dioxide (**NO_2**) is given off, a qualitative reaction for nitric acid and nitrate ions. As with concentrated sulfuric acid, the reaction proceeds in two stages:

$$Cu + 2HNO_3 \longrightarrow CuO + 2NO_2\uparrow + H_2O$$

$$\begin{array}{llll|l|l|l} Cu^{0} & - 2e^- & \longrightarrow & Cu^{+2} & 2 & & 1 \\ & & & & & 2 & \\ N^{+5} & + e^- & \longrightarrow & N^{+4} & 1 & & 2 \end{array}$$

$$CuO + 2HNO_3 \longrightarrow Cu(NO_3)_2 + H_2O$$

Sum:

$$Cu + 4HNO_3 \longrightarrow Cu(NO_3)_2 + 2H_2O + \underset{\text{red-brown gas}}{2NO_2\uparrow}$$

Phosphoric acid and **its soluble salts can be proved** with silver ions, with which it forms a yellow-colored precipitate of silver phosphate.

$$\underbrace{3\,Na^+ + PO_4^{-3}} + \underbrace{3\,Ag^+ + 3\,NO_3^-} \longrightarrow Ag_3PO_4\downarrow + \underbrace{3\,Na^+ + 3\,NO_3^-}$$

Functional analysis is an important step in the study of organic compounds. It is used to determine what functional groups are contained in the molecules of the compounds under study. The analysis data indicate to which class of organic compounds the product belongs.

Compounds containing a multiple bond (**double** or **triple carbon-carbon - alkenes, dienes, alkynes**) decolorise **bromine water** or **2% aqueous potassium permanganate**.

If ethylene is passed through an aqueous solution of bromine, the solution is discolored.

Ethylene reacts under ordinary conditions with bromine. 1,2-dibromoethane is obtained.

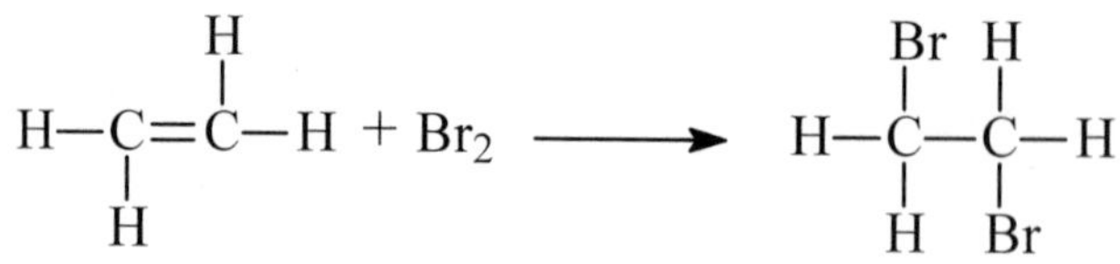

1,2-dibromoethane

$$C^{-2} - e^{-} \longrightarrow C^{-} \quad \left|\, 1 \,\right| \; 2 \; \left|\, 2 \right.$$

$$Br_2 \longrightarrow 2\,Br^{0} + 2 \times 1\,e^{-} \longrightarrow 2\,Br^{-} \quad \left|\, 2 \,\right| \qquad \left|\, 1 \right.$$

When the bromine joins the ethylene molecule at the same time as the new carbon-bromine bonds are formed, the double bond is broken and becomes a simple bond.

Alkynes join bromine to give bromine derivatives of alkenes and alkanes in turn:

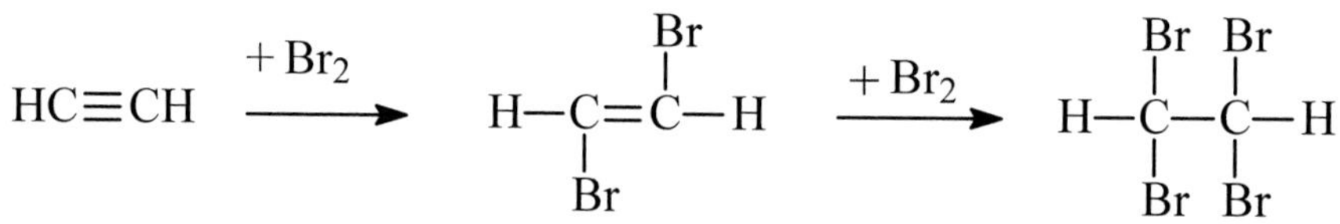

1,2-dibromoethylene 1,1,2,2-tetrabromoethane

$$C^{-} - 2\,e^{-} \longrightarrow C^{+} \quad \left|\, 2 \,\right| \; 2 \; \left|\, 1 \right.$$

$$Br_2 \longrightarrow 2\,Br^{0} + 2 \times 1\,e^{-} \longrightarrow 2\,Br^{-} \quad \left|\, 2 \,\right| \qquad \left|\, 1 \right.$$

The reaction with bromine is qualitative to prove a fold relationship.

The decolorisation of bromine water is a qualitative reaction for the presence of a fold bond in the molecule of the compound under study.

Alkenes oxidize at ordinary temperature. Their oxidation with **hydrogen peroxide** and **potassium permanganate** (moderate oxidation) is characteristic. Oxidation produces different products depending on the type of alkene and oxidant. Under mild oxidation conditions (H_2O_2 and a 2% solution of $KMnO_4$), the alkenes yield divalent alcohols called glycols. The decolorization of the potassium permanganate solution is another qualitative reaction for the presence of a multiple bond in the molecule of the compound under study.

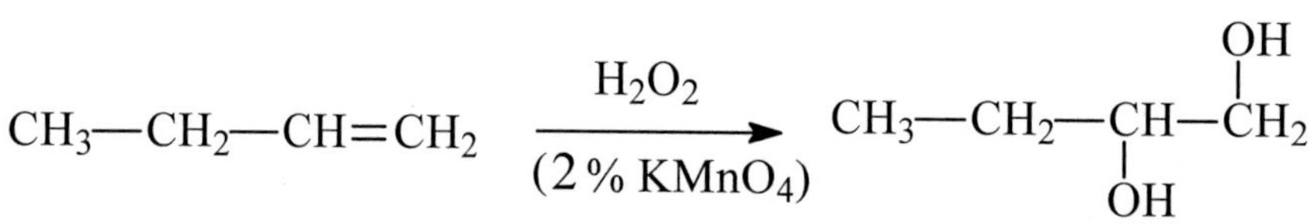

1-butene 1,2-butanediol

Under conditions of vigorous oxidation with a concentrated solution of potassium permanganate in the presence of concentrated sulfuric acid, the carbon chain of the unsaturated hydrocarbons is broken at the complex bond and mixtures of carboxylic acids are obtained.

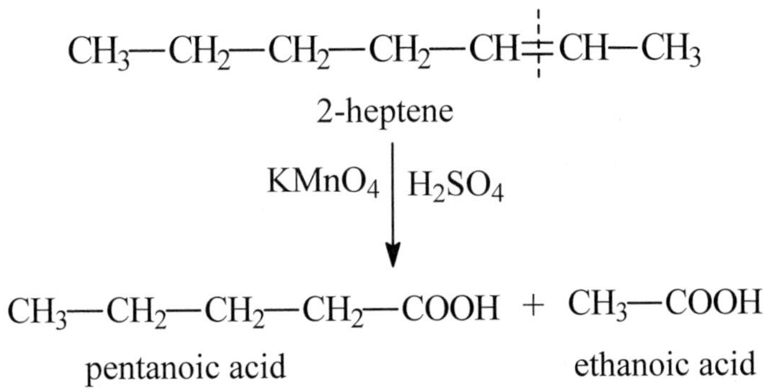

As with alkenes, the complex bond between carbon atoms can also be demonstrated with potassium permanganate solution (reduction to manganese dioxide). The carbon chain is broken at the triple bond and a mixture of carboxylic acids is obtained:

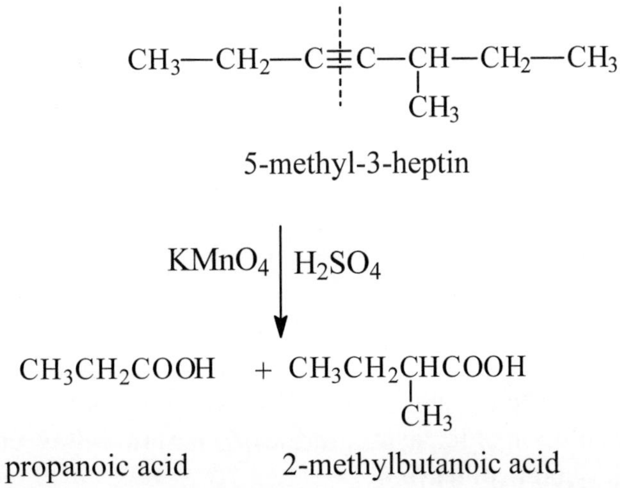

Alcohols are proven by the **iodoform reaction**. To crystals of iodine (I_2) is added a solution of sodium base of definite concentration until the solution is completely decolored. The test substance is added and the reaction mixture heated. A golden yellow precipitate of iodoform (triiodomethane) is separated. The process is represented by the following equation:

$$C_2H_5OH + 4\,I_2 + 6\,NaOH \longrightarrow CHI_3\downarrow + HCOONa + 5\,NaI + 5\,H_2O$$

The iodoform (haloform) reaction is used to prove ethyl alcohol, both in traffic accidents and in forensic practice.

Multivalent alcohols, in which the hydroxyl groups are bonded to adjacent carbon atoms, interact with freshly precipitated copper dihydroxide with a **sky-blue color**. A water-soluble complex compound of **dark blue color** is obtained.

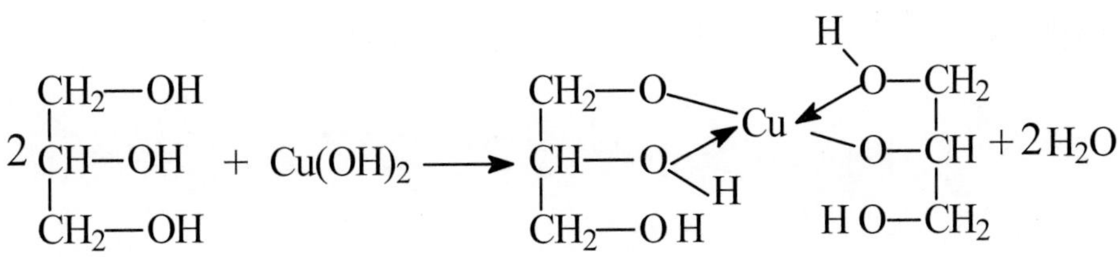

copper glycerate

Monosaccharides (glucose and fructose) each contain five hydroxyl groups bonded to adjacent carbon atoms. These react with freshly precipitated copper dihydroxide to give a complex compound of **dark blue color**.

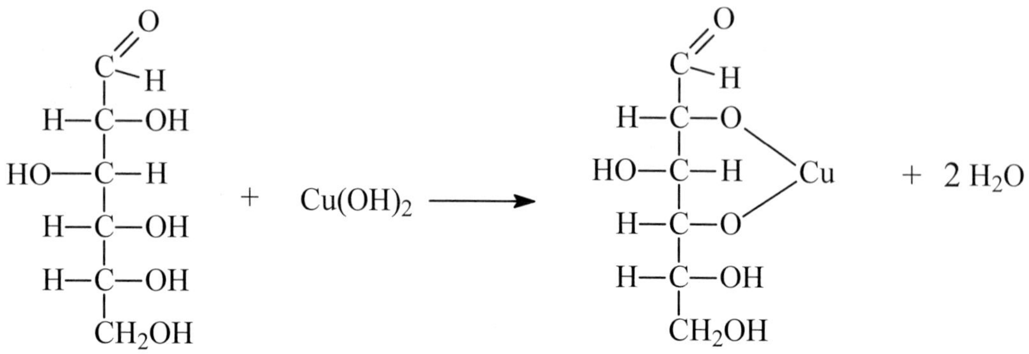

It has been found that copper dihydroxide can also react with two molecules of glucose to give the following complex compound:

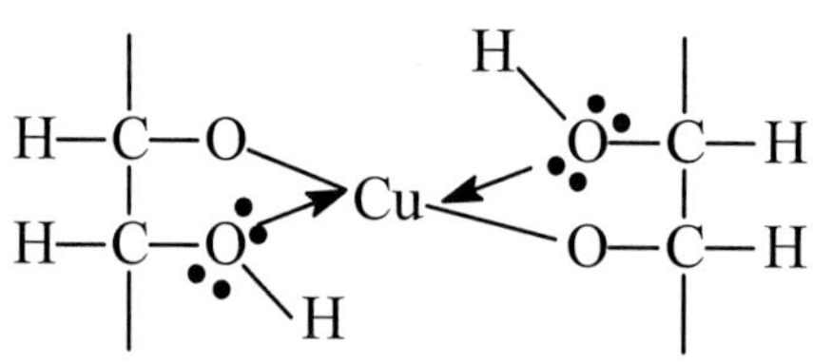

In **phenol** molecules, the functional hydroxyl group is bonded to a carbon atom from the aromatic nucleus. With a dilute solution of ferric trichloride of a **lemon-yellow color**, phenol forms a complex compound with a **blue-violet coloration**.

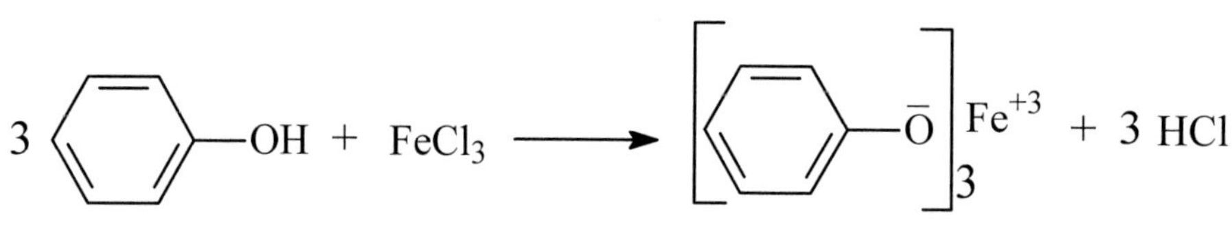

Aldehydes are good organic reducing agents (see functional group structure) and are oxidized by weak oxidizing agents such as silver-ammonia hydroxide (**Tolens reagent, silver mirror** reaction), Fehling's solution, and fresh-earth copper dihydroxide**.**

The increased reactivity of the hydrogen atom of the aldehyde group determines the reducing action of aldehydes. Specific oxidizing agents for the aldehyde group (including monosaccharides) are silver-ammonia hydroxide, Fehling's solution and freshly precipitated copper dihydroxide. These are used to prove the aldehyde group.

If an aldehyde is added to a silver-ammonia hydroxide and the mixture is heated, a shiny silver is given off on the walls of the vessel (silver mirror reaction):

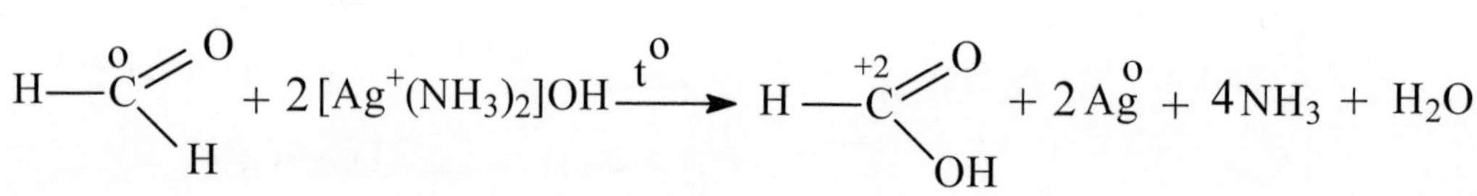

$$C^{0} - 2\bar{e} \longrightarrow C^{+2} \quad \left|\, 2 \,\right| \quad 2 \quad \left|\, 1 \right.$$

$$Ag^{+} + e^{-} \longrightarrow Ag^{0} \quad \left|\, 1 \,\right| \quad \quad \left|\, 2 \right.$$

Copper dihydroxide is a sparingly soluble precipitate. For a smoother redox process, a Fehling solution is used in which the divalent copper ions are bound in a soluble complex compound. A Fehling solution is prepared by mixing copper sulfate and an alkaline solution of a segnet salt (potassium sodium tartrate, potassium sodium salt of tartaric acid. The systematic name of the acid is 2,3-dihydroxy-1,4-butanedicarboxylic acid). The result is a water-soluble complex of divalent copper ions with the two hydroxyl groups of tartaric acid. The oxidation-reduction process proceeds by heating a Fehling solution with aldehyde, which produces a tile-red precipitate of dicopper oxide:

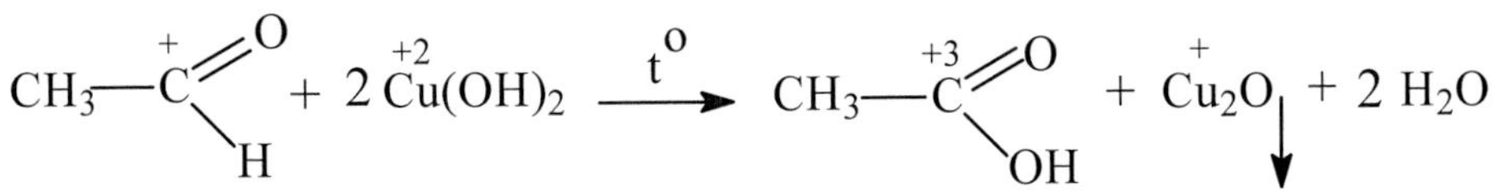

$$C^{+} - 2e^{-} \longrightarrow C^{+3} \quad | \ 2 \ | \quad | \ 1$$
$$\qquad\qquad\qquad\qquad\qquad\qquad\ \ 2$$
$$Cu^{+2} + e^{-} \longrightarrow Cu^{+} \quad | \ 1 \ | \quad | \ 2$$

The presence of an **aldehyde group** in the structure of glucose is proved by interaction with **Fehling's solution** and with a silver-ammonia hydroxide **(Tolens reagent)** - the "silver mirror" reaction.

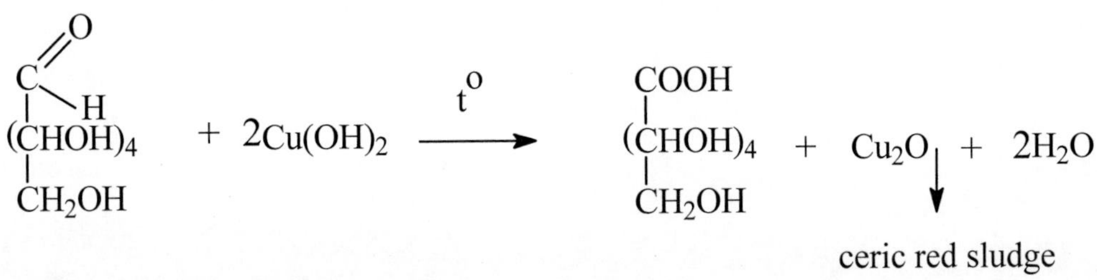

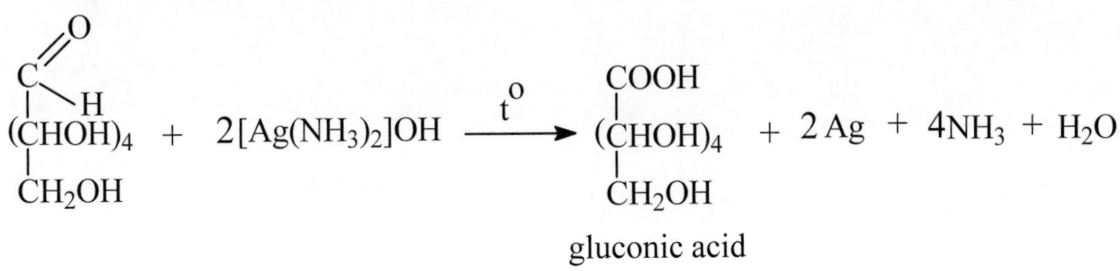

Ketones are resistant to weak oxidants. They are oxidized only by strong oxidants, for example, **concentrated potassium permanganate solution in the presence of concentrated sulfuric acid**. The carbon chain is broken adjacent to the carbonyl group and a mixture of carboxylic acids is obtained.

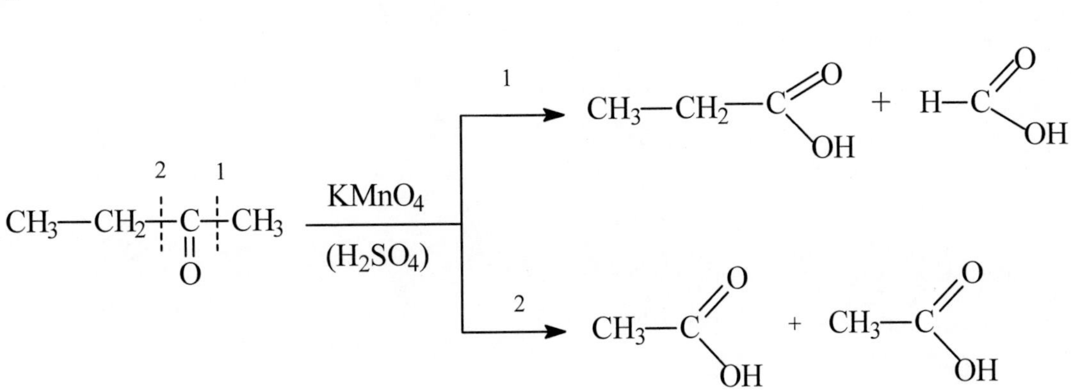

Formic acid contains an aldehyde group in its structure and under the condition of vigorous oxidation is converted into the weak and **perishable** carbonic acid, which decomposes.

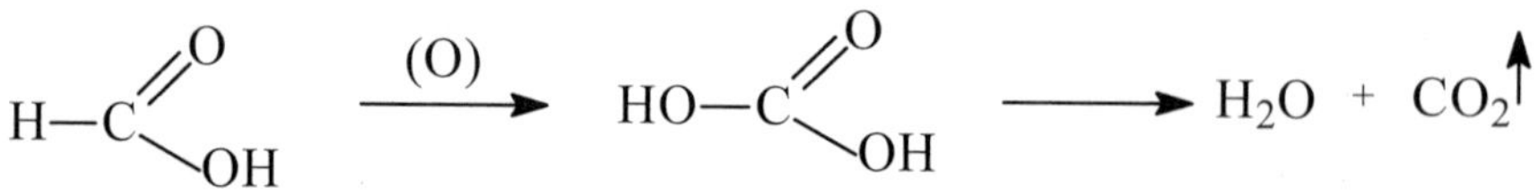

(H_2CO_3)

Carboxylic acids stain blue and violet litmus red.

Aniline on treatment with acidified solution of potassium bichromate ($K_2Cr_2O_7$) produces a black coloration of **aniline black**.

Starch treated with **an alcoholic solution of iodine (iodine tincture)** forms a compound with a **blue color** which disappears on heating and reappears on cooling. This is a **qualitative reaction for starch detection.**

Protein substances - A typical qualitative reaction for peptide bond detection is the **biuret reaction**. In an alkaline solution of a peptide under the action of copper sulfate, depending on the size of the peptide molecule, a characteristic coloration is produced - blue-violet to red- violet.

Xanthoprotein reaction - used to detect aromatic nuclei in the side chains of amino acid residues, e.g. in phenylalanine, tyrosine, etc. A golden yellow coloration is obtained by treating with conc. nitric acid.